Manganese in Powder Metallurgy Steels

Manganese in Powder Metallurgy Steels

Andrej Šalak and Marcela Selecká

Viva Books
New Delhi | Mumbai | Chennai | Kolkata
Bengaluru | Hyderabad | Kochi | Guwahati

First Indian Edition 2013
Reprinted 2015, 2017

VIVA BOOKS PRIVATE LIMITED

- 4737/23 Ansari Road, Daryaganj, New Delhi 110 002
  E-mail: vivadelhi@vivagroupindia.net, Tel. 42242200
- 76, Service Industries, Shirvane, Sector 1, Nerul, Navi Mumbai 400 706
  E-mail: vivamumbai@vivagroupindia.net, Tel. 27721273, 27721274
- Jamals Fazal Chambers, 26 Greams Road, Chennai 600 006
  E-mail: vivachennai@vivagroupindia.net, Tel. 28290304, 28294241
- B-103, Jindal Towers, 21/1A/3 Darga Road, Kolkata 700 017
  E-mail: vivakolkata@vivagroupindia.net, Tel. 22836381, 22816713
- 7, Sovereign Park Apartments, 56-58, K. R. Road, Basavanagudi, Bengaluru 560 004
  E-mail: vivabangalore@vivagroupindia.net, Tel. 26607409, 26607410
- 101-102 Mughal Marc Apartments, 3-4-637 to 641, Narayanguda, Hyderabad 500 029
  E-mail: vivahyderabad@vivagroupindia.net, Tel. 27564481, 27564482
- First Floor, Beevi Towers, SRM Road, Kaloor, Kochi 682 018, Kerala
  E-mail: vivakochi@vivagroupindia.net, Tel: 0484-2403055, 2403056
- 232, GNB Road, Beside UCO Bank, Silpukhuri, Guwahati 781 003
  E-mail: vivaguwahati@vivagroupindia.net, Tel: 0361-2666386

www.vivagroupindia.com

Published by arrangement with
Cambridge International Science Publishing
7, Meadow Walk, Great Abington
Cambridge Cb1 6AZ, UK

ISBN 978-81-309-2297-3

Published by Vinod Vasishtha for Viva Books Private Limited, 4737/23, Ansari Road, Daryaganj, New Delhi - 110 002.

Printed and bound in India.

Contents

1

Introduction

The manufacture of machine parts by powder metallurgy processes and their application has an irreplaceable role in this continually evolving industry, particularly in the automotive industry. By the diversity of methods and iron-based materials powder metallurgy makes it possible to achieve the properties of parts that meet demanding requirements for strength, toughness and other parameters only in the sintered-porous state, in some cases also with appropriate heat treatment. To achieve these properties of the parts, sintered steels are alloyed with larger amounts of alloying elements (this trend will also continue in future), in some standardized steels with up to ~10%, which is due to porosity and partly due to the heterogeneity of the structure of sintered steels compared with conventional steels. Despite this, the sintered parts are competitive. Therefore, the production of highly stressed components with high static and dynamic characteristics are the focus of a permanent basic and applied research.

Copper, nickel and molybdenum are alloying elements which have been used from the beginning of the production of sintered alloyed steels for the manufacture of structural components. The primary reason for using these elements is their low affinity for oxygen and thus the modest requirements on the reduction potential of the atmospheres and, to a lesser extent, also metallurgical reasons. A disadvantage of copper, nickel and molybdenum as alloying elements is their high price. A special group of sintered steels is formed by phosphorus-alloyed steels.

Manganese, chromium and partly silicon, among them especially manganese, have not been used for the preparation of sintered alloy steels. The main reason from a thermodynamic point of view has been and still is their high affinity for oxygen and stringent requirements on the purity of sintering atmospheres with an extremely low oxygen potential for the reduction of oxides covering the surfaces of these powders, particularly manganese powders, despite the fact that they are the cheapest alloying elements. Among these elements, manganese has the highest affinity for oxygen and, according to thermodynamic consideration, its oxides can be reduced with difficulties by solid carbon only at high temperatures.

Manganese, which in classical metallurgy steels occurs as a basic element, is used for manufacture of high strength structural steels and for design and manufacture of special steels with high wear resistance, has become a factor, despite its high affinity for oxygen, the subject of special attention as an alloying element in powder metallurgy. The aim of studies concerned with manganese is the effort partly to replace nickel and copper by cheap manganese for the production of sintered alloy steels. Studies of the application of manganese as an alloying element for the preparation of sintered steels started early in the fifties of last century. Since then, special attention has been paid to manganese in science and research up to present time. This is evidenced by several hundreds of published works. It should be noted that in all the published papers presented the positive properties of sintered manganese steels, with the properties depending on the selected material and technological conditions applied in their preparation. The achieved values of the mechanical and other properties of manganese steels clearly confirm and show unambiguously that the iron powder matrix of the compacts was alloyed with manganese during sintering independently of the purity of the sintering atmosphere or the type of manganese carrier and sintering conditions.

In addition to high affinity for oxygen, manganese has a high vapour pressure, which is its physical property that can not be controlled. The manifestation of the high vapour pressure of manganese is its sublimation from manganese particles in the compact, depending on the sintering temperature. When sintering manganese steels, its high affinity for oxygen already becomes evident in the form of vapour, thereby reducing during sintering the oxides covering the surface of base plain iron and/or prealloyed powder in the compact, and the moisture content of the sintering atmosphere securing the effective sintering of manganese steels.

In this case, the impact of manganese in the form of manganese vapour on sintering and alloying is characterised in the literature as the self cleaning / protection / reduction effect with respect to the sintering atmosphere, regardless of its purity. This physical feature of manganese allows sintering of manganese steels.

Manganese as an alloying element in powder metallurgy and the basic alloying element in classical metallurgy, where the effect of manganese vapour as a reduction agent is also evident, is also characterised by a high hardening effect on the iron matrix.

Another such element is chromium with a slightly lower affinity for oxygen compared with manganese, which in conjunction with carbon is also characterised by high hardening effect on the iron matrix. Consequently, they can be used together for alloying of sintered steels. Hybrid steels prepared on the basis of chromium- and/or molybdenum-prealloyed powders with the addition of manganese can form a new group of sintered material systems to produce statically and dynamically highly stressed sintered parts.

Chromium and manganese are in this case the two cheapest alloying elements with the highest hardening effect. It should be noted here that the hybrid sintered steels are characterised by greater or even complete homogeneity of the structure resulting in technical and economic advantages of their application. These are the benefits of alloying sintered steels with chromium and manganese and small additions of molybdenum.

The aim of this monograph is to present a comprehensive overview of the factors that influence the properties of manganese-alloyed sintered steels listed in the cited papers. They are the physical and metallurgical characteristics of manganese, chromium and other alloying elements. Types of ferrous and prealloyed powders, various kinds of manganese carriers, including various master alloys, different conditions of pressing and forming high density compacts, sintering conditions, *i.e.* under laboratory short- and long-term sintering up to about 180 min at temperatures of 700°C to 1300°C, purity of sintering H/N, N/H and N atmospheres with a dew point of –30°C to –70°C, different protection of charges of sintered samples and manganese-alloyed components against their possible oxidation according to the thermodynamics by insufficiently clean atmospheres, with no account of the reduction effect of the manganese vapour in an atmosphere containing moisture. The results achieved at sintering of manganese steels under industrial conditions accomplish those attained under laboratory sintering conditions. All studies in this monograph are presented without discussion, which are published in nearly 289 references cited, to avoid influencing the reader in research or practice with such discussion.

The presented findings with tangible results provide a very broad overview of the effect of the above mentioned factors on the studied properties of manganese steels. This has created suitable conditions for evaluating their effectiveness in relation to the achieved characteristics so that in future it will be possible to focus on research detailing the impact of the investigated or other factors which would result in the highest desired properties of sintered manganese steels with relatively the lowest production costs. The findings for the sintering of manganese and hybrid steel in industrial furnaces confirm their possible application in the mass production of sintered parts.

The results of all cited papers are processed in three parts. Parts I and III present the results of the works of the authors of this monograph which were also achieved with coauthors. Part II presents the results of other authors, so that together they constitute the above-mentioned general overview of the current knowledge of sintered manganese steels published in the cited papers. By considering the concrete results the reader will be able to make a picture of the current state of knowledge of manganese in powder metallurgy.

This book is intended for anyone who in any way deal with manganese in powder metallurgy, scientific and research and development workers, manufacturers of powders and of finished parts that represent the final products of the presented results. To some extent, the book can also be used for teaching purposes.

Part I

2

Thermodynamic conditions for the Mn–O system in sintering of manganese steels

2.1. Manganese in steelmaking

Manganese is the most widely used alloying element in the production of classic steels. Because of its possible application as an alloying element in the production of steels (this has not as yet been realised), it is necessary to discuss its physical–metallurgical characteristics which determine its irreplaceable role in classic steel production. In this connection, it is important to stress the fundamental difference between the production of classic manganese-alloyed steels and the production of sintered steels which influences the properties of final products. The classic and also manganese-alloyed steels are produced by melt metallurgy with the unlimited solubility of manganese in the melt. The production of a solid skeleton from a dispersed iron powder and its alloying with manganese and also with other alloying elements by sintering take place by diffusion mechanisms only in the solid state. This production is affected by a considerably larger number of mutually interacting factors in comparison with the production of classic steels. These factors must also be taken into account in the application of manganese as an alloying element in powder metallurgy with its physical–metallurgical characteristics, some of which are discussed later.

Due to its physical–metallurgical properties, manganese has a special position in ingot steel metallurgy, *i.e.* alloying and deoxidation. At the same time, manganese is the cheapest alloying element. The density of manganese is 7.21 g cm^{-3}, melting point 1244°C.

Manganese is found in every classic structural steel. It comes from ore, and steels with the content of up to 0.4% Mn are not regarded as alloyed steels. The steels with the manganese content of 0.8–1% and with 0.4–1% C are characterised as low-alloy steels, the steels with 0.8–1.7% Mn and with 0.6-0.1% C belongs to quenched and tempered steels with the hardness of 300–350 HV and the steels with 0.9–1% C are classified in the group of steels with no deformability. Higher alloyed steels with 6–12% Mn are characterised as steels with high wear resistance, with the Hadfield steel (12% Mn, 1.2% C) being the main representative of this group of steels.

The highest solubility of manganese in α-iron is 3% and in γ-iron it is up to 55%. Manganese is present in four modifications, *i.e.* α, β, γ and δ. From the viewpoint of sintering it is necessary to ensure that manganese at the transformation points, *i.e.* at 727°C (α→β), 1100°C (β→γ) and 1136°C (γ→δ) expands and this expansion is the main reason for the growth of the dimensions of sintered manganese-alloyed compacts.

Like nickel, manganese forms the open γ-region and by this decreases the γ→α transformation temperature. Diffusion of manganese in α- and γ-iron is considerably slower than carbon diffusion. The manganese atoms can replace iron in the carbide and, consequently, it may be expected that the iron carbide will be greatly enriched with manganese. This phenomenon is more evident in low-carbon steels and vice versa.

When the manganese content is increased the pearlitic transformation point is shifted to the side with lower carbon content. Increasing manganese content also reduces the rate of transformation in both the pearlitic and bainitic ranges, and the maximum rate of transformation in the bainitic range and also the martensitic transformation are shifted to lower temperatures. A special feature of the diffusion processes in the manganese steels is that in certain systems the steel can contain simultaneously ferrite, bainite, martensite and austenite.

Alloying of pure iron with manganese has almost no effect on its notch toughness. At a carbon content of up to 0.2% no quenching of the steels takes place even after rapid cooling of the manganese steels from the γ-region in water. Consequently, these steels change their properties only slightly even after long-term tempering.

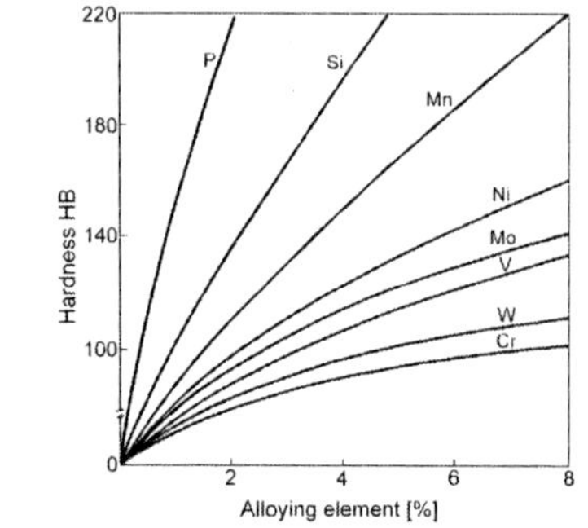

Fig. 2.1. Effect of alloying elements on hardness of ferrite [1].

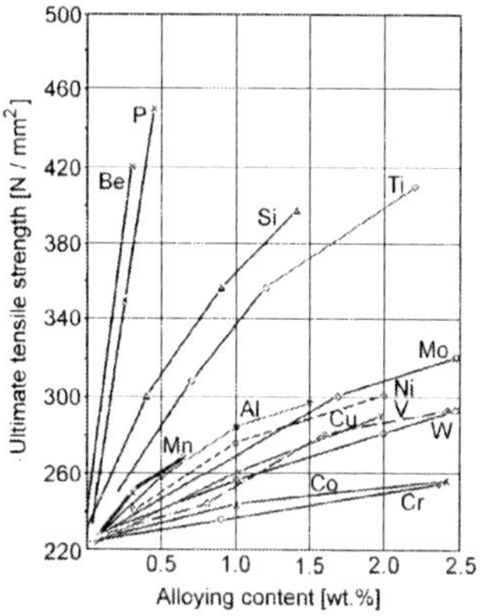

Fig. 2.2. Effect of alloying elements on tensile strength of ferrite.

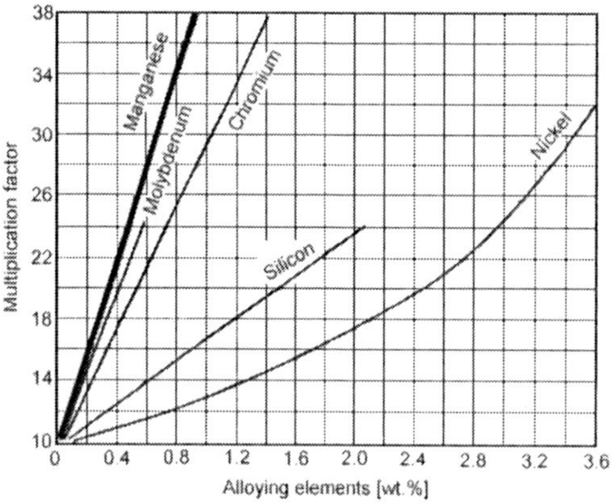

Fig. 2.3. Multiplication factor of alloying elements.

Manganese has a higher hardening effect on ferrite compared with other alloying elements, for example, nickel and, in particular, chromium. The hardening effect of manganese has been and still is the main aim of research also in powder metallurgy in the production of sintered structural steels. Figure 2.1 shows the hardening effect of manganese on the hardness of ferrite in comparison with other elements, and Fig. 2.2 shows the effect on ultimate tensile strength. These data are valid only for carbon contents of up to 0.1%. The high hardening effect of manganese on ferrite is also shown by the multiplication factor in Fig. 2.3.

The parameters important for hardenability are the Grossman factor for 1 wt.% alloy, oxidation reaction enthalpies and diffusion coefficients as the additional data presented in the diagrams in the previous section for the elements used and researched most

Table 2.1. Grossman factor for 1 wt.% alloy (GF), diffusion coefficient of alloying element in γ-iron (D_x), self-diffusion coefficient of γ-iron (D_{Fe}) and free enthalpy of formation of oxide (ΔG) (kJ·g·atom) oxygen) [1]

Alloying element	GF	D_x / D_{Fe}	ΔG
Copper	1.3	1	–150
Nickel	1.4	0.5	–250
Molybdenum	3.7	5	–310
Iron			
Chromium	3.1	5	–540
Manganese	4.5	2.5	–500
Vanadium	–	5	–620
Boron	–	–	–650
Silicon	1.7	10	–680

extensively in powder metallurgy, are presented in Table 2.1. (The changes in the free enthalpy ΔG of oxide formation is a measure of how readily the material oxidises assuming that metal and oxide are present as the solid phase).

The deoxidation effect of manganese in the production of steel is evident as a result of its high affinity for oxygen. Manganese in the melt reacts with oxygen and also sulphur with the formation of oxides and sulphides which easily transfer into the slag and also evaporate and react with oxygen in the atmosphere above the melt surface.

2.2. Basic thermodynamic characteristics of protective atmospheres for sintering steels alloyed with manganese and other elements

The sintering of materials based on ferrous powders, which represent almost 80% in mass of total production of powder metallurgy, must be perfected under a protective atmosphere. The task of the protective atmosphere under the effect of the heat at elevated temperatures is the formation of a solid body from a dispersed powder system. The effect of the protective atmosphere expresses itself during the entire sintering process, *i.e.* yet in yearly stages of heating of the compacts at dewaxing, which adversely affects the quality of the atmosphere in spite of it that in most cases it occurs only during a relatively short time during this sintering stage.

The most important function of the atmosphere is the reduction of the oxide films covering the surface of the used powder components used for the preparation of the compacts (base and alloying powders stored in air), including the surface of interior pores in the compacts. The reduction of the atmosphere containing some amount of humidity to the equilibrium conditions for the given material system must guarantee satisfactory clean metal surfaces of the metal powder components in the compacts to form contacts (necks, bonds), so that the material transport during sintering between powder particles is not inhibited under contemporary protection of the compacts against the oxidation [2].

The further role of the atmosphere is to protect the system being sintered against oxidation during preheating, isothermal sintering and cooling up to the exit of the sintered parts from the furnace. The flowing gases used as protective atmospheres transport away also the gaseous products of the reduction. The protective atmosphere also guarantees during the isothermal sintering hold the growth of interparticle bonds as well as a substantial extent of alloying of the austenite matrix with additional elements with the impact on the homogeneity of the structure and required physical and mechanical properties of the material.

In the protective atmospheres used in powder metallurgy, pure hydrogen exerts a high reducing effect in most cases but due to some of its drawbacks (highly explosive, high cost), cracked ammonia is often used (75% H_2/25% N_2) as a proper protective atmosphere. At present, modern nitrogen-based atmospheres (nitrogen + hydrogen) are used in most cases. They are formed as a mix of two gases over a wide ratios up to (5–3)% H_2/(95–97)% N_2. Therefore, the ratio of these gases is prepared in loco adapted to the material systems being sintered. Nitrogen, N_2, as molecules, does not react with iron nor with oxygen. It is basically neutral or even beneficial for many ferrous alloys [3].

The carrier of oxidation of the sintered materials is the water vapour H_2O in the H_2/H_2O atmosphere. Due to a porous structure, pressed powder components react more readily with the surrounding atmosphere than fully dense materials and this affects the choice of the gas and its optimum mix and gas flow rate.

The susceptibility of an element to oxidation is indicated by the change of the free enthalpy of oxide formation. The temperature dependence of that and of oxygen potential for different metal oxides is shown in Fig. 2.4. This dependence is used to determine the partial

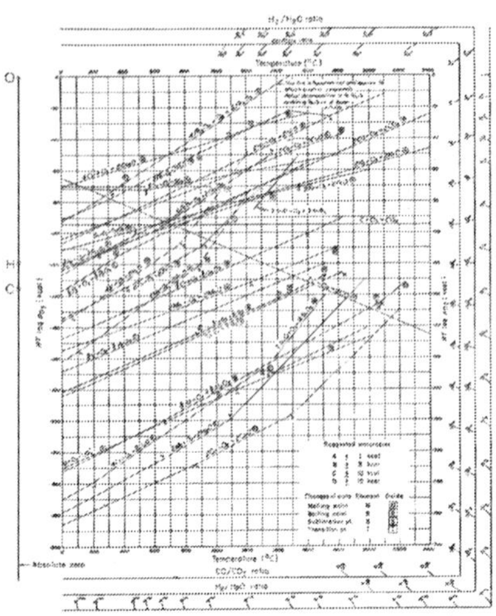

Fig. 2.4. Temperature dependence of enthalpy changes and oxygen potential of metal oxides (Richardson–Ellingham diagram) with the line for Mn at 1150°C [4].

pressure of oxygen which is in equilibrium with the oxide and the corresponding H_2/H_2O ratios at the given temperature for a specific metal. (The CO/CO_2 equilibrium ratios for the metals are in this case not considered because this atmosphere was not used for sintering of manganese steels described in this work). The values of this oxygen pressure for several metals, including corresponding dew points, are shown in more detail in Fig. 2.2. In general, the fraction of water vapour in a furnace atmosphere in a H_2/H_2O mixture is determined by the dew point measurements. The dew point is a measure of the water to hydrogen partial pressure ratio and tells the temperature at which water vapour will condense. It is a common measure of the atmosphere reduction potential. The dependence of the H_2/H_2O ratio in an atmosphere and the dew point is shown in Fig. 2.6.

The equilibrium data for interesting metals are those used in selecting the atmosphere for the given material system and sintering conditions. The use of equilibrium data to calculate compositional requirements for the protective atmosphere implicitly assumes that the system in question may be brought to an equilibrium or steady state condition. The steady state differs from a true equilibrium here in that only a partial, or local, equilibrium is established at the metal–vapour (atmosphere) interface. Supplying the atmosphere at a rate sufficient to overpower any compositional changes that might occur does this by reaction of the atmosphere with metal surfaces. It is readily obvious that this situation is not possible with porous metal materials since the surface extends well into the interior volume, the volume of pores and continuous rapid replenishment of atmosphere

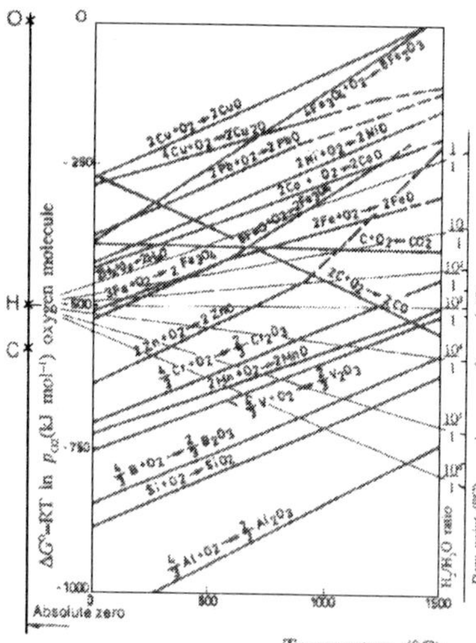

Fig. 2.5. Temperature dependence of enthalpy changes and oxygen potential of some selected metal oxides and corresponding dew points (Richardson–Ellingham diagram) [1,4].

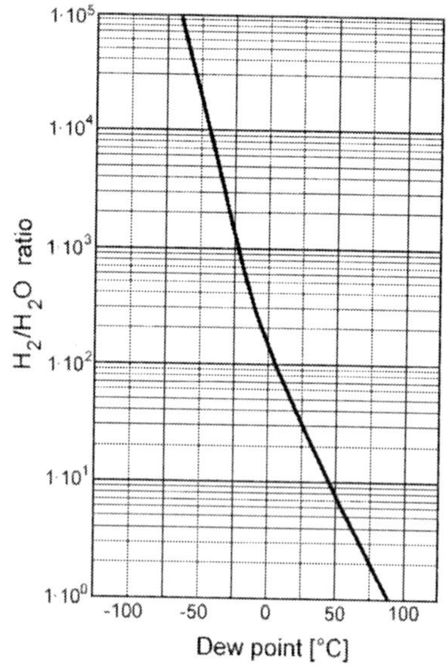

Fig. 2.6. Correlation between H_2/H_2O ratio and dew point [5].

in contact with their interior surfaces is physically impossible under normal furnace conditions.

Additionally, any minor change of local temperatures can affect equilibrium ratios. The situation inside the parts is very different, indeed, in the regions equilibrium can be established only between local carbon content and in discussed case also local manganese and local oxygen or residual water vapor formed by reaction of hydrogen with iron oxide. The effect of the atmosphere inside the parts can be important at relatively low densities. For this reason, the ideal stages should be considered almost completely valid only for the 'skin' – surface of PM parts. What really happens, which reactions proceed in the core, strictly depends on porosity, pore structure, part volume and part size, and perhaps on the flow rate of the atmosphere. The porosity increases in a large amount the total reaction surface with the active gas of the sintered compact. For instance, at the total porosity of 12% of a sample the surface area of interconnected pores is around 130 times larger as its geometrical surface area [6]. For a better understanding of actual sintering conditions, definitely different from the classical equilibrium concept, it may be necessary to discuss from the thermodynamic point of view the different

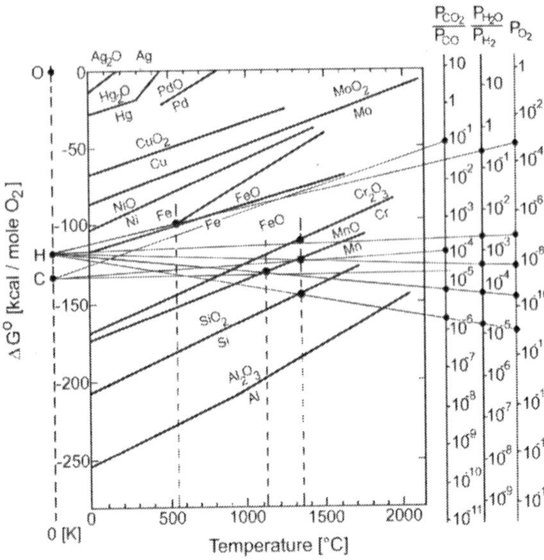

Fig. 2.7. Graphical determination of the equilibrium dissociation pressure P_{O2}, $P_{H2O/H2}$ and $P_{CO2/CO}$ for FeO/Fe, Cr_2O_3/Cr, MnO/Mn and SiO_2/Si systems at various temperatures [1,2,7].

possible atmospheres separately for individual material systems, what the researchers tried and try to do especially for the Fe–Mn system.

Figure 2.7 illustrates the equilibrium oxygen dissociation pressure for FeO (550°C), MnO at 1120°C, the most common sintering temperature, and for Cr_2O_3, MnO and SiO_2 at 1350°C. The difference between these dissociation pressures is larger than by 6 orders of magnitude, *i.e.* it is very low for MnO. Figure 2.8 shows for comparison the determination of equilibrium temperatures for Cu, Mo, Fe, Cr, Mn and Si in the H_2/H_2O atmosphere.

Manganese belongs to the elements, which form the oxides reducible only with carbon not with hydrogen. The sintering of manganese and of manganese containing alloys and the reduction of its oxides with hydrogen should be possible only under the presence of some other element, which forms with it a solid solution. Iron is an excellent element for this application [9,10]. Manganese with oxygen forms the compounds Mn_2O_3, MnO_2, MnO_3, Mn_2O_7, Mn_3O_4 and MnO. The majority of them have also different modifications. MnO_2 and Mn_2O_3 easily decompose to Mn_3O_4 and oxygen and by this reason are easily reducible. MnO_2 (manganese dioxide, manganese ore – pyrolusite) dissociates at the temperatures of 400–500°C [11], and is reducible also with hydrogen was used therefore for

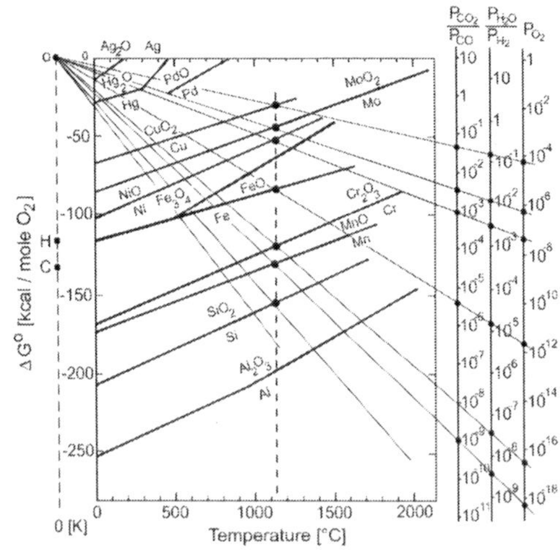

Fig. 2.8. Graphical determination of equilibrium temperatures for Cu, Mo, Fe, Cr, Mn and Si in the H_2/H_2O atmosphere [8].

alloying of sintered manganese steels. The oxides Mn_3O_4 and MnO exist at temperatures higher than 800°C, and should be reducible only with carbon at presence of the element with high solubility of manganese [12]. These Mn-oxides prove to be of greatest importance for powder metallurgy. Iron in this case can act as an activator at temperatures higher than the ordinary sintering temperature of 1120°C and, therefore, the analysis is always concentrated on MnO as the most hardly reducible Mn-oxide. The oxidation of metallic manganese in air starts at ~400°C under the formation of an active bright blue-grey Mn-oxide, which at 500°C transforms in air to a stable inactive green in colour MnO. At the increased temperature, over 900°C forms also an inactive dark grey form of MnO of non-stoichiometric composition [13].

All requirements concerning the formation and reduction of manganese oxides during the sintering of mixed manganese steel according to the Richardson–Ellingham diagram shown in Figs. 2.4 and 2.5 are related to MnO only. The basic reaction of manganese and oxygen expresses the equation:

$$2Mn + O_2 = 2MnO.$$

The thermodynamic conditions for reduction of MnO oxide by various hydrogen-containing atmospheres or by solid carbon are described below. They only particularize the oxygen potentials in

Table 2.2. Equilibrium data of Mn/MnO in O_2 (partial oxygen pressure) gas mixture of H_2/H_2O and corresponding water contents; d.p. – dew point [3,8]

	Temperature [°C]								
	600	700	800	900	1000	1100	1150	1200	1400
P_{O2} [Pa]	10^{-38}	10^{-34}	10^{-30}	10^{-26}	10^{-24}	10^{-22}	10^{-21}	10^{-20}	10^{-16}
H_2/H_2O	$1\cdot10^{8}$	$1\cdot10^{7}$	$3\cdot10^{6}$	$7\cdot10^{5}$	$2\cdot10^{5}$	$9\cdot10^{4}$	$5.0\cdot10^{4}$	$4.0\cdot10^{4}$	$2.7\cdot10^{3}$
P_{H2O} [Pa]	0.01	0.05	0.13	1.06	2.35	2.06	3.89	10.10	37.67
dew point [°C]	102	–90	–80	–75	–60	–57	–50	–40	–30
H_2O vol.[%]	$1\cdot10^{-6}$	$9\cdot10^{-6}$	$5.2\cdot10^{-5}$	$1.4\cdot10^{-4}$	$1.1\cdot10^{-3}$	$1.3\cdot10^{-3}$	$2.0\cdot10^{-3}$	$3.8\cdot10^{-3}$	$3.7\cdot10^{-2}$

the atmosphere under some sintering condition for MnO and partly for FeO. The calculated minimal dew points and oxygen potentials for H_2/H_2O atmospheres in the temperature range of 600 to 1400°C to reduce MnO or to prevent oxidation of manganese are listed in Table 2.2. These data correspond to the free enthalpy changes shown in Figs. 2.4 and 2.5 demonstrating high affinity of manganese for oxygen and hard reducibility of its oxides. It is possible to deduce from this why manganese as a PM alloying element did not find up to the time none large scale application in production of sintered alloyed steels. These thermodynamic data were and in some cases are considered as a decisive factor against manganese in spite of the results obtained at sintering of manganese steels. It would be very hard up to impossible to fulfill these requirements for the purity of the atmosphere for sintering of manganese steel, namely under industrial conditions.

The thermodynamic requirements for the purity of a sintering atmosphere should be not constant during the entire sintering temperature range. The total preheating period and the cooling stage under laboratory and, in particular, under industrial sintering conditions need the atmosphere with higher purity than that for isothermal sintering. This is often decisive for the determination of the dew point of the atmosphere used. The preheating period in the sintering process is that in which dewaxing of the compacts occurs at which the lubricant melts and evaporates followed by burn-off requiring a humid atmosphere at the temperature below ~600°C [14]. On the other hand, the Mn-containing parts require for elimination of oxidation first the manganese carriers in the compacts an atmosphere with an extremely low dew point, as for instance of –102°C at 600°C, what is not realizable in practice.

In general, the fraction of water vapour in a furnace atmosphere in H_2/H_2O mixtures is determined by the dew point measurements. The relation between the dew point and moisture is listed in Table 2.3. However, on cooling, the compacts will again recross the oxidation-reduction boundary and be oxidized during cooling. If that occurs at a high temperature, then the compacts will be discoloured and possibly attacked.

A detailed view on the oxygen potential in the sintering atmospheres previously and nowadays commonly used in powder metallurgy, *i.e.* endothermic gas (endogas), dissociated ammonia, hydrogen and on nitrogen-based atmospheres at different dew points for Cr and Mn is shown in Fig. 2.9. It follows from the diagram that H_2/H_2O and temperatures are the critical factors determining whether oxidation or reduction will occur. It was stated in [15] that none of the analyzed atmospheres can, according to these calculations, reduce the oxides or avoid further oxidation of Cr and Mn at 1120°C. The equilibrium values of the H_2/H_2O ratio and the dew points in H_2/H_2O atmospheres for selected metal–oxide systems are better shown in Fig. 2.6. It is necessary to note the differences in oxygen

Table 2.3. Relation between dew point and moisture content of gases [1]

Dew point		Moisture		Dew point		Moisture	
(°C)	(°F)	mg/l	vol.%	(°C)	(°F)	mg/l	vol.%
30	86	30.2	4.18	–15.6	4	1.54	0.18
25.6	78	23.7	3.22	–20	–4	1.08	0.125
20	68	17.3	2.31	–25.6	–14	0.675	0.076
15.6	60	13.3	1.75	–30	–22	0.455	0.0503
10	50	9.4	1.21	–35.6	–32	0.272	0.0294
7.8	46	8.15	1.04	–40	–40	0.178	0.0189
5.6	42	7.05	0.895	–45.6	–50	0.107	0.0112
0	32	4.8	0.6	–50	–58	0.064	0.0065
–5.6	22	3.28	0.4	–56.7	–70	0.029	0.0028
–10	14	2.35	0.282	–62.2	–80	0.014	0.0014
–13.3	8	1.83	0.216	–65	–85	0.009	0.0008

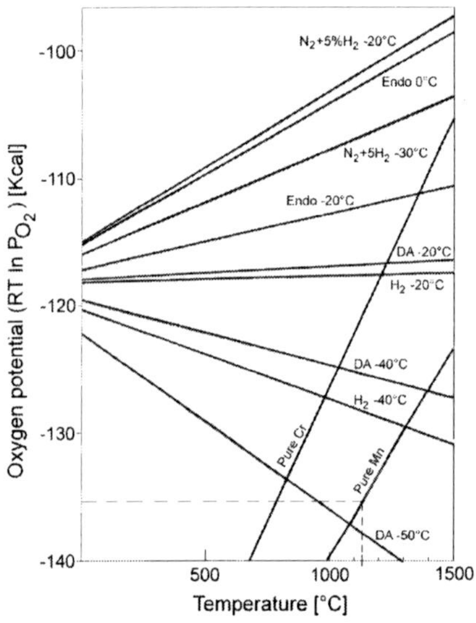

Fig. 2.9. Oxygen potential in endogas, dissociated ammonia, hydrogen and N_2-based atmospheres at different dew points (indicated temperatures refer to the dew point of the atmosphere [8,15,16].

potentials in H_2/H_2O atmospheres for sintering with Mo and Ni alloyed steels compared to that alloyed with Cr and Mn. Low requirements for the purity of the atmosphere for industrial sintering of pressed iron based parts were perhaps the main starting reason for alloying of sintered steels mainly with Ni and Cu as elements

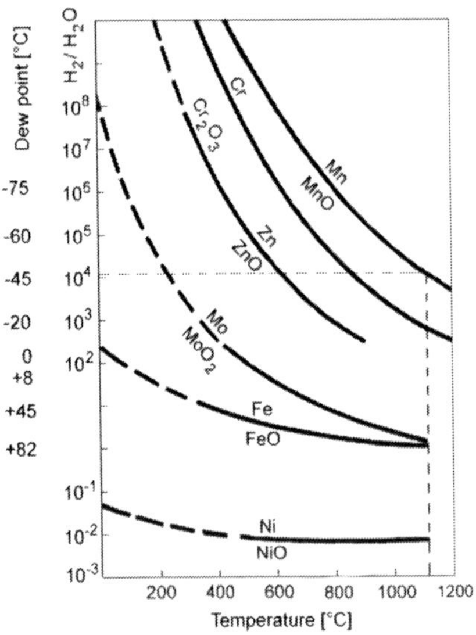

Fig. 2.10. Equilibrium values of H_2/H_2O ratio and the dew points in H_2/H_2O atmospheres for selected metal-oxides systems [4,7].

with low affinity for oxygen, *i.e.* low requirements o the purity of the sintering atmospheres [2], Fig. 2.10.

All mentioned reactions for Mn–O–H systems from the thermodynamic point of view are calculated for the unit activity of the system. They refer to chemically pure compounds, including hydrogen. In the real systems, the chemical activity of, for example, all manganese carriers as, for instance, ferromanganese grades and/or some special Mn-containing master alloys, is much lower because they contain different impurities, mainly iron. Accordingly, FeO and vice-versa (unlimited solid solubility) can substitute the places in the lattice of MnO.

The phase stability diagrams Mn–O–H and Mn–O–C for 800, 1200 and 1300°C, elaborated on the basis of mentioned thermodynamic analyses for reduction and conversion reactions of manganese oxides with $H_{2(g)}$, $CO_{(g)}$ and $C_{(s)}$ for the Mn–O–H and Mn–O–C systems, made it possible to deduce comprehensively the following results [8]:

- reduction of MnO → Mn by neither $H_{2(g)}$, $CO_{(g)}$ nor $C_{(s)}$ occurs below 1280°C
- reduction of MnO is possible only with $C_{(s)}$ at temperature above 1280°C
- conversion of MnO_2 to Mn (MnO_2 is not stable) is possible, the highest step of conversion reaches the reaction with $C_{(s)}$
- reduction both of Mn_2O_3 → Mn at >1000°C and of Mn_3O_4 → Mn at >1100°C with $C_{(s)}$ can take place
- reduction of $MnO_{(g)}$ in $H_{2(g)}$ and $CO_{(g)}$ can take place.

According to [17], the reduction of MnO to Mn occurs with carbon (solid) as graphite at a temperature over 1200°C evenly only through the Mn-carbide (Mn_5C_2) as follows from the overall reduction reactions for this process:

$$5MnO + 7C = Mn_5C_2 + 5CO$$
$$2MnO + Mn_5C_2 = 7Mn_{(g)} + 2CO.$$

During the sintering of manganese steel, the conditions for the reduction with solid carbon do not form because at the temperature over 900°C, solid carbon (graphite) in sintered steel cannot be present due to its dissolution in austenite.

In term of the mentioned thermodynamic calculations, the presence of MnO in being admixed sintered Fe–Mn compacts is apriori assumed because the all calculations start with this Mn-oxide. The presence of some manganese oxides introduced or during sintering formed in sintered manganese containing systems in the literature was not observed and therefore also not analyzed. The surface of the starting manganese carrier particles (electrolytic manganese, ferromanganese grades, and/or special Mn-containing master alloys) should be covered with some layer formed at milling in which could be some manganese oxide or Mn–O oxide film. The MnO, which is green in colour and by this easy discernible also by look, is possible to exclude due to its reducibility only at the temperature exceeding the common sintering temperatures as follows from the previous data. Beside it, there are no data in the literature regarding the Mn-oxides on the particles of the manganese carriers. There is only the possibility for oxidation of manganese particles in the compacts during preheating period as well under laboratory as under industrial sintering conditions up to the temperature at which the equilibrium occurs at the dew point of the atmosphere used. For instance, by the use of the H/N sintering atmosphere with a dew point of –50°C, the equilibrium conditions for the Mn–O system corresponds to the temperature of 1150°C. This purity of the atmosphere does not fulfill the requirements for the preheating period over the temperature 400°C, at which manganese starts to oxidize, Table 2.2.

These data show that the sintering of manganese-alloyed steels in accordance with the thermodynamic requirements on the purity of the protective atmosphere cannot be realised without taking into account the stage of preheating to sintering temperature. In addition to this, all the thermodynamic criteria for the Mn–O system take into account only the possible reduction of MnO. There are no data indicating whether this manganese oxide should be introduced into the Fe–Mn system as a layer on the manganese carrier powder, or whether the manganese particles in the compact should oxidise in the atmosphere with insufficient purity. Under no conditions can the MnO be reduced to such an extent that it would be capable of alloying the iron matrix of the compact in some actual sintering conditions. On the basis of these data it should be accepted that the sintering of manganese-alloyed steels takes place under different physical–metallurgical conditions.

2.3. Influence of protective atmospheres on proper sintering of carbon containing steels

In general, the task of the protective gas is to prevent oxidation of the compacts during sintering and reduce any oxide skins remaining from the powder processing, so that the contact between particles, and therefore, material transport during sintering are not inhibited [5]. In the sintering of carbon-containing steels, the protective atmosphere should prevent the decarburisation of the compacts.

To obtain the required high strength and other functional properties for greatly differing technical applications, the sintered steels always contain carbon mostly in combination with the additions of suitable alloying elements. It is well-known that the optimum properties of the sintered steels in combination with the additions of suitable alloying elements can be achieved only by a precise control of carbon control, what is a further requirement for the atmosphere. A comparison between the ranges of carbon content of fully dense wrought steel versus sintered porous steel is strongly unfavourable to PM. A precise control of carbon content is always a critical factor for elimination the decarburisation and by this for achieving, systematically, the highest required properties. The controlling role of the shielding atmosphere is discussed briefly in Table 2.4.

Taking into account manganese, to these tasks with the special reference to the purity atmosphere it is necessary to add protection against oxidation of manganese particles as the alloying element during preheating, including the dewaxing zone, and also throughout the entire sintering process. This requirement follows from the thermodynamic considerations for the Mn–O system.

The mentioned 'unit' stages of sintering can proceed partly in other temperature ranges in relation to the starting base and alloy elements used, they can overlap each other. This can be, for example,

Table 2.4. 'Unit' stages occurring during the sintering of steels under the influence of a controlled atmosphere [18]

No.	Unit stage	Temperature range [°C]
1	Lubricant removal in the liquid state	150–200
2	Lubricant removal by gaseous decomposition	300–600
3	Refining (reduction of iron oxides)	750–1000
4	Carbon diffusion	900–1050
5	Diffusion of alloy additions	1050–1300
6	Carbon pick-up by atmosphere (in some cases)	1000–1300
7	Carbon restoration	1050–800
8	Microstructure formation	950 – 600
9	Final cooling under reducing or inert conditions	600 – 50

reduction of iron oxides. All starting powder particle surfaces used in powder metallurgy are commonly clean, not covered with stoichiometric oxides. It can therefore be assumed that the powders are covered with chemisorbed oxygen. Similarly, the alloying of iron matrix with some alloy element can proceed at lower temperature in dependence on mutual diffusivity of the elements.

The possible atmospheres, considering only the sintering of carbon-containing steels, which depend also on the type of alloy addition, may be:

- endogas from gaseous hydrocarbons (usually methane or propane)
- synthetic endogas (mixtures of methanol and nitrogen)
- cracked ammonia
- nitrogen-based mixes (N_2 + H_2) without carburising addition
- nitrogen-based mixes (N_2 + H_2) with carburising addition
- highly diluted endogas
- vacuum.

The most often used carburising addition is 0.5% of natural gas (methane, CH_4) to the atmosphere. This is especially important if the sintering atmosphere is required to have a certain 'carbon potential' to prevent decarburisation, *e.g.* during the sintering of steels containing C, or in other cases to promote carburisation. The water will combine with the carbon in the methane to form CO or CO_2, leaving the carbon in the iron unattacked.

Laboratory sintering of carbon-containing steels under a getter (alumina + graphite) is a common method to prevent the decarburisation of the parts. The amount of graphite addition must eliminate supercarburisation. The use of some controlled atmosphere depends on the local source of the base gas. The same atmospheres and methods to protect the sintered steels against the decarburisation are valid also for manganese steels.

In respect to carbon in the compacts, pure hydrogen as a protective atmosphere at sintering is the most common of commercially pure gases because it provides the most effective reducing atmosphere, but does not ensure decarburizing. The same character has also pure nitrogen as a cheap and explosion-safe inert gas or purging gas, especially for metals such as iron and copper which do not react readily with the furnace gas. On contrary, in this case carbon acts as a reduction agent to the oxygen introduced in the system with the base powders. Nitrogen-based protective atmospheres are produced by mixing inert, dry nitrogen with small, carefully controlled amounts of 'active' ingredients to control oxide reduction, carbon potential and strict dimensional accuracy, microstructure and mechanical and other required properties. Cracked (dissociated) ammonia is relatively a low-cost reducing atmosphere to sinter many metallic materials, is free from CO, CO_2 and water, and does not contain another oxygen or sulphur compound. The dissociation reaction occurring near 1000°C gives a pure, low moisture content atmosphere. The residual ammonia content is typically below 250 ppm, and as long as the moisture content is low the atmosphere is nearly neutral with respect to carbon. Due to its high H_2 content, the gas is decarburising and is therefore not suitable for sintering carbon-containing steels [5].

Decarburisation of manganese steels could be prevented by the same methods as for other steels, but no experimental knowledge applied in practice is available. Regarding the effect of some alloying element on the carbon activity in austenite, for instance Si or Ni, this element originates a decrease of the carbon content of saturated austenite and also of pearlite, whereas alloying additions, like Cu, Mn and Cr, decrease carbon activity. For this reason, the addition of more than 0.7% C to nickel-alloyed PM steels should be avoided to prevent any possible formation of brittle carbides at the grain boundaries [18]. As can be deduced from this, the addition of more than ~0.6% C to the sintered manganese steels should be studied in more detail regarding the characteristics of manganese in wrought steels, Chapter 2.1.

3

Alloying and sintering of manganese steels in terms of high manganese vapour pressure

3.1. Vapour pressure of elements

Every chemical element is characterised by its physical–chemical property – vapour pressure – the actual value of which depends only on temperature. The temperature dependence of the vapour pressure of some metals used in industry, including those used in powder metallurgy as alloying elements or base metals, is shown in Fig. 3.1.

Some crystalline materials exhibit a relative high vapour pressure which attains the barometric pressure at the temperature lower than

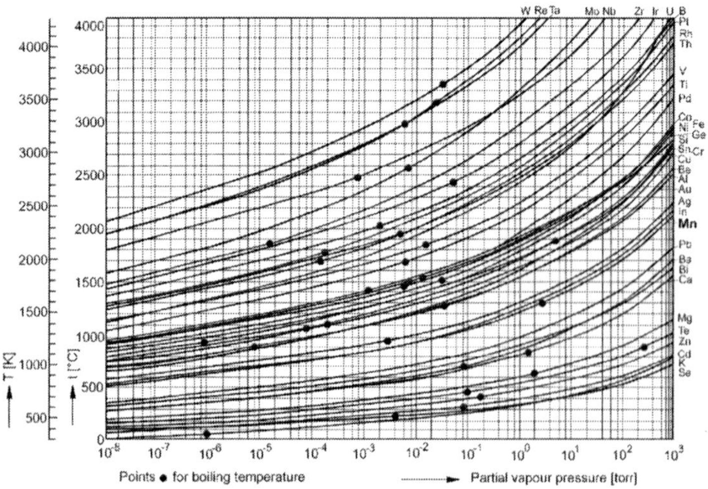

Fig. 3.1. The temperature dependence of vapour pressure of various metals.

their melting point. For this reason it is not possible to attain the melting point by heating at the pressure of one atmosphere because at such a pressure they proceed directly into the gas phase. This state change is sublimation and the temperature at which the vapour of the solid material reaches the pressure of one atmosphere is the normal sublimation point. The vapour pressure of a liquid (the same also for solid sublimating matter) does not depend on the presence of an inert gas in the space over the liquid which will be saturated by the gas of the liquid phase in that space should therefore be evacuated. This is limited only to small pressures of an inert gas. In the case that these pressures are high then also the indifferent gas influences the equilibrium vapour pressure above the liquid. The pressure which the saturated vapour reaches is independent of the amount of liquid or vapour and depends only on temperature [19].

Manganese belongs to the group of crystalline materials which on the basis of their physical–chemical substance sublimate at the barometric pressure at temperatures lower than their melting point. In the case of manganese these temperatures form practically the main range of sintering temperatures. The knowledge regarding the factors affecting the sublimation of manganese will be used to analyze its special effect on sintering and alloying if we want to use it as an alloying element in production of sintered steels.

3.1.1. Basic formulas characterising the sublimation of manganese from solid manganese

To explain the real conditions of manganese sublimation with the impact on sintering of manganese steels, it is important to investigate the dependence of manganese vapour on temperature, the vapour volume formed, and the sublimation rate and time. The processes occurring during sublimation of the solid phase are the same as at previously mentioned evaporation of the liquid phase. The difference lies only in the metal surfaces from which the evaporation or sublimation occurs. The manganese vapour pressure can be calculated from the relation valid for α- and δ-manganese including the constants according to [12,20,21]:

$$\log p = A \cdot T + B \cdot \log T + C \cdot T + D$$

where A, B, C and D are the constants (–14920, –1.96, 0, 18.32), T is temperature (K), p the vapour pressure (Pa) for manganese. According to these data, the corresponding equation for manganese is:

$$\log p = -14920 \cdot T^{-1} - 1.96 \cdot \log T + 18.315 \ [\text{Pa}].$$

The vapour volume of manganese generated by sublimation or evaporation is calculated according to the standard equation used for an ideal gas, which with some simplification can be applied even for manganese vapour:

$$p \cdot v = m / M \cdot R \cdot T$$

where R is the gas constant (9.31441 J·K^{-1} mol^{-1}), p vapour pressure in Pa, T temperature in K, v gas volume in m^3, m mass in g, M molecular mass (54.93 g·mol^{-1} for manganese).

The rate of atom sublimation from a surface unit can be expressed according to equation:

$$\log W = -3.358 + 0.5 \log M + \log p - \log T$$

where W is the sublimation rate in g·cm^{-2}·s^{-1}, M the molecular mass in g·mol^{-1}, p the vapour pressure in Pa, and T the metal surface temperature in K.

The sublimation time of certain metal quantities from the surface at a certain temperature can be expressed by the equation:

$$\tau = m / v \cdot F$$

where τ is the sublimation time in s, m the metal mass in g, v the sublimation rate in g·cm^{-2}·s^{-1} and F the free metal surface in cm^2.

The temperature dependence of manganese vapour pressure and sublimation rate for solid manganese is listed in Table 3.1. Table 3.2 gives the calculated data for sublimation rates and time of manganese from cube-shaped manganese particles in dependence on temperature up to 1250°C. The calculation refers to the geometrical surface of the particles of elemental manganese and to the temperature range of 400–1250°C covering the temperatures at which the oxidation of manganese in air starts (400°C), the preheating phase in sintering and the isothermal sintering temperature range applied mostly also for sintering the manganese steels (1100–1200°C) regarding the melting point of manganese of 1244°C. It is necessary to expect that at given sublimation rates the sublimation of larger Mn particles, *e.g.* 40–50 μm in size, will also finish during the preheating period. The longer sublimation time compared to that for pure manganese can be expected when using ferromanganese or some special Mn-containing master alloys as manganese carriers due to their lower chemical

Table 3.1. Manganese vapour pressure and sublimation rate in dependence on temperature [22–24]

Temperature [°C]										
400	500	600	700	800	900	1000	1100	1150	1200	1250
Manganese vapour pressure [Pa]										
$4.0 \cdot 10^{-10}$	$2.3 \cdot 10^{-7}$	$2.9 \cdot 10^{-5}$	$1.3 \cdot 10^{-3}$	$2.9 \cdot 10^{-2}$	0.38	3.2	20.0	45.0	99.0	188.0
Sublimation rate [$g \cdot cm^2 \cdot s^{-1}$]										
$1.0 \cdot 10^{-14}$	$2.6 \cdot 10^{-11}$	$2.2 \cdot 10^{-9}$	$4.0 \cdot 10^{-9}$	$8.7 \cdot 10^{-8}$	$9.8 \cdot 10^{-7}$	$8.1 \cdot 10^{-6}$	$4.8 \cdot 10^{-5}$	$1.03 \cdot 10^{-5}$	$2.2 \cdot 10^{-4}$	$4.0 \cdot 10^{-4}$

Table 3.2. Calculated sublimation time of manganese vapour from the cube particles 10, 20 and 50 mm in size and vapour volume generated from 1 g of manganese in dependence on temperature at corresponding sublimation rates [8,20,23] (Corresponding sublimation rates are listed in Table 3.1)

Temperature [°C]		400	500	600	700	800	900	1000	1100	1150	1200	1250
Sublimation time [s]	10 μm	$2.4 \cdot 10^{10}$	$4.6 \cdot 10^{10}$	$3.8 \cdot 10^{5}$	$8.7 \cdot 10^{3}$	$4.2 \cdot 10^{2}$	30.3	4.1	0.9	0.6	0.15	0.1
	20 μm	$4.8 \cdot 10^{10}$	$9.1 \cdot 10^{7}$	$7.6 \cdot 10^{5}$	$1.7 \cdot 10^{4}$	$8.2 \cdot 10^{2}$	66.7	8.2	1.8	1.5	0.3	0.2
	50 μm	$1.2 \cdot 10^{11}$	$21.3 \cdot 10^{8}$	$1.9 \cdot 10^{5}$	$4.4 \cdot 10^{4}$	$2.1 \cdot 10^{3}$	$1.7 \cdot 10^{2}$	20.4	4.5	3.5	0.8	0.5
Vapour volume [l]		1.00	1.15	1.30	1.45	1.60	1.75	1.90	2.1	2.1	2.2	2.5

Remark: the particles 10 and 20 mm in size can be accepted as mean in a select charge, probably 50 mm particles in size as maximum.

activity (lower Mn content). For instance, in one cubic centimetre of a Fe–3Mn compact with a density of 7.0 g cm^{-3} manganese added in the amount of 0.21 g contains $8.6 \cdot 10^6$ particles 15 μm in size or $4.5 \cdot 10^5$ particles 40 μm in size.

The influence of the manganese particles (specific) surface area on the time unit of the sublimated manganese volume is evident. Also, by raising the temperature by 100°C the sublimation rate increases by approximately two powers. The rate of manganese sublimation from fine particles at high temperatures takes place almost immediately due to the low specific surface area. These data must be taken into account in the evaluation of the alloying of sintered manganese steels. The calculated data show that manganese can be effectively sublimated even at low temperatures. The real manganese carrier particles exhibit mechanical, structural and lattice defects which increase the volume of manganese sublimated in unit time.

It is evident from the given data that the manganese vapour volume, generated by sublimation from Mn particles in some compact, is considerably higher than the pore volume, *i.e.* the manganese vapour fills the total open volume in the compact already in early stages of sublimation. In real systems, the manganese carrier is a milled powder with the highest size limit only, for instance 40–45 μm. Sublimation starts at some temperature of all manganese particles. Since the manganese particle size distribution is in the given extent markedly large, the sublimation continues up to completion of sublimation from the largest particles and during this process is also affected by the physical–mechanical characteristics of the individual particles and the heating rate of compacts. The vapour pressures of alloying elements commonly used in powder metallurgy compared with that for manganese are listed in Table 3.3.

3.1.2. *Effect of manganese vapour in laboratory sintering of Fe–Mn–C samples*

The effect of high manganese vapour pressure formed by sublimation in the form of gaseous manganese was firstly observed at laboratory sintering of mixed iron–manganese–(carbon) compacts in porcelain boats.

Fe–4.5Mn–0.36C cylindrical samples were sintered with the aim to study in greater detail the microstructure characteristics and hardness of the samples. The samples were prepared on the basis of SC100.26

Table 3.3. Vapour pressure of manganese and of some alloying elements used in powder metallurgy in dependence on temperature in the temperature range 900–1300°C [8,22,24–26]

Element	Temperature [°C]				
	900	1000	1100	1200	1300
	Vapour pressure [Pa]				
Mo	$2.03\cdot10^{-17}$	$4.07\cdot10^{-15}$	$3.73\cdot10^{-13}$	$1.85\cdot10^{-11}$	$5.66\cdot10^{-10}$
Si	$2.25\cdot10^{-17}$	$5.40\cdot10^{-6}$	$8.13\cdot10^{-5}$	$8.44\cdot10^{-4}$	$6.49\cdot0^{-3}$
Ni	$3.95\cdot10^{-6}$	$1.17\cdot10^{-15}$	$2.11\cdot10^{-4}$	$2.68\cdot10^{-3}$	$2.25\cdot10^{-2}$
Fe	$2.99\cdot10^{-6}$	$6.47\cdot10^{-15}$	$8.85\cdot10^{-4}$	$8.40\cdot10^{-3}$	$5.93\cdot10^{-2}$
Cr	$1.08\cdot10^{-5}$	$2.36\cdot10^{-4}$	$3.26\cdot10^{-3}$	$3.13\cdot10^{-2}$	$2.24\cdot10^{-1}$
Cu	$4.23\cdot10^{-4}$	$6.10\cdot10^{-3}$	$5.94\cdot10^{-2}$	$4.23\cdot10^{-1}$	$2.33\cdot10^{0}$
Mn	0.38	3.23	20.0	95.0	367.0

sponge iron powder and manganese was added in the form of high-carbon ferromanganese (76% Mn, 6.6% C, FeMnC). The samples, casually placed in porcelain boats instead of commonly used alumina (oxal) boats, were sintered in a laboratory Mars furnace in a quartz tube 35 mm in diameter at a temperature of 1200°C in technical purity hydrogen from gas cylinders. After finishing sintering, the porcelain boats blackened compared to the previous sintering trials in the alumina boats, as shown in Fig. 3.2(a),(b). The sintered samples became metallic clean. No similar features were observed if the sintering of Fe–Mn–(C) samples was carried out in a laboratory retort furnace in steel boxes. This was also the case in the treatment of pieces of electrolytic manganese, Fig. 3.2(c). The ends of the boats

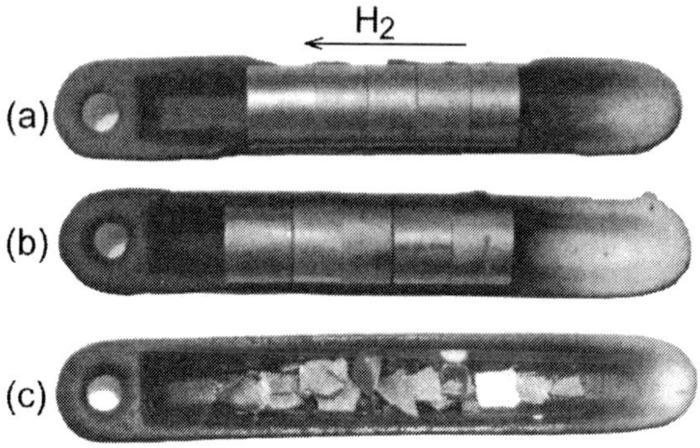

Fig. 3.2. Fe–4.5Mn–0.36C steel compacts (10 mm diameter, 10 mm long), sintering in porcelain boats at 1200°C in hydrogen for: (a) – 30 min, (b) – 10 min, (c) – pieces of electrolytic manganese, annealing 10 min at 1130°C [20,22–24].

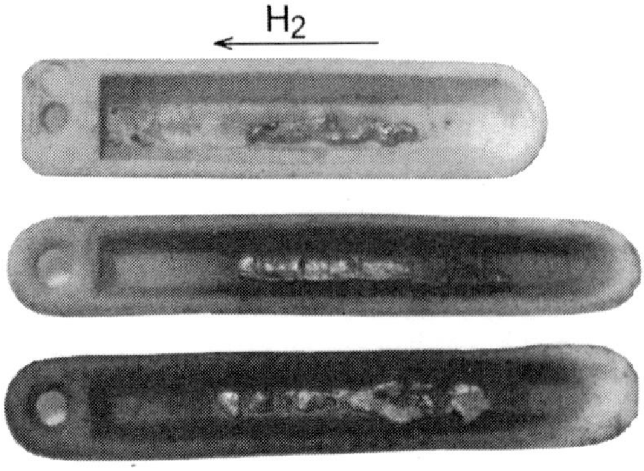

Fig. 3.3. High-carbon ferromanganese powder after annealing in hydrogen at 1120°C for: (a) –10 min, alumina boat, (b) – 3 min, porcelain boat, (c) – 10 min, porcelain boat. Note: The melting point of 1060°C was determined for high-carbon ferromanganese [20,22,23,25,27–29].

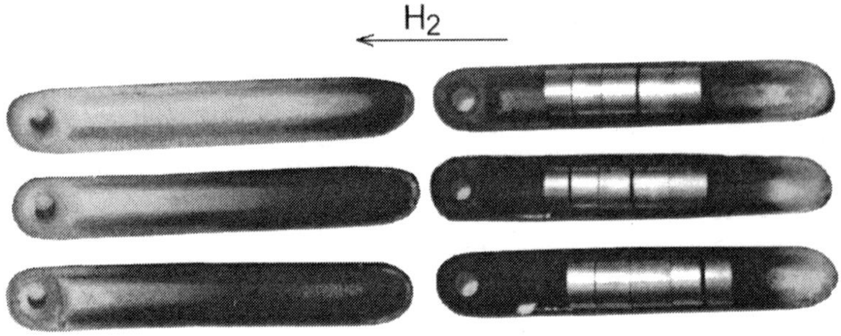

Fig. 3.4. Fe–4.5Mn–0.36C compacts (10 mm diameter, 10 mm long) as in Fig. 3.2, sintered in porcelain boats at 1100°C in hydrogen for: (a) – 10 min, (b) – 30 min, (c) – 60 min; (left) – empty boats in front of the active boats [20,23,24–26,29]

at the entry of hydrogen to them (right sides) were white and the left sides in the direction of the hydrogen flow became darker.

Figure 3.3 shows the boats with high-carbon ferromanganese annealed at 1120°C. Again, the appearance of the alumina boat did not change in this case and no blackening was detected, Fig. 3.3(a). As shown in Figs. 3.3(b),(c), the blackening of the boats also expressed itself in dependence on annealing time. Longer processing time resulted in blacker boats. The liquid phase of high-carbon ferromanganese was formed at 1060°C [20] and, therefore, the appearance of the ferromanganese charge processed in all three boats corresponds to the liquid phase at this temperature.

Figure 3.4 shows the blackening of the empty porcelain boats which were in the furnace tube placed in front of the boat with

Fig. 3.5 Porcelain boats with powder electrolytic manganese annealed in hydrogen for 10 min at temperature (°C): (a) – 900, (b) – 1020, (c) – 1050, (d) – 1100, (e) – 1250.

the compacts being sintered. All porcelain boats with the sintered compacts became darker in dependence on the sintering time as was shown previously. The empty boats also became in dependence on sintering time. More intense blackening of both groups of the boats occurred at longer sintering time. This shows that the manganese vapour, escaping from the sintered samples, was transported by hydrogen away over a longer distance through the tube. The manganese vapour transported by hydrogen also condensed on the surface of the boat where it reacted with the porcelain. The consequence of this only possible reaction was the previously mentioned blackening of the porcelain boats.

Figure 3.5 shows the boats with powder electrolytic manganese annealed at various temperatures. With increasing temperature the blackening of the walls of the boats became greater and covered a larger surface of the boats, more intensively in the direction of the hydrogen flow. The blackening of a boat is demonstrated also after treatment at 900°C, Fig. 3.5(a). The liquid phase of manganese formation at 1250°C is shown in Fig. 3.5(e). The side views of the boats, shown in Fig. 3.6, prove that the gaseous manganese filled the entire tube volume and by this also condensed on the outer sides of the boats.

Figure 3.7 compares the appearance of the heat treated alumina and porcelain boats with electrolytic manganese in two forms and with powder high-carbon ferromanganese (liquid phase formed).

These features, blackening, observed only at sintering of the Fe–Mn samples as well as at annealing of electrolytic manganese

Fig. 3.6. Side view of porcelain boats. (a), (b), (d), (e) as in Fig. 3.5 [20,23–25,28].

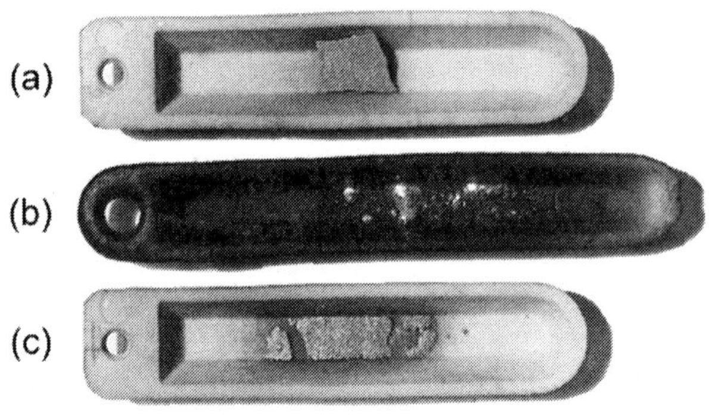

Fig. 3.7 Appearance of the boats with electrolytic manganese (EMn) and high-carbon ferromanganese (FeMnC) heat treated for 30 min at 1100°C in hydrogen: (a) – piece of EMn, alumina boat, (b) – FeMnC, powder, porcelain boat, (c) – EMn powder, alumina boat.

and high-carbon ferromanganese, used as manganese carriers, in the porcelain boats, proved the manganese sublimation and its reaction with porcelain only, Fig. 3.2(a). These features occurred under various annealing times and temperatures in the range 900–1250°C. The presented photographs prove realistically the manganese sublimation partly escaping from the samples being sintered into the flowing atmosphere from the compacts being sintered. In processing of manganese samples in the alumina boats, the entire amount of manganese vapour, sublimating from the manganese powder charge of the samples, was transported away with flowing hydrogen, which then burnt out at exit from the tube.

It was also established that the formation of a black layer on the porcelain boats under all tested conditions is the consequence of the only possible reaction of gaseous manganese with the porcelain condensed on its surface. In sintering of Fe–Mn samples it is necessary to note that the hydrogen used was only of technical purity.

These results were used to find a physical–chemical explanation of these observations and by this to study their consequences in the sintering and alloying of manganese steels. Since the samples sintered in the atmospheres which did not fulfill the thermodynamic requirements did not oxidise, they remained metallic clean, Figs. 3.2 and 3.4.

The presented knowledge about manganese sublimation and its known high affinity for oxygen was used initially as a method for increasing the purity of the sintering atmosphere. For this purpose, the tube of the Mars laboratory furnace in the hot zone was filled with pieces of electrolytic or ferromanganese (5–10 mm) and heated to 800°C. The hydrogen flowing through this furnace was used as a protective atmosphere for sintering of Fe–Mn–C samples in the furnace installed in front of the previous furnace in the direction of the hydrogen flow. The outlet dew point of hydrogen was –80°C [30]. These tests proved intense sublimation of manganese already at 800°C. Laboratory sintering of Fe–Mn–(C) compacts in hydrogen with the dew point considerably higher than the dew point for sintering at 1100°C, Table 2.2, was not carried out because the compacts produced by this procedure did not show better characteristics than those sintered in the hydrogen of normal purity.

3.1.3. Manganese sublimation and condensation in free space

In this case, the free space was the tube of the sintering furnace (diameter 30 mm), in which the previously described experiments were carried out, in contrast to the iron compact with a manganese addition. The surfaces of the pieces and powders of electrolytic manganese and carbon ferromanganese after annealing in boats, as presented in the photographs in the previous sections and produced under different conditions, were studied by scanning electron microscopy (SEM). Figure 3.8 shows different manifestations of the condensation of manganese vapour formed by sublimation in the given thermal conditions. These figures show that the sublimation of manganese from the manganese carrier did not occur spontaneously from the entire surface of individual particles or from the surface of the partially sintered powder charge, as shown in Fig. 3.5 [20,23–25,29,31].

Figure 3.8(a) shows needle-shaped formations extending from the edge of a piece of electrolytic manganese formed by condensation of manganese vapour. The background of the figure shows further formations of fine needles. Figure 3.8(b) shows, at a high magnifications, typical needless formed by the condensation of manganese vapour at a high temperature. X-ray analysis of a single 'needle' detected only pure manganese.

Sublimation takes place preferentially from defective areas of the manganese particle (particles), represented by sharp edges, needle-shaped formations with point ends (especially active), mechanical and lattice defects and the specific surface area of the heat-treated

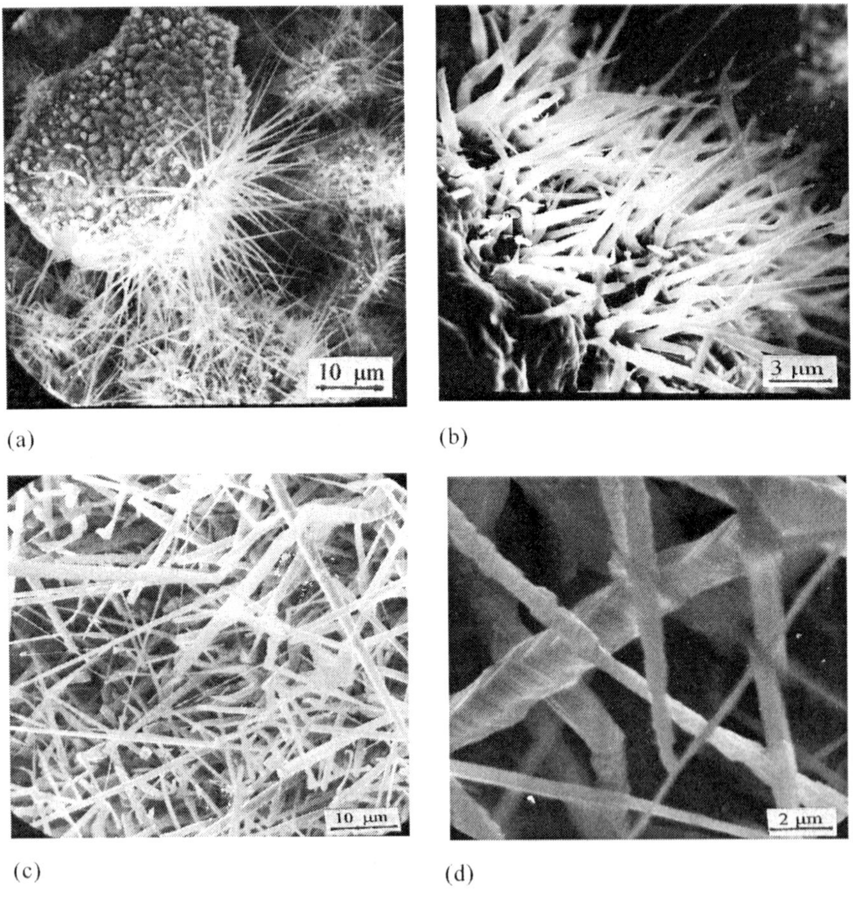

Fig. 3.8. Surface appearance of: electrolytic manganese in piece and powder form heat-treated under various conditions in hydrogen: (a) – pieces of electrolytic manganese, 15 min, 950°C, (b) – as (a) 10 min, 1100°C, (c) powder charge of EMn, 60 min, 1000°C, (d) as (c) 120 min, 1000°C.

manganese piece. The processed powder consists of individual particles of greatly differing shapes and size in the range of the selected particle size distribution. Figure 3.8(c) shows needles formed by the condensation of manganese vapour at heating of electrolytic manganese powder. No special sharp 'needles' formed on the surface of the charge, instead there was a network of fine 'fibres' of different thickness and shape which were connected together and did not extend to any large distance from the surface of the charge as the previously mentioned individual sharp needles. It can be concluded that each fibre forms independently points for sublimation by active points on two particles of the powder charge. These fibres, formed by the condensation of metal vapour, manganese in this case, have the form of whiskers. They are ductile and can be deformed to different degrees. Figure 3.8(d) shows the fibres under a high magnifications indicating formation of the granules of the final shape and thickness in condensation.

Figure 3.9 shows the surface of a piece of electrolytic manganese after annealing at two temperatures in air and in hydrogen.

Annealing in air resulted in the formation of individual short needles on the spherical formations of the surface. These needles are the product of sublimation and condensation of manganese vapour because no such formations were found in the initial condition. As shown by Fig. 3.9(b), after annealing in hydrogen there were clusters of needles formed by sublimation and condensation of manganese vapour from the previously mentioned obviously highly defective spherical formations on the surface. The processed pieces

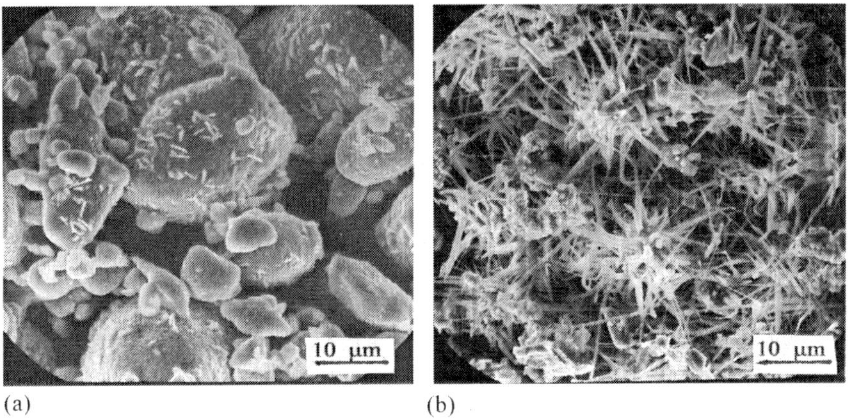

(a) (b)

Fig. 3.9. Surface of original uncrushed piece-electrolytic manganese annealed at: (a) 15 min, 800°C in air, (b) 10 min, 1000°C in hydrogen. SEM.

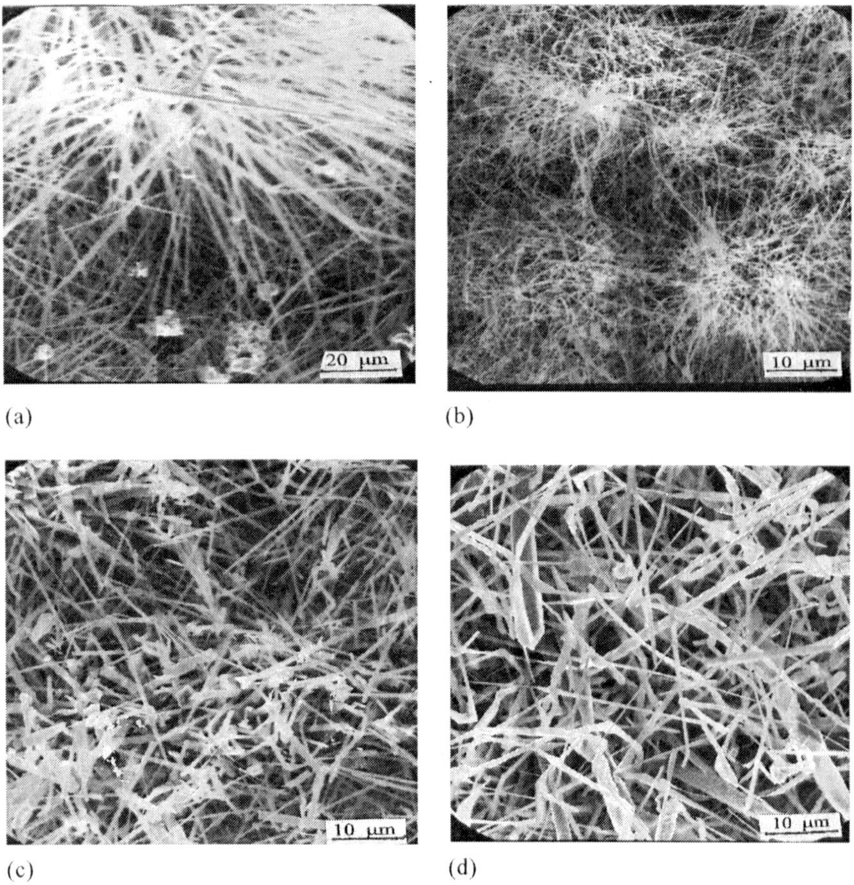

Fig. 3.10. Surface appearance of high-carbon ferromanganese powder annealed under various conditions: (a) 15 min, 1050°C, (b) 15 min, 1000°C, (c) 30 min, 1000°C, (d) 120 min, 1100°C; all in hydrogen, SEM. Coiled needle on (d) – EDX analysis: 98.15% Mn, 1.75% Si.

of electrolytic manganese were stored for a long period of time in air prior to processing.

Figure 3.10 shows the surfaces of samples of the high-carbon ferromanganese powder annealed in different conditions. Individual fibres formed on the surface of all samples as a result of condensation of manganese vapour, Fig. 3.10(a),(b), and a network of fibres shown in Fig. 3.10(c),(d) was also found. X-ray analysis detected in one of the fibres 19.15% Mn and 1.75% Si. In other cases, the fibres contained 98.83% Mn, 0.62% Si, 0.03% S, 0.11% K and 0.3% Al. This means that the accompanying, mostly contaminating, elements which can be found in ferromanganese, were carried away

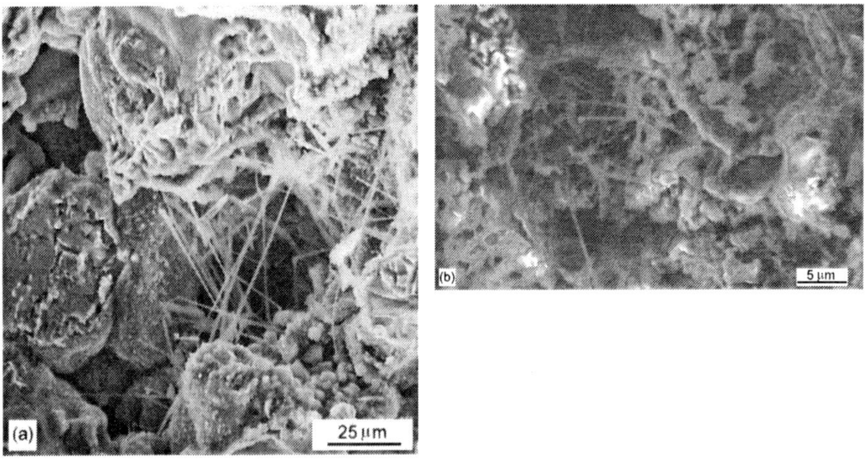

Fig. 3.11. (a) Fracture of Fe–4.5Mn–0.36C steel sample, SC100.26 iron powder, FeMnC, 600 MPa, induction sintering 3 min at 1200°C in hydrogen; (b) fracture of Fe–3Cr–0.5Mo–3Mn–0.24C steel sample, CrM prealloyed powder, FeMn, 600 MPa, industrial sintering 40 min at 1180°C in $70N_2/30H_2$ atmosphere, SEM.

by the manganese vapour from high-carbon ferromanganese during sublimation of manganese.

These results show that at the given conditions of heat treatment of the investigated samples of electrolytic manganese and carbon ferromanganese only a small proportion of the manganese vapour was carried away by flowing hydrogen which subsequently reacted with the porcelain or was carried away from the tube of the furnace. The manganese vapour condensed preferentially on the surface of the samples in the form of individual needles and/or fibres. It should be stressed that all the manifestations of sublimation and condensation of manganese have been and are influenced, in addition to the initial form (individual pieces or powder), also by the working space, the temperature and duration of heat treatment and the type of atmosphere.

These manifestations of the sublimation and condensation of the manganese vapour indicate that the rate of sublimation of manganese from the actual particles of the manganese carrier is higher and its extent is also greater than the values calculated for elementary manganese and presented in Table 3.1.

Figure 3.11 shows the manganese whiskers on the fracture surfaces of two sintered Fe–Mn–C steels. In preparation of the mixtures of iron powder and manganese powder, insufficient mixing, usually in the laboratory conditions, may result in the formation of large

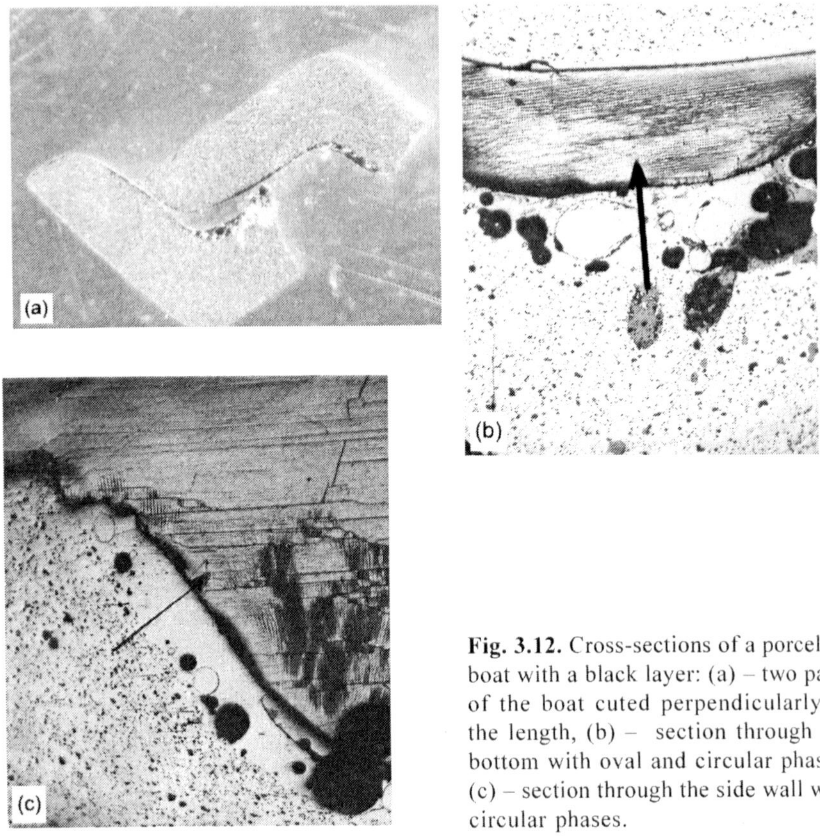

Fig. 3.12. Cross-sections of a porcelain boat with a black layer: (a) – two parts of the boat cuted perpendicularly to the length, (b) – section through the bottom with oval and circular phases, (c) – section through the side wall with circular phases.

pores in the compact during compacting together with a cluster of manganese particles. In this case, the fracture surface of the sintered specimen shows, in large pores, fine fibres connecting together areas that are active in sublimation.

3.2. Reaction of manganese vapour with porcelain

The black layers formed on the walls of the porcelain boats at sintering of Fe–Mn–C samples and at annealing of manganese carriers are a product of the reaction of gaseous manganese with SiO_2. Porcelain is formed by about 70% of SiO_2, 24% of Al_2O_3 and rest are the oxides of other elements. Gaseous manganese did not react with alumina boats.

The porcelain boat with a black layer was sectioned in the transverse direction to produce samples for analysis. Analysis was carried out by optical microscopy and spectral analysis. Figure 3.12

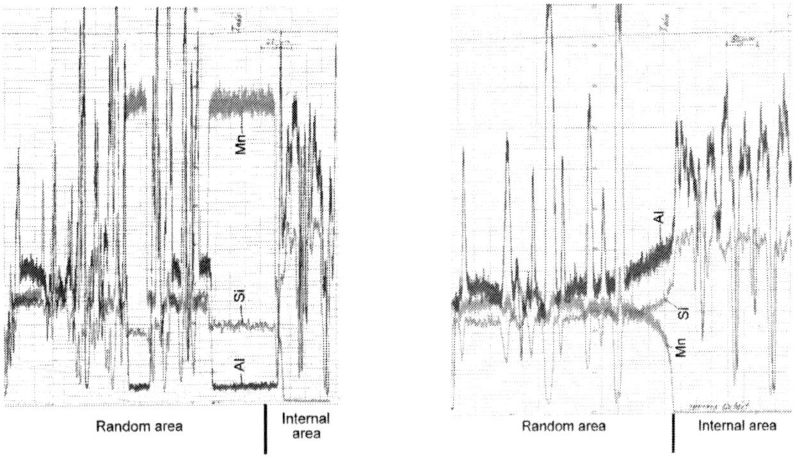

Fig. 3.13. Line spectral analysis (spectral analyzer EMX-SM/ARL) on the perpendicular cross-sections: left – through the bottom, Fig. 3.12(b),(d), right – through the lateral wall, Fig.3.12(c).

shows the cross-sections of the boat in three positions. The first view, figure (a), shows both halves of the boat with the traces of a black layer at the bottom and sidewalls. Figure 3.12(b) shows the section through the bottom of the boat and the direction of microanalysis, and Fig. 3.12(c) shows the same section through the side wall of the boat. It can be seen that extensive changes took place in the structure of the porcelain in the investigated conditions. One of these changes is the formation of a subsurface layer with full density in comparison with the porous initial state of the material. Another manifestation associated with the formation of the regions with full density is the formation of large spherical pottery of different sizes in the subsurface layer on the bottom of the boat. The third new manifestation is the formation of oval and circular particles in the subsurface pore-free layer of the boat. Spectral analysis, carried out in both parts of the boat, as shown in Fig. 3.13, confirms the presence and local changes in the concentration of Si and Al and also the presence of Mn. Analysis of the sample in accordance with Figs. 3.12(b),(d), which was performed through the circular phase, shows that it is an Mn + Si metallic phase with a higher Mn concentration.

3.3. Summary

The basic formulas, characterising the condition for manganese sublimation from solid manganese, are described.

The sublimation of manganese in the form of vapour from the manganese carrier particles in a sintered Fe–Mn–C compact, carried away by the atmosphere, was experimentally documented for the first time as the blackening of the surface of porcelain boats due to the condensation of manganese vapour on their surface. The surface of the sintered compacts retained its metallic appearance. No blackening was detected on the boats produced from aluminium oxide.

Blackening of the surface of the porcelain boats is the consequence of the reaction of manganese vapour with SiO_2. The reaction of the manganese vapour with this oxide in the porcelain was also reflected also in the formation of a pore-free layer and of large pores in the subsurface layer of the boat and the Mn + Si metallic phase.

The sublimation of manganese from the investigated individual pieces of the particles of the manganese carrier (electrolytic ferromanganese and ferromangaese grades) and subsequent condensation of the manganese vapour after annealing at elevated temperatures was reflected in the formation of whisker-type needles.

The sublimation of manganese from the manganese powder particles annealed separately at elevated temperatures was reflected in the formation, on the surface of the powder charge, of a network of whiskers connected together in the areas on the surface with respect to sublimation (sharp points, sharp edges, including lattice defects, etc).

The existence of manganese whiskers in the form of very fine fibres can also be detected on the fracture surfaces of the sintered samples, if a large pore with a large range of manganese particles remained in the compact as a result of insufficiently effective mixing.

4

Alloying and sintering of manganese steels by manganese vapour

4.1. Microstructure formation

4.1.1. Conventional radiation sintering

Chapter 3 describes the effects of sublimation and condensation of the manganese vapour initially in the form of blackening of the surface of porcelain boats and then in the formation of three-dimensional manganese whiskers after annealing pieces of powder electrolytic manganese and/or carbon ferromanganese in a free space, in the shaft of a sintering laboratory furnace. This chapter describes how these phenomena are reflected in the formation of the structure of manganese-alloyed steels and, in the final analysis, in the properties of sintered components under the effect of gaseous manganese.

The alloying of powder particles by the manganese gas phase was studied on cylindrical samples with a diameter of 10 mm and 10 mm long produced from a Fe–4% Mn–0.32% C powder mixture compacted at a pressure of 600 MPa and prepared on the basis of two types of iron powder with different structural characteristics. The following powder iron was used: a) air-atomised iron powder produced by air atomisation and referred to as RZ (Rohzunderverfahren method), annealing at a temperature of ~1100°C, fairly large spherical particles and a well recrystallised structure, coarse-grained structure, coded here R; b) iron powder referred to as Hametag (eddy milled, platelet-shaped particles with a highly disordered lattice, dislocation density of $8.4 \cdot 10^9$ cm^{-2}, after annealing at 800°C, fine-grained structure) which is therefore characterised as a powder with high structural activity for sintering as well as for alloying, coded here H [32–37]. Manganese was added to the powder iron in the form of high-carbon ferromanganese (75% Mn, 6% C, 1% Si) with the particle size of <40 μm. (Both RZ and Hametag iron powder grades are no longer available).

The samples were sintered at temperatures of 1100 and 1200°C for 3, 10, 30 and 60 min in alumina (oxal) boats in a laboratory Mars furnace in a 18 mm tube in commercial purity hydrogen. Boats with the samples were inserted directly into the isothermal zone of the furnace and cooled at a rate of ~200–300°C·min^{-1} by rapid withdrawal of the boat into the cool part of the tube. The thermocouple was placed on the sintered samples. The microstructures of selected samples providing information on the formation of the structure of the sintered manganese steel, based on the previously mentioned iron powder grades, are discussed below.

The microstructures of the samples prepared on the basis of RZ iron powder, sintered at 1100°C, are shown in Fig. 4.1.

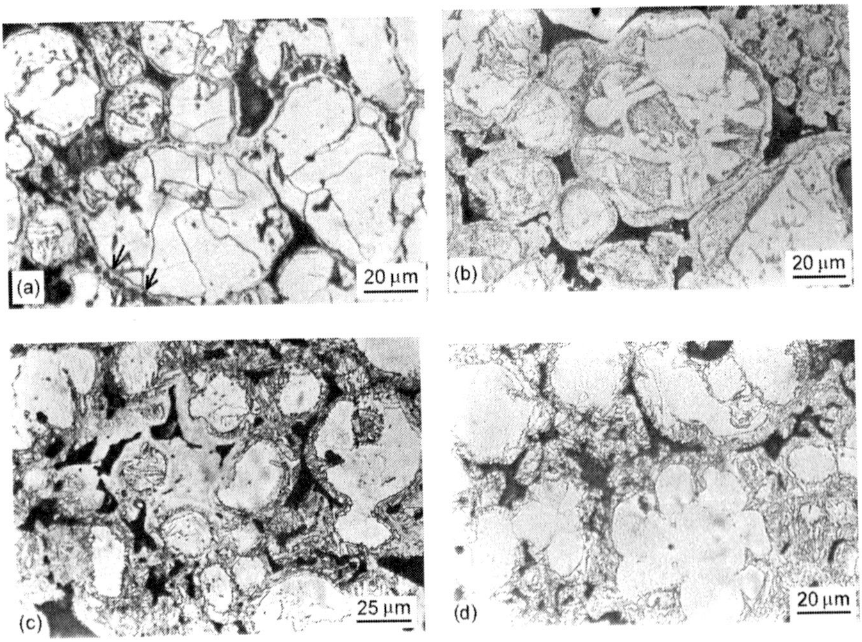

Fig. 4.1. Microstructure of Fe–4Mn–0.32C samples; RZ iron powder, sintering at 1100°C in hydrogen: (a) – 3 min, (b) – 10 min, (c) – 30 min, (d) – 60 min. Nital etched.

Figure 4.1(a) shows the basic structure of the particles of RZ iron powder and on the surface of all particles a thin uniform layer formed as a result of condensation of manganese vapour combined with high-rate surface diffusion. In the area indicated by arrows one can see the original surface of the powder particle without any signs of diffusion of manganese from this layer into the core of the particles. This means that the surface of the powder particles was alloyed with the gas phase when the sintering time and temperature were increased. The thickness of the layer increased and microvolumes of the powder particles were also gradually alloyed as a result of diffusion. This resulted in the formation of a manganese-alloyed structure of the iron powder particles in the compact by solid-state diffusion. It may be assumed that the area in the centre of Fig. 4.1(c) is a solidified liquid phase which formed in the cluster of high-carbon ferromanganese particles which formed the liquid phase at a temperature of 1060°C [20]. After sintering for 60 min, the manganese-alloyed layer became less distinctive, Fig. 4.1(d). Large areas of the particles remained ferritic as a result of the highly coarse-grained structure of this type of powder.

Figure 4.2 shows the characteristic microstructures of the sample sintered at 1200°C. Sintering at a higher temperature resulted in the formation of more uniform alloyed layers on the surface of the individual powder particles, Fig. 4.2(a). This photograph also shows that typical 'funnels' of the start of grain boundary diffusion which have the form of V-shaped regions where diffusion-induced grain boundary migration (DIGM) along the prior grain boundaries has penetrated [38,39]. This process of grain boundary diffusion from the condensed manganese layer corresponds to the scheme in Fig. 4.3. As indicated by the microstructure shown in Fig. 4.2(b), after sintering

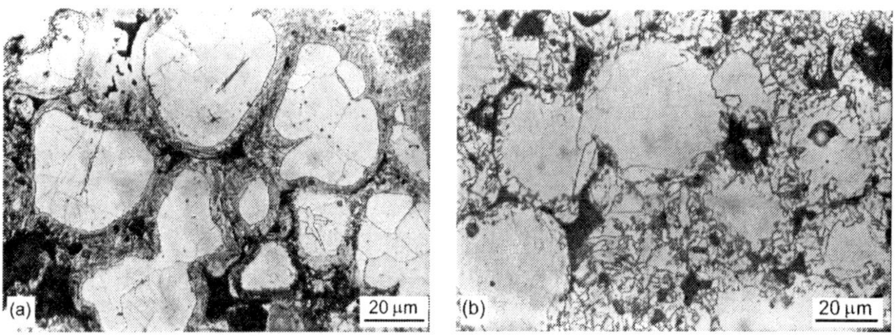

Fig. 4.2. Microstructure of Fe–4Mn–0.32C samples; RZ iron powder, sintering at 1200°C in hydrogen for: (a) – 3 min, (b) – 60 min. Nital etched.

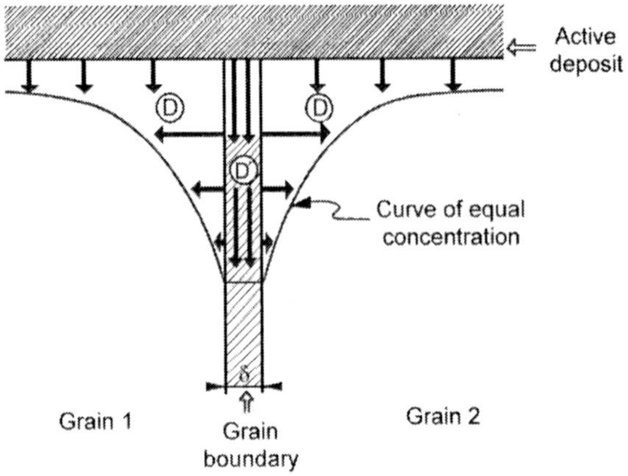

Fig. 4.3. Model of grain boundary diffusion from the deposited (condensed) layer [40]. *d* – boundary width, $D' > D$ – diffusion coefficients.

for 30 min the alloyed layers with new very fine grains became less distinctive. This is similar to the structure shown in Fig. 4.1(d). These phenomena are a consequence of reversed sublimation of manganese from the sintered compact into the atmosphere. The effect of the sintering time and temperature on the process of alloying of the matrix by diffusion of manganese from the surface layer and the structural activity of the iron powder is quite evident here.

These structures shows that in the compact based on this type of powder the particles showed sharply defined alloyed layers, from internal ferritic regions of isoconcentrations which do not correspond to the expected uniform alloying by bulk diffusion.

To confirm the extent of alloying of the iron powder particles with manganese, measurements were taken on all the investigated conditions to determine the microhardness of structural components in four areas of the sintered samples, *i.e.* the ferrite in the core of the particles (A), the ferrite in the immediate vicinity of the boundary of the alloyed layer (B), the alloyed layer (C), and completely alloyed small particles (D). The dependences of the microhardness of the individual regions on the sintering time at both temperatures are shown in Fig. 4.4.

A large difference was found in the values of microhardness of ferrite between the regions A and B and the phase-alloyed regions C and D in relation to sintering time, with the difference between these regions being only minimum. In the ferritic areas the microhardness increased after sintering for 3 and 10 min to 150 and 180 HV 0.05

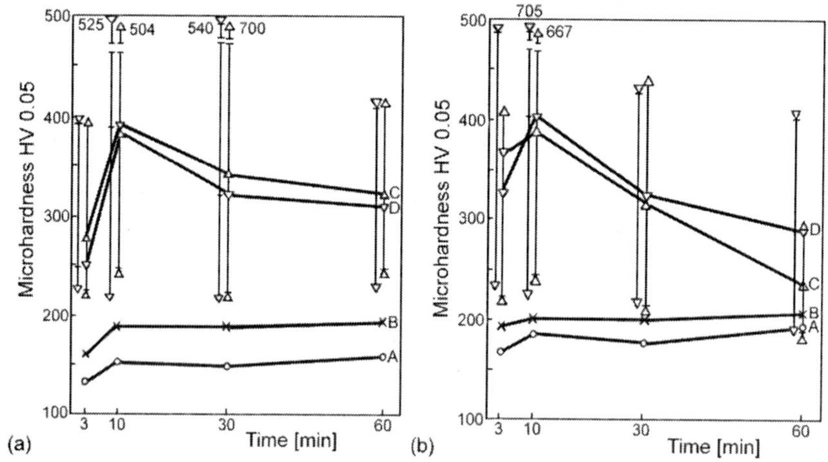

Fig. 4.4. Microhardness HV 0.05 of structural components of Fe–4Mn–0.32C samples measured in regions A–D in relation to sintering time at: (a) – 1100°C, (b) – 1200°C; A – ferritic core of the particle, B – ferrite at the boundary of the alloyed layer, C – legovaná vrstva, D – complete alloyed fine particles (every value is a mean value of 50–210 measuremets) [41,42].

and remained constant with increasing sintering time. Therefore, the 'white' area in these cases are already formed by the solid solution of manganese in ferrite. This is caused by hardening of the ferrite by the substitutional solution of manganese. The highest microhardness in the regions C and D was recorded after sintering for 10 min, with the average value being 390 HV 0.5 (range 220–625 HV 0.05). With increasing sintering time the microhardness of the alloyed regions decreased to the average value of 330 HV 0.05 (range 200–420 HV 0.05). This shows that after sintering for 10 min the alloyed regions in the microvolumes were formed by pearlite, bainite and martensite, and after sintering for 60 min by fine pearlite and bainite.

Sintering at 1200°C increased the microhardness of ferrite in the regions A and B to 180–200 HV 0.05 (solid solution), again without any change with increasing sintering time. After sintering for 3 min, the microhardness of the region C was higher than that of the region D. This corresponds to the large amount of manganese sublimated from the ferromanganese particles and subsequent condensation on the surface of the iron powder particles at a higher temperature. The average microhardness of the components in the regions C and D after sintering for 10 min was slightly higher than after sintering at 1100°C. An increase of sintering time resulted in a more rapid decrease of the microhardness of the components in comparison with

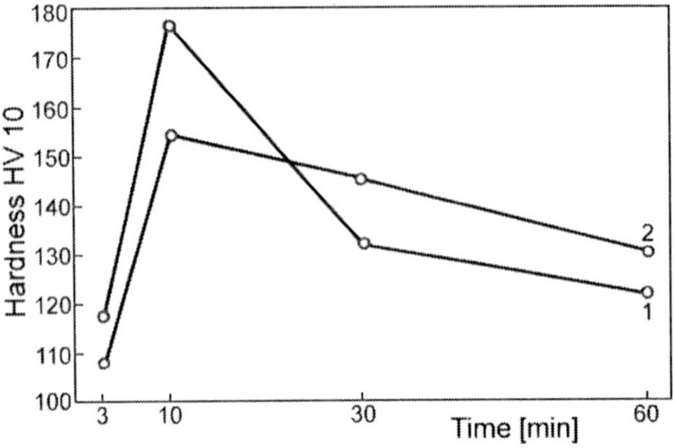

Fig. 4.5. Dependence of apparent hardness of Fe–4Mn–0.32C samples on sintering time at: 1 – 1100°C, 2 – 1200°C as in Fig. 4.4. (Every value is a mean value of 25 measurements) [41,42].

the previous case, in the region C to a value slightly higher than 200 HV 0.05, Fig. 4.4(b). This is the result of reversed sublimation of manganese (demanganisation) from these areas into the atmosphere, as indicated in Fig. 4.1(d).

The dependence of apparent hardness of the tested steel samples on sintering time and temperature, shown in Fig. 4.5, exhibits approximately the same character as that of microhardness. According to hardness values, sintering at 1200°C resulted in more homogeneous microstructures with prolonged sintering time.

The presented micrographs and diagrams showing uniform alloyed surface layers on all iron particles in the compacts formed by manganese prove that it was and is possible only by gas-phase (manganese vapour) – solid-phase (iron powder particle) system in contrary to conventional alloying in the solid-phase (alloy element) – solid phase (iron particle) system affected by the iron powder grade with its special characteristics and by sintering conditions.

Figures 4.1 and 4.2 show the initial stages of alloying the surfaces of the powder particles in a compact by the condensation of manganese vapour and subsequent gradual changes in the thickness of the layer with increase of the sintering time at the given temperature up to reversed sublimation of mangaese from the layer. This process is also confirmed in Fig. 4.4 by the changes of the microhardness of the structural components, and by Fig. 4.5 by the changes of hardness in relation to sintering time. On the basis of these data it may be concluded that the highest manganese concentration of the alloyed layer in the given sintering conditions was obtained after the first 10 min of sintering.

These results show that the formation of phase boundaries with the manganese isoconcentration of the defined alloyed layer by diffusion takes place at a high rate as a result of a higher concentration gradient of the manganese gas phase. This phenomenon is based on the high chemical driving force (100% Mn in the atomic layer on the surface of the particles in relation to the zero manganese concentration in the iron powder particles).

The previously mentioned driving force gradually decreases by successive completion of the sublimation of manganese from the finer ferromanganese particles, subsequent increase of the thickness of the surface layer formed as a result of condensation and progressing diffusion of manganese from the surface layer into the core of the particles. This concerns mostly the previously mentioned diffusion-induced grain boundary migration without precipitation, with the controlling role played by the chemical driving force, *i.e.* a very sharp discontinuity in composition at the grain boundary. The grain boundary migration can stop after some growth if the temperature is high enough. It was stated that the alloyed surface layer may consist of fine grains and that a steep gradient is to be expected between the layer and the substrate, and that the diffusivity in the migrating boundary is faster by as much as a factor of 3000 [43–45]. Grain boundary migration as a key to full density materials and microstructural processes, such as recrystallisation, are regarded as essential to the creation of new materials, as stated in [46]. This is

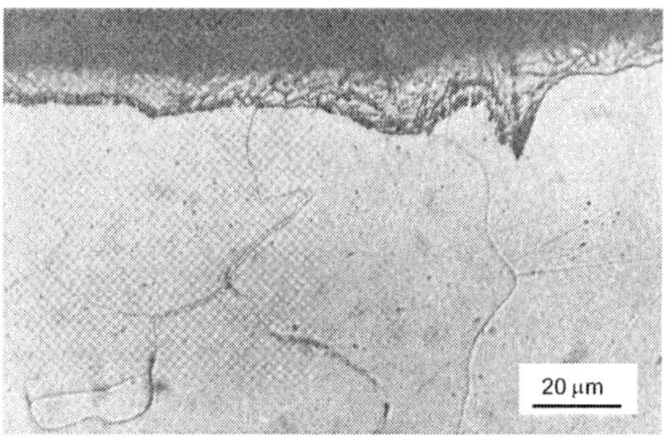

Fig. 4.6. Cross-section of a low-carbon steel plate heated 30 min at 1120°C in a capsule with low-carbon ferromanganese powder indicates some grain boundary migration in course of heat treatment. Nital etched [47].

also the case of sintered steels alloyed by manganese as follows from the shown microstructures.

The diffusion grain boundary migration in the Fe–Mn system is presented in Fig. 4.6. In this case, heating of the closed cylindrical capsules, containing small plates of low-carbon steel, partly immersed in ferromanganese powder, resulted in the 'plates' surface manganizing and carburising. The manganized layer appeared not only on the immersed part of the plates, but also on surface with no solid contact with ferromanganese. Although the structure of the manganized layer is indistinct, there is an indication of a grain boundary movement. The tips of the layer do not always concide with the visible grain boundaries in ferrite [47].

After sublimation of manganese from the individual particles and condensation of Mn vapour on the particle surface and subsequent diffusion into the bulk of the iron particles in the compact, a reverse process starts after a certain sintering time, *i.e.* sublimation of manganese preferentially from the alloyed layer of the material into the atmosphere. This is due to the change of the manganese concentration gradient, which in this case is significantly smaller than in sublimation and condensation of particles of manganese (concentration of manganese in the surface layer of sintered samples and zero concentration in the atmosphere). Manifestations of this process can be seen on the microstructures of sintered samples which are shown in Fig. 4.1(d) and Fig. 4.2(b) respectively with the formation of a large number of new fine grains in the previously alloyed areas of the iron powder particles. The same character of the microstructures was found for Fe–4 Mn steel sintered at 1200°C and 1280°C for 50 min but without explanation of insufficient alloying with manganese [48].

The diffusion trajectories in alloying of the iron matrix with by only bulk diffusion with the coefficients of bulk diffusion of $2.33 \cdot 10^{-11}$ $cm^{-2} \cdot s^{-1}$ at 1100°C and $1.7 \cdot 10^{-10}$ $cm^{-2} \cdot s^{-1}$ at 1200°C were calculated for comparison. The results showed that during these sintering periods the iron matrix particles would have been alloyed by bulk diffusion to a lesser distance from the surface. In doing so, manganese should be located only in the solid solution [49,50].

Figure 4.7 shows the linear distribution of manganese in the compacts sintered for 30 to 60 min at 1200°C. Sharp peaks after sintering for 3 min show the highest concentrations of manganese on the very surface as a result of the lack of diffusion inside the particles. The alloyed layer itself has a uniform concentration of

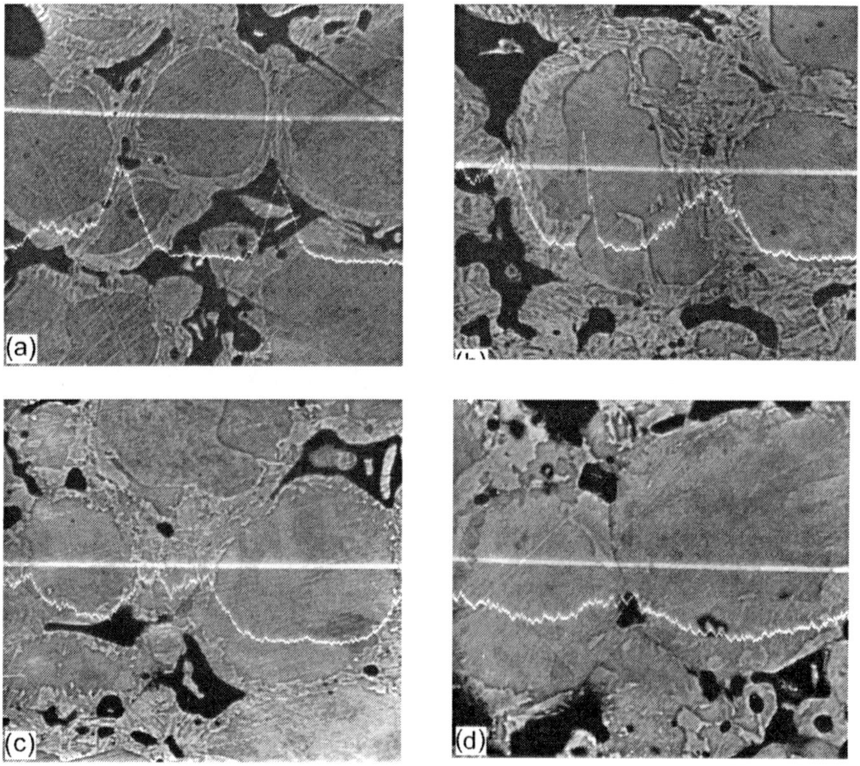

Fig. 4.7. Line distribution of manganese in Fe–4Mn–0.32C samples; RZ iron powder, <100 μm, 600 MPa, FeMnC, 1200°C in hydrogen for: (a) 3 min (b) 10 min, (c) 30 min, (d) 60 min. Nital etched [45–49].

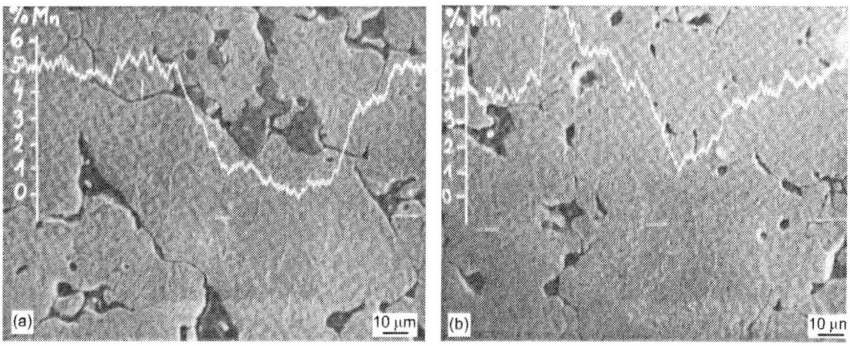

Fig. 4.8. SEM of a sintered compact (Fe–6.5Mn) with the Mn profile scans (wave dispersion line analysis); ASC100.29 iron powder, FeMnC, <0.315 mm, 600 MPa, 30 min, 1120°C, 95Ar/5H$_2$ atmosphere [45].

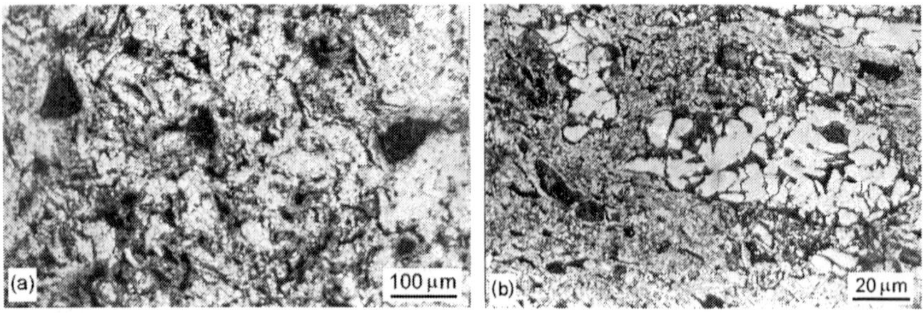

Fig. 4.9. Microstructure of Fe–4Mn–0.32C samples; Hametag iron powder, sintering at 1100°C in hydrogen for: (a) – 3 min, (b) – 10 min. Nital etched.

manganese which is reflected in the downward trend curve and also in the microhardness values. A sharp difference in alloying between the alloyed surface layers and the cores of the particles is clearly shown after sintering for 3 and 10 min. It can be seen that extending sintering time increases the thickness of the layer and the concentration of manganese becomes more uniform so that after 60 min sintering the particle surface has the minimum concentration of manganese and the curves in this case are characterised by only manganese in solid solution, which confirms the previously mentioned demanganisation.

Figure 4.8 shows the electron micrographs with manganese concentration profiles along a line of scan, indicated by the white abscisses realized by [47] on the basis of the knowledge about the sublimation and manganese presented [26,34]. The analysed samples were prepared using iron powder ASC100.29 with admixture of 8% of high-carbon ferromanganese. The scan line in figure (a) is responsible for the high peak similar in the interparticle contact plane like in Fig. 4.8(a). In figure (b) there is evidence of the transverse alloyed of the iron matrix with manganese.

Figure 4.9 shows the microstructure of samples prepared on the basis of iron powder Hametag with fine particles and a larger number of lattice defects as compared to coarse-grained RZ powder. In both states in the same conditions larger areas of the particles are alloyed with strong manganese diffusion inside the particle and also at the grain boundaries in the Hametag powder structure.

The extent of matrix alloying with manganese also in smaller microvolumes of powder particles suggests that the DIGM process was also crucial in this case.

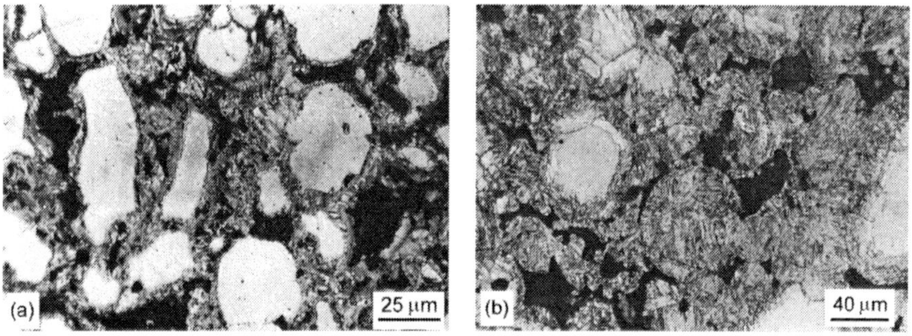

Fig. 4.10. Microstructure of Fe–4Mn–0.32C samples; RZ iron powder, induction sintering in hydrogen for: (a) – 1 min, 1100°C, (b) – 10 min, 1200°C. Nital etched.

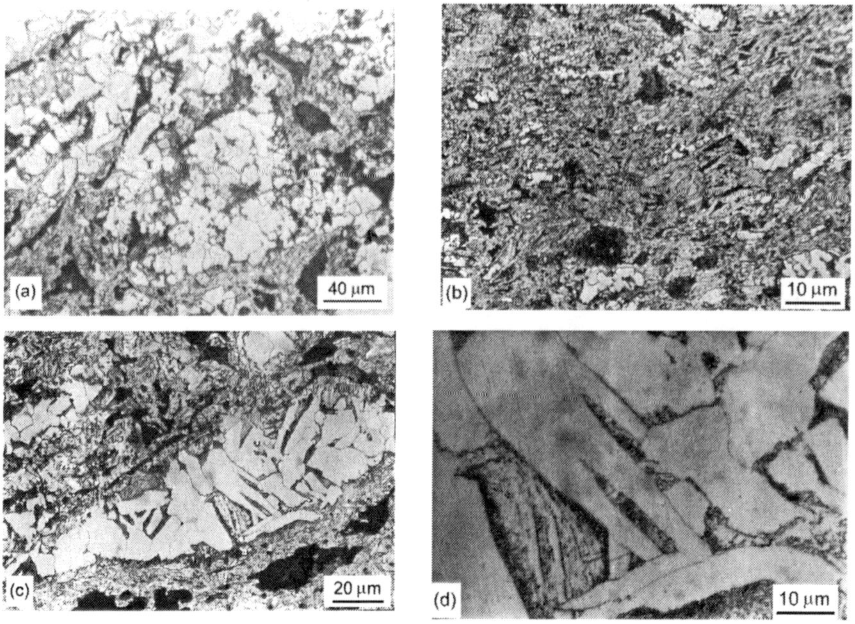

Fig. 4.11. Microstructure of Fe–4Mn–0.32C samples; Hametag iron powder, induction sintering in hydrogen for: (a) – 3 min, 1100°C, (b) – 3 min, 1200°C, (c) – 1 min, 1300°C, (d) – as (c) – detail. Nital etched.

4.1.2. Induction sintering

Induction sintering of the same samples as above was performed with a frequency of 1000 MHz with a heating rate of 17 $K \cdot s^{-1}$ at temperatures of 1100, 1200 and 1300°C for 1, 3 and 10 min. In this case, unlike radiant heating, hydrogen flowing through a quartz tube with an induction coil around it was not heated.

Figure 4.10 shows the microstructure of induction sintered samples prepared from RZ iron powder. After sintering for only 1 min an alloyed layer, separated by a phase boundary from the ferritic core of the particles was formed by the DIGM process on the surface of all the particles. After sintering for 10 min at 1200°C the majority of the particles in the area of the photograph was already alloyed through the entire thickness. The cores of larger particles remained ferritic [51,52].

The microstructures of the induction-sintered samples prepared on the basis of Hametag iron powder are shown in Fig. 4.11. Even in this case the higher structural activity of the powder was reflected in more extensive alloying of the internal part of the particles. The particle in Fig. 4.8(c) shows the alloyed surface layer as in Fig. 4.2(a) as evidence of diffusion by DIGM. A detail of the microstructure in Fig. 4.8(d) shows the heterogeneity of structures and substructures of the starting powder particles and austenite. This has a beneficial effect on the diffusion of manganese into the iron powder particles [52].

The nature of these microstructures shows that sublimation of manganese from the carrier, in this case the particles of carbon ferromanganese, with subsequent condensation on the particle surfaces of the iron powder in the compact, took place at 1100°C already after a short period of time and that gaseous manganese condensed on the surface of iron powder particles and then in the entire volume of the compact.

The process of alloying with manganese was strongly affected by the structural, substructural and technological properties of the iron powders determined by their production method. The RZ coarse-grained powder structure has proved suitable for studying alloying with gaseous manganese. Both methods of heating, radiant and induction, were identical from the physico-metallurgical point of view of their effect on the structures of the samples. This alloying process can be described as gas-phase (manganese vapour) – solid-phase (iron particle) alloying of iron-based materials with manganese.

4.2. Nucleation of new grains in base iron powders in Fe–Mn materials

Studies dealing with the diffusion-induced grain boundary migration in the solid phase–gas phase systems showed nucleation of new grains on the sample surface [43,44].

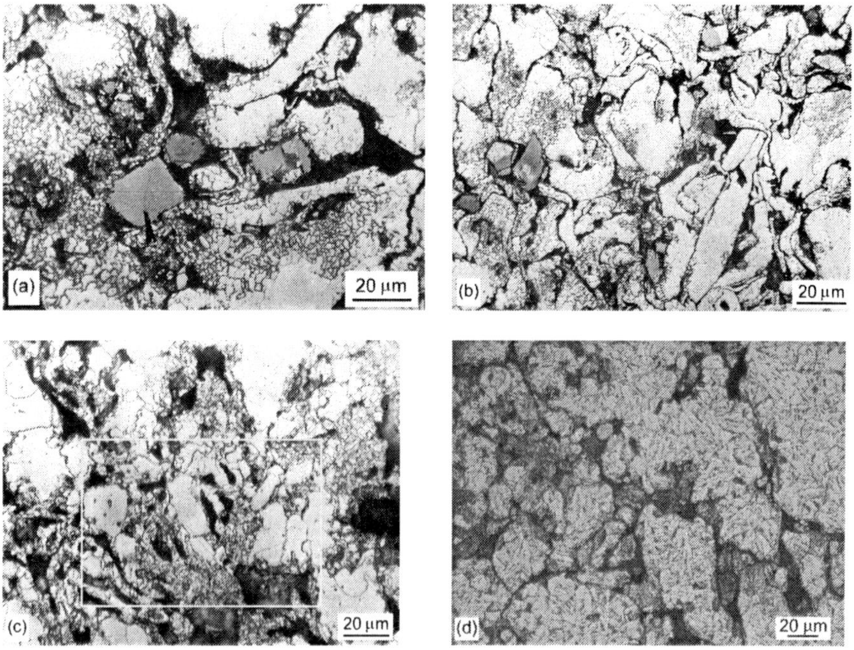

Fig. 4.12. Microstructures of Fe–4Mn–0.32C and of Fe–0.85Mo–3Mn–0.5C samples indicating the nucleation of new grains under the effect of manganese vapour; (a), (b), (c) – Hametag iron powder; (a), (b) – sintering 30 min at 750°C, (c) – sintering 3 min at 1100°C; (d) – hybrid Fe–0.85Mo–3Mn–0.5C mixed powder, 600 MPa, sintering 3 min at 875°C, cracked ammonia. Grey particles in (a) and (b) – FeMn, (c) and (d) FeMnC particles. Nital etched.

Figure 4.12 shows the nucleation of new fine grains in the samples based on Hametag iron powder and in the samples based on prealloyed Ast85Mo powder sintered under different temperature conditions in the gaseous phase–solid phase ($Mn_{(g)}$–Fe) system. As shown by Figs. (a) and (b), sublimation of manganese from ferromanganese particles (grey particles) takes place already at 750°C, as shown by 'gaps' at the edges of the particles that constitute a failure and are therefore active sites for sublimation. New very fine grains were not found in the powder particles in these powders in the initial state nucleated in the vicinity of these new particles.

Nucleation of new grains is also evident in Fig. 4.12(c) as well as in Fig. 4.11(a). New fine grains contribute to the accelerated and uniform diffusion of manganese and to alloying a greater volume of the individual powder particles by solid state diffusion. It can be assumed that the nucleation of new grains in the presence of gaseous manganese is also supported by the presence of a larger number of

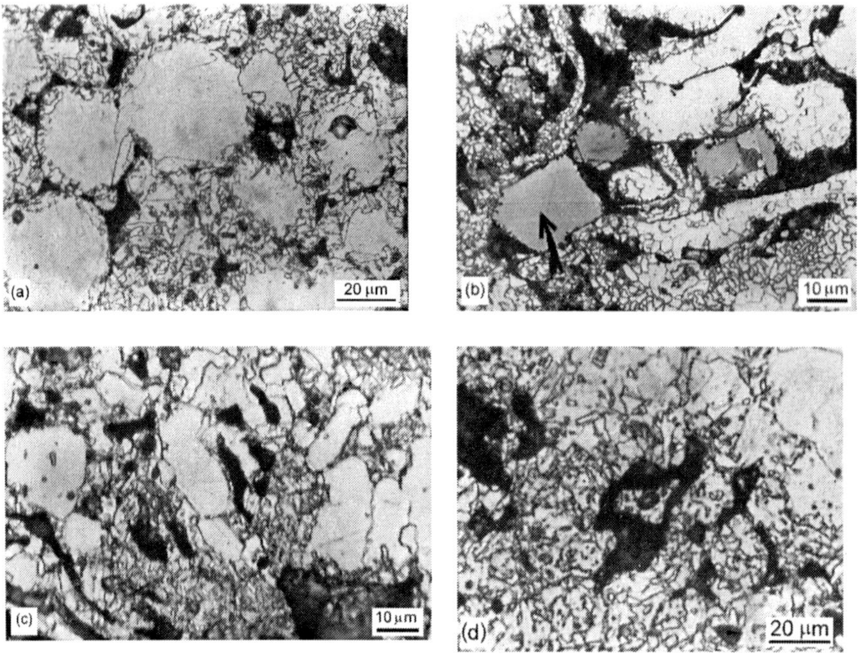

Fig. 4.13. Micrographs in detail; (a) – of Fig. 4.2(b), 60 min, 1200°C, RZ iron powder, (b) of Fig. 4.12(a), 5 min, 750°C, H iron powder, (b) – of Fig. 4.12(a), 30 min, 750°C, H iron powder, arrow – grey FeMnC particle, (c) – of Fig. 4.12(c), 3 min, 1100°C, H iron powder, (d) – demanganized microstructure of Fe–4Mn–032C sample, H, 30 min at 1200°C, H – Hametag iron powder. Nital etched.

lattice defects and, to some extent, this also leads to the alloying of the particles of the base powder, Fig. 4.10(d).

Nucleation of new grains at the early stages of sintering is shown in greater detail in Fig. 4.13. New fine grains can be found on the surface of the coarse RZ powder particles after demanganisation of the alloyed layer. These new fine grains nucleated before or during alloying by condensation of gaseous manganese and its subsequent diffusion into the surface layer of powder particles, Fig. 4.13(a). Figures 4.13(b) and Fig. 4.13(c) show the creation of fine grains after short-term sintering of Fe–4Mn–032C samples based on Hametag iron powder. Figure 4.13(d) shows the microstructure with fine grains of the same sample after sintering for 30 min at 1200°C. Very fine grains of powder in the starting powder remained after demanganisation and did not grow even at such high temperatures. This shows the highly fine-grained microstructures of the manganese-alloyed steels which may therefore alloyed in microvolumes with different concentrations of manganese and carbon.

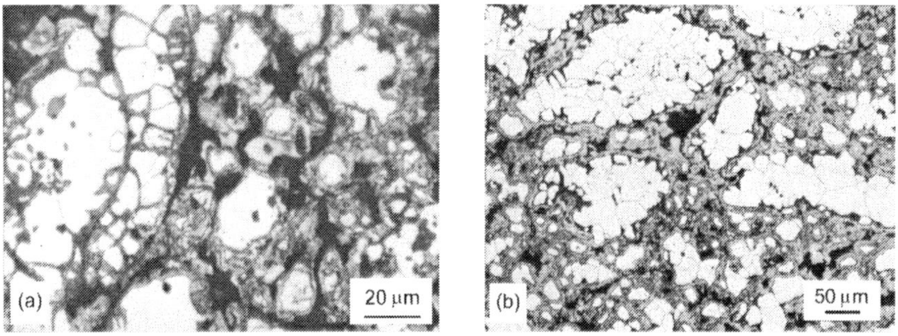

Fig. 4.14. (a) Microstructure of Fe–4Mn–0.3C samples; SC100.29 iron powder, 600 MPa, 30 min, 875°C, cracked ammonia, (b) microstructure of Fe–4Mn–0.3C sample, ASC100.29 iron powder, 600 MPa, 10 min, 1120°C, cracked ammonia. Nital etched [53].

The nucleation of new grains in the examined system iron powder–manganese vapour system is inevitable, but is influenced to some extent by different structural activity of the iron powders.Figure 4.14 shows the formation of new fine grains also in particles of iron powder SC100.26 and ASC1010.29 powder. These two types of iron powders are located in terms of structural activity between the RZ powder on the one hand (coarse-grained) and Hametag powder on the other hand (fine-grained structure, larger number of lattice defects).

4.3. Sintering and alloying of manganese steels analysed by the dilatometric tests

The dilatometric tests on PM materials for the characterisation of the behaviour of the compacts during sintering, mainly the dimensional changes which are crucial for the dimensional stability and in some cases the behaviour of the alloyed systems at the α–γ phase transformation are commonly carried out. The following dilatometric tests of Fe–Mn–(C) steels were perfected in the frame of the project [54] with the aim to study the effect of all variables affecting the final dimensional change of the samples.

In this case also, the entire sintering process takes place in three stages: initial, intermediate and final. The initial stage of sintering is characterised by rapid growth of point interparticle necks. In the intermediate stage, the pore structure becomes smoother and has an interconnected, cylindrical nature, with pore rounding, densification and grain growth. The final stage of sintering is a slow process wherein isolated, spherical pores shrink by a bulk diffusion mechanism [55].

Dimensional changes of compacts recorded on dilatometric graphs are determined by the heat conductivity of iron, initially through non-metallic contacts formed between the particles during pressing of the compacts. The initial sintering stage is in reality characterised by rapid formation of metallic bonds – of interparticle necks always at temperature higher than 400°C, but the majority of them form at temperatures below 800°C, *i.e.* in the α-iron region. In this process, the effect of the temperature is stronger than that of time. The characteristics of the necks in form of facets are characterised by specific surface (mm^2/mm^2), specific number (mm^{-2}), specific size (mm) and can be measured also by the density, tensile strength and coercivity changes. During the final sintering stage the number of necks decreases due to coalescence forming bigger necks [1,54]. The start of the formation of the necks is determined in [2] for the temperatures >500°C.

These features, affected by sintering and material conditions applied, express themselves on the dilatometric graphs as will be shown in following.

4.3.1. Effect of base materials and processing variables

The effect of manganese and carbon content, of manganese carriers and of iron and of some prealloyed powders as common variables acting in the preparation of sintered steels became for this reason the object of these dilatometer tests.

Specimens 4 × 4 × 15 mm in size with the density of 6.5 g cm^{-3} were prepared from two iron powder grades and Mo- and Cr+Mo-base prealloyed powders. Electrolytic manganese (EMn) and medium- (FeMn) and high-carbon ferromanganese (FeMnC) were used as manganese carriers. The parameters of the test specimen must be taken into account in the evaluation of the dilatometer graphs.

All manganese carriers contained oxygen in amount of 0.6–0.7% after milling in nitrogen. The as-milled powder particles were covered with a dark layer as a product of the abrasion of cast milling balls with ferromanganese.

Sintering was carried out in a horizontal pushrod dilatometer under the following conditions: isothermal sintering for 30 min at 1150°C in the $75H_2/25N_2$ atmosphere with a dew point of –30°C, heating rate of 10°C·min^{-1} and cooling rate of 30°C·min^{-1}. Plain iron

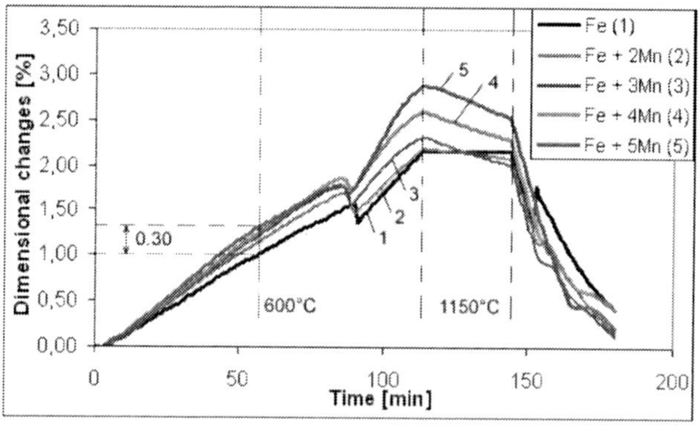

Fig. 4.15. Dilatometric graphs for Fe and Fe–(2,3,4,5)Mn specimens; SC100.26, FeMn (80% Mn, 1.1% C, 0.67% O, <45 μm) [57–60].

specimens were tested under the same conditions. (Deviations from these values are given in the captions).

4.3.1.1. Effect of manganese content

Figure 4.15 shows the dilatometric graphs for Fe and Fe–(2–5)Mn specimens. The dimensional behaviour of the materials gives a good indication of the homogeneous process. With addition of Mn, the deviation from the base line given by plain iron material used as a reference, starts to be noticeable already at temperatures well below the α–γ transition temperature. This is clearly marked in the graphs at 600°C as total higher expansion by 0.3% for the Fe–5Mn compact compared to the Fe specimen. According to the graphs, the swelling of Mn specimens can be shifted to lower temperature. To eliminate manganese oxidation, the sintering atmosphere at 600°C should have a dew point of –102°C (chapter 2, Table 2.1). The increasing expansion is affected only by continuous manganese sublimation from ferromanganese particles and by manganese vapour condensation on the surfaces of iron particles in the compact, as shown in Fig. 4.1(a). This effect causes the larger growth of the dimensions in relation to that of the Fe-specimen. The recorded dimensional changes of the Fe-specimen reflect already the previously mentioned formation and growth of the interparticle necks in the initial sintering stage.

Solid-state diffusion of manganese from the surface layers into the interior of iron particles starts in this initial sintering stage increasing the swelling by the solid solution and by the formation of new phases, chapter 4.1. (Fe is not soluble in manganese – one way diffusion of manganese into iron).

The characteristic marked dimensional changes of the Fe-Mn specimens in dependence on manganese addition at the α–γ transformation are the demonstration of an intense alloying of the iron matrix by manganese expressed also in a decrease of the phase transformation temperature by Mn addition compared to that of the plain iron compact. The dimensional changes on the graphs show that the preheating period is the very important stage of sintering and alloying of the iron matrix by manganese. The samples attained the higher swelling by ~0.7% compared to the Fe sample with the completed preheating period at 1150°C. The expansion in the γ-period is also higher with increasing Mn content which is partly compensated during isothermal sintering by shrinkage.

The shrinkage of the Fe–Mn sample length was recorded in the dilatometric graphs during isothermal sintering compared to the constant dimensions of the Fe specimen. This appearance is caused by two factors: a) reverse sublimation of manganese from the samples alloyed with manganese as mentioned previously, *i.e.* demanganisation characterised by manganese loss in the sintered compacts due to finishing of the supply of manganese vapour (finish of the manganese sublimation from all manganese particles in the compact) and the change of the manganese gradient (relatively high in the alloyed compact and zero in the atmosphere), b) activated sintering by manganese demonstrated in the decrease of the dimensional change of the Fe–2Mn specimen at the finish of the hold below that of the Fe specimen. The shrinkage line of the the Fe–Mn specimens is parallel during the holding. It means that the pores in the compacts are filled with manganese vapour firstly up to the finish of the sublimation from the coarsest particles and then due to demanganisation.

The demanganisation of the samples was also demonstrated in some extent in the course of the cooling part in the graphs compared to that of the Fe specimen graph. The wave-form flat changes in the cooling graphs of the Fe–Mn specimens were recorded. Consequently, according to this, the final as sintered manganese content in the specimens due to demanganisation will be lower than the starting Mn addition. The final dimensional changes of the manganese-carbon containing steel specimens recorded by dilatometry cannot be therefore accepted as representative for the specific sintered Mn alloy steel.

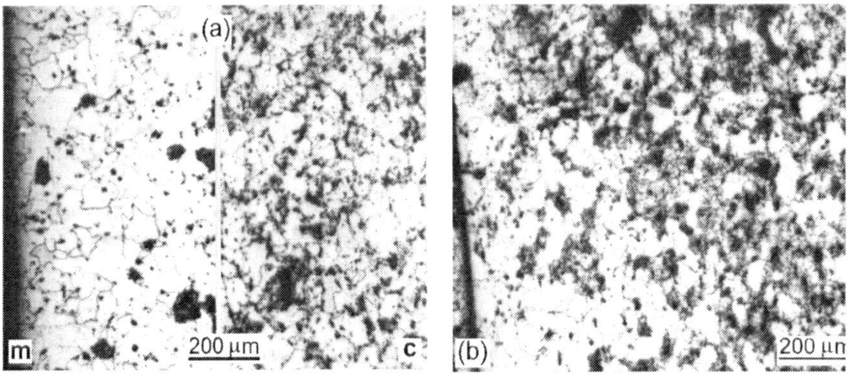

Fig. 4.16. Microstructures of: (a) – Fe–2Mn sample; (m – margin, c – core); (b) – margin of Fe–5Mn sample. Nital etched [57].

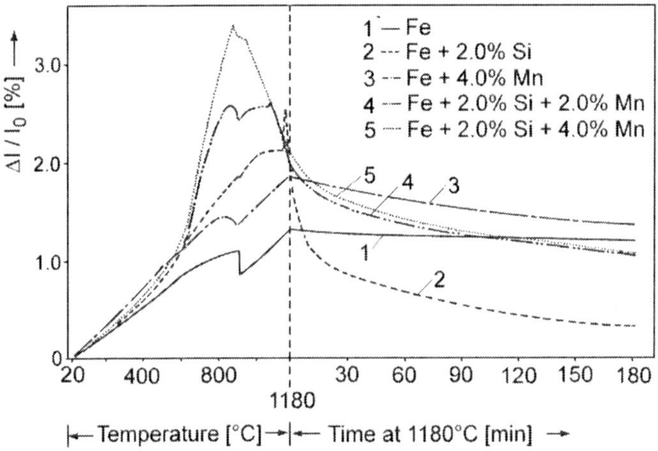

Fig. 4.17. Dimensional change of iron powder compacts (80% T.D.) containing FeMn and/or FeSi as alloying components; heating rate 10°C min^{-1}, 3 h, 1180°C, dry hydrogen [61].

The results show that the Mn sublimation followed by homogenisation of the iron matrix already starts as fairly low sintering temperatures.

The micrographs shown in Fig. 4.16 prove significant demanganisation of the samples during sintering at dilatometric tests, especially in random layers in dependence on the manganese addition. At higher manganese addition the micrographs indicate a smaller degree of demanganisation.

Figure 4.17 shows dilatometric curves of samples of the Fe-Mn–Si system prepared from an iron powder by adding low-carbon industrial

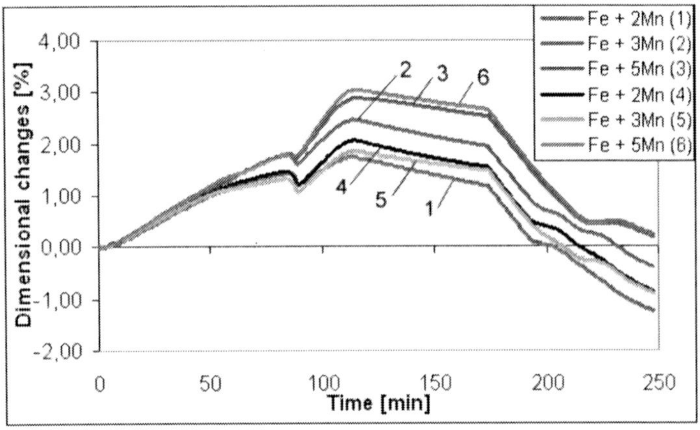

Fig. 4.18. Dilatometric graphs for Fe–(2,3,5)Mn samples based on two iron powder grades: 1,2,3,4 – SC100.26 iron powder, 5,6,7 – ASC100.29 iron powder, FeMn [57,60].

ferromanganese (0.11% C, FeMn) and/or ferrosilicium (15.4% Si, 0.4% C, FeSi). Both elements individually and collectively caused more growth in size during heating compared with the Fe sample and characteristic changes in the α–γ transition and shrinkage during isothermal sintering in comparison with Fig. 4.15.

The binary Fe–Mn compacts show also increased expansion in the preheating stage and the beginning of the α–γ transformation at lower temperature than the Fe-compact.

The Fe–Si compact expands more during preheating and the transformation is less visible. The ternary Fe–Si–Mn compacts are characterised by excessive swelling just below the transformation temperature due to formation of new phases.

It is important to note here that the sintering of the mixed Fe–Si–Mn system was successful despite the fact that Si has a higher affinity for oxygen than manganese (Section 3.2).

4.3.1.2. Effect of iron powder grade

The effect of the iron powder grades used is also demonstrated in graphs in relation to the manganese content clearly in the highest swelling values for the specimens at the start of isothermal sintering at 1150°C, as shown in Fig. 4.18.

The Fe–2Mn specimen based on ASC100.29 powder manganese vapour exhibited larger swelling compared to that based on SC100.26 iron powder and conversely the Fe–3Mn specimens. The effect of the iron powder grade was not clearly evident in the graphs for the Fe–5Mn specimens due to more homogeneous through alloying of the Fe matrix by manganese. It follows from this that alloying of these iron powder grades. The particles with manganese formed by a different mechanism during preheating.

The shown effect of iron powder grade on sintering and alloying of manganese steels is given by the specific characteristics of every iron powder grade caused by the production method. As in other sintered alloyed steels, this can cause marked differences in the final properties of the compacts.

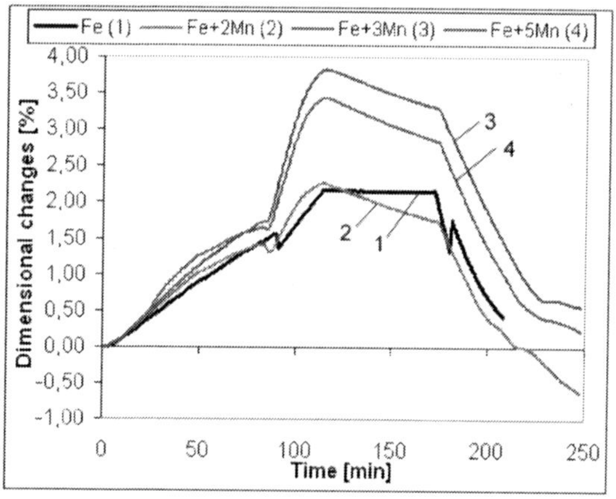

Fig. 4.19. Dilatometric graphs for: Fe, Fe–(2,3,5)Mn–(0.16,0.24,0.40)C_c samples; Mn and C as FeMnC (76% Mn, 6.6% C, 0.77% O, <45 μm) [57–59,62].

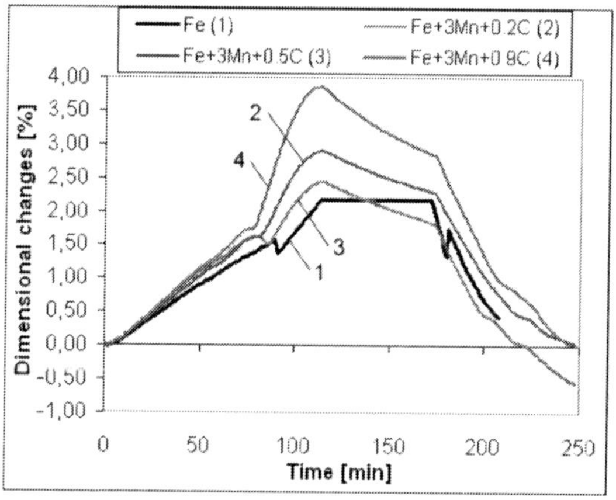

Fig. 4.20. Dilatometric graphs for: Fe, Fe–3Mn–(0.2,0.5,0.9)C samples, FeMn, C as natural graphite; SC100.26 iron powder in both cases [57–59,62].

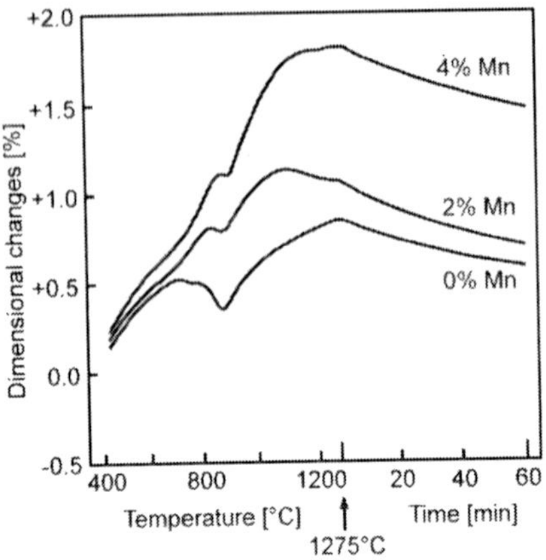

Fig. 4.21. Dilatometric graphs for Fe–xMn–0.3C (x = 0, 2, 4); 600 MPa, heating rate 10K min^{-1}, 60 min, 1275°C, vacuum [53].

4.3.1.3. Effect of carbon content

Carbon was added to the Fe–(2,3,5)Mn compacts as a combined addition in the form of high-carbon ferromanganese and as graphite to the Fe–3Mn compacts. Figures 4.19 and 4.20 show the dilatometric graphs of Fe, Fe–(2,3,5)Mn–(0.16,0.24,0.40)C and Fe–3Mn–(0.2,0.5,0.9)C specimens characterised by various manganese and carbon contents and carbon carrier.

The mentioned differences in the base material characteristics were demonstrated clearly by swelling in the γ-region up to 1150°C in both steel groups due to solubility of carbon in austenite. The highest swelling was shown by the Fe–3Mn–0.24C sample and differed from that of the Fe–3Mn–0.5C sample. On the other side, the Fe–5Mn–0.4C steel exhibited the same swelling of the dimensions as the Fe–3Mn–0.9C steel (3.9%). The markedly higher swelling of the Fe–Mn–C samples compared to that of the Fe–Mn sample is also caused by the manganese increasing solubility of carbon in the γ-solid solution. Higher swelling and by this increased solubility of carbon in austenite were recorded for the samples with the combined addition of high-carbon ferromanganese and carbon. Increased sintering activity compared with the previous case (Fe–2Mn) was recorded for the Fe–2Mn–0.16C compact compared also to the Fe–3Mn–0.2C specimen (manifested in larger shrinkage below the Fe line at the finish of the hold). The form and amount of carbon as well as the manganese carrier type and manganese addition were expressed in shrinkage during the hold and in small differences in cooling parts of the curves.

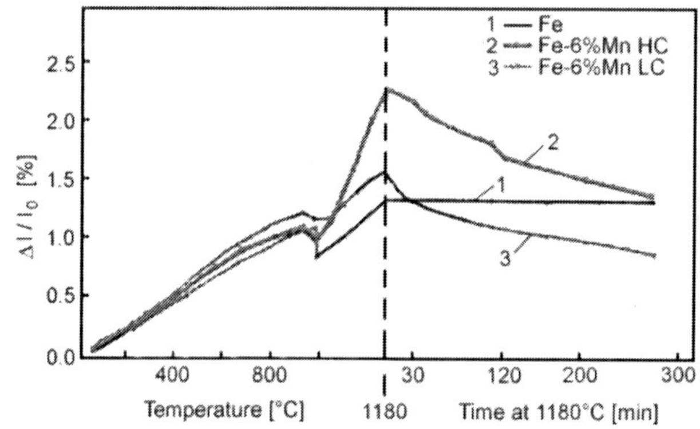

Fig. 4.22. Dimensional changes during sintering of Fe–6% Mn sample with high (HC) / low (LC) carbon ferromanganese and of pure Fe; sintering at 1180°C up to 300 min [63].

According to [53], the dilatometric graphs in Fig. 4.21 show also the homogenisation process of Fe–Mn steels by manganese starting noticeably already at the temperatures well below the α–γ transition. It should be added here that the maximum expansion of the Fe–4Mn-0.3C specimens is significantly lower than that shown in Fig. 4.15.

Figure 4.22 shows the dilatometric curves of the Fe–6Mn system with high and low carbon content after prolonged isothermal sintering. The characteristic dimensional changes during the α–γ transformation, caused by manganese, are clearly evident. Also, greater growth of the size of the sample with the higher carbon content in comparison with the lower C content, as demonstrated by Figs. 4.19 and 4.20, was recorded. In this case, no dimensional changes were observed in the Fe samples during isothermal sintering, as shown in Fig. 4.15.

4.3.1.4. Effect of manganese carrier

The testing of the effect of medium-carbon ferromanganese and electrolytic manganese as manganese carriers on alloying of sintered steels, including previous results, has a scientific-technical and economical meaning. The ferromanganese grades are the cheapest industrial manganese carriers and, on the other hand, the electrolytic manganese is the cleanest one but more expensive compared to the ferromanganeses.

The graphs in Fig. 4.23 show clearly a different effect of manganese addition in the form of electrolytic manganese (EMn) and

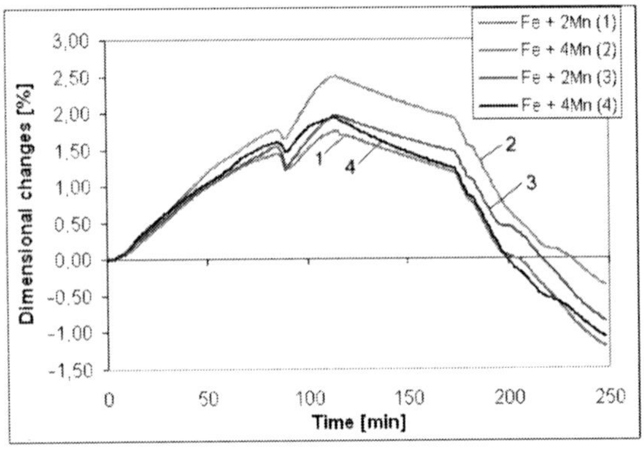

Fig. 4.23. Dilatometric graphs for Fe–(2,4)Mn samples; SC100.26 iron powder, manganese as: 1, 2 – FeMn, 3, 4 – EMn [57,62].

of medium-carbon ferromanganese (FeMn) on dimensional expansion of the compacts. The Fe–2Mn sample alloyed with FeMn exhibited the lowest swelling than that alloyed with EMn and the Fe–4Mn sample showed the highest swelling in the investigated systems. On the contrary, the difference in swelling between the Fe–2Mn and Fe–4Mn samples alloyed with EMn is minimal. The curves exhibit the same character during the cooling period.

The differences between the two manganese carriers consisted in particle size (medium-carbon ferromanganese, <45 μm, electrolytic manganese, <20 μm) and in manganese concentration (lower chemical activity of ferromanganese compared to the electrolytic manganese particles). The finer manganese particles sublimated early at lower temperatures and manganese vapour in such a case escaped through the open pores into the atmosphere (demanganisation), due to relatively low temperature for diffusion alloying of the Fe matrix (Chapter 3, Table 3.2). This caused a lack of manganese for alloying due to the different range of the particle size distribution (<20 μm → <45 μm).

In spite of this, the uniform distribution of individual particles of the same size cannot be ensured in any of the milled manganese carrier powders. Heat treatment of electrolytic manganese and of high-carbon ferromanganese powder batches with the particle size <20 μm after milling in air was carried out with the aim to complete previous results. The results as the manganese loss in dependence on treatment temperature and time are shown in Figs. 4.24 and 4.25.

The initial larger manganese loss from the FeMnC charge was caused by decarburisation. The electrolytic manganese powder charge

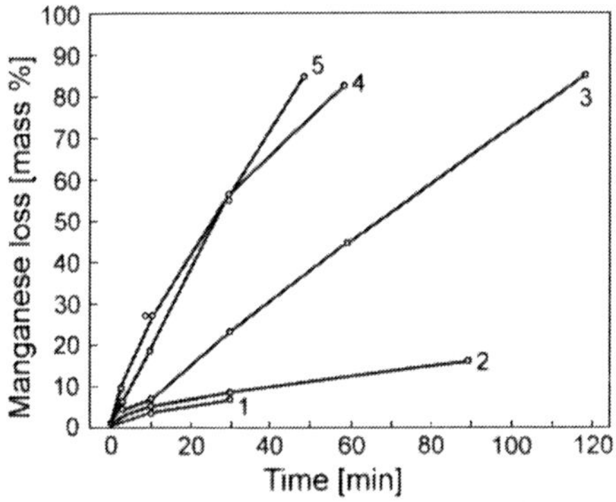

Fig. 4.24. Manganese loss from loose manganese powder in heat treatment in a laboratory Mars furnace in an alumina boat in high-purity hydrogen in dependence on temperature and time; EMn, mass 0.03 g, specific surface area 0.54 $m^2 \cdot g^{-1}$; 1 – 1000°C, 2 – 1100°C, 3 – 1130°C, 4 – 1200°C, 5 – 1220°C.

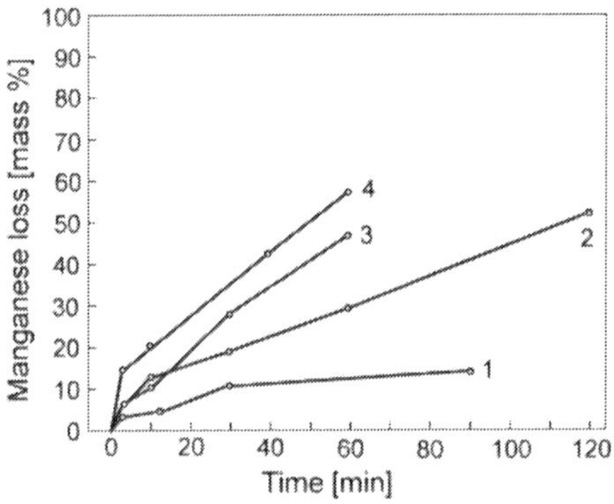

Fig. 4.25. As in Fig. 4.24 – FeMnC; mass 0.04 g, specific surface area 0.62 $m^2 \cdot g^{-1}$; 1 – 1030°C, 2 – 1120°C, 3 – 1200°C, 4 – 1220°C [20,23,25,57].

exhibited a larger loss of manganese at all processing conditions compared to that of high-carbon ferromanganese, *e.g.* after treating for 60 min. The manganese loss of the EMn particles after treatment for 60 min at 1200°C was 85% and that of the FeMnC particles 50%. This was also the case for FeMn. It is important to take into account the increasing of the manganese vapour pressure with increasing temperature. It may be assumed that this is also caused by the even lower chemical activity of FeMnC in comparison with that of FeMn and also by bonding of manganese with carbon in FeMnC.

The effect of the charge mass of electrolytic manganese on the manganese loss by sublimation of two standard sintering temperatures is shown in Table 4.1.

Table 4.1. Mass and percentage loss of manganese by sublimation from electrolytic manganese (particle size <20 μm) at annealing for 10 min at 1120°C and 1220°C in a laboratory Mars furnace in an alumina boat in high-purity hydrogen in dependence on the mass of the batch [25,56]

Temperature [°C]					
1120			1220		
Test batch	Mn loss		Test batch	Mn loss	
[g]	[g]	[%]	[g]	[g]	[%]
0.03050	0.00300	9.80	0.03156	0.00723	22.91
0.30725	0.00463	1.50	0.30068	0.00770	2.56
2.99415	0.01260	0.42	3.02810	0.01825	0.61

As expected, the results show an increased manganese loss by sublimation in all cases at increased annealing temperature. On the other hand, the manganese loss at both temperatures decreased with increasing mass of the test batch of EMn particles. It was the effect of the 'thickness' of the test batch; the lower mass of the test batch – the thinner layer – more particles took part in sublimation in relation to the mass of the batch compared to that in a thicker layer (gravitation compaction of the batch). This shows the need to secure the maximum homogeneous distribution of Mn particles in the mixture at mixing with the iron powder (elimination of clusters). This can explain the formation of whiskers in the fractures of the steels shown in chapter 3, Fig. 3.11.

4.3.1.5. Effect of prealloyed powders

The prealloyed FeMo and FeCrMo(V) powders with the addition of manganese can be used to produce new hybrid nickel- and partly also copperless or separately new high-performance sintered steel groups forming hybrid steels. The test samples were prepared from the prealloyed Fe–1.5Mo (Astaloy Mo, AstMo), Fe–0.85Mo (Astaloy 85Mo, Ast85Mo) and Fe–3Cr–0.3Mo–0.3V powders.

Figure 4.26 shows the dilatometric graphs for samples based on Ast85Mo powder with the addition of manganese and carbon isothermally sintered at 1200°C.

The graphs proved also an intense dimensional expansion of Mn-containing specimens compared to the plain Ast85Mo specimen already at the temperature of 470°C. This was caused by the same effect of manganese as in the case of the Fe–Mn specimens. This makes it possible to shift the start of the sublimation of manganese to this low temperature, as already carried out previously. The

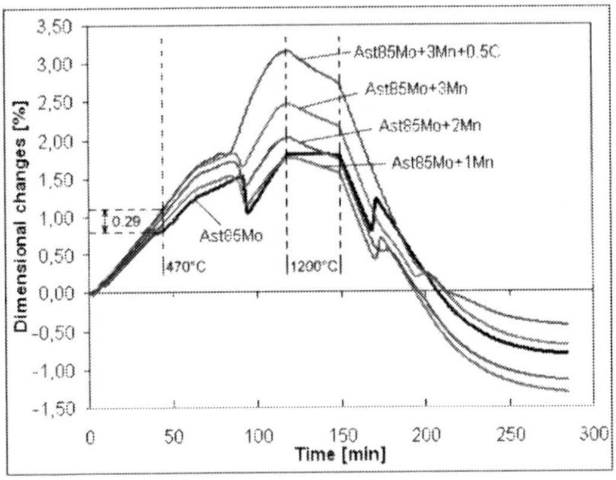

Fig. 4.26. Dilatometric graphs for hybrid Fe–0.85Mo–(0,1,2,3)Mn and Fe–0.85Mo–3Mn–0.5C steel samples; Mn as FeMn, C as graphite, 30 min, 1200°C, heating rate 10°C min^{-1}, cooling rate 15°C min^{-1}, 30H_2/70N_2 atmosphere, dew point –40°C [57,64].

swelling of the Fe–0.85Mo–1Mn specimen during preheating up to 1200°C is similar in its extent to the sample with no Mn addition, *i.e.* both samples have the same thermal expansion with minimum changes in the microstructure which could have been caused by the Mn addition. In this case, the Mn addition would also be reflected in higher sintering activity manifested by greater shrinkage during isothermal sintering, lower than that of the sample with no Mn addition. The same form of the graphs during holding was obtained for the samples with the addition of 2 and 3% Mn, *i.e.* Mn has an enhancing effect on sintering of this system. The cooling curve of the graph for the sample with 1% Mn showed the γ–α transformation similar to the graph for the sample with no Mn addition, contrary to the samples with 2 and 3% Mn.

The addition of 0.5% carbon resulted in a relatively large increase of swelling in the γ-range compared to the Fe–3Mn steel without carbon. Also in this case it is necessary to allow for some demanganisation of the specimens demonstrated by the same character of shrinkage as that for the Fe–Mn samples.

Figure 4.27 shows the dilatometric graphs for the hybrid Fe-1.5Mo and Fe–1.5Mo–3Mn–0.25C steel specimens based on the prealloyed AstMo powder. The linear swelling of the hybrid Fe–1.5Mo–3Mn-0.25C steel according to the graph is approximately at the same level at 1150°C as that shown in Fig. 4.26 for the Fe–0.85Mo–3Mn-0.5C specimen, considering higher sintering temperature and carbon content. It is possible to assume the same course of alloying of both Fe–0.85Mo and Fe–1.5Mo prealloyed matrices by manganese

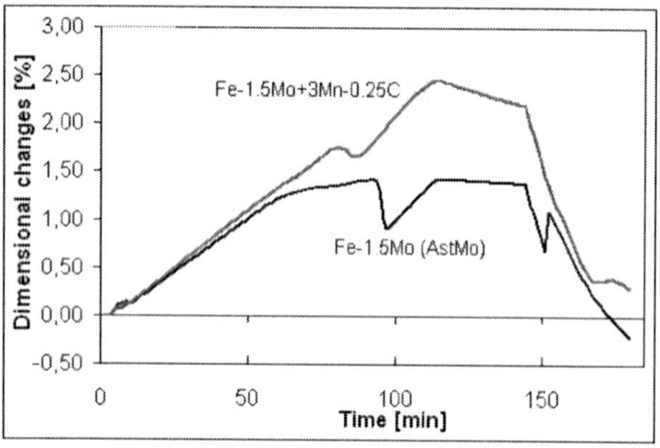

Fig. 4.27. Dilatometric graphs for: Fe–1.5Mo (AstMo) and for hybrid Fe–1.5Mo–3Mn–0.25C (graphite) samples; Mn as FeMn.

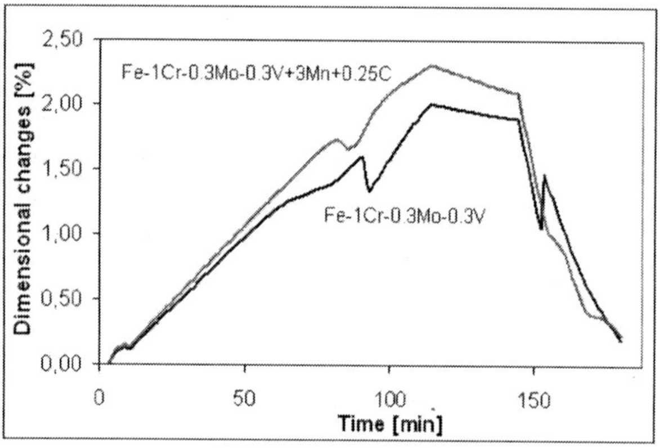

Fig. 4.28. Dilatometric graphs for Fe–3Cr–0.3Mo–0.3V (KIP 30CRV) and hybrid Fe–3Cr–0.3Mo–0.3V–3Mn–0.25C samples; Mn as FeMn, C as graphite; 30 min, 1150°C, heating and cooling rates as at processing of Fe–Mn samples [57].

including the shrinkage caused by demanganisation and by the sintering enhancing effect of manganese.

The graphs for Fe–3Cr–0.3Mo–0.3V samples without and with addition of 3% Mn and 0.25% C (graphite), shown in Fig. 4.28, differ from those based on Mo prealloyed powders by markedly lower final swelling at 1150°C followed by lower shrinkage. This is caused by different compositions of the prealloyed matrices. On the contrary, higher swelling was exhibited by the Fe–Cr–Mo–V prealloyed matrix compared to the Fe–0.85Mo and Fe–1.5Mo matrices. The dilatometric graphs of all three systems based on these prealloyed powders showed extensive alloying with manganese.

The dilatometric tests were also carried out to study the impact of medium- and high-carbon ferromanganese (<40 μm) in particular to achieve consistent dimensional changes of the Fe–4Mn–0.8C and Fe–4Mn–0.8C compacts based on the NC100.24 iron powder at temperatures of 1120°C and 1300°C in hydrogen with a dew point of –60°C. In this case, it was also found that sintering occurs after some time by reversed sublimation of manganese (demanganisation) from the sintered samples. According to these results, the final dimensional changes in the tested systems were affected by the isothermal sintering temperature and the composition of the added ferroalloy. In doing so, was also recorded differential impact of combined and free carbon on dimensional changes of the compacts. Differences in the effect of bonded and free carbon on the dimensional changes of the compacts were also noted [59,65].

4.3.2. Effect of manganese on isothermal sintering and cooling process according to the dilatometric graphs

4.3.2.1. Enhancing effect of manganese in solid-state sintering

The linear shrinkage of the Fe–Mn sample length recorded in the dilatometer graphs during isothermal sintering was in previous chapters explained and analytically determined by demanganisation. The dimensional swelling at the beginning of isothermal sintering when this was for all steels maximal, changed to 'uniform' dimensional shrinkage with extending holding time, as shown in previous figures.

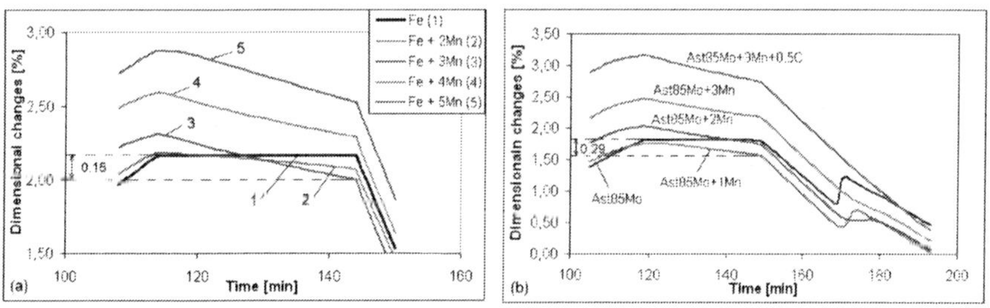

Fig. 4.29. Sections of the isothermal sintering part of the dilatometer graphs shown in: (a) – Fig. 4.15 (Fe, Fe–(2,3,4,5)Mn), (b) – Fig. 4.26 (Fe–0.85Mo, Fe–0.85 Mo–(1,2,3)Mn and Fe–0.85Mo–3Mn–0.5C steel samples [57].

The sections of Mn and Fe–Mo steel sample dilatometric graphs as representative steels based on plain iron and prealloyed powders, shown in Fig. 4.29, provided additional information on the shrinkage of the manganese-containing steels compared to the dimensional changes of plain iron. The shrinkage of the specimens with 'lower' manganese content (Fe–(2,3)Mn and Fe–0.85Mo–(1,2)Mn) at the finish of the isothermal sintering was below the constant dimensional change of the Fe compacts. This level was exeeded at different times of isothermal sintering.

The shrinkage in the case for the Fe–3Mn specimen shown in section (a), and for the Fe–0.85Mo–1Mn sample in section (b), was lower than for those containing carbon. The shrinkage of steel samples with the addition of Mn and C in form of FeMnC was 0.44% smaller than that for the Fe sample (Fig. 4.19) and the shrinkage of the samples with the addition of carbon in the form of graphite was 0.36% lower than the values for the Fe sample. This feature was a result of the progressive effect of manganese on alloying, demonstrating also as an enhancing factor for solid-state sintering of manganese-containing steels. As follows from all graphs, the course of the shrinkage of the specimens in dependence on Mn content during isothermal sintering was parallel. It means that the effect of manganese as a sintering activator was constant and that the total shown shrinkage was the result of both factors, *i.e.* demanganisation and the enhancing effect of manganese on sintering. This phenomenon is also clearly evident in the Fe–5Mn system in relation to the carbon content, Fig. 4.29. The result of this feature was the increased density of the sintered Mn-containing steel specimens compared to the Fe specimen. Also, the higher shrinkage of the Fe–Mn–C specimens, compared to that without carbon, demonstrated the enhancing effect of manganese + carbon on solid-state sintering. Some data for this feature follow from Mn microanalysis listed in Table 4.2.

4.3.2.2. Transformation of austenite characterised by cooling curves of the dilatometer graphs

Figure 4.30 shows in detail the cooling curves in the previously discussed dilatometric graphs, Fig. 4.29. All the cooling curves of the Mn-containing samples show in a certain temperature range the effect of manganese and carbon by a dip as compared to the straight line for the Fe and Fe–0.85Mo samples. In the cases examined in relation to the temperature at which these changes occurred, the bainitic

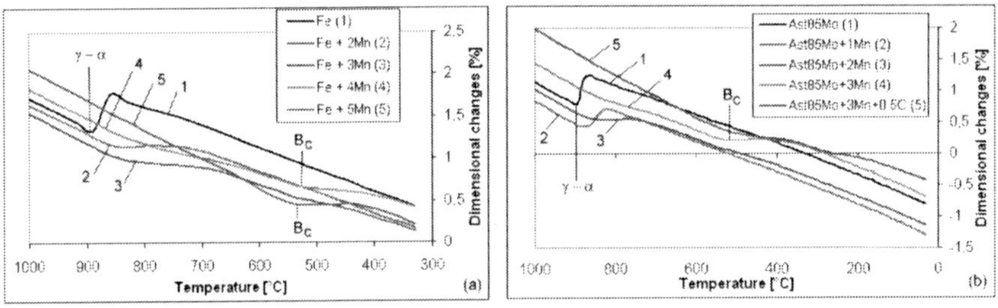

Fig. 4.30. Cooling curves in the dilatometric graphs showing the transformation of austenite in dependence on 'residual' manganese and carbon content in the tested steel samples presented in: (a) – Fig. 4.15 (Fe, Fe–(2,3,4,5)Mn), (b) – Fig. 4.26 (Fe–0.85Mo, Fe–0.85Mo–(1,2,3)Mn, Fe–0.85Mo–3Mn–0.5C).

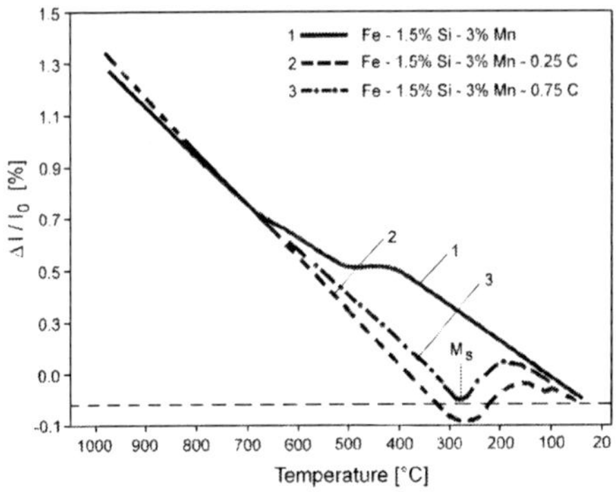

Fig. 4.31. Dilatation during cooling of Fe–Si–Mn–C samples showing the martensitic transformation of the main part of samples volume [65].

transformation took place at temperature from 600 to 400°C during cooling at a given rate, as indicated. In these case these data can only be regarded as indicative, because they must be related to the actual content of manganese which is in this state substantially lower than the added amount due to the previously mentioned demanganisation and possible decarburisation. For this reason, the curves did not change their course in the martensitic transformation temperature range. These curves showed the transformation of Mn-containing austenite present in relatively small amounts only in the middle of the samples, as indicated by Fig. 4.16.

In this way, the cooling curve of the dilatometric graph of the Fe-Mn–Si–C samples indicates the martensitic transformation, as shown in Fig. 4.31. Figure 4.32, characterising the Fe–1.5Cr–3Mn-0.2Mo-0.8C system, records martensitic and banitic transformations.

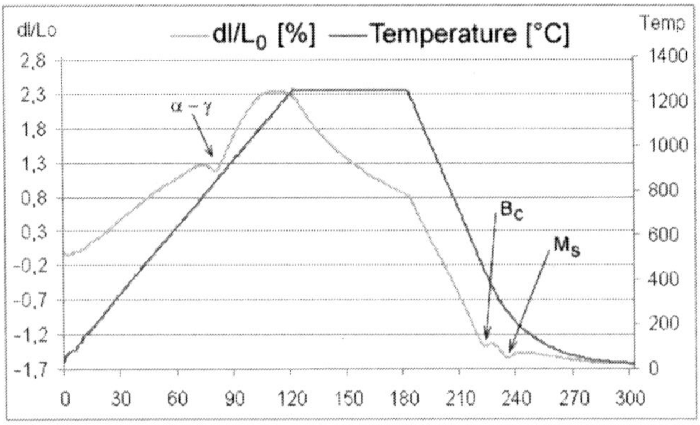

Fig. 4.32. Dilatometric graph of sintered Fe–3Mn–1.5Cr–0.2Mo–0.8C steel: CrL base prealloyed powder, specimen 5 × 5 × 15 mm, density 6.5 g cm^{-3}, 60 min, 1250°C, $5H_2/95N_2$ atmosphere, dew point –60°C [66].

4.3.2.3. Effect of cooling rate on the transformation of austenite in sintered steels as indicated by dilatometric graphs

Despite this seemingly negative result, the cooling part of the dilatometric graph appears to be an appropriate method for research into the impact of the continuous cooling rate on the structural characteristics and thus on the final properties of sintered material. This need is very topical, for example, in rapid cooling, in general, but particularly in the research of the sinter hardening process. Such dilatometric tests should be carried out on sintered and efficiently analysed (in the sintered state) bending strength samples (10 × 5 × 55 mm). Four dilatometric samples can be prepaed from every bend test sample. The sample should be heated in nitrogen only to the austenitizing temperature of the alloy with the appropriate holding time and the sample should then be cooled from this temperature at the selected rate. All analytical procedures can then be conducted on the resultant dilatometric sample – microstructure with detailed measurement of the microhardness of structural components, apparent hardness, and the three-point bending strength at a span of 10 mm, and fractures for completing the results. The bending strength of the dilatometric samples was tested and some results are published in [57,67]. The proposed method ensures the stability of the chemical composition of the samples and reliable data are recorded on the transformation of austenite, depending on the cooling rate, particularly on the bainitic and martensitic transformation of the investigated system.

This type of research of the impact of cooling rate on the austenite transformation of sintered steels or of the effects of various additives on hardenability, can also be carried out in the reversed direction, *i.e.* determination of the actual cooling rate of components operating in a continuous furnace. Bend test samples would be sintered together with components and then would be dilatometrically and metallographically tested at different cooling rates. At some rate this would yield identical data on austenite transformation and the the corresponding properties of the compact which were determined after sintering in a furnace. Consequently, the actual cooling rate of components in an industrial furnace would be determined.

In both cases, the initial standard samples with defined chemical, structural, mechanical and fracture characteristics would be used.

5

Effect of base materials and of various processing methods on mechanical and some special properties of manganese steels

The aim of this chapter is to present the detailed view on the properties of manganese steels in terms of their practical use regarding only sublimation of manganese.

5.1. Electrolytic manganese and ferromanganese grades - physical–metallurgical and technical characteristics

Electrolytic manganese and ferromanganese grades are characterised by the highest manganese contents and therefore are often used as manganese carriers in the investigation of sintered manganese steels. On the other hand, because of their high Mn content, their application in research is largely affected from the thermodynamic point by the fear of their oxidation in the milling and sintering of Mn compacts and therefore are not widely used. There is a continued effort to replace by different master alloys with a lower manganese content.

Given the previously mentioned and documented evidence it is evident that the sintering of manganese alloy steel takes place and is possible only by the reaction of gaseous manganese and oxygen from the sintering atmosphere and the already mentioned condensation of manganese on the surface of the basic powder particles.

Description and analysis of their physico-chemical and technical characteristics is necessary for their optimal use as carriers of manganese as an alloying element for sintered Mn steels. It is appropriate if these Mn carriers are used with a particle size distribution with a specific limited maximum fraction.

As regards the manganese carriers, it should be noted in advance that they are the cheapest source of manganese for powder metallurgy, since they are made for conventional steel production and for other purposes, and for powder metallurgy only milling and screening is necessary. The following are the basic characteristics of these carriers of manganese in the state of production and conditions of milling to produce a powder with the characteristic properties. More detailed information about these three manganese carriers in some applications was presented in [68].

5.1.1. Electrolytic manganese

The melting point of manganese is 1244°C, and density 7.20 g cm^{-3}. Metallic manganese exhibits the highest technical quality as regards the content of accompanying elements P, C, Si and Fe. Therefore, due to its highest chemical purity, electrolytic manganese is the most expensive manganese grade. Cheaper metallic manganese with lower chemical purity is produced by the silicothermic method [69]. Electrolytic manganese (two grades) contains 99.7 to 99.95% Mn, 0.05 to 0.01% P, 0.1 to 0.02% C and 0.010 to 0.015% S. The as supplied electrolytic manganese is brittle, it has the form of irregular plates with a thickness of ~1.0–1.5 mm. The bottom of the cathode surface of the plates is flat silvery metallic clean. The upper surface of the plates is rough, brightly grey in colour and formed by non-uniform sphere-like formations. Figures 5.1(a),(b) show the characteristic surface 'roughness', Fig. 5.1(c) the side view and Fig. 5.1(d) the cross-section of a plate. The structure of electrolytic manganese is mechanically heterogeneous and, as shown in Fig. 5.2, contains also phases formed by manganese with accompanying elements. The shown character of the structure indicates easy milling to powder. It is also necessary to note that manganese and manganese alloys contain a 'large' amount of gases. Electrolytic manganese contains 300–500 ppm O, 100 ppm N and 100–150 ppm H. The solubility of oxygen in manganese is several orders lower than that in iron [70].

The behaviour of solitary pieces and powder electrolytic manganese in heat treatment at elevated temperature as the consequence of the sublimation and condensation of manganese is shown in detail and explained in Chapter 3.

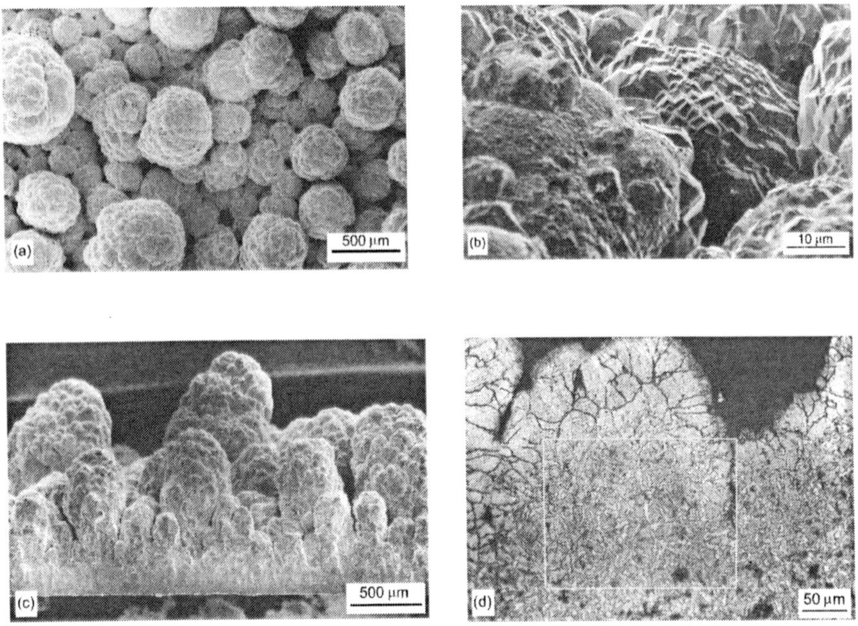

Fig. 5.1. Characteristic morphology of electrolytic manganese; (a), (b) view on the top of a plate, (c) a side view on a plate, (d) cross-section of a plate [68]. Nital etched.

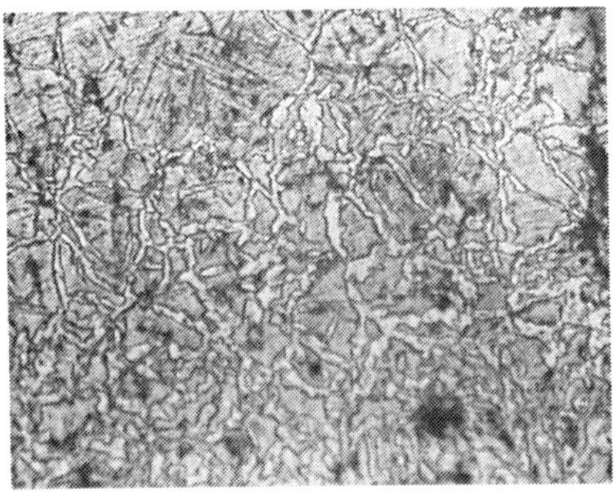

Fig. 5.2. Detail of the microstructure of electrolytic manganese shown in Fig. 5.1(d) [68]. Nital etched.

5.1.2. Ferromanganese grades

The alloying of the wrought steels with some elements occurs through the ferroalloys. Ferromanganese grades used in the steelmaking industry are produced in electric furnaces. Lumpy ferromanganese is a practically odourless material when dry, with a silvery metallic surface. On the basis of the carbon content, they are divided to high-carbon ferromanganese (max. 7% C), medium-carbon ferromanganese (max. 1–1.5% C), and low-carbon ferromanganese (max. 0.5% C).

Industrial ferromanganese is a multicomponent alloy and its melting point is 1200–1250°C and density 6.8–7.1 g cm^{-3}. The liquid phase of high-carbon ferromanganese forms by heating in air at 1230°C. On the other hand, the liquid phase of the same grade of ferromanganese forms at 1060°C by heat treatment in hydrogen [7]. The concentration of silicon in ferromanganese grades is usually 0.8–1.5% causing the formation of a hard solid solution of silicon in the iron–manganese matrix. Phosphorus is found in the quaternary Mn–Fe–C–P system (carbon ferromanganese) as a phosphide (60% Fe, 20% Mn and 20% P). The total amount of non-metallic inclusions in ferromanganese is ~0.04 to 0.40% [70]. The ternary Mn–Fe–C system is characterised by many phases the areas of which are shown in Fig. 5.3. According to this diagram, Mn_3C could be formed in ferromanganese with carbon content >1%. As stated in [71], the existence of this basic carbide Mn_3C in cast ferromanganese was not confirmed. The industrially produced ferromanganese grades contain at 1300°C hydrogen in the amount of 60 cm^3/100 g of the alloy, 0.025 vol. % of oxygen and 0.8 vol. % of nitrogen both cast at 1500°C [70–72].

The chemical compositions of industrially manufactured carbon ferromanganese grades are given in Table 5.1.

The manufacture of nitro-medium carbon ferromanganese could be considered as the cheapest alloying element in powder metallurgy, especially in steels alloyed with Mn and Cr [72,73]. The use of nitro-medium carbon ferromanganese as an Mn–N carrier in connection with Cr-prealloyed powders could be interesting despite

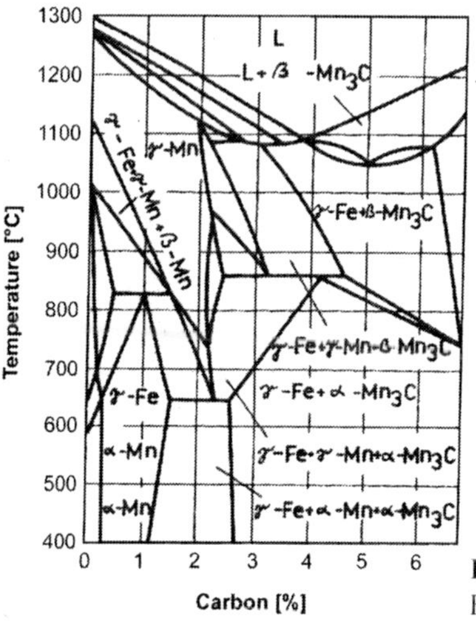

Fig. 5.3. Equilibrium diagram of ternary Mn-Fe–C alloy [20,68,71].

Table 5.1. Chemical composition [mass %] and density of carbon ferromanganese grades; 1 – high-carbon ferromanganese – HC, 2 – medium-carbon ferromanganese (affinné) – MC, 3 – low-carbon ferromanganese (suraffinné) – LC, 4 – as to 3 – nitro. (Further HC – FeMnC, MC – FeMn, LC – up to the present time has not probably been used as a manganese carrier in PM)

Mn grade	Mn		C		Si		P		N_2	Density [g cm^{-3}]
	Spec.	Typ.	Max.	Typ.	Max	Typ.	Max.	Typ.	Typ.	
1 – FeMnC	75–77	76	8	6.7	2.0	0.2	0.35	0.2	–	7.3
2 – FeMn 82C10	80–83	81	1.0	0.9	0.6	0.45	0.20	0.18	0.12	7.4
3 – FeMn 83	80–83	81	0.5	0.45	0.6	0.5	0.20	0.18	0.045	7.4
4 – FeMn 82C10	80–83	81	1.0	0.9	0.6	0.45	0.20	0.18	0.80	7.4

Note: 1 – OFZ Istebné, 2,3,4 ERAMET Manganese (formerly ELKEM Sauda)

the small nitrogen content. Manganese, like chromium, belongs to the elements increasing the solubility of nitrogen in iron. The maximum solubility of nitrogen in 80Mn–Fe alloy at 1500°C is 0.8% [70]. The investigation of the effect of hydrogen, nitrogen and dissociated ammonia on the mechanical properties of iron compacts sintered at 1120°C for various time confirmed their increase after sintering in nitrogen and in cracked ammonia compared to that sintered in hydrogen. This was also proved by increased nitrogen content in sintered compacts and by the presence of nitrides in the microstructures [74]. Introduction of nitrogen to sintered iron compacts in the atmosphere $5NH_3/95N_2$ at 600°C and 920°C was investigated [75–77].

5.1.2.1. Microstructure of ferromanganese

The microstructure of high-carbon ferromanganese is shown in Fig. 5.4. The HC ferromanganese structure was evaluated according to [71]. It was stated there that HC ferromanganese contains a grey phase,

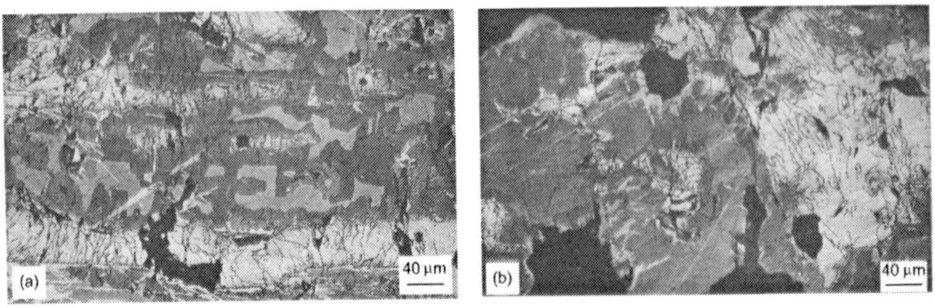

Fig. 5.4. Light microstructures of high-carbon ferromanganese [68,71]. Nital etched.

which is a solid solution of silicon (~4% Si) in the iron–manganese matrix (Mn/Fe = 75/25), and an acicular phase. It was further stated that, according to X-ray analysis, the acicular phase can be identified as a complex iron–manganese $(Mn,Fe)_7C_3$ carbide in which 10% of Mn is replaced by iron. Phosphorus in the microstructure was also identified by X-ray analysis. The decrease of carbon content in medium- and in low-carbon ferromanganese grades compared to that in high-carbon ferromanganese conduces to the decrease of the content of the carbide phase in the structure [2].

The fracture surfaces of the tested crushed ferromanganese moulds contained in all hollows ('large closed pores') relatively large aggregations of acicular shapes next to the geometrical form shown in Fig. 5.5. Such relatively large needles were formed by sublimation and condensation of manganese from sharp edges orientated against them in the hollows during the cooling of the ferromanganese mould (called hold). The formation of the hollows during solidification is perhaps the consequence of high hydrogen content in ferromanganese [68,70,71,73]. The microstructure of the needles is heterogeneous, *i.e.* the gaseous manganese took along also accessory elements present in HC ferromanganese [73] did not explain the mechanisms of the formation of such manganese carbides in the needle form interconnected with the surface of the hollows. It is can be supposed that according to [71] the mentioned acicular phase is that shown in Fig. 5.6, *i.e.* as the above mentioned complex carbides. The thickness of the needles was 0.1 to 0.25 mm measured on optical micrographs. These phases are the components of the milled ferromanganese powder.

The formation of the acicular phase as in the previous case was also observed in the fractures of an experimentally produced mould of the Fe–40Mn–30Cr–5Mo–5C master alloy with a markedly

Fig. 5.5. Fracture of the crushed high-carbon ferromanganese mould with acicular carbide (needle) phase at two magnifications [68]. SEM.

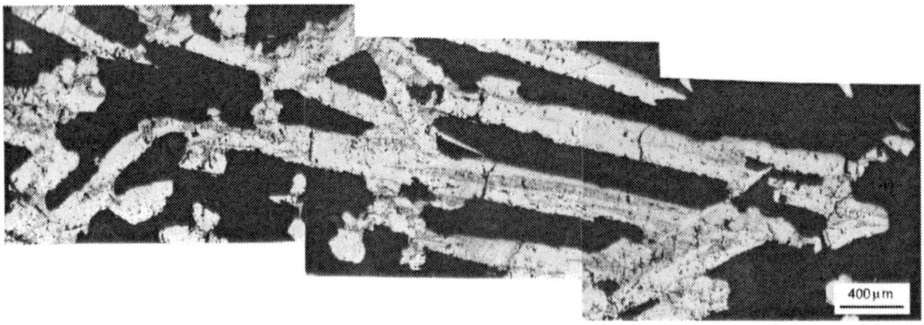

Fig. 5.6. Metallographic cross-section of acicular (needle) phase – manganese carbides shown in Fig. 5.5 at higher magnification [68]. Nital etched.

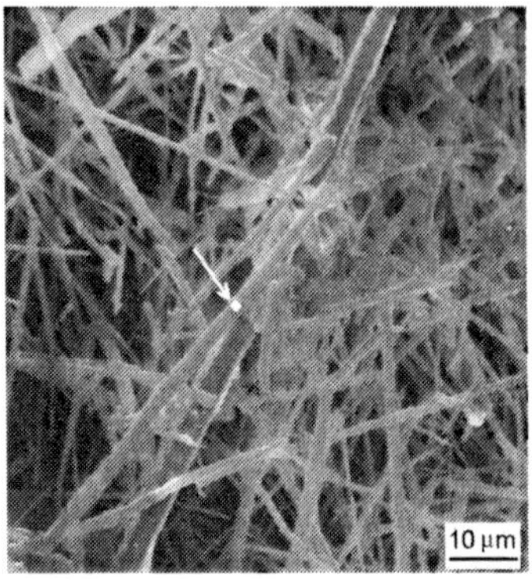

Fig. 5.7 (left). Fracture of a Fe–40Mn–30Cr–5Mo-5C master alloy crushed mould [68,75]. SEM.
Fig. 5.8 (right). Manganese needles (whiskers) formed on the surface of high-carbon ferromanganese powder charge heat treated 120 min at 1000°C in hydrogen in and alumina boat; arrow • X-ray analysis: 98.25% Mn, 1.75% Si.

lower manganese content, as shown in Fig. 5.7. This master alloy should be milled to the powder used as the Cr–Mn–Mo alloy carrier. The acicular manganese carbides formed in the hollows of the ferromanganese mould are basically similar to those formed by sublimation of manganese from the high-carbon ferromanganese powder shown in Fig. 5.8. They differ in thickness, shape and length in dependence on composition and different heat treatment conditions under which they formed. According to EDX analysis, the composition of the needles corresponds to the base composition of the tested ferromanganese and documents that silicon also sublimates with manganese.

5.1.3. As milled characteristics of manganese carriers

The milling tests of the mentioned manganese carriers under various conditions and as milled specific powder characteristics regarding their possible use for alloying in powder metallurgy were conducted in the framework of a project described in [56]. The relatively high hardness of all three manganese carriers makes possible their milling to powder to the desired size in the commonly known mills [1]. The aim of the milling tests was to characterise the properties of manganese carrier powders with which they are used for alloying of manganese steels.

The crushed pieces ~10 to 30 mm in size of both ferromanganese grades (HC and MC) seem to be optimal as the starting size for milling to powder (without slag). Both ferromanganese grades were milled in a ball mill and some tests were done in a high performance two-drum ball vibratory mill (rotation frequency 980–1450 min^{-1}, amplitude of vibrations in the range 1–4 mm) [68]. The milling tests were carried out in air and in nitrogen due to the fear of permanent manganese oxidation for an unreal adverse effect on sintering under the thermodynamic requirement of the purity of the sintering atmosphere. The results of the milling tests were evaluated by the particle size distribution and by the oxygen content of as milled powders.

Particle size distribution

Figure 5.9 shows the particle size distribution curves of FeMnC and FeMn ferromanganese milled in a ball mill and of electrolytic manganese supplied as milled to the required maximum size of 20 μm which can be regarded for representative results attained under applied laboratory milling conditions. The curves (a) and (b) show a marked increase of finer particles with the culmination to an average value of ~20 μm in size. This was proved also by the analysis of FeMn ferromanganese particles <100 μm in size with an average size of ~4 μm. According to these results, the as milled FeMnC and FeMn ferromanganese powders contained the minimal fraction of the coarser particles close to maximum <45 μm and <100 μm in size. The average size of electrolytic manganese (EMn) was <1 μm. The milling conditions of EMn were not declared by the manufacturer. The milling tests of all three manganese carriers proved their easy milling followed by sieving to the required fine powders.

Considering that all ferromanganese grades contain some amount of slag which during crushing of the moulds becomes a very fine powder and by this gets into the undersize powder,

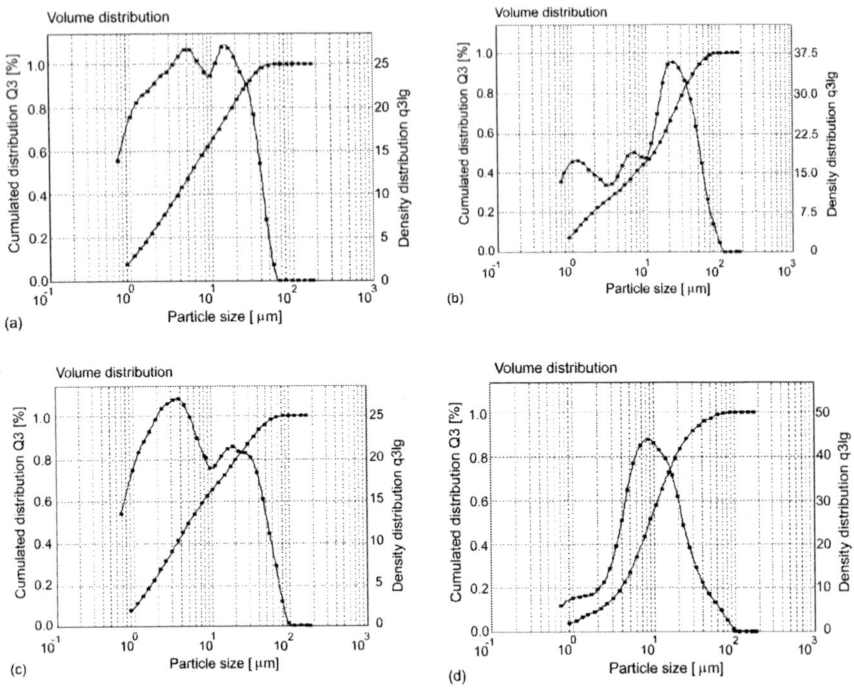

Fig. 5.9. Particle size distribution curves of as milled: (a) FeMnC ferromanganese (<45 μm), (b) FeMn ferromanganese (<45 μm), (c) FeMnC ferromanganese (<100 μm), (d) as supplied electrolytic manganese (EMn) (<20 μm) [54,68].

it can be recommended to use for milling the sizing of high-carbon ferromanganese of 10–20–(30) mm in size. The possible amount of powder slag in the milled powder decreases as a result of the formed finest particles. From this point of view, the use of an undersize powder formed, for instance, at milling of high-carbon ferromanganese for the production of welding electrodes is disadvantageous for its use in powder metallurgy. FeMn ferromanganese is chemically cleaner and therefore it would be possible to use the undersize (<45 μm) powders formed by milling for the electrodes or for other purpose. The presented particle size distribution curves depend on the milling conditions, including the mass of the charge, on the ratio to the ball size and amount in the ball and on the time period of the sieving to the required maximum size of the powder. The size of the powders does not decrease uniformly from the maximum up to not defined minimal size.

Figure 5.10 shows in more detail the morphology of as milled manganese powders, fine and some larger particles in the range of the particle size distribution. Figures 5.10(c),(d) show the morphology

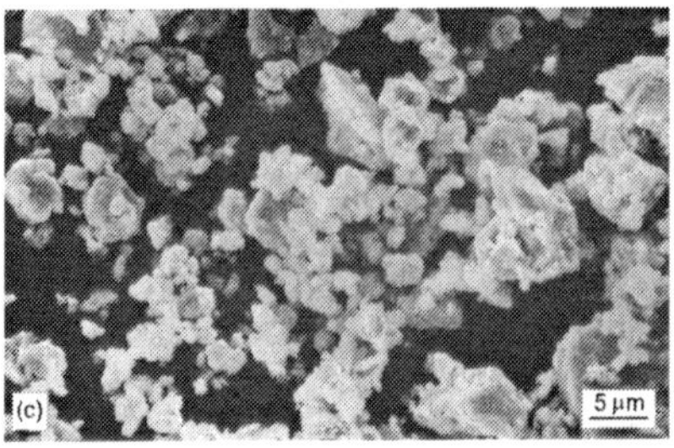

Fig. 5.10. Morphology of as milled: (a) electrolytic manganese, (b) HC ferromanganese, (c) nitro-MC ferromanganese [68,73]. SEM.

Table 5.2. Oxygen concentration ranges in electrolytic manganese (EMn) and medium- (FeMn) and high-carbon (FeMnC) ferromanganese with particle size <45 µm milled in a ball mill in air and in nitrogen (in brackets – rare). (EMn in piece form for milling test was used) [54]

Mn carrier	EMn	FeMn	FeMnC
	Oxygen content [mass %]		
Air	0.84 – 1.10	0.81 – 1.35 (2.3)	0.72 – 1.43
Nitrogen	0.63 – 0.90 (1.4)	0.33 – 0.57 (1.5)	0.25 – 0.68

of nitro-MC ferromanganese powder particles. The margins of them seem to be modestly rounded compared to that of EMn and/or HC and MC ferromanganese powders.

Oxygen content in as milled manganese carriers

The oxygen content in as milled powders was determined by two analyzers (L, B) conducted in four laboratories. Table 5.2 gives the ranges of the analyses of the oxygen concentrations determined by 97 analyses. (The repeated analyses of the same batch of a powder by the same analyzer showed only negligible differences). The oxygen concentrations in all three manganese carriers were on the same level after milling in air, but after milling in nitrogen the concentration was lower.

Table 5.3. Oxygen concentration of various manganese carriers milled in air or in nitrogen in a ball mill with the particle size <40 μm determined by two analyzers (L, B) [68,75]

Analyzer O content	Manganese carrier					
	*EMn (air)	FeMn (N)	FeMn (air)	**FeMn (air)	FeMnC (air)	FeMnC (N)
L [%]	0.84	0.47	0.80	1.33	0.89	0.25
B [%]	1.14	0.64	1.19	2.28	1.43	0.45

Note: *as external milled shipment, ** particle size formed by milling of ferromanganese for production of the welding electrodes

Table 5.4. Oxygen concentration (O) in medium-carbon ferromanganese powders (83.1% Mn, 1.1% C) milled in a ball mill (before milling 0.18% O)

Particles size [μm]	100 – 160	<100	63 – 100	45 – 63	<45
O [mass %]	0.17	0.64	0.25	0.68	1.32

The differences in oxygen concentrations of manganese powders determined by different analyzers can always be expected as being due to sublimation up to sublimation of manganese from the analysed powder. The previously mentioned differences in oxygen concentrations determined by different analyzers are given in Table 5.3. The adverse effect of the manganese powder oxygen content on the sintering process under laboratory as well as industrial conditions was not observed [1,76].

The effect of the particle size on the oxygen concentration of as milled medium carbon ferromanganese is given in Table 5.4. The oxygen concentration in the powder increases with decreasing particle size.

The surface of all three as milled manganese carrier powders was covered by a dark grey (black) layer which was the source of the measured increased oxygen concentration compared to that before milling. The black layer was a product of the abrasion between the milled manganese carrier and the grinding balls (cast manganese steel). X-ray analysis of as milled HC ferromanganese powder detected manganese silicide (Mn_5Si_3), iron carbide (Fe_2C), iron manganese ($FeMn_3$) and in small diffractions unstable manganese MnO_2 oxide in the surface black layer. The MnO_2 oxide (pyrolusite) was also used [9,10] as a manganese carrier in sintering of manganese steels [8]. This black layer can be wiped away, *e.g.* by hands which become black. This means that the dry milling of the manganese carriers did not result in the formation of a stable manganese oxide on the powders.

5.1.3.1. Effect of ferromanganese powder addition to various iron powder grades on ejection pressure

Ejection pressure is an approximate characteristics of the dry-die wall friction of the compacts ejected from the die. The possible effect of hard manganese powder particles on the wear of the die and the ejection pressure in dependence on iron powder grade and on the added manganese amount in the form of FeMnC and FeMn ferromanganese powder was determined. The results are given in Table 5.5 [68,77].

No increase of the ejection pressure was recorded with increasing manganese addition or with changes of the ferromanganese grade. This behaviour of ferromanganese powders in the green compacts was caused by the black surface layer formed by dry milling. This layer acted as a lubricant at ejection of the samples. It is possible in accordance with this statement to assume a similar effect on compressibility of the powder mixtures with the milled manganese powder addition independently on the kind of the milled manganese carrier. On the contrary, the base iron powder grade has a relatively strong effect on the ejection force was recorded. The lowest ejection pressure was recorded for the samples based on ASC100.29 iron powder.

Table 5.5. Ejection pressure for Fe–(2–5)% Mn samples (∅ 10 × 10 mm) compacted at 600 MPa in a steel die in dependence on iron powder grades and manganese addition in the form of FeMnC and FeMn ferromanganese (mean values from 10 measurements). 0.8% HW wax as lubricant [68,77]

	Ejection pressure [MPa]											
Iron powder	NC100.24				SC100.26				ASC100.29			
Mn [%]	2	3	4	5	2	3	4	5	2	3	4	5
FeMnC	100	100	100	90	60	70	70	60	60	60	70	60
FeMn	100	100	90	100	80	80	80	80	70	60	50	50

5.2. Low-carbon low-alloy sintered steel

This chapter discusses the mechanical properties of sintered steels low-alloyed with manganese, chromium, molybdenum and copper with a very low carbon content in dependence of the cooling method.The aim of research of low-alloy sintered steels alloyed with two elements with high affinity was to produce a bainitic structure at different sample cooling rates after sintering, because bainite is characterised by an appropriate ratio of strength and toughness properties. In doing so, Mn, Cr and Mo are characterised by a relatively high hardening effect on ferrite.

The test specimens (tensile strength bars (ISO 27400) and impact strength bars (ISO 5754) were prepared on the basis of sponge iron powder produced by reduction of mill scale [78,79]. 1.4% Mn in the form of FeMnC (76% Mn, 6% C) was added to the iron powder and also carbon in the amount of 0.11%, 0.52% of Cr in the form of ferrochrome, 0.55% Mo in the form ferromybdenum and in one case 0.6% Cu in the form of electrolytic copper. 0.5% zinc stearate as a lubricant was added to all samples. The particle size of all alloying additions was <40 μm. At the time of testing the addition of copper should contribute to the formation of a liquid phase to produce a more homogeneous structure and inrease the density of the compacts. The powder mixture was pressed to produce samples of both types at a pressure of 600 MPa. To achieve a higher density, green compacted samples were annealed for 30 min at 750°C and again at 600 MPa. The second process was repeated and by this way the three levels of density were attained by single, double and triple press-sinter technique. Final sintering of all the samples for each process was performed in a laboratory retort furnace for 2 h at 1120°C in the cracked ammonia with a dew point of –30°C.

Different cooling rate of samples were achieved by cooling all the samples in the furnace to a temperature of 900°C and then: a) in the retort removed from the furnace in air at 8–10°C min^{-1} which corresponds to the actual term sinter hardening and therefore this cooling mode will be termed as sinter hardening, b) quenching in oil at a temperature of 30°C (oil quenched), and c) quenching in water at 25°C (water quenched). In one case, the cooling of the samples was carried out only in the furnace – furnace cooling.

The composition of the sintered compacts was (mass %): 0.07 C, 1.36 Mn, 0.52 Cr, 0.53 Mo, 0.63 Cu, 0.09 Si. The samples cooled only in the furnace attained a tensile strength of 279 MPa and the elongation of 9.1% [79,80].

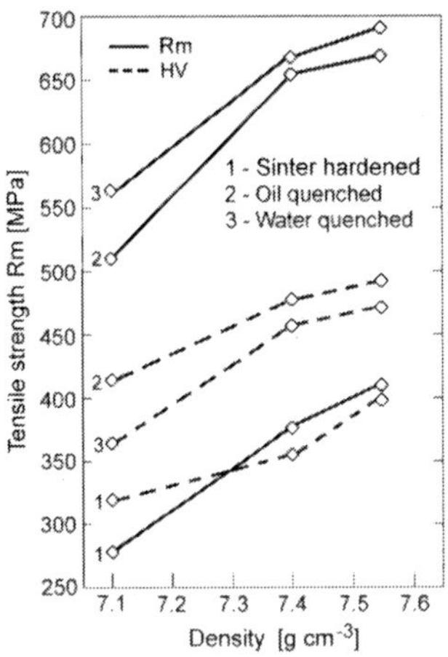

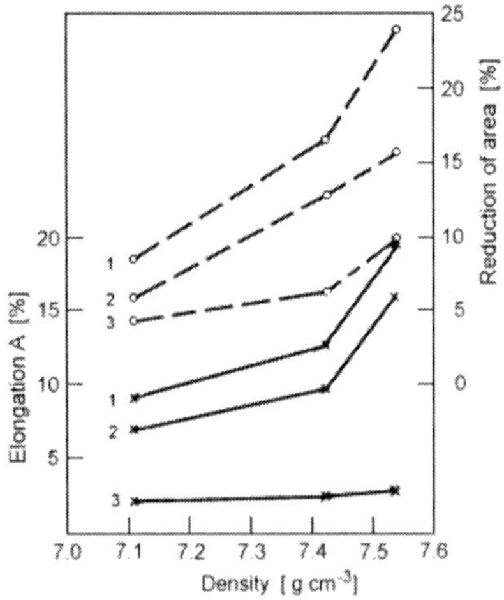

Fig. 5.11 (left). Tensile strength and hardness of sintered of low-carbon low-alloyed Fe-0.1C–1.4Mn–0.50Cr–0.50Mo–0.6Cu steel in dependence on density and cooling mode. 1 – sinter hardened, 2 – oil quenched, 3 – water quenched [83,85].
Fig. 5.12 (right). Elongation and reduction of area of sinter hardened samples as in Fig. 5.11 in dependence on density and cooling mode [83].

5.2.1 Mechanical properties

Figure 5.11 shows the tensile strength and hardness of the low-carbon low-alloy FeMnCrMoCuC steel in dependence on the density and cooling mode. Both properties increased first with increasing density of the samples and also after all three cooling modes. It is necessary to stress especially the increase of the tensile strength of oil- and water-quenched samples in spite of the very low carbon content.

The as sinter hardened samples had the tensile strength in the rage of 283 to 406 MPa and the water-quenched one in the range of 570 to 690 MPa. The density as well as the cooling mode of the samples expressed themselves in the elongation and reduction of area by decreasing values as shown in Fig. 5.12. The highest values of both properties were obtained for the sinter hardened samples.

The ratio of yield and tensile strength values of the tested steels listed in Table 5.6 also at highest strength and hardness values provides for some level of the toughness values.

The impact energy (KCV) and the unnotched (KC) values of the previously specified low-carbon low-alloyed FeMnCrMoCuC steel samples in dependence on density and cooling mode are listed in Table 5.7.

In connection with the mentioned impact energy values, especially the KCV values of the samples in dependence on density and oil and

Table 5.6. Ratio of yield and tensile strength of the tested low-carbon low-alloyed FeMnCrMoCuC steel samples in dependence on density and cooling mode

Density [g cm^{-3}]	Yield strength/tensile strength ($R_p0.2/Rm$ [%]		
	Cooling mode		
	Sinter hardening	Oil quenching	Water quenching
7.12	77.5	99.0	91.3
7.27	69.5	94.2	85.3
7.55	61.0	94.2	85.5

Table 5.7. Impact energy (KCV) and unnotched (KC) values of the tested steel samples in dependence on density and cooling mode

Density [g cm^{-3}]	KCV [J]			KC [J] unnotched
	Cooling mode			
	Sinter hardening	Oil quenching	Water quenching	Sinter hardening
7.12	11.8	7.2	9.0	24.0
7.27	18.0	21.1	14.0	33.0
7.55	25.7	12.7	14.8	61.0

water-quenched values compared to those sinter hardened. The large difference in the values obtained for the smooth specimens (without notches) should also be noted. Minimum differences were obtained in the yield and tensile strength ratios for the samples water- and oil-quenched in dependence on the density as compared with the sinter hardened specimens.

In a separate test, the compacts with same base composition with 0.27% C as the water-quenched samples had a tensile strength of 880 MPa and elongation of 2.5%.

Increased addition of copper to 1.6% to the composition of the steels with 0.6% Cu caused a decrease in tensile and impact energy values of the samples.

5.2.2. Microstructure and fracture

The microstructure of the sample cooled slowly in the furnace is formed by polyhedral ferrite and pores and by small areas of pearlite, Fig. 5.13(a). Increased cooling rate caused marked changes in the microstructure character. The fraction of ferrite decreased and the fraction of the bainitic phases increased. The water- and/or oil-

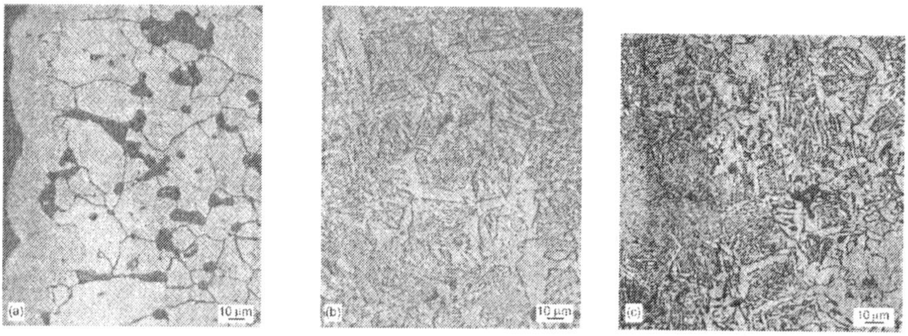

Fig. 5.13. Microstructures of sintered FeMnCrMoCuC steel samples; (a) ρ = 7.12 g cm^{-3}, furnace cooling, electroless nickel-coated fracture surface, (b) ρ = 7.37 g cm^{-3}, sinter hardening, (c) ρ = 7.55 g cm^{-3}, water quenching. Nital etched [83].

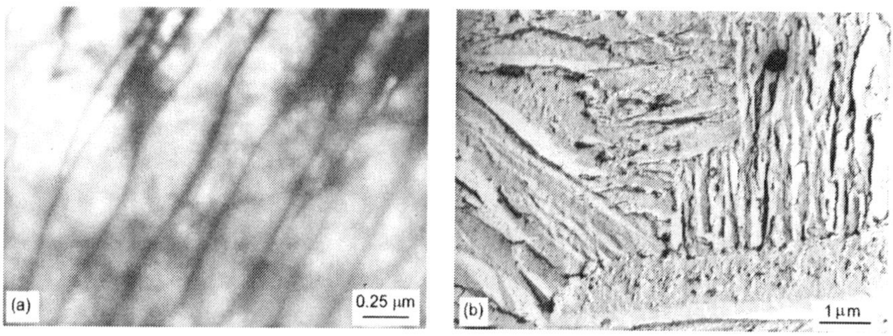

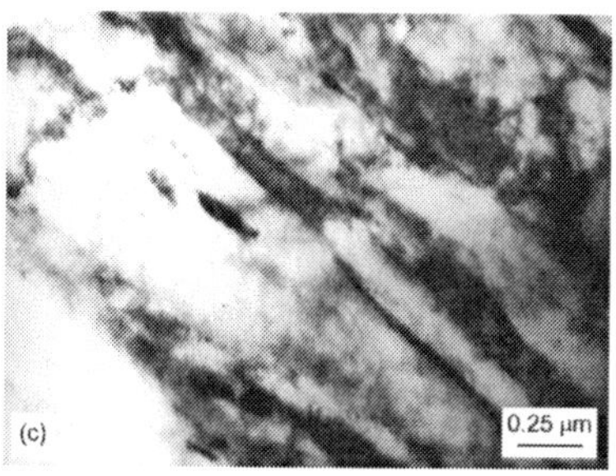

Fig. 5.14. Bainitic structures of sintered mixed Fe–1.4Mn–0.5Cr–0.5Mo–0.6Cu–0.07C steel, sintered density 7.37 g cm^{-3}. (a) thin foil – bainite and lath ferrite, (b) extraction carbon replica shadowed with Au, (c) as (a) with marked areas of dislocation of high density. TEM.

quenched microstructures are formed by bainite, often by low-carbon martensite and bainite. The Widmanstätten structure was also observed in these microstructures, Figs. 5.13(b),(c) [86]. The bainitic character of the microstructures resulted in the KCV impact energy values of 9 to 14 J. The impact energy (unnotched) KC values in the range of 24 to 61 J were obtained for the as sinter hardened steel samples.

According to the known knowledge, the alloying elements retard and by this decrease the temperature of the start of the bainitic transformation. To explain this, the characteristic bainite structures

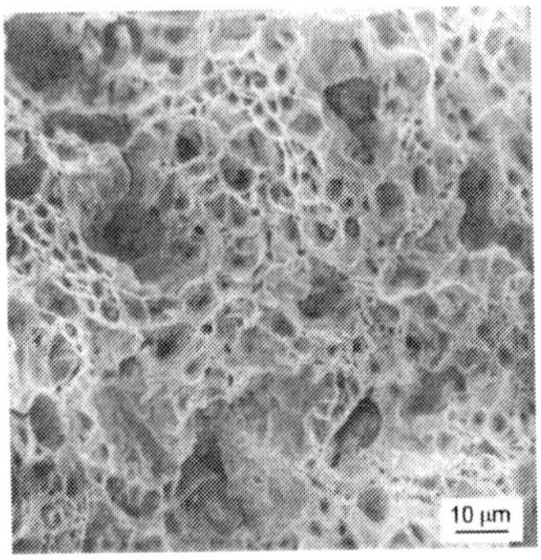

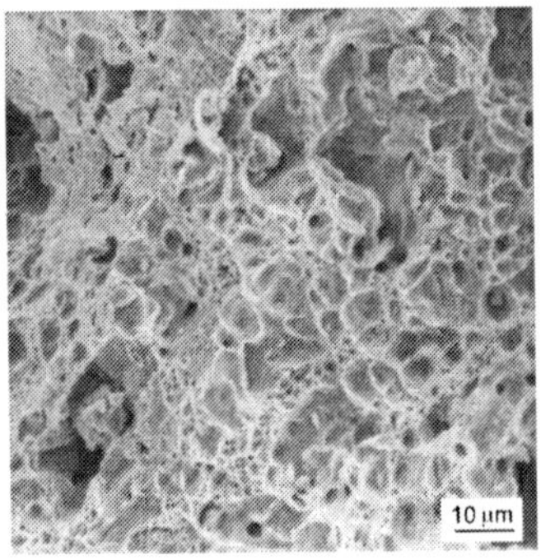

Fig. 5.15 (left). Fracture of the sinter hardened Fe–1.4Mn–0.5Cr–0.5Mo–0.6Cu–0.07C steel samples; ρ = 7.37 g cm^{-3}, Rm = 352 MPa.
Fig. 5.16 (right). Fracture of oil-quenched steel samples as in Fig. 5.15; ρ = 7.55 g cm^{-3}, Rm = 612 MPa, A = 5.7%.

of a tested mixed sintered steel cooled from 900°C in nitrogen are shown in Fig. 5.14.

In spite of high cooling rates and the corresponding high strength and hardness values the fractures of the sinter hardened and oil-quenched samples shown in Figs. 5.15 and 5.16 have a typical dimple morphology. This secures, under the given processing conditions, the optimum combination of strength and toughness properties of these sintered steels.

5.3. Properties of induction sintered and upset-forged manganese steels

Induction high-frequency (HF) sintering tests were performed mainly with an Fe–4.5Mn mixed steel to study the starting states of the formation of the microstructures during short-time sintering regarding the skin effect and corresponding mechanical properties of the tested steels. The basic question of this investigation was whether induction sintering – in spite of short diffusion times – promotes material properties that can be compared with those of conventional long-time sintering [85]. Induction sintering was conducted with individual tensile test bar samples (ISO 2740) and impact strength bars (10 × 10 × 55 mm).

Hametag (eddy milled) and RZ (air-atomised) iron powder were used for the preparation of the test samples. Manganese in amount of 4.5% was added in the form of high-carbon ferromanganese (75% Mn, 5.5% C) and this also included 0.33% C and in one case 1% Cu. The samples from powder mixtures with a lubricant were compacted at 600 MPa. Induction sintering (1 MHz) of the samples took place in a quartz tube surrounded by an inductor in hydrogen from bottles, *i.e.* in hydrogen of technical purity for 1, 3, 10 and 30 min at temperature of 1100, 1200 and 1300°C at a heating rate of ~1000 K min^{-1}. Upset forging took place after reheating (5 min, 1100°C, N_2) of the samples to 50% reduction of the original thickness. The standard test specimens from the forged billets were produced by machining.

5.3.1. Mechanical properties of induction sintered Fe–4.5Mn–0.33C steel

The steels prepared on the basis of RZ iron powder

The mechanical properties of these steels are listed in Table 5.8. Increase of all properties was observed already after 1 min HF sintering as the proof of active alloying by manganese. The elongation of all samples was <1%.

The changes in the density of the sintered samples in dependence of sintering temperature and time were relatively small. The strength

Table 5.8. Mechanical properties of induction sintered Fe–4.5Mn–0.33C steel based on RZ iron powder, high-carbon ferromanganese, 600 MPa [86,88]

HF sintering conditions		Density [g cm^{-3}]	C_c [mass %]	UTS (Rm) [MPa]	Hardness HV 20	KC [J]
°C	min					
1100	1	6.93	0.31	273	160	3.9
	3	6.94	0.29	296	159	5.5
	10	6.94	0.24	315	179	5.7
	30	6.93	0.19	420	180	5.8
1200	1	6.93	0.30	238	185	4.4
	3	6.94	0.27	372	189	5.7
	10	6.95	0.19	415	177	6.0
	30	6.92	0.15	427	133	4.8
1300	1	6.94	0.27	339	137	6.0
	3	6.94	0.20	377	161	6.8
	10	6.98	0.15	394	149	8.0
	30	6.97	0.14	407	152	6.4

properties were affected by insufficient alloying of the RZ iron matrix by manganese during the short-time sintering (time, temperature).

The steels prepared on the basis of Hametag iron powder
The mechanical properties of these steels are listed in Table 5.9.

The rise in sintered density corresponding to the starting green density of 6.57 g cm^{-3} during induction sintering is the result of the higher structural activity of Hametag iron powder. The tensile strength values lie also above those based on RZ iron powder. The lower starting density of the samples based on Hametag iron powder should be kept in mind.

The comparison of the tensile strength values attained by induction sintering of the samples of the same composition based on both iron powder grades with different structural activity is shown in Figs. 5.17 and 5.18.

Figure 5.19 shows the decrease of carbon content during sintering due to decarburisation caused by the reaction with H_2O so that accurate evaluation of the mechanical properties attained in relation to the combined carbon content can be carried out.

5.3.2. Microstructure

Figures 5.20 and 5.21 show the microstructures of induction sintered

Table 5.9. Mechanical properties of induction sintered Fe-4.5Mn-0.33C steel. Hametag iron powder, high carbon ferromanganese, 600 MPa [52,86,89]

HF sintering conditions		Density [g cm^{-3}]	UTS (Rm) [MPa]	Hardness HV 20	KC [J]
°C	min				
1100	1	6.58	294	133	7.3
	3	6.58	359	151	6.9
	10	6.63	465	166	-
1200	1	6.60	413	158	9.8
	3	6.61	542	165	11.8
	10	6.65	650	156	-
1300	1	6.62	565	157	10.1
	3	6.63	589	157	11.7
	10	6.75	600	149	-

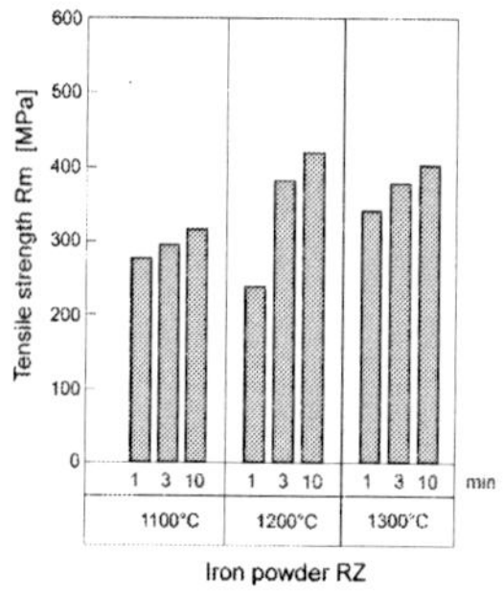

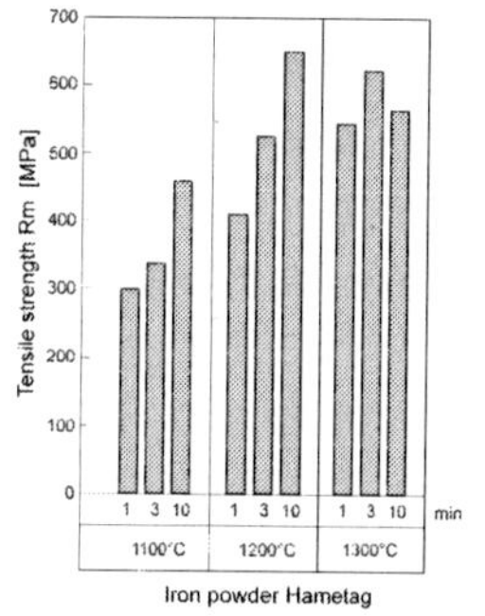

Fig. 5.17 (left). Tensile strength of induction sintered Fe–4.5Mn–0.33C steel samples based on RZ iron powder in dependence on sintering time and temperature [52,87]. **Fig. 5.18.** (right). Tensile strength of induction sintered Fe–4.5Mn–0.33C steel samples based on Hametag iron powder in dependence on sintering time and temperature [52,86,89].

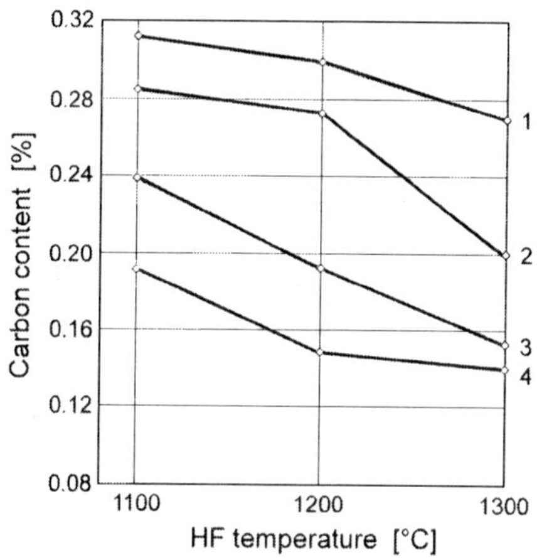

Fig. 5.19. The decrease of the combined carbon content during HF sintering of Fe–4.5Mn–0.33C steel samples in dependence on sintering time and temperature. 1 – 4 sintering time [min] [51].

samples based on RZ and Hametag iron powder grades. After sintering for 1 min at 1300°C, the extent of alloying of the microstructure is higher than that after sintering at 1200°C for 10 min. The same phenomenon, determined by the differences in the extent of alloying, in relation to the sintering conditions is also shown on the Hametag powder samples after sintering for 1 minute at two temperatures, Fig. 5.21.

The materials based on RZ iron powder were alloyed with manganese to a significantly lower extent. The cores of the starting particles remained ferritic. The microstructures of the samples based on Hametag iron powder are more homogeneous already after 1 min

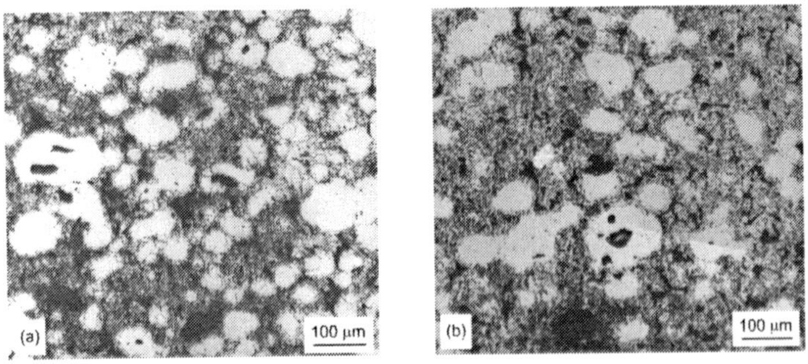

Fig. 5.20. Microstructure of Fe–4.5Mn–0.33C samples based on RZ iron powder induction sintered: (a) 10 min, 1200°C, (b) 1 min, 1300°C. Nital etched.

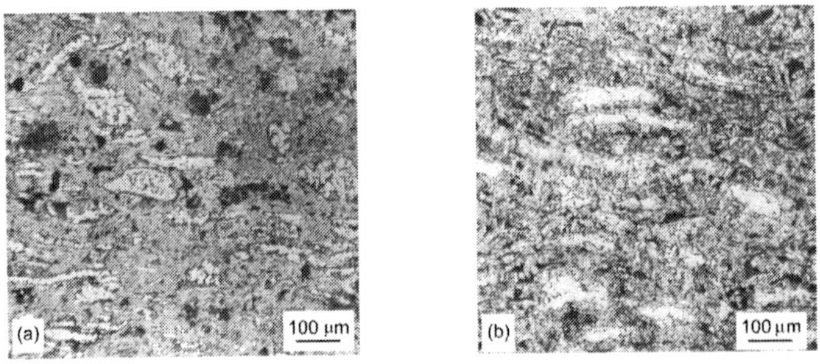

Fig. 5.21. Microstructure of Fe–4.5Mn–0.33C steel samples based on Hametag iron powder induction sintered: (a) 1 min, 1100°C, (b) 1 min, 1300°C. Nital etched.

HF sintering. The differences in mechanical properties are also the consequence of characteristics of both iron powder grades.

5.3.3. Mechanical properties of induction sintered and upset-forged Fe–4.5Mn–0.33C steel prepared on the basis of both iron powder grades

Table 5.10 lists the mechanical properties of induction sintered and upset-forged Fe–4.5Mn–0.33C steels based on the Hametag iron powder.

The upset-forged steels based on RZ iron powder processed under the same sintering conditions showed tensile strength values in the range of 600 to 730 MPa [51,88].

Fe–4.5–Mn–0.33–C steel with addition of 1% Cu

The effect of 1% copper addition on the properties of induction-sintered and upset-forged Fe–Mn–C steels was investigated. The results are listed in Table 5.11.

Table 5.10. Density and mechanical properties of upset-forged Fe–4.5Mn–0.33C steel based on Hametag iron powder in relation to the induction sintering conditions [86,88,89]

HF sintering conditions		Density [g cm^{-3}]	UTS (Rm) [MPa]	Elongation A [%]	Hardness HV 20
°C	min				
1100	1	7.59	555	0	212
	3	7.59	681	1.6	247
1200	1	7.56	622	0.8	248
	3	7.58	737	1.2	257
1300	1	7.57	741	1.6	239
	3	7.61	805	1.2	240

Table 5.11. Mechanical properties of induction sintered and upset-forged Fe–4.5Mn-1Cu–0.33C steels based on Hametag iron powder [85]

HF sintering conditions		Density	UTS (Rm)	Elongation A	Hardness
°C	min	[g cm^{-3}]	[MPa]	[%]	HV 20
As sintered					
1200	1	6.58	445	0	153
	3	6.57	475	0	151
1300	1	6.53	518	1.1	157
	3	6.56	561	2.0	145
As upset-forged					
1200	1	7.58	576	1.0	296
	3	7.58	710	1.0	329
1300	1	7.59	849	1.9	300
	3	7.62	737	1.3	231

It follows from the results, Tables 5.10 and 5.11, that addition of 1% copper to the tested induction sintered Fe–4.5Mn–0.33C material system does not contribute to any improvement in the mechanical properties.

5.4. Effect of various material and processing conditions on mechanical and some specific properties of manganese steels

This chapter presents the properties of sintered manganese steels prepared on the basis of: a) two types of iron powders with very different structural activities – the previously presented Hametag and RZ powders, b) three types of commercial iron powders, *i.e.* sponge powders NC100.24 and SC100.26 and atomised powder ASC100.29

partly in order to compare their effects on the properties of manganese steels. The impact of different sintering and cold and hot forming conditions on the mechanical and friction properties of manganese steels prepared based on these powders with different amounts of added manganese and various kinds of manganese carrier is evaluated.

5.4.1. Effect of iron powder grades of markedly different structural activity and manganese addition on properties of sintered manganese steel

The test samples were prepared from the previously mentioned Hametag iron powder, coded here H (particle size <0.10 mm) and RZ iron powder, coded here R (particle size <0.15 mm). Hametag iron powder is eddy milled (self-communition) powder which is characterised by the fine-grain structure formed by optimal recrystallisation annealing and by the large number of lattice defects (powder of high structural activity). The starting materials were chopped into pieces of annealed wire lengths 11 mm long and 2.3 mm in diameter.The iron powder was produced by gradual, dynamic, non-uniform deformation of wire particles by mutual collisions and impact on the walls of the mill. Impacts in the longitudinal axis are more intense, so that the milling material gradually passes from cylindrical to spherical shape. Changing the shape of particles to the powder formation takes place in various stages of milling [90].

The RZ iron powder was manufactured by air atomisation of a melt of synthetic pig iron of approximately eutectic concentration (~4% C, 0.1% Si), followed by annealing at ~1100–1150°C in its own atmosphere. The iron powder particles are spherical and therefore coarse-grained [1,32,35,41,91]. Both iron powder grades are no longer available but they can be accepted as the examples of the iron powders for their best and the worst starting structure and substructure character in terms of alloying by solid-state diffusion with some powder alloying element. Given that these types of iron manganese alloy powders have been discussed in previous chapters, particularly their alloying with manganese, their influence on the mechanical and friction properties of manganese steels is liste in a later chapter. Figure 5.22 presents their basic significantly different characteristics of structures whose effects have already been studied extensively

Figure 5.22(a) shows the leaf form of the particles of Hametag iron powder prior to sieving and annealing carried out at a temperature of ~800°C. These powder particles acquired the characteristic fine-grained

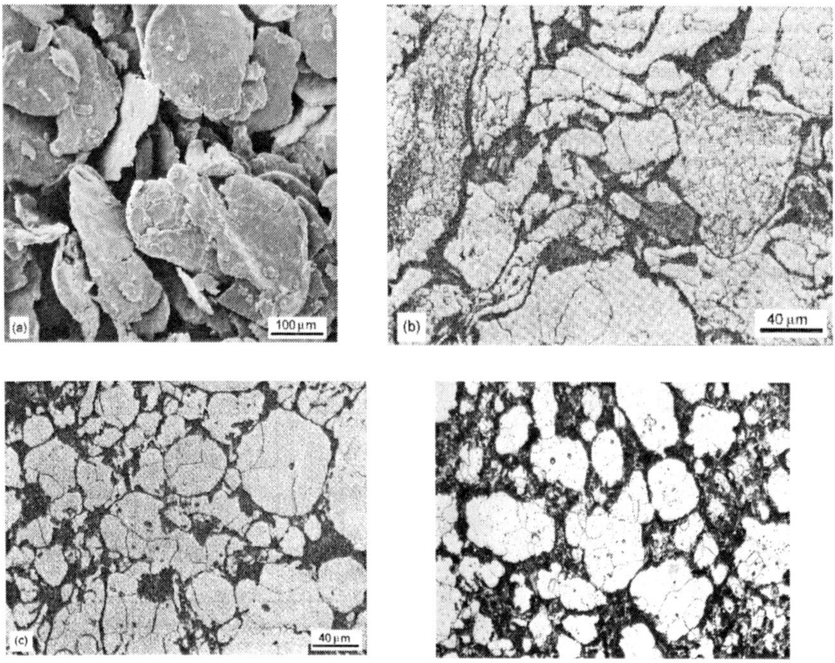

Fig. 5.22. Characteristic structures of: (a) and (b) Hametag iron powder, (a) as eddy-milled, (b) microstructure of a sintered compact with addition of ferromanganese (dark particle); 600 MPa, 20 min, 800°C, cracked ammonia, (c) green compact based on RZ iron powder with addition of ferromanganese, 600 MPa, (d) microstructure of a sintered Fe–4.5Mn–C compact; 600 MPa, 30 min, 1120°C, cracked ammonia, (b), (c), (d) – Nital etched [1].

structure and in this case also created new very fine grains, Fig. 5.22 (b). Fig. 5.22(c) shows the spherical shape of RZ powder particles with the characteristic coarse-grained structure. The consequence is that in alloying this type of powder by the manganese gas phase, which takes place mostly by slow bulk diffusion, particles inside the sintered compact remain ferritic, Fig. 5.22(d). As a result of the spherical shape of these particles one can expect even less extensive formation of interparticle necks already during preheating.

Tensile strength bars were prepared to verify the effect of the two types of powder with greatly different structural activity (Hametag, RZ) on the mechanical properties of sintered manganese steels. Manganese in the amount of 2.5–5.5% in the form of air-milled high-carbon ferromanganese (75% Mn, 7.75% C, <45 μm) and a lubricant in the amount of 0.85% was added. The specimens of both powder mixtures were compacted at 600 MPa and sintered in a retort laboratory furnace in dissociated ammonia with the dew point

Table 5.12. As sintered density (ρ), carbon and manganese contents added and as sintered values (mass %) in the steels; Mn and C as FeMnC [37,92–94]

Iron powder	Hametag (H)				RZ (R)			
Designation	H1	H2	H3	H4	R1	R2	R3	R4
ρ [g cm^{-3}]	7.26	7.19	7.12	7.08	7.08	7.08	6.95	6.93
Mn added	2.5	3.5	4.5	5.5	2.5	3.5	4.5	5.5
*C added	0.20	0.28	0.36	0.44	0.20	0.28	0.36	0.44
C_c	0.14	0.15	0.20	0.26	0.11	0.16	0.28	0.33
Mn	2.2	3.2	4.2	4.5	2.4	3.3	4.1	4.8

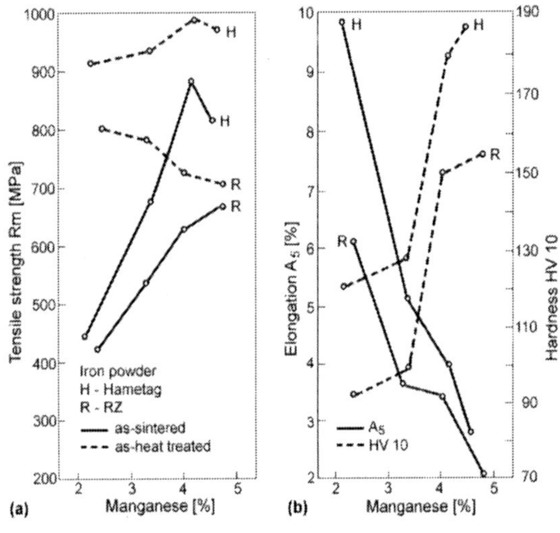

Fig. 5.23. Effect of as sintered manganese content and of iron powder grade (H, R) on: (a) sintered and heat treated tensile strength, (b) elongation and hardness; heat treatment – quenching 850°C, N_2, oil, 200°C, 30 min, air [37,92–94].

of –30°C for 3 h at 1120°C. The specimens were tested as sintered and as heat treated (oil-quenching and tempering 30 min at 200°C).

A part of the specimens was oil-quenched and tempered and a part upset-forged (10 min, 1100°C, N_2) by the reduction of the thickness of ~40% and then machined by turning to the standard tensile test bars.

The as sintered density values and manganese and carbon contents of the tested steels are listed in Table 5.12. The effect of the manganese content of both iron powder grades on the as sintered and as heat-treated mechanical properties of Fe–(2.5–5.5)Mn–xC steels is shown in Fig. 5.23.

These sintering conditions resulted in a decrease in the manganese content and a relatively large decrease in the carbon content.

A large difference in the mechanical properties due to the influence of the types of iron powder is clearly visible. In both cases the tensile strength increased almost linearly with the Mn content increasing up to 4–4.5%. The highest tensile strength of the sintered steels of 886 MPa was obtained for the Fe–0.2C–4.2Mn steel H3. On the other hand, the steel R4 showed the highest tensile strength of 672 MPa. The same characteristic effect of the iron powder grade and manganese content on the elongation and hardness of the tested steels is shown in Fig. 5.23(b) bearing in mind in all cases also the as sintered carbon contents. Given that the carbon content in the sintered state in both systems ranged from 0.11 to 0.33%, in the case of the steels in the system H it was in the range of 0.14–0.26% and, therefore, these material systems can be regarded as low-carbon. The effect of manganese to about 4% resulted in an almost linear increase or decrease of these properties. All the properties of the steels prepared on the basis of the RZ powder were lower.

Heat treatment increased the tensile strength of both steel groups. The steels based on the H iron powder attained the tensile strength in the range of 910 to 996 MPa without any marked effect of manganese addition. The steels based on the R iron powder had the tensile strength in the rage of 800 MPa with decreasing tendency to 700 MPa at increasing manganese content. The increase of the hardness of both groups of Mn steels differed in relation to the base iron powder.

The type of iron powder was also reflected on the fracture surfaces as shown in Fig. 5.24. The fracture surface of sample H1 is ductile nature with fine-dimple morphology. In contrast, the fracture surface

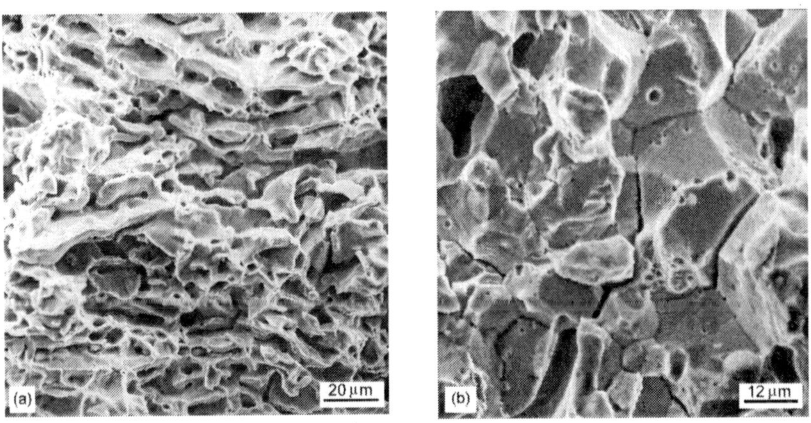

Fig. 5.24. Fractures of sintered Mn steel prepared on the basis of H iron powder. (a) Fe–2.2Mn–0.14C, (b) Fe–4.5Mn–0.26C, (Table 5.12, Fig. 5.22). SEM.

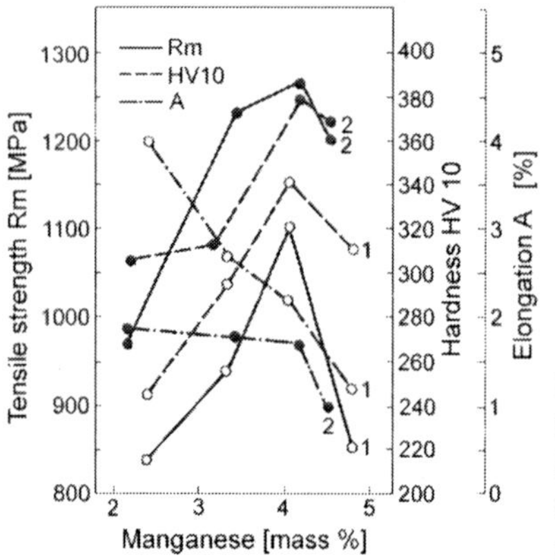

Fig. 5.25. As upset-forged tensile strength, hardness and elongation in dependence on iron powder grade (H, R) and on manganese content of Fe–Mn–C steels. As upset-forged density of the samples ~7.75 g cm^{-3} [91,94].

Fig. 5.26. Microstructure of an as upset-forged (reduction 40%) sintered manganese steel sample. (a) H2 – 3.4% Mn, 0.15% C, (b) H4 – 4.5% Mn, 0.26% C; H2, H4 see Table 5.12. Nital etched.

of the samples H4 is distinctively brittle intergranular due to the expected significant fraction of martensite in the structure. This explains the decrease in the tensile strength of the compacts with addition of 5.5% Mn.

Figure 5.25 shows the mechanical properties of steels presented in the previous figure as upset-forged with reduction of 40%. The H specimens reached the highest tensile strength value of 1264 MPa, while the R specimens 1100 MPa. The highest values of upset-forged and tensile strength as well as of elongation were obtained for the steel with 4% Mn followed by a rapid decrease for 5% Mn. Hardness shows a maximum at the same manganese content as strength does. The difference in the effect of both iron powder grades is again clearly discernible also in elongation values.

The uniform and fine microstructure is characteristic for upset-forged manganese steels prepared on the basis of Hametag iron

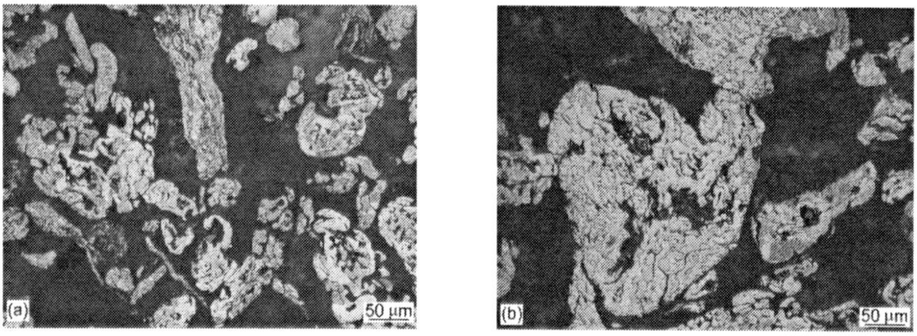

Fig. 5.27. Microstructure of particle of MH80 sponge iron powder, fraction >0.315 mm, after milling for 15 min in a vibration mill and annealing for 30 min at 720°C in hydrogen. Nital etched.

powder. Figure 5.26 shows the microstructure of samples with two different manganese contents; fine carbides can be clearly seen. The hardness values indicate that the structures of these steels are composed of bainite and martensite.

The fine-grained iron powder was also produced by milling the sponge iron powder MH80, coarse fraction >0.315 mm (and also SC100.26 iron powder, fractions >0.16 mm) in a vibration mill for 15 min followed by annealing for 30 min at 720°C. An example of the structure of this powder is shown in Fig. 5.27.

5.4.2. Effect of various hot forming processes on mechanical properties of manganese steels

The tested hot forming processes for the attainment of high density of manganese steel and by this of high mechanical properties were: extrusion, rolling, upset forging and closed-die forging (form of the die – standard test bar).

Extrusion. Samples for testing extrusion were prepared from Hametag iron powder. Manganese, including the corresponding amount of carbon, was added to the iron powder in the form of high-carbon ferromanganese (<40 µm) in an amount: 2.5% (H1), 3.5% (H2), 4.5% (H3) and 5.5% Mn (H4). The carbon and manganese content in the samples in the sintered condition is shown in Table 5.13. The preforms for extrusion from the powder mixture with the addition of lubricant were pressed at 300 and 600 MPa and sintered 60 min at 1120°C in cracked of ammonia with a dew point of –30°C.

Extrusion of samples with a reduction of 80 and 90% after 10 min heating at 1120°C in nitrogen was performed in a device

Table 5.13. Compaction pressure for extrusion preforms and as sintered density (ρ) of the bars extruded by two degrees of reduction (R) for manganese steels H1, H2, H3 and H4 [95–97]

Ser. No.	Steel	Compact. pressure [MPa]	As sintered			As extruded ρ [g cm^{-3}] Reduction R [%]	
			Mn [%]	C [%]	ρ [g cm^{-3}]	80	90
1	H1	300	2.6	0.13	6.15	7.40	–
2	H1	600			7.0	7.37	7.40
3	H2	300	3.2	0.16	6.12	7.42	–
4	H2	600			6.95	7.28	7.45
5	H3	300	4.0	0.23	6.1	7.39	–
6	H3	600			6.9	7.43	7.50
7	H4	300	5.0	0.26	6.1	7.40	–
8	H4	600			6.9	7.46	7.57

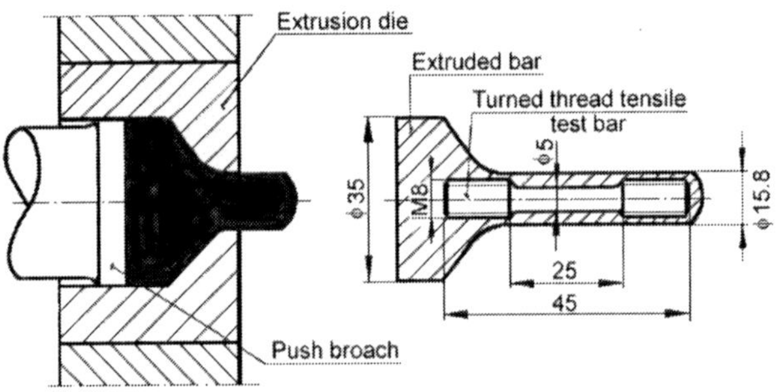

Fig. 5.28. The die for forward hot extrusion and the dimensions of the extruded bar and of the turned tensile strength bar (dimensions in mm) [95].

shown in Fig. 5.28. The samples for determining the tensile strength of extruded bars were manufactured by turning.

Figure 5.29 shows the tensile strength values of H1, H2 and H3 steel samples manufactured by turning extruded bars to the standard form by 80% reduction. Before machining to tensile strength bars a part of the extruded bars was annealed 60 min at 600°C in dissociated ammonia.

The same tensile strength of 886 and 873 MPa was approximately obtained for the annealed H1 and H2 steels in spite of the fact that the H1 preform was compacted at 300 MPa compared to that of the H2 steel preform compacted at 600 MPa. The elongation of the

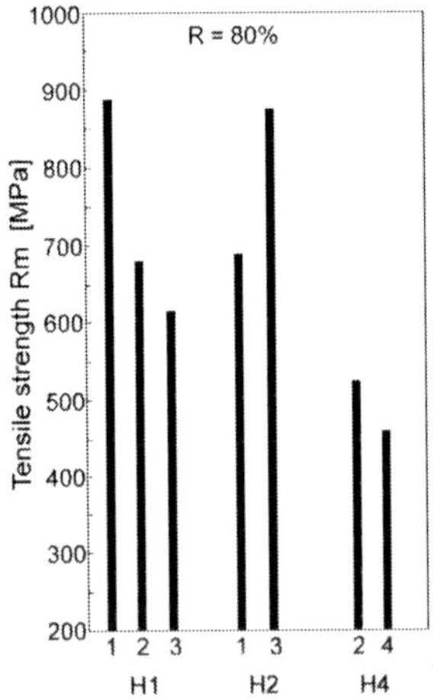

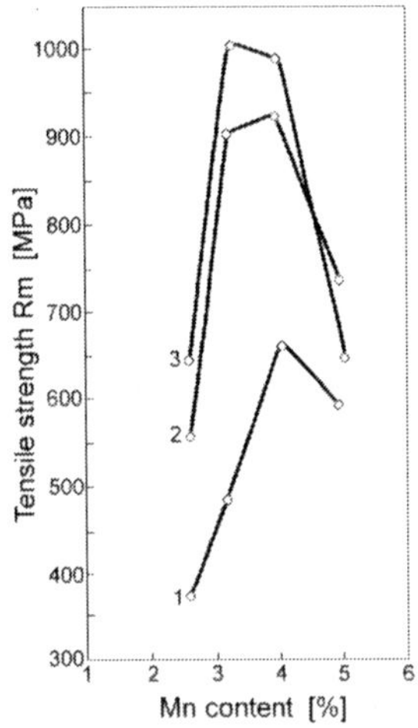

Fig. 5.29 (left). Tensile strength of manganese steels H1, H2 and H4 hot extruded with 80% reduction. 1 – 300 MPa, annealed, 2 – 300 MPa, unannealed, 3 – 600 MPa, unannealed, 4 – 600 MPa, unannealed (Table 5.13) [95–97].

Fig. 5.30 (right). Tensile strength of sintered manganese steels in dependence on manganese content and forming method. 1 – as sintered, 2 – as upset-forged by reduction of 50%, 3 – as extruded by reduction of 90% (unannealed) [98,99].

H1 steel from the preform compacted at 600 MPa was 13%. Figure 5.30 shows the dependence of as sintered, as upset-forged and as extruded tensile strength samples on manganese addition. The highest tensile strength under given processing conditions attained the steels extruded by reduction of 90%, Table 5.13. The extruded H1 steel attained the elongation of 17% and reduction of area of 33%.The ductility of the steels H2–H4 could not be measured. It is important to stress the reduction of the tensile strength of the tested steels in all the conditions at a Mn content higher than 4%.

Rolling and upset and closed-die forging. These tests were conducted on the tensile strength bars based on a Hametag iron powder iron (particle size <0.315 mm, 12% of the fraction <0.1 mm) and the following materials: 1 – plain iron powder (Fe), 2 – Fe–0.5% graphite, 3 – Fe–3.5% Mn–0.5% C (total). Manganese was added to powdered iron in the form of high-carbon ferromanganes (<40 μm). Samples for this purpose were pressed at a pressure of 600 MPa and sintered for 60 min at 1100°C in cracked ammonia, dew point –30°C.

Sintered samples were tested in the state sintering and in state after hot forming processes referred to various reducing the thickness of the sample. Samples from the processing of were heated for 10 min at 1100°C in nitrogen. Real measurements of the temperature

Table 5.14. Density (ρ), tensile strength (Rm), elongation (A) and hardness (HV10) of sintered iron and carbon and manganese steels after hot forming by rolling, upset forging a closed-die forging with different reductions; R – reduction [100,101]

Hot forming process	Material	R [%]	ρ [g cm^{-3}]	Rm [MPa]	A [%]	HV10
Rolling	Fe	30	7.5	254	7.4	125
		50	7.6	396	12.4	139
		70	7.6	444	17.0	156
	Fe–0.5C	30	7.5	412	7.0	146
		50	7.6	404	15.0	163
		70	7.6	429	17.0	169
	Fe–3Mn–0.5C	30	7.6	825	1.7	225
		50	7.7	866	1.6	278
		70	7.7	971	2.4	398
Upset forging	Fe	30	7.5	353	6.6	146
		50	7.7	399	15.6	144
	Fe–0.5C	30	7.6	391	8.9	172
		50	7.7	429	10.5	149
	Fe–3Mn–0.5C	30	7.6	818	2.0	266
		50	7.6	914	1.4	317
Closed die forging	Fe	70	7.7	450	10.1	178
	Fe-0.5C	70	7.7	530	9.3	180
	Fe-3Mn-0.5C	70	7.7	913	2.6	356

of the samples under the processing conditions from 800 to 900°C were perfected.

The mechanical properties of materials in the state after the above processes for heat treatment are shown in Table 5.14.

All the three hot-forming methods resulted in a large increase of tensile strength and hardness depending on the reduction.The largest increase of tensile strength to 913–971 MPa in comparison with the sintered state (279 MPa) was recorded for the manganese steel with the ductility decreasing in all the three conditions.

5.4.3. *Effect of various sintering conditions, powder grade and manganese carrier*

5.4.3.1. Sintering in α-phase and during preheating and hold time

From general knowledge about sintering of iron based materials it emerges that, from the technological point of view, for the course of the metal body forming process from a dispersed body and its alloying not only with manganese, the knowledge of the lowest temperature of manganese sublimation and temperature stability of iron powder structure properties, which determine its sintering activity, was and is important.

Regarding this aspect for sintering and alloying through the manganese vapour, the effect of sintering at 875°C and during preheating, which seems to be extreme in relation to the conventional sintering conditions, followed by holding at sintering temperature, on the mechanical properties of manganese steels was investigated [102].

For this purpose, in addition to the Hametag iron powder, specimens (tensile strength bars) were also prepared from sponge SC100.26 and NC100.24 and iron powder and water-atomised ASC 100.29 iron powder. Manganese in the amount of 4.5% Mn was added in the form of high-carbon ferromanganese to the iron powders.

The tensile strength of manganese steels based on different iron powder grades sintered at 875°C is shown in Fig. 5.31. The as sintered density of the specimens was 6.75–6.82 g cm^{-3}. Sintering in the α-region resulted in the formation of interparticle metal bonds

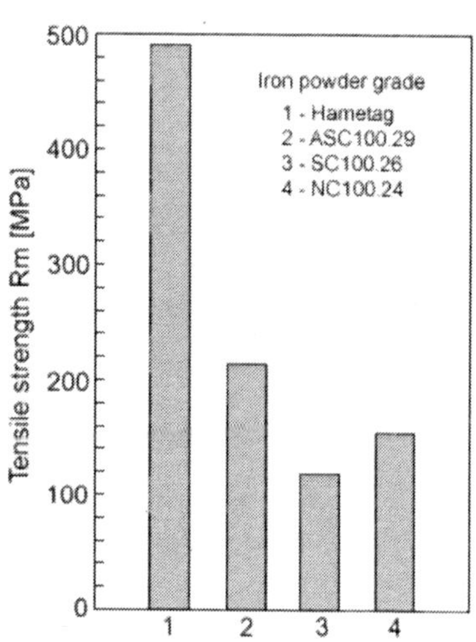

Fig. 5.31. Tensile strength of Fe–4.5Mn–0.4C steels prepared from four iron powder grades. 600 MPa, 80 min, 875°C, cracked ammonia, dew point –30°C, heating rate 35°C min^{-1}, FeMnC <45 μm [8,53,94,102].

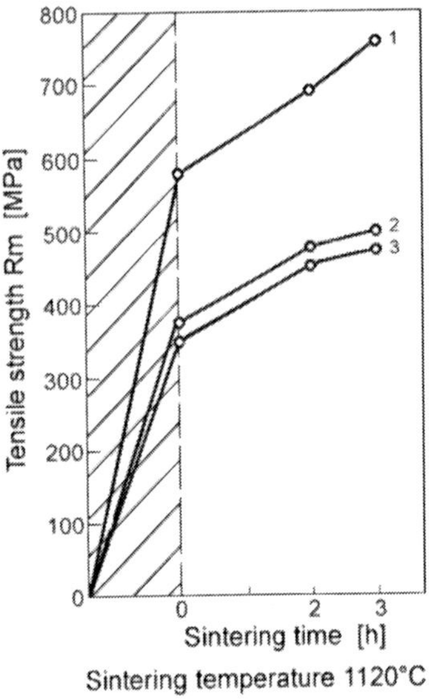

Fig. 5.32. Tensile strength of Fe–4.5Mn–0.4C steel samples in dependence on heating time to sintering temperature for: 0 – preheating, heating rate 17°C min^{-1}, 2 and 3 h hold time, cracked ammonia, dew point –30°C. 1 – Hametag, 2 – SC100.26, 3 – NC100.24 iron powder. Other as in Fig. 5.31 [8,94,102].

between the iron particles in the compacts and in some extent in alloying by manganese.

The samples based on Hametag iron powder had a tensile strength of 496 MPa at the hardness of 205 HV 10. Despite the lower tensile strength, the steels based on the two sponge iron powder grades exhibited the hardness in the range of 153 to 171 HV 10. The data prove that there was no oxidation of ferromanganese particles in the 'insufficiently' clean sintering atmosphere for the tested material systems under the given sintering conditions.

The results demonstrate the continuous sublimation of manganese from the ferromanganese particles in the compacts at this temperature and alloying of iron matrices by manganese and its effect on the tensile strength of the commercially produced iron powders.

Sintering of the iron powder based materials in the α-region has been recommended as the method for studying and evaluating the physical–structural and technological properties of the iron powder grades in relation to the formation of the metal bonds and the start of alloying with some selected alloying elements [103–105].

Figure 5.32 shows the tensile strength of the Mn steels of the same composition as in Fig. 5.31 with special attention given to the preheating time to 1120°C. The preheating time resulted in the relatively largest increase in tensile strength caused by manganese compared to that after 2 and 3 h sintering with the preservation of

Table 5.15. As sintered density and carbon and manganese content (mass %) for Fe–(1–5)Mn–xC steels determined on tensile strength samples. SC100.26, 600 MPa, 60 min, 1120°C, cracked ammonia, dew point −30°C. Mn and C as FeMnC

As sintered	Manganese addition [%]				
	1	2	3	4	5
Density [g cm^{-3}]	6.98	7.01	6.92	6.86	6.81
C [%]	0.10	0.17	0.21	0.23	0.32
Mn [%]	0.99	2.01	2.93	3.34	4.39

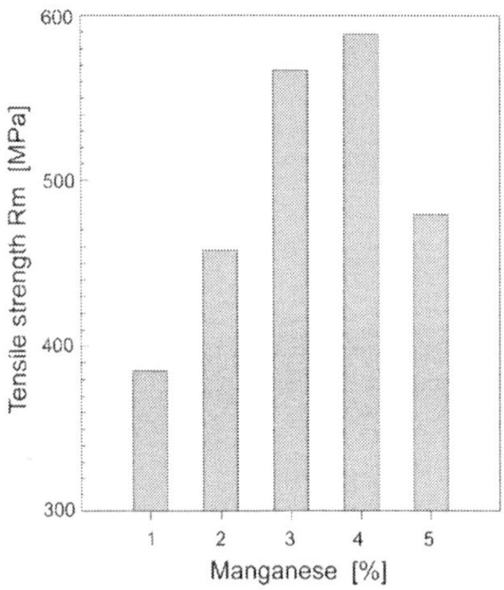

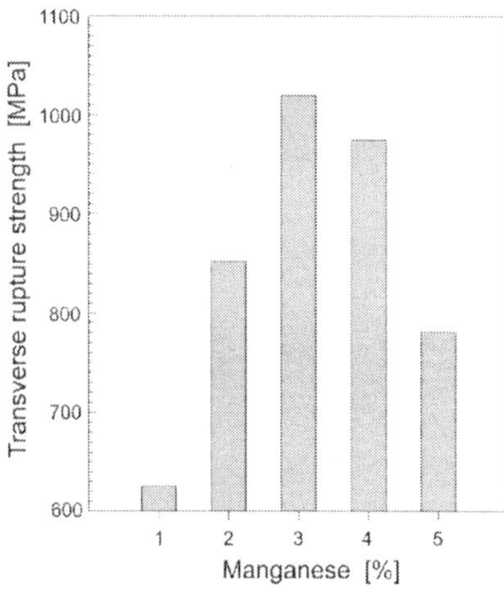

Fig. 5.33 (left). The dependence of tensile strength on manganese addition of sintered Fe–(1–5)Mn–xC steel. SC100.26, 600 MPa, 60 min, 1120°C, cracked ammonia, dew point −30°C [77,94].

Fig. 5.34 (right). The dependence of transverse rupture strength on manganese addition of Fe–(1–5)Mn–xC steel. SC100.26, 600 MPa, 60 min, 1120°C, cracked ammonia, dew point −30°C [77,94],

the differences in tensile strength between the materials based on the two sponge iron powder grades [103,104,106].

5.4.3.2. Effect of manganese addition

The powder mixtures based on SC100.26 iron powder with a manganese addition of 1, 2, 3, 4 and 5% Mn were prepared to study the effect of manganese addition. The corresponding combined carbon content in the powder mixtures coming from the high-carbon ferromanganese was 0.08, 0.16, 0.24, 0.32 and 0.40%, respectively. The standard tensile and transverse rupture strength and impact energy bars were compacted at 600 MPa and sintered for 60 min at 1120°C in

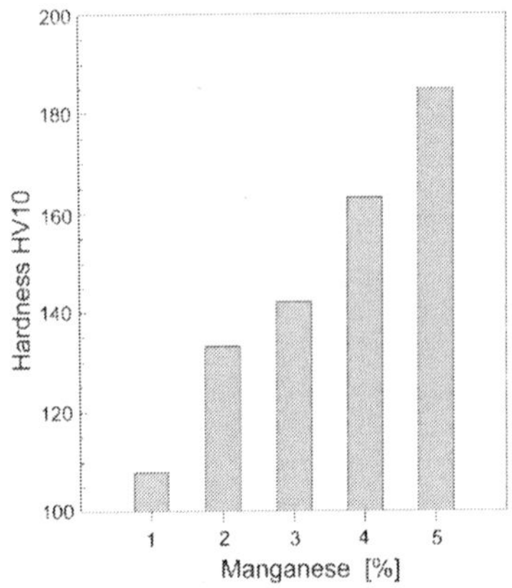

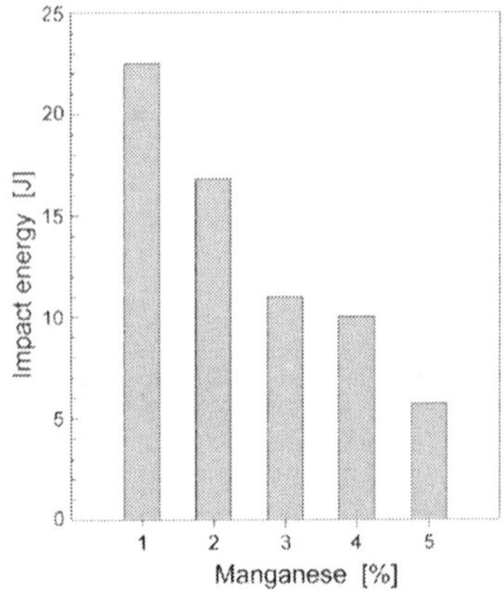

Fig. 5.35 (left). The dependence of hardness on manganese addition for sintered Fe–(1–5)Mn–xC steels. SC100.26, 600 MPa, 60 min, 1120°C, cracked ammonia, dew point –30°C [77,94].

Fig. 5.36 (right). The dependence of impact energy on manganese addition for sintered Fe-(1-5)Mn-xC steels. SC100.26, 600 MPa, 60 min, 1120°C, cracked ammonia, dew point –30°C [77,94].

a laboratory retort furnace. As sintered carbon and manganese contents are listed in Table 5.15.

The loss of manganese content during sintering was caused by demanganisation. (This effect is more marked at sintering of small batches of samples under laboratory conditions compared with sintering of larger batches also in industrial furnaces).

The dependence of the tensile strength and transverse rupture strength values on manganese content is shown in Figs. 5.33 and 5. 34 and of the hardness and impact energy values in Figs. 5.35 and 5.36. Tensile strength increased with increasing manganese content up to 589 MPa for 4Mn steel, *i.e.* by ~98 MPa/1% Mn. In the Fe-5Mn–C steel, the tensile strength became substantially lower. The similar course was also recorded for the dependence of transverse rupture strength on manganese addition with the maximum of 1019 MPa for 3Mn steel. Hardness increased to 185 HV10 for 5Mn steel. The impact energy compared to hardness decreased from 22.5 J for 1Mn to 5.7 J for 5Mn steel with increasing manganese addition.

5.4.3.3. Effect of sintering temperature, compacting pressure, sintering time and iron powder grade

Effect of sintering temperature. The mechanical properties of the tested

Table 5.16. Sintered density (ρ) and mechanical properties of Fe–(1–5)Mn–0.3C steels. ASC100.29, electrolytic manganese <5 μm, 600 MPa, 75 min, 1120°C and 1280°C, cooling rate 50–70°C min^{-1} [58,63]

Mn added [mass %]	Sintering temperature [°C]	ρ [g cm^{-3}]	Rm [MPa]	A [%]	KC [J]	Hardness HV 30
1	1120	7.21	384	11.2	50	93
	1280	7.25	403	12.4	52	103
2	1120	7.11	542	6.2	36	117
	1280	7.16	593	6.1	49	116
4	1120	6.98	594	1.2	8	227
	1280	7.01	966	2.5	23	275
5	1120	6.91	462	0.9	6	258
	1280	6.94	896	1.4	21	269

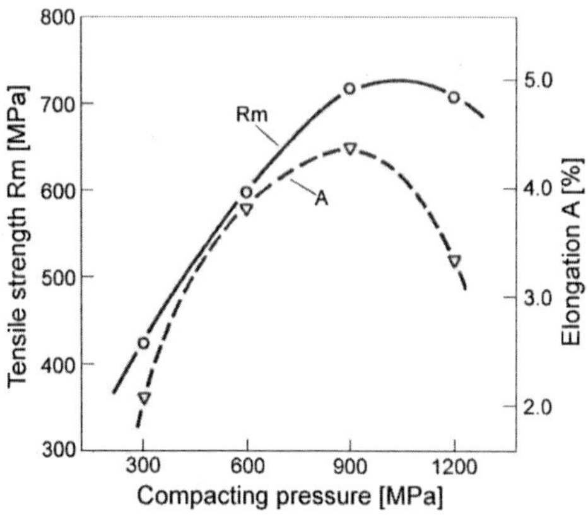

Fig. 5.37. Tensile strength and elongation of Fe–3Mn–0.3C steel in dependence on compacting pressure. 75 min, 1160°C. Other as in Table 5.16 [53,94,107,108].

Fe–(1–5)Mn–0.3C steels sintered in technical purity hydrogen under a getter – alumina + 5% graphite – at two temperatures are presented in Table 5.16. The testing of the mechanical properties revealed that their improvement, except that of elongation, as a result of sintering at higher temperatures is more pronounced at higher Mn content. While at 2% Mn the effect of sintering temperature is moderate, at 4% Mn a decisive improvement of both tensile strength and impact energy is observed.

Effect of compacting pressure. Investigations carried out with differently compacted Mn steel compacts showed that in the case of Mn steels the usual steady increase of the mechanical properties with increasing compacting pressure (some density increase) is found only

Table 5.17. As sintered density of Fe–(1,2,4)Mn–(0.08, 0.16, 0.32)C steel samples (tensile strength bars) in dependence on iron powder grade and manganese addition. 60, 90 and 120 min, 1120°C, cracked ammonia, dew point –30°C. Mn as FeMnC, <45 μm [103,109]

Iron powder grade	As sintered density [g cm^{-3}]								
	Sintering time [min]								
	60	90	120	60	90	120	60	90	120
	1Mn			2Mn			4Mn		
N	6.77	6.79	6.81	7.02	7.00	7.01	6.89	6.88	6.98
S	6.79	6.79	6.84	7.06	7.06	7.07	6.89	6.89	7.01
A	6.86	6.88	6.92	7.09	7.12	7.10	6.89	6.89	7.06

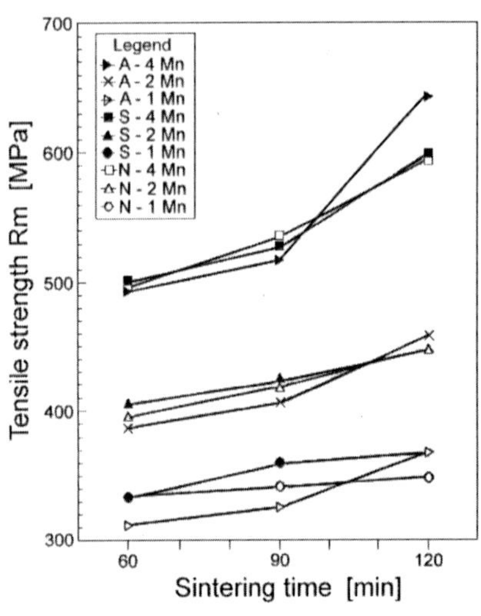

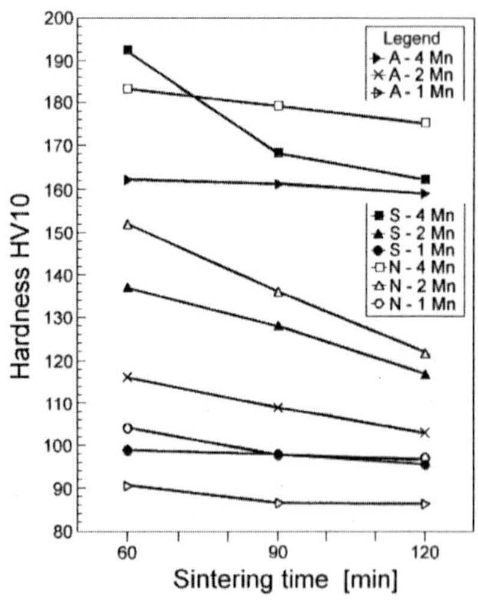

Fig. 5.38 (left). Tensile strength of sintered Fe–(1,2,4)Mn–xC steels in dependence on sintering time at 1120°C, manganese addition and iron powder grade (N, S and A). 600 MPa, 1120°C, cracked ammonia, dew point –30°C [54,103,108].
Fig. 5.39 (right). Hardness of the steels as in Fig. 5.38 [54,103,108].

up to a certain compacting pressure level, at still higher densities as shown in Fig. 5.37.

Effect of sintering time and of iron powder grade. The test specimens on the basis of the sponge NC100.24 (N) and SC100.26 (S) iron powder and of water atomised ASC100.29 (A) iron powders were prepared. Manganese in amount of 1, 2 and 4% was added as high-carbon ferromanganese to the powders. The tensile strength and impact energy bars from the powder mixtures were compacted at

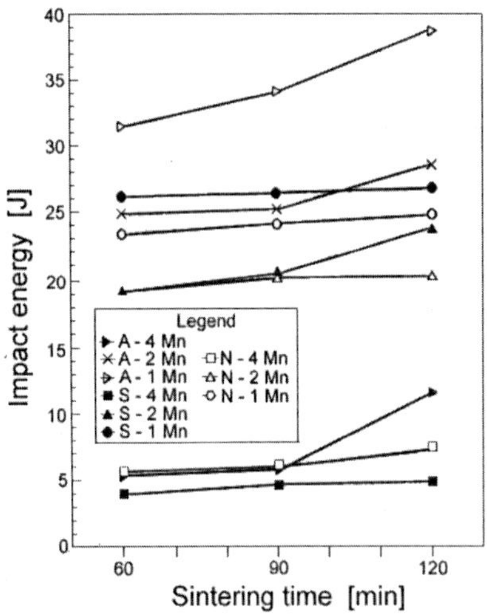

Fig. 5.40. Unnotched impact energy of the steels as in Fig. 5.37 [54].

600 MPa and sintered at 1120°C in a laboratory retort furnace. The as sintered density of the tested steel samples is listed in Table 5.17.

The dependence of tensile strength and hardness on sintering time, iron powder grade and manganese addition is shown in Figs. 5.38 and 5.39 [103].

The result show the almost linear increase of tensile strength with increasing sintering time and minimal differences caused by different iron powder grades, except for A4 steel sintered for 120 min. The 1Mn steels attained the tensile strength in the range of 312 to 334 MPa (60 min) and of 348 to 367 MPa (120 min), and the tensile strength of the 4Mn steels was 492 to 499 MPa (60 min) and 593 to 644 MPa (120 min), respectively. The hardness of the investigated steels decreased with increasing sintering time.

Sintering time resulted in minimum changes of impact energy, except for the steels based on ASC100.29 iron powder after sintering for 120 min (Fig. 5.40). The impact energy values of 1Mn steel samples based on A powder are higher by 17 to 34% compared to those based on N and S powders. The relatively lowest impact energy values were recorded for the steels based on N powder for 1Mn effect on hardness and in a minor extent on impact energy contrary to 2Mn steel and to those based on S powder for 4 Mn steel.

The effect of iron powder grade on transverse rupture strength (TRS) was also studied. The TRS of the sponge iron powder grades was higher than that of the atomised iron powder. The highest TRS value of 1110 MPa was obtained for the steel with the addition of 4%

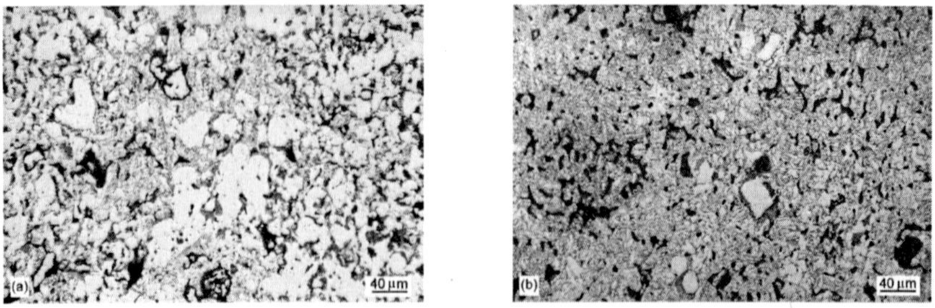

Fig. 5.41. Microstructure of sintered 2Mn steel specimens based on: (a) NC100.24 iron powder, (b) ASC100.29 iron powder. 600 MPa, 60 min, 1120°C, cracked ammonia, dew point –30°C. Nital etched.

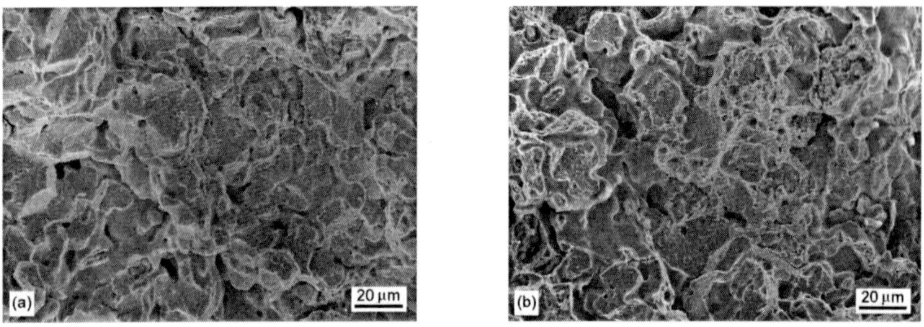

Fig. 5.42. Fracture surface of sintered 4Mn steel tensile strength specimens based on: (a) NC100.24 iron powder, (b) ASC100.29 iron powder. 60 min, 1120°C. SEM.

Mn for sponge iron powder SC100.26 (600 MPa, 60 min, 1120°C, cracked ammonia).

Figure 5.41 shows the micrographs of the 2Mn steel specimens based on N and A iron powder. The number of ferritic regions in Fig. 5.40(a) is greater than in Fig. 5.40(b).

The fracture surfaces of the 4Mn steel specimens based on NC100.24 (N) and ASC100.29 (A) iron powders are shown in Fig. 5.42. The fracture surface of 4Mn steel specimen based on N iron powder is characterised by a marked fraction of intergranular brittle facets and a minor fraction of transgranular brittle facets. On the contrary, the fracture surface of 4Mn steel specimen based on A iron powder is characterised mostly by transgranular brittle facets.

Figure 5.43 shows the yield, tensile strength and impact energy values of sintered Fe–(2,3,4)Mn–xC steels based on ASC100.29 iron powder with manganese addition in the form of electrolytic manganese obtained by sintering for different times. Sintering was

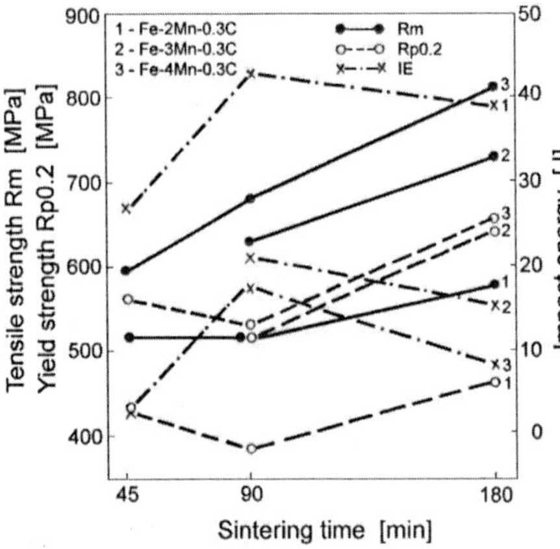

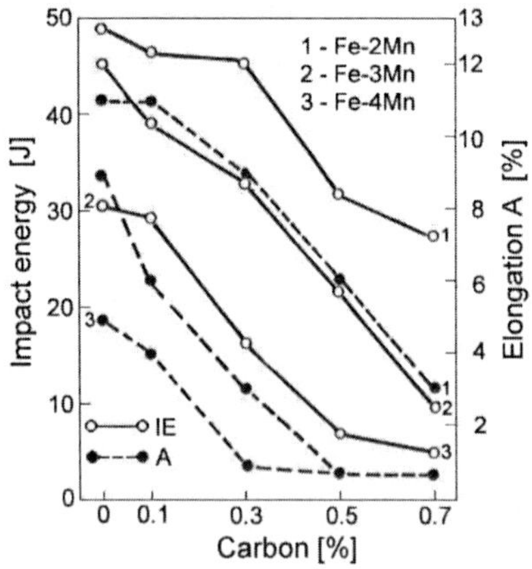

Fig. 5.43 (left). Yield and tensile strength and impact energy of sintered Fe–(2,3,4) Mn–0.3C steels in dependence on sintering time at 1120°C. ASC100.29, 600 MPa [55–57,60].

Fig. 5.44 (right). Impact energy (unnotched) and elongation of sintered Fe–(2,3,4) Mn–C steels in dependence on carbon addition (graphite). 90 min, 1160°C. Other as in Fig. 5.4.2 [54–56,60].

performed in hydrogen of technical purity under an alumina + graphite getter. The highest tensile strength of 810 MPa was recorded for the Fe–4Mn–0.3C steel sintered for 180 min.

According to Fig. 5.44 the impact energy and elongation values of Fe–(2,3,4)Mn–xC steels sintered for 90 min decreased with increasing carbon and manganese addition and by this in general with increasing hardness of these steels. It is important to stress here the relatively high values of both characteristics recorded for the sintered Fe–Mn–C steels caused probably by longer sintering time and a slightly higher sintering temperature in comparison with, for example, sintering at 1120°C.

5.4.3.4. Effect of iron powder grade, manganese addition and manganese carrier on transverse rupture strength of manganese steel

Effect of manganese addition and manganese carrier type on transverse rupture strength. The specimens for transverse rupture strength testing were prepared on the basis of N, S and A iron powders. Manganese was added to the powder mixtures in the amount of 1 to 5% in the form of electrolytic manganese (EMn) and medium- (FeMn) and high-carbon ferromanganese (FeMnC). The starting carbon contents in the mixtures with manganese addition as EMn and FeMn

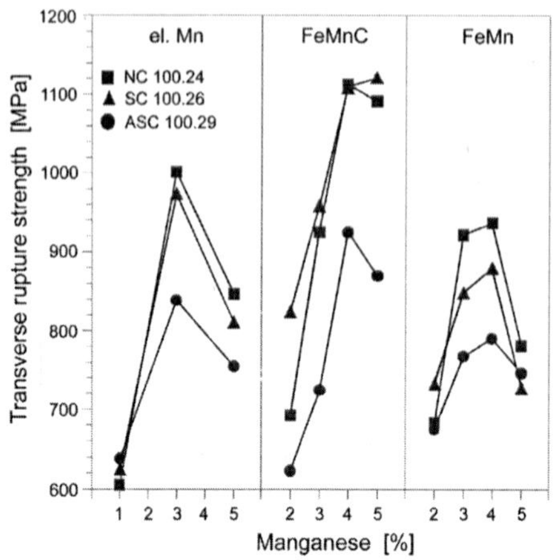

Fig. 5.45. The dependence of transverse rupture strength (TRS) on iron powder grade, manganese addition and manganese carrier type for sintered Fe–(1–5)Mn–xC steels based on NC, SC and ASC iron powder grade. 600 MPa, 60 min, 1120°C, cracked ammonia, dew point –30°C [77].

Table 5.18. Characteristics of powder mixtures and sintered density (ρ) of the Fe–2Mn–C and Fe–4Mn–xC steels. SC100.26, Mn as FeMnC, FeMn and EMn, starting carbon content for 2Mn steel was 0.16% and for 4Mn steel 0.32%, 60 min, 1120°C, cracked ammonia.

Mn [%]	2					4				
Manganese carrier	FeMnC		FeMn		EMn	FeMnC		FeMn		EMn
	a	n	a	n	a	a	n	a	n	a
*Oxygen [%]	0.95	0.57	0.89	0.67	0.70	0.95	0.57	0.89	0.67	0.70
ρ [g cm^{-3}]	6.92	6.95	6.96	6.94	6.97	6.86	6.89	6.90	6.90	6.92

*Milling of ferromanganese in: a – air, n - nitrogen, *in manganese carrier.

were adapted by the graphite addition to that introduced by FeMnC (0.08% C for 1Mn, 0.16% for 2% Mn, 0.24% for 3% Mn, 0.32% for 4% Mn and 0.40% for 5Mn steel.

As shown in Fig. 5.45, the manganese carrier type has a strong effect on the TRS of the tested steels. The highest strengthening effect was recorded using the high-carbon ferromanganese, as documented by the TRS values of 879 to 1113 MPa for the 4Mn specimens. Further, a larger effect of iron powder grade on the transverse rupture strength of the manganese steel was observed. The drop in TRS was also recorded in this case for the Fe–5Mn steels.

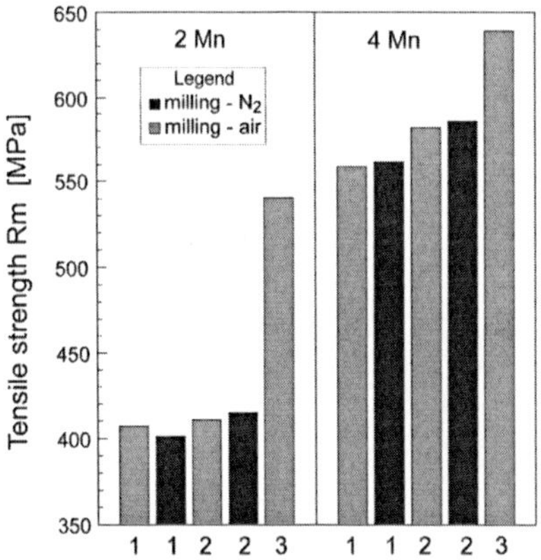

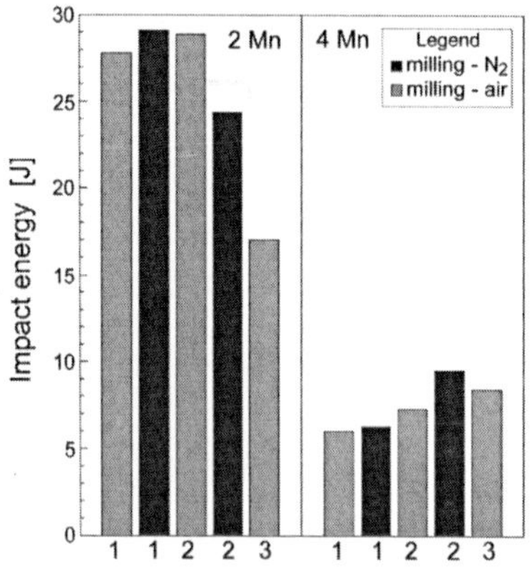

Fig. 5.46 (left). Tensile strength of sintered Fe–2Mn–C and Fe–4Mn–C steel samples. Mn as: 1 – FeMnC, 2 – FeMn, 3 – EMn. (oxygen content in manganese carriers is in Table 5.18) [77].

Fig. 5.47 (right). Impact energy of sintered Fe–2Mn–C and Fe–4Mn–C steel samples. Other as in Fig. 5.45 [77].

Effect of manganese carriers milled in air and in nitrogen on oxygen content, tensile strength and impact energy. Regarding the thermodynamic requirements for the purity of the atmosphere for sintering of manganese steels, the milling tests of the base manganese carriers in air and under nitrogen were carried out with the aim to decrease the oxygen content introduced into the systems by these materials.

To decrease the oxygen content brought by the milled manganese powder particles in the iron–manganese system, in some cases it was recommended to mill the manganese carrier under nitrogen. To consider this effect, the powder mixtures reported in Table 5.18 were prepared and tensile strength and impact toughness bars were compacted and sintered.

The corresponding tensile strength and impact energy values of the sintered Mn steels are shown in Figs. 5.46 and 5.47.

Minimum or no differences in the tensile strength and impact energy values were recorded when the manganese carrier, milled in air or under nitrogen, was used. A marked increase in tensile strength and a small decrease in impact energy values were recorded when electrolytic manganese was used instead of ferromanganese as the manganese carrier was used.

Regarding the possible effect of cooling rate and of the manganese carrier on the mechanical properties of Mn steels, the effect of

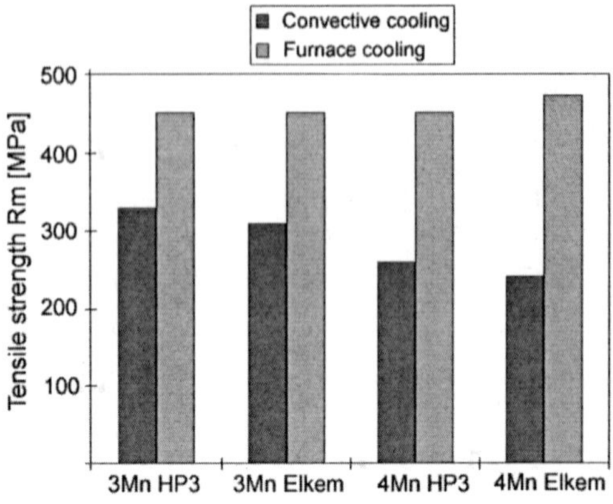

Fig. 5.48. Histogram of tensile strength of Fe–(3,4)Mn–0.8C steels; SC100.26 iron powder, 600 MPa, 60 min, 1180°C, HP3 – high-carbon ferromanganese, Elkem – medium-carbon ferromanganese, both <40 μm [94,110].

high- and medium-carbon ferromanganese grades and of two cooling rates, *i.e.* 3–4°C min^{-1} (furnace, slow) and ~40°C min^{-1} (convective, rapid) on the tensile strength of Fe–(3,4)Mn–0.8C steel was investigated [110]. The samples were sintered in hydrogen with the dew point better than – 40°C. The results are show in Fig. 5.48. Both ferromanganese grades resulted in lower tensile strength of the 4Mn steel after conventional (slow) cooling. The same tensile strength of 460–480 MPa was obtained for both rapidly cooled steels regardless of the ferromanganese grade.

Sintering of specimens under industrial conditions. The sintering of Fe–Mn–C steels with different Mn and C content under industrial conditions commonly used for sintering the components alloyed with Ni, Cu and Mo steels was performed. No special measures were taken to adapt the sintering conditions for Mn steels according to the thermodynamics. High-carbon ferromanganese was used as a manganese carrier and sintering was carried out in an industrial $70N_2/30H_2$ atmosphere with the inlet dew point of –50°C for 40 min at 1180°C. The data are supplemented by the tensile strength values after sintering at 1200°C and by the effect of adding 0.5% Mo to 3Mn steel. The tensile strength values are shown in Fig. 5.49. The maximum tensile strength of 589 MPa was recorded for the Fe–4Mn–0.7C steel and the tensile strength of the Fe–3Mn–1C steel was 555 MPa. The effect of Mo addition resulted in higher tensile

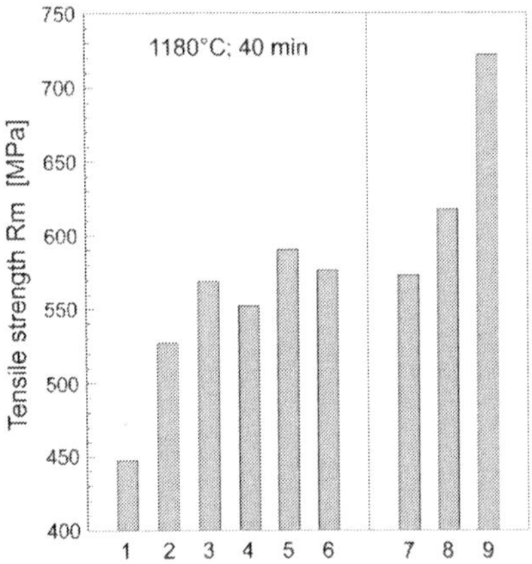

Fig. 5.49. Tensile strength of sintered Fe–xMn–xC steels; SC100.26 iron powder, 600 MPa. Composition: 1 – 2Mn–0.4 gr., 2 – 2.5Mn–0.7 gr., 3 – 2.5Mn–1.0 gr., 4 – 3Mn-0.7 gr., 5 – 3Mn – 1.0 gr., 6 – 4Mn–0.4 gr., 7 – 3Mn–0.2Mo–0.7gr.,7 – 2.5Mn–0.5Mo–0.7 gr., 8 – 3Mn–0.2Mo–0.7 gr., 9 – 3Mn–0.5Mo–0.5 gr., 1 to 6 and 8 – industrial sintering 40 min, 1180°C, 7, 9 – laboratory sintering: 60 min, 1200°C, H_2; gr. – graphite [77].

strength after industrial sintering (steel 8) and mainly after sintering at 1200°C (steel 9, 745 MPa) [53,77].

Effect of iron powder grade and manganese carrier on homogeneity of the Weibull modulus for sintered Mn steels. The reliability of Fe–Mn–C sintered steels is an important characteristic and was evaluated on by hardness measurements correlated with the iron powder grade and manganese carrier type.

The samples were produced from sponge iron powders NC100.24 (0.22% O) and SC100.26 (0.18% O) and water-atomised iron powder ASC100.29 (0.12% C) to which 0.8% of HW wax was admixed as a lubricant. The manganese carrier was high-carbon ferromanganese (76.6% Mn, 6.5% C, 0.77% O), medium-carbon ferromanganese (78% Mn, 1.4% C, 0.2% O) and electrolytic manganese. The powders were blended to obtain a nominal Mn content of 5%. In the case of medium-carbon ferromanganese and electrolytic manganese the carbon content was corrected by addition of graphite to obtain the same carbon input as in the case of high-carbon ferromanganese for 5% Mn, *i.e.* 0.40%. Specimens for bending tests (10 × 5 × 55 mm) were compacted at 600 MPa and sintered for 60 min at 1120°C in cracked ammonia. Transverse rupture stress was measured on five samples of each variant. Weibull moduli *h* were calculated by means of linear regression between 30 measured hardness values and the estimation of the '*failure probability*' [10,11]. The density, transverse rupture strength, hardness and Weibull moduli *h* values of the tested Mn steel variants are given in Table 5.19.

The transverse rupture strength values were significantly higher in the samples made from sponge iron powders. This may have been

Table 5.19. Density (ρ), transverse rupture strength (TRS), hardness HV10 and homogeneity of the Weibull moduli *h* calculated on the basis of hardness measurements of sintered Fe–5Mn–(0.4)C steels for three different iron powder grades and manganese carriers: FeMnC, FeMn, EMn [111,112]

Iron powder grade	Manganese carrier	ρ [g cm^{-3}]	TRS [MPa]	HV 10	*h*
NC100.24	FeMnC	6.71	1092	161	47
	FeMn	6.61	781	180	7
	EMn	6.67	847	192	23
SC100.26	FeMnC	6.78	1124	154	28
	FeMn	6.69	706	199	13
	EMn	6.80	811	192	10
ASC100.29	FeMnC	6.84	870	201	6
	FeMn	6.68	746	167	9
	EMn	6.75	754	173	10

caused by a higher degree of alloying with manganese resulting in higher hardening due to the higher specific surface area of sponge iron. The samples were based on water-atomised powder with a smaller specific surface area. Comparing the effect of manganese carriers based only on TRS values, the samples can be arranged in the following order: high-carbon ferromanganese, electrolytic manganese and medium-carbon ferromanganese.

According to the hardness values in the samples made from both sponge iron powders, alloying with high-carbon ferromanganese resulted in lower surface hardness than in the samples made from the water-atomised powder. This can be related to the fine pores present in the materials, which cede more easily during the indentation.

Comparing the Weibull moduli *h*, it can be seen that the materials produced from the sponge iron powders exhibit higher homogeneity, *i.e.* reliability was ensured using high-carbon ferromanganese. The use of medium-carbon ferromanganese resulted in low Weibull moduli in samples based on all the three iron powder grades. The poor performance of the water-atomised powder ($h = 6 - 10$) in spite of the highest hardness coincides with the low transverse rupture stresses. The homogeneity modulus *h* calculated from the hardness

Table 5.20. Ejection pressure of Fe–(2–5)% Mn samples (diameter 10 mm, 10 mm height) in dependence on iron powder grade, manganese addition and manganese carrier. Mn as FeMnC and FeMn, both <45 μm, 600 MPa, 30 min, 1120°C, cracked ammonia, dew point –30°C [77]

Iron powder grade	Ejection pressure [MPa]											
	NC100.24				SC100.26				ASC100.29			
Mn [%]	2	3	4	5	2	3	4	5	2	3	4	5
FeMnC	100	100	100	90	60	70	70	60	60	60	70	60
FeMn	100	100	90	100	80	80	80	80	70	60	50	50

values serves as a good means for comparison from the view point of the reliability of these materials.

Effect of iron powder grade and ferromanganese addition on ejection pressure. Ejection pressure is a measure for the dry-die wall friction of the sample ejected from the die. The cylindrical samples based on NC100.24, SC100.26 and ASC100.29 iron powder grades with addition of manganese in amount of 2–5%, graphite to the total amount of 0.5% of carbon and of 0.8% HW wax as a lubricant were prepared for the tests. The results (mean values of 10 samples) are given in Table 5.20.

No marked difference as observed between the two manganese carriers in relation to the ejection pressure. Not even the increasing amount of the ferromanganese powder addition caused any substantial increase in ejection pressure. The effect of powder grade on ejection pressure was recorded. The lowest ejection pressure was measured for the ASC100.29 iron powder, while the highest one exhibited the compacts based on NC100.24 iron powder. The ejection pressure for samples without ferromanganese based on NC100.24 iron powder was of 85 MPa and for SC100.26 and ASC100.29 powder of 70 MPa.

The surface of both as milled manganese carrier powders was covered by a dark grey (black) layer which was the source of the measured increased oxygen concentration in them compared to that before milling (Chapter 5.1). This black layer acted as a lubricant at ejection of the compacted samples. For this reason, dry milling of these carriers in air reduces friction between green compacts and increases their density [77].

5.4.4. Effect of manganese addition and of iron powder grade on friction properties

5.4.4.1. Friction and mechanical properties of sintered steels of various composition

To test the friction properties, the samples from the steels with various carbon content and with the addition of manganese in the form of FeMnC and FeMn and for comparison with other common steels without manganese addition were prepared in three groups:

group 1: the Fe–(0.3, 0.7)% C, Fe–(1.75, 4)% Ni–1.5% Cu-0.5% Mo–(0.3, 0.7)% C steels, without and with 0.5% MnX (machining aid, solid lubricant) addition, based on SC100.26 and ASC100.29 iron powder, Fe–1.75Ni–1.5Cu–0.5Mo (Distaloy SA), Fe–4Ni–1.5Cu-0.5Mo (Distaloy SE) and Fe–0.85Mo (Astaloy 85Mo) powder,

group 2: the Fe–(1–5)% Mn–(0.1–0.4)% C based on three iron powder grades,

group 3: the Fe–(2.5, 3)% Mn–(0.1–0.4)% C based on three iron powder grades,

– manganese carrier: a) FeMnC (76% Mn, 6.5% C), b) FeMn (80% Mn, 1.25% C), particle size of both <45 μm;

– natural graphite CR12 [113].

The standard tensile strength bars and impact toughness bars were prepared by compaction at 600 MPa from the powder mixtures with 0.8% HW wax as a lubricant. The specimens were sintered for 60 min at 1120°C in cracked ammonia. The steels of all three groups listed in Tables 5.23, 5.25 and 5.27 are numbered 1–37.

Testing of friction properties by the disc–prism method (Amsler method) at the load of 100 N at a sliding speed 0.5 m s^{-1} for a

Fig. 5.50. (a) – Scheme of the disc-block friction test method, (b) – sketch of the test specimen [113].

Table 5.21. Combined C content (C_c), density (ρ), tensile strength, hardness and impact toughness values of tested steels. S – SC100.26, A – ASC100.29, SA – Distaloy SA, SE – Distaloy SE, 0.3C and 0.7C = 0.3% and 0.7% graphite addition

Steel No.	Steel	C_c [%]	ρ [g cm^{-3}]	Rm [MPa]	*HV10	K [J]
1	S-0.3C	0.33	6.97	189	76	20.8
2	S-0.3C-MnX	0.29	6.82	214	82	14.3
3	S-0.7C	0.61	6.94	234	110	13.6
4	S-0.7C-MnX	0.63	6.84	248	111	11.4
5	A-0.3C	0.29	7.04	220	74	26.5
6	A-0.3C-MnX	0.32	6.99	229	80	21.5
7	A-0.7C	0.56	7.08	253	90	22.2
8	A-0.7C-MnX	0.59	7.02	247	96	21.4
9	SA-0.3C	0.28	6.96	533	158	22.5
10	SA-0.3C-MnX	0.31	6.87	496	169	18.9
11	SA-0.7C	0.55	6.94	598	191	15.5
12	SA-0.7C-MnX	0.67	6.92	555	182	14.6
13	SE-0.3C-MnX	0.29	6.90	572	187	20.1
14	SE-0.7C-MnX	0.53	6.89	623	202	19.5

*measured on cross-section of tested bars

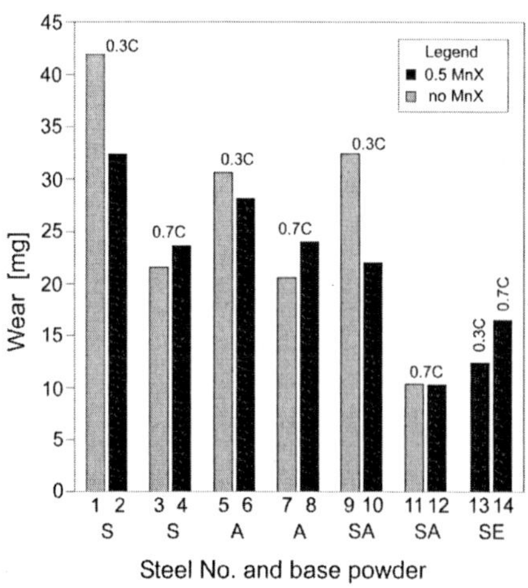

Fig. 5.51. Dry wear of steels No.1–14, Table 5.22 [113].

sliding distance of 1000 m was carried out, as shown in Fig. 5.50. A tool steel (ASTM D2) disc (54 HRC, diameter 50 mm, width 10 mm) with a freshly ground surface (Ra = 0.15 mm) was used for the manufacture of the die and as prisms and rectangular broken impact toughness bars were used for testing. They were tested

Table 5.22. Dry wear, coefficient of friction and temperature in the contact zone at friction tests of steels listed in Table 5.21

Steel No.													
1	2	3	4	5	6	7	8	9	10	11	12	13	14
Wear [mg]													
41.9	32.4	21.6	23.4	30.6	28.1	20.6	24.0	32.4	22.0	10.4	10.3	12.4	16.5
Friction coefficient μ													
0.57	0.56	0.42	0.58	0.50	0.63	0.40	0.66	0.65	0.42	0.63	0.42	0.41	0.45
Temperature [°C]													
60	59	48	59	52	62	50	63	56	53	57	51	60	52

on both sides perpendicular to the pressing direction. Two steel samples of each steel were tested for friction and the mean value of four results is presented. The friction properties of the steels are characterised by dry wear as mass loss, the friction coefficient calculated from the friction moment and the temperature measured at a distance of 5 mm from the contact place.

Properties of Fe–C and FeNiCuMo-C steels – group 1. For this test series, samples with the addition of 0.3 and 0.7% graphite without and with 0.5% MnX (machining aid, lubricant [114]) addition were prepared. The main characteristics and mechanical properties of the investigated steels are listed in Table 5.21.

Tribological properties of steels – group 1. The results of the friction tests are listed in Table 5.22 and the values of the wear are shown separately in Fig. 5.51.

The lower wear values were obtained for the steels with 0.7% graphite addition. The marked decrease in wear was recorded for the steels of the highest hardness and tensile strength (No.11–14). A decrease in wear of the 0.3C steels with 0.5% MnX addition and the lowest wear for the SA-0.7C steel (No.11, 12) were recorded. The relative difference in the wear values between the highest and lowest wear is larger than for the corresponding hardness and impact energy values. A decrease in the friction coefficient and of temperature in the samples at minimum wear of the steel was recorded.

Properties of the Fe–(1–5)Mn–(0.1–0.4)C steels – group 2. In this test series the samples based on three iron powder grades with the addition of manganese in the amount of 1 to 5% added in the form of high-carbon ferromanganese were tested. The main characteristics and mechanical properties of these steels are shown in Table 5.23.

Table 5.23. Main characteristics and mechanical properties of the steels (group 2). Iron powder grade: S – SC100.26, N – NC100.24, A – ASC100.29. Starting carbon content coming from high-carbon ferromanganese for steel (mass %): 1Mn – 0.08, 2Mn – 0.16, 3Mn – 0.24, 4Mn – 0.32, 5Mn – 0.40

Steel No.	Iron powder	Mn [%]	ρ [g cm^{-3}]	Rm [MPa]	TRS [MPa]	KC [J]	HV 10
15	S	1	7.01	382	625	26.7	99
16		2	6.94	454	824	19.1	137
17		3	6.90	545	1025	12.5	158
18		4	6.85	582	982	4.0	192
19		5	6.80	470	770	3.8	202
20	N	1	6.88	334	605	23.3	108
21		2	6.84	394	692	19.0	147
22		3	5.93	458	925	10.5	162
23		4	6.78	525	1113	6.0	179
24		5	6.72	495	1092	5.5	195
25	A	1	6.97	312	602	31.3	91
26		2	6.95	386	643	24.8	116
27		3	6.92	451	902	15.2	150
28		4	6.85	492	937	5.4	162
29		5	6.78	480	781	4.7	198

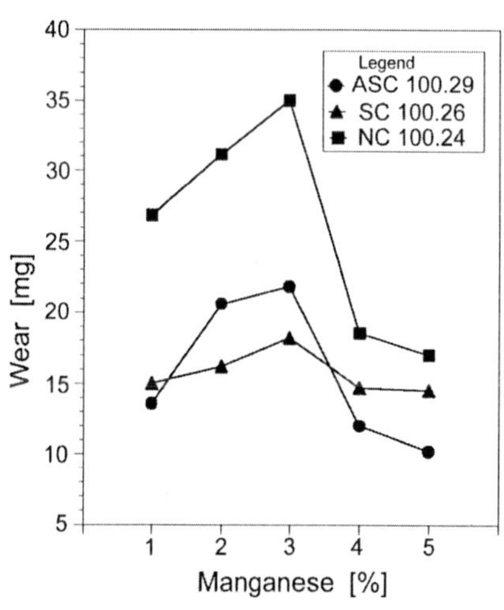

Fig. 5.52. Dry wear of sintered Fe–(1–5)Mn–xC steels in dependence on iron powder grade and manganese addition as FeMnC [77,113].

Table 5.24. Dry wear, friction coefficient and temperature at friction tests of the steels (No.15 – 29), group 2

Steel No.														
15	16	17	18	19	20	21	22	23	24	25	26	27	28	29
Wear [mg]														
15.0	16.2	18.2	14.7	8.5	26.8	31.1	29.8	18.5	13.8	13.6	20.6	21.8	12.0	10.2
Coefficient of friction μ														
0.60	0.57	0.56	0.66	0.65	0.61	0.54	0.59	0.60	0.62	0.63	0.58	0.51	0.49	0.56
Temperature [°C]														
60	59	59	58	59	55	59	61	59	59	60	57	55	52	55

Table 5.25. Main characteristics and mechanical properties of industrially sintered Fe–Mn–C steels (group 3) in dependence on base iron powder and ferromanganese grade. C_c = 0.83–0.91% [113]

Steel No.	Mn [%]	Powder	Mn-carrier	*ρ [g cm^{-3}]	Rm [MPa]	TRS [MPa]	*KC [J]	*HV 10
30	2.5	S	FeMnC	6.88	566	1025	9.2	191
31	2.5	S	FeMn	6.86	569	1054	10.4	192
32	2.5	A	FeMnC	6.88	593	979	7.1	193
33	2.5	A	FeMn	6.88	681	1005	10.8	192
34	2.5	A85Mo	FeMn	6.81	565	948	6.9	193
35	3.0	S	FeMnC	6.80	575	993	6.9	193
36	3.0	S	FeMn	6.86	590	1042	10.6	190
37	3.0	A85Mo	FeMn	6.87	613	1051	8.5	202

*valid for impact energy bars

Friction properties of 2nd group steels. The results of the friction tests are listed in Table 5.24 and the dependence of dry wear on manganese content and iron powder grade is shown in Fig. 5.52.

Effect of iron powder grade on the wear of the 2nd group of Fe-Mn-C steels was recorded. Differences in the degree of alloying of different iron powder grades as a consequence of alloying by manganese in the vapour form were the reason for these results [10]. The highest wear was recorded for the steels based on NC100.24 iron powder. A relatively small difference was found in the wear for the steels based on the SC100.26 and ASC100.29 iron powders. The wear of the Fe–(1–3)Mn samples (except the NC100.24 powder) was lower compared with those shown in Fig. 5.51 (No. 1–10). A

Table 5.26. Dry wear, friction coefficient and temperature at friction tests of steels (No. 30–37), group 3

Steel No.							
30	31	32	33	34	35	36	37
Wear [mg]							
9.7	6.1	6.9	5.6	7.0	6.8	12.6	10.8
Coefficient of friction μ							
0.33	0.30	0.26	0.28	0.37	0.35	0.48	0.41
Temperature [°C]							
47	49	45	42	47	48	43	51

decrease in wear for the 4 and 5% Mn-containing samples up to 10 mg was recorded at the decreasing tendency with increasing Mn and C content. Relatively small differences were recorded in the friction coefficient and in temperature.

Properties of industrially sintered Fe–(2.5,3)Mn–1C steels – group 3. In this test series, the Mn steel samples with equal starting carbon content of 1.0% based on SC100.26 (S) and ASC100.29 (A) iron powder were prepared. Molybdenum in the amount of 0.2% in the form of Astaloy 85Mo (Ast85Mo) powder for the steels No. 34 and 37 was added. The samples were sintered 40 min at 1180°C in an industrial pusher furnace in a $70N_2/30H_2$ atmosphere with a dew point of –50°C. The main characteristics and mechanical properties of these steels are given in Table 5.25.

The minimum difference in the density and hardness values is a characteristic feature of these steels. The tensile strength values for the steels based on ASC100.29 iron powder were higher compared with those based on the SC100.26 powder. The addition of 0.2% Mo increased the tensile strength and hardness of the Fe–3Mn–1C steel.

Friction properties of group 3 steels. The friction properties of the industrially sintered steels are given in Table 5.26. The steels of this test group show a marked decrease in wear <10 mg, in the friction coefficient and temperature when compared with the previous results (groups 1 and 2). The lowest wear of all the tested steels in this chapter, *i.e.* 5.6 mg at the lowest friction coefficient (0.28) and the lowest temperature (42°C) were recorded for steel No. 33 (2.5Mn, ASC100.29 iron powder). The wear of the steels based on the ASC100.29 iron powder alloyed with medium-carbon ferromanganese was lower than the wear of the steels based on the SC100.26 iron powder. The optimum friction properties in the whole

Table 5.27. Density (ρ), hardness and friction coefficient (μ) and temperature (*T*) of tested Fe–(2.5, 3)Mn–1% graphite steels. SC100.26 iron powder; 40 min, 1180°C, $75N_2/25H_2$ atmosphere, dew point –50°C [77,113]

Steel Fe-	ρ [g cm^{-3}]	HV10	Wear [mg]	μ	T [°C]
*2.5Mn–0.88C_c	6.82	191	9.7	0.33	47
**2.5Mn–0.85C_c	6.87	192	6.1	0.30	49
*3Mn–0.91C_c	6.81	193	6.8	0.35	48
**3Mn–0.84C_c	6.89	190	12.6	0.48	53

Mn as *high-carbon ferromanganese, **medium-carbon ferromanganese

investigated steel range were obtained for the Fe-2.5Mn-1C steel No. 33. The addition of 0.2% Mo did not affect the friction properties.

The results of the friction tests conducted on the selected industrially sintered steel samples (group 3) were as follows. The marked decrease in the wear of the manganese steel was obtained at a higher carbon content, as shown in Table 5.27. Markedly lower wear and friction coefficient values and partly also the temperature in these steels were measured. Compared with Fig. 5.52, the wear and friction coefficient and partly also temperature in these steels were considerably lower compared to the previous results. The effect of ferromanganese on the wear seems to be connected with the manganese content. No ferrite was found in the microstructures by optical microscopy.

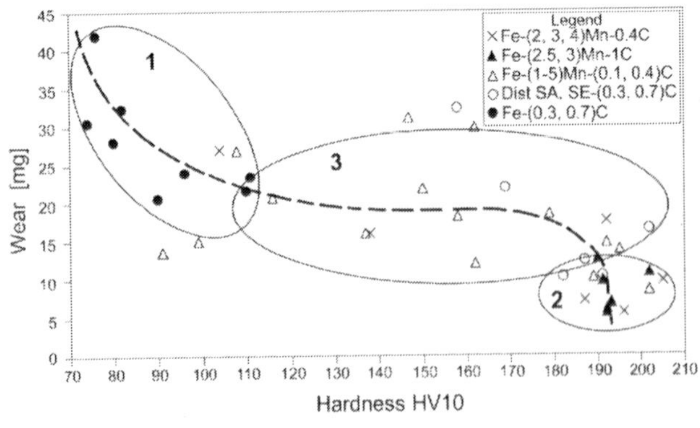

Fig. 5.53. The dependence of wear on hardness values of the steels No.1–37 according to the above mentioned tables [113].

The friction coefficient should be in the range 0.2–0.5 and should depend only slightly on the friction speed. All the materials meet these values and most of them have even higher values, as indicated by the tables presented previously.

5.4.4.2. Wear vs. hardness of sintered steels

The wear values obtained under the given friction test conditions of all three steel groups (Tables 5.24, 5.26 and 5.28) in relation to the hardness values are shown in Fig. 5.53.

In the diagram there are three areas of the data. The first one (1) is for Fe–C materials (1–7) with the wear values >20 mg at the hardness <110 HV 10, the third one (3) is for FeNiCuMoC steels (11–14) and mainly for FeMnC steels (29–37) characterised by the wear values <10 mg and by the hardness of the samples in a small range of ~190–200 HV 10. The second area (2) is characterised by the wear values of ~10 to about 25 mg but with a large hardness range of the specimens of ~110–200 HV 10. It follows from this that hardness is not a sufficiently accurate characteristic for dry wear of steels. The difference should be exlained on the basis of the microstructure with its singularities in dependence on alloying and processing of the steels.

Materials characterised by 14 types of microstructure having a microhardness from 131 to 1142 HV 0.01 were prepared to test the friction properties (resistance to bonding of structural components) of the carbon steel with the hypoeutectoid composition by quenching in water and tempering by different methods. The friction coefficient was measured and recorded in special device in which the sample with a defined microstructure was moved at a low speed and a ball (1890 HV 0.05) was pressed to it with a pressure of 10 g. It was found that the results obtained in the friction tests of the examined

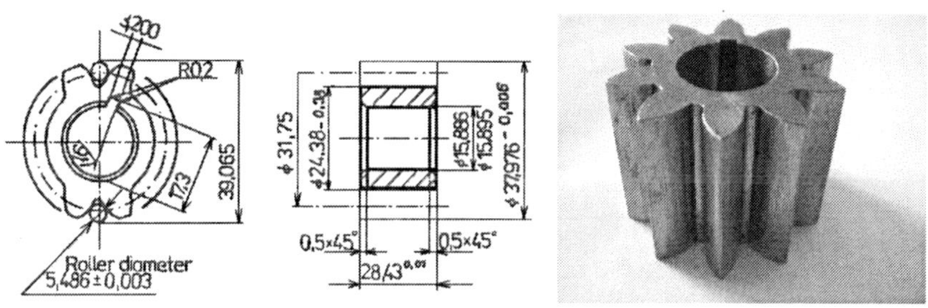

Fig. 5.54. (left) Dimensions in mm of a spur gear for hydrogenerators; (right) Photo of the gear [116–118].

Table 5.28. Base characteristics of the gears fulfilling the test life time 1000 h

Steel	Density [g cm^{-3}]	Rm [MPa]	Hardness HV10/HRC
Fe–4Ni–1C	7.15	449	165/49
Fe–3.5Mn–0.5Mo–0.3C	6.95	692	238/55

Remark: Manganese added as FeMnC, <45 μm, Rm determined on tensile strength bars sintered together with the gears

microstructures are not in any respect related to their microhardness and are closely linked with the nature of the microstructure [114,115].

5.4.5. Industrially sintered prototype structural parts prepared from manganese steel

Gears for hydrogenerators. The gears for oil hydrogenerators are produced from wrought steel by machining, through hardening and grinding. The object of the interest was to prepare such gear shown in Fig. 5.54 used with the service pressure of 1.5 MPa with the aim to replace the Fe–4% Ni–1.0% C sintered steel (compaction to the density ~6.9 g cm^{-3}) finally through hardened (850°C/0.5 h, oil, 200°C/1 h, air) to the hardness of 35 to 45 HRC with manganese steel and to try to replace their production finally also from wrought steel [116–118].

Several Mn steels were investigated for this application. The best results were attained with Fe–3.5Mn–0.5Mo–0.3C steel. The gears were compacted at 600 MPa and industrially sintered for 40 min at 1180°C in a 75N_2/25H_2 atmosphere with the dew point –50°C.

The required test life time of the gears was 1000 h at the pressure of 3 MPa at frequency 2800 rpm without damage. The tested gear was used as driven and the gear from hardened wrought steel as driving. The Fe–4Ni–1.0C steel gear and the Fe–3.5Mn–0.5Mo-0.3C as sintered steel gear fulfilled this requirement. Their base characteristics are given in Table 5.28.

These gears were further subjected to the tests under cyclic oil pressure changing every 10 s in the range of 0.7 to 3 MPa during 1000 h. This test condition fulfilled the as sintered Fe–3.5Mn–0.5Mo-0.3C steel gear which was subjected further for long time test under cyclic pressure in the range of 0.7 to 5 MPa. The test after 5000 h was interrupted due to capacity reasons. The as sintered Fe–3.5Mn-0.5Mo–0.3C steel gear was undamaged after this test without the change of the dimensional tolerances.

Fig. 5.55. (left) Components plate (65 g), two as sintered and one broken, (right) components cube (90 g) [54,113,118,119].

The results attained prove that the gears for the hydrogenerators can be produced by the powder metallurgy method from Fe–Mn–Mo–C steel as sintered only without some heat treatment also as replacement of wrought steel gears.

The components plate and cube. From production of some structural components on the basis of Distaloy SE powder there were selected the prototype components *plate* and *cube* shown in Fig. 5.55 with the aim to produce them from manganese steel. The components were prepared from 2, 3 and 4% Mn steels on the basis of SC100.26 powder and for comparison those prepared from DistSE powder were also tested. Manganese was added to the iron powder in form of medium-carbon ferromanganese (81.2% Mn, 1.5%C) and also HW lubricant with 0.8% and 0.4% graphite. The tensile strength and impact energy bars were also pressed and sintered. The components and specimens were compacted at 600 MPa. Sintering of the components and specimens was carried out under: (a) a laboratory continuously working furnace with manual shifting of the boxes (F) for 60 min at 1130°C in cracked ammonia with the dew point –23°C at heating rate of ~8.5°C/min and cooling rate ~6.5°C/min, (b) components in industrial pusher furnace (C) for 60 min at 1100°C in a $70N_2/30N_2$ atmosphere with the dew point –50°C at heating rate ~ 6.5°C/min and cooling rate ~4°C/min [54,113,118,119].

The component *plate* was tested in the three-point bending test and the component *cube* by compression up to destruction. The results, supplemented by the hardness and impact energy values, are presented in Table 5.29.

The bending loading force for components *plate* was higher by 13% for 2Mn, by 6.7% for 3Mn and by 3.2% for 4Mn steel when compared with FeNiCuMo steel components sintered together. The height of the component was in the range of 11.5 to 11.6 mm. The compression force of components *cube* was higher by 16% for 2Mn,

Table 5.29. Characteristic properties of tested plate and cube components. DistSE – Fe–4% Ni–1.5% Cu–0.5% Mo, 2Mn – 2% Mn, 3Mn – 3% Mn, 4Mn – 4% Mn
(a) plate

Steel	HV 10	K [J]	Bending force [kN]	HV 10	K [J]	Bending force [kN]
	F furnace			C furnace		
DistSE	115	–	16.2	114	–	14.5
2Mn	125	12.0	18.4	135	16.5	15.0
3Mn	140	6.10	15.8	156	11.0	17.5
4Mn	210	4.40	12.2	200	10.0	21.0

Remark: The density of Mn-steel components was in the range of 6.41 to 6.67 g cm^{-3}

(b) *cube*

Composition	HV 10	K [J]	Compression force [kN × 10]	HV 10	K [J]	Compression force [kN × 10]
	F furnace			C furnace		
DistSE	138	–	16.2	123	–	18.3
2Mn	140	12.0	18.4	132	17.0	10.2
3Mn	172	6.20	15.8	158	11.0	11.8
4Mn	210	4.10	12.2	202	10.2	12.0

Remark: The density of Mn-steel components was in the range of 6.86 to 7.05 g cm^{-3}

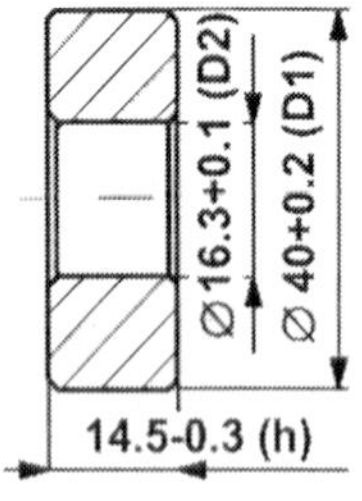

Fig. 5.56. Three types of the rollers in size: 22, 30 and 40 mm in diameter, G = 14 g for 22 mm, 105 g for 30 mm, 170 g for 40 mm and their dimensions in mm presented with tribological properties [113,118,119].

by 23% for 3Mn and by 19% for 4Mn steel compared with those for FeNiCuMo steel components. The shrinkage of the height for Mn steel components was in the range of –0.2% to –1.2% and for FeNiCuMo steel from –0.4% to –0.8%.

The compression force for components *cube* was higher by 16% for 2Mn, by 23% for 3Mn and by 19% for 4Mn steel compared with FeNiCuMo steel components.

Component roller. The photo and drawing of the component *roller* developed from Mn steel are shown in Fig. 5.56 as possible replacement of their production from Fe–4%Ni–1.5% Cu–0.5% Mo –0.6% graphite powder mixture (DistSE) and from wrought steel. Manganese steel *rollers* were prepared from Fe–3.5% Mn–0.7% graphite steel on the basis of ASC100.29 powder. The components were compacted at 600 MPa and sintered under industrial conditions in a pusher furnace (1180°C/40 min, $75N_2/25H_2$, entry dew point –50°C).

The tribological tests of the Mn *rollers* and of those prepared from DistSE powder were performed by the disc (roller)-prism method as shown in Fig. 5.49 at two loads. The results are given in Table 5.30.

The *roller* from Fe–3.5Mn–0.7C steel exhibited the lowest wear (96 mg) compared with FeNiCuMoC steel (344 mg), and significantly lower wear (50→2405 mg) when tested against a hardened counterpart. The *rollers* from manganese steel were successfully tested by the user. The results proved the possibility of replacing their production from wrought steel.

Table 5.30. Tribological characteristics of the rollers prepared from Mn (Fe–Mn–C) and FeNiCuMo steels. *W* – wear, μ – friction coefficient, *T* – temperature in the contact area. (Mn-roller: ρ = 6.89 g cm^{-3}, 170 HV 100; Ni-roller: ρ = 6.99 g cm^{-3}, 206 HV 10). Test load: 140 and 100 N, sliding rate 0.5 $m \cdot s^{-1}$, sliding distance 1000 m; *mild steel – counterpart, **hardened steel – counterpart

Steel	*Load 100 N			*Load 140 N			**Load 140 N		
	W [mg]	μ	T [°C]	W [mg]	μ	T [°C]	W [mg]	μ	T [°C]
DistSE-0.6C	344	0.40	63	596	0.41	54	2405	0.36	58
3.5Mn-0.7C	96	0.39	56	104	0.60	64	50	0.19	57

6

Effect of additional elements on the properties of manganese steels

This chapter provides information and knowledge about the wider impact of the additional elements, *i.e.* molybdenum, copper, tin, phosphorus and boron, on the mechanical and other properties of sintered mixed manganese steels and some steels alloyed with manganese, chromium and other elements. Particular attention in this chapter is devoted to boriding of sintered steels to produce hard coatings on their surfaces resistant to wear under the technical and technological conditions in which they were produced. Samples of these steels were prepared under different forming and sintering conditions with respect to the possible application of such materials for the production of sintered parts with the characteristics of each system achieved under the particular conditions.

6.1. Effect of molybdenum

The effect of 0.5% Mo addition and of compacting pressure on the properties of sintered iron powder and manganese steels was investigated regarding the high strengthening effect of molybdenum in this case for sintered manganese steel and for comparison also for plain iron. The following materials were used for the preparation of the tensile and impact energy bars: Hametag iron powder, high-carbon ferromanganese (75% Mn, 7.7% C <40 μm) and ferromolybdenum (Fe55Mo <15 μm). The specimens from the powder mixtures with addition of 0.8% of HW wax as a lubricant were compacted at 150,

Table 6.1. Designation and initial chemical composition of the investigated steels [120]

Alloying element mass [%]	Designation and composition of the investigated steels									
	A	B	C	D	E	F	G	H	K	L
*Mn	Fe	2.5	3.5	4.5	5.5	–	2.5	3.5	4.5	5.5
*C	–	0.25	0.36	0.46	0.56	–	0.25	0.36	0.46	0.56
Mo	–	–	–	–	–	0.5	0.5	0.50	0.5	0.5

*Mn and C added as FeMnC

200, 400 and 600 MPa and sintered together in a laboratory retort furnace for 3 h at 1120°C in dissociated ammonia with a dew point of –30°C.

6.1.1. Properties of Fe–Mn–C steels with molybdenum addition

Table 6.1 gives the designation of the tested materials and the additions of Mn, C and Mo.

6.1.1.1. Density

The dependence of the density of the materials on compacting pressure is shown in Fig. 6.1. In particular, the results show a steep increase of the density of the samples compacted at 200 MPa compared to those compacted at 150 MPa, except of the plain iron powder compact A. A further increase of the compacting pressure resulted in a linear increase of the density of all tested materials. The addition of 0.5% Mo caused a considerable increase of the density of material F (Fe–0.5Mo) at lower pressures. The density of Mo-containing materials attained at the compacting pressure of 200 MPa was higher by 1.5 g cm^{-3} compared to Fe–Mn–C steel samples. All Fe–Mn–Mo–C steels had a higher density than the Fe–Mn–C steels.

Table 6.2 lists the as sintered manganese and carbon contents in the investigated materials for the detailed characterisation of the presented mechanical properties.

6.1.1.2. Chemical composition of the tested steels

The as sintered carbon and manganese contents of the specimens compacted at 150 and 600 MPa are listed in Table 6.2.

Table 6.2. Manganese and carbon contents of the Fe–Mn–Mo–C steels compacted at 150 and 600 MPa and sintered for 3 h at 1120°C in dissociated ammonia [120]

Compacting pressure [MPa]	Alloying element mass [%]	Steel									
		A	B	C	D	E	F	G	H	K	L
150	Mn	0.30	1.92	2.84	3.80	4.60	0.32	1.98	2.84	3.85	4.75
	C	0.07	0.07	0.08	0.10	0.12	0.08	0.07	0.11	0.12	0.14
600	Mn	0.31	2.33	3.40	4.30	4.85	0.32	2.25	3.22	4.28	4.93
	C	0.07	0.18	0.19	0.23	0.26	0.07	0.11	0.11	0.13	0.16

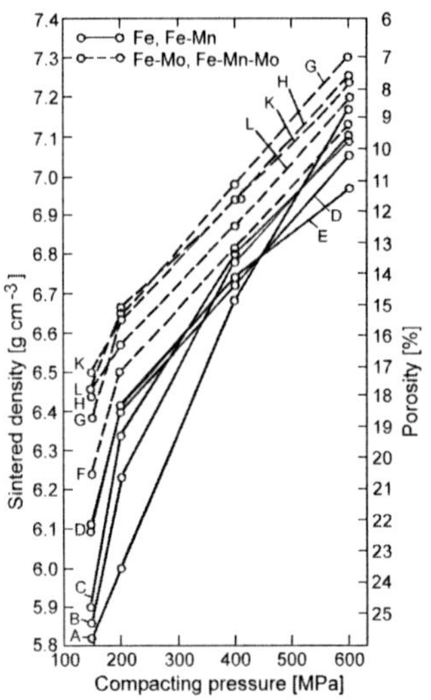

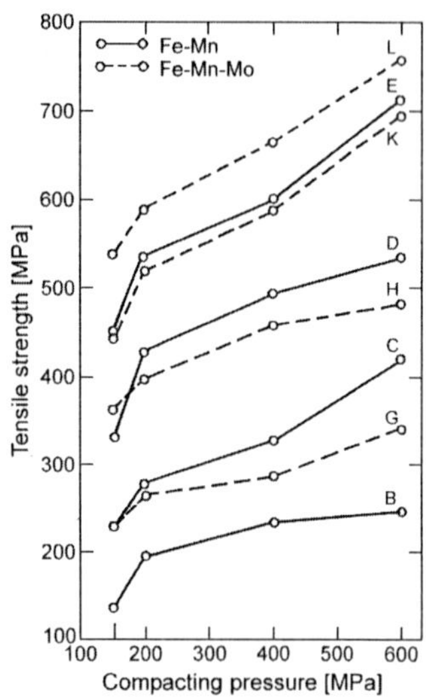

Fig. 6.1 (left). The dependence of as sintered density on compacting pressure for Fe–Mn–C and Fe–Mo–C steel samples (A, B, C, D, E, F, G, H, K, L) [120].
Fig. 6.2 (right). The dependence of the density on compacting pressure for sintered Fe–Mn–C and Fe–Mn–Mo–C steel samples (B, C, D, E, G, H, K, L) [120].

6.1.1.3. Dependence of the mechanical properties on density

Figure 6.1 shows the large inrease of the density of the compacts with the increase of compacting pressure from 160 to 200 MPa. A similar increase in dependence on compacting pressure was recorded for the tensile strength values, Fig. 6.2. A further increase of the compacting pressure raised the tensile strength in a nearly linear fashion. Compared to the starting manganese addition in the steels, the Fe–Mn–0.5Mo–C steels attained a higher tensile strength at a lower manganese content than the Fe–Mn–C steels. The highest

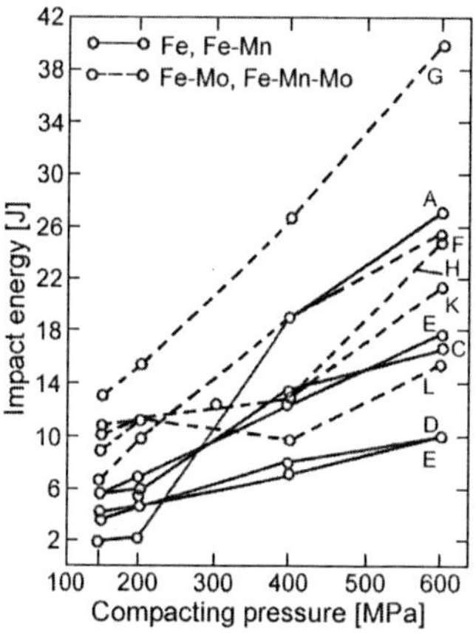

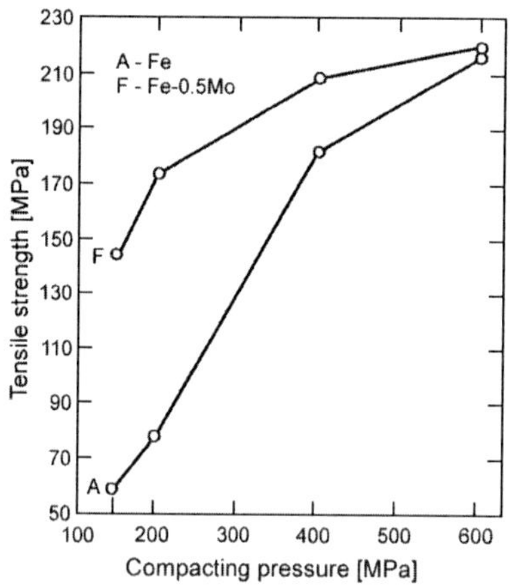

Fig. 6.3 (left). Dependence of impact energy values (unnotched) upon compacting pressure of the steels A–L [120].

Fig. 6.4. (right). Dependence of tensile strength upon compacting pressure for sintered iron (A) and Fe–0.5Mo alloys (F) [120].

tensile strength of 750 MPa was recorded for the Fe–4.75Mn–0.5Mo–0.14C steel (L) compacted at 600 MPa. The tensile strength of the Fe–4.85Mn–0.5Mo–0.2C (E) under the same conditions was 700 MPa.

The impact energy values dependent upon the compacting pressure are shown in Fig. 6.3. In all cases, the impact energy also increased with increasing compacting pressure. The impact energy of the Fe–Mn–C steels decreased with increasing manganese and corresponding carbon content. The highest impact energy values of 40 J was obtained for the Fe–2.25Mn–0.5Mo–0.11C steel (G) compacted at 600 MPa. The lowest impact energy value of 8.2 J was shown by the steels Fe–(4.30, 4.85)Mn–0.5Mo–(0.23, 0.26)C (D, E).

Figure 6.4 shows the tensile strength of sintered iron (A) and of Fe–Mo (F) materials without carbon addition. Here was observed considerable increase in the tensile strength of Fe–0.5Mo steel, especially when compacted at lower pressure. The tensile strength was in step with the relevant density changes of these specimens. The highest strength of ~210 MPa was attained for both materials compacted at 600 MPa. The highest hardness was recorded for the steel Fe–4.85Mn–0.26C (E) in the range of 128 to 231 HV 10 with increasing compacting pressure. The steel Fe–4.93Mn–0.5Mo–0.36C showed, under the same conditions, the hardness of 108 – 190 HV 10.

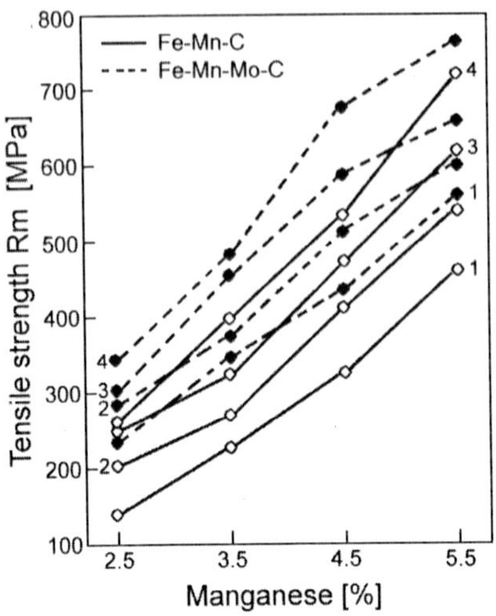

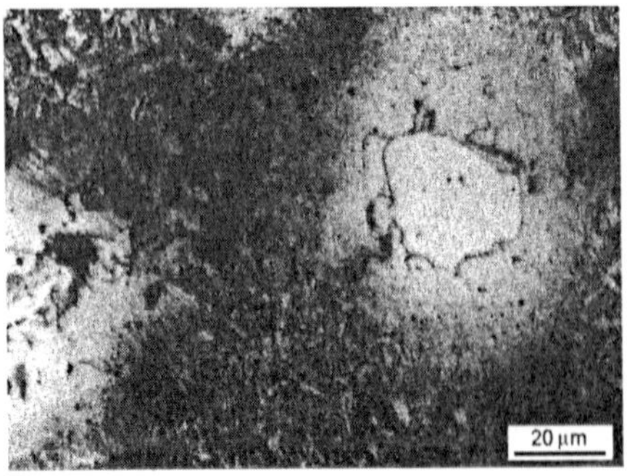

Fig. 6.5 (left).Tensile strength of Fe–xMn–xC and Fe–xMn–0.5Mo–xC steels in dependence on manganese addition and compacting pressure. 1 –150 MPa, 2 – 200 MPa, 3 – 400 MPa, 600 MPa [60].

Fig. 6.6 (right). Microstructure of sintered Fe–4.5Mn–0.5Mo–C steel. 600 MPa, 60 min, 1120°C. Nital etched [47].

Table 6.3. Microhardness (HV 0.01) of sintered A – L steel microstructure constituents. Samples compacted at 600 MPa [120]

Steel	A	B	C	D	E	F	G	H	K	L
HV 0.01	129–196	142–167	149–236	148–469	314–789	135–196	149–236	175–363	252–394	335–425

The dependence of the tensile strength of the two compared steels on the Mn addition is shown in Fig. 6.5.

6.1.1.4. Microstructure and microhardness

Figure 6.6 shows the microstructure of Fe–4.5Mn–0.5Mo–C steel sintered 1 h. As shown at sintering such a compact for 1 h it was found that the coarser ferromolybdenum particles under sintering conditions used were not totally dissolved and that they formed with base iron powder particles the poreless joint what contributed to the increase of density. Initial stages of alloying the iron matrix with Mn and diffusion of Mo from ferromolybdenum particles are clearly visible.

The microhardness values characterizing the microstructures are given in Table 6.3.

The as sintered manganese and carbon content must be regarded in the evaluation of these microhardness values. According to the microhardness values, the microstructures of the steels A and F were formed by ferrite and fine pearlite. The microstructure of the steel G was formed by bainite and martensite. The microstructure of the steel L was formed by upper and lower bainite.

As follows from the results, the addition of 0.5% Mo in the form of ferromolybdenum increased the density and strength properties of the Fe–C and Fe–Mn–C steels. The steels Fe–4.5Mn–C and Fe–4.5Mn–0.5Mo–C compacted at 200 MPa reached the tensile strength of 536 and 591 MPa, respectively. The highest increase of density and strength due to 0.5% Mo addition attained the Fe–3.5Mn–C steel. The highest impact energy values of the whole range on compacting pressure attained the Fe–2.5Mn–0.5Mo–C steel.

6.1.2. Manganizing of powder steels combined with sintering

Manganese surface alloying leads to case layers of austenitic manganese steel whose behaviour is similar to that of Hadfield's steel. This process thus offers the potential for producing sintered wear-resistant parts because of increasing surface layer after work-hardening with the same accuracy that is obtained from sintered parts.

Two examples of manganizing of Mo-containing powder steels with manganese vapour will now be discussed. The sintering atmosphere in this case served for manganese vapour transport. Figure 6.7 shows the microstructure of a specimen of mixed Fe–

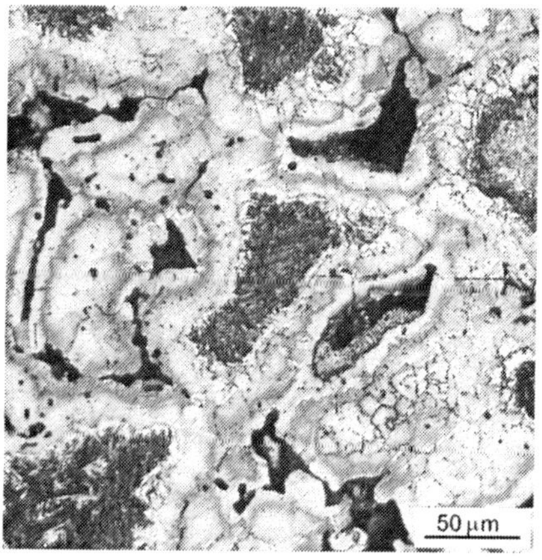

Fig. 6.7. Microstructure of Fe–6Mn–1Mo–1.2C mixed powder steel. Sintering for 30 min at 1250°C in cracked ammonia (dew point −30°C) under a getter of FeMnC powder mixed with Al_2O_3 as diluent. Density 6.1 g cm^{-3}, ASC100.29 iron powder, high-carbon ferromanganese, ferromolybdenum. Nital etched [1].

Fig. 6.8. Microstructure of Fe–2Ni–0.55Mo–0.3C steel. Sintering for 30 min at 1250°C in cracked ammonia (dew point –30°C). Density 7.1 g cm^{-3}, Astaloy A powder. Specimens were at sintering only surrounded with ferromanganese powder. Nital etched [1].

6Mn–1Mo–1.2C (graphite) powder steel sintered for 30 min at 1250°C in cracked ammonia under manganese powder and alumina as diluent. The composition of this steel corresponds to that of wrought steel replacing partly the Hadfield steel. As a result of high porosity of the specimen hard layers were also formed inside of the specimen on the surfaces of initial iron powder particles in the compact or on larger grains originated by sintering.

The manganese-alloyed surface layer was also formed on a specimen of Fe–2Ni–0.55Mo–0.3C (graphite) of high density. In this case the specimens were placed in a steel box surrounded only by FeMnC powder without any contact with ferromanganese. Manganizing of the specimens occurred only by transfer of manganese vapour from the surrounding ferromanganese powder inside the compact, Fig. 6.8. A white hard layer also formed on the surface of this specimen as the proof of the formation of hard surface layers only through manganese vapour – manganizing. The specimens could be cut only by the electroerosion method. There are simplified procedures which enable manganizing to be carried out over a considerably shorter period of time possibly together with sintering.

6.1.3. Sintering of Fe–Mn–C and Fe–Mn–Mo–C steels in α-region and at 1120°C

The effect of molybdenum on the properties of manganese alloy steels is described by the following mechanical properties of Fe–4.5Mn-0.39C and Fe–4.5Mn–0.5Mo–0.39C steels prepared on the basis of Hametag (H) and RZ iron powders. Manganese and carbon were added

Table 6.4. As sintered density, tensile strength, elongation and hardness of Fe–Mn–C and Fe–Mn–Mo–C steels sintered under given conditions in cracked ammonia, dew point –33°C, H – Hametag, R – RZ iron powder [104]

Iron powder	Steel Fe–	Density [g cm^{-3}]	UTS [MPa]	A [%]	HV 10
	Sintering 80 min at 875°C				
H	4.5Mn–0.39C	6.73	496	1.0	205
	4.5Mn–0.5Mo–0.39C	6.73	445	1.5	170
R	4.5Mn–0.39C	6.71	143	1.4	171
	4.5Mn–0.5Mo–0.39C	6.73	138	1.5	167
	Sintering during preheating to 1120°C, 0 h hold				
H	4.5Mn–0.39C	6.82	579	0.6	177
	4.5Mn–0.5Mo–0.39C	6.87	619	2.0	184
R	4.5Mn–0.39C	6.78	373	1.5	197
	4.5Mn–0.5Mo–0.39C	6.80	441	1.5	189
	Sintering 1 h at 875°C + 2 h at 1120°C				
H	4.5Mn–0.39C	7.08	718	3.0	171
	4.5Mn–0.5Mo–0.39C	7.07	758	3.0	188
R	4.5Mn–0.39C	6.90	468	2.2	135
	4.5Mn–0.5Mo–0.39C	6.91	548	1.8	125

to the iron powder in the form of FeMnC <40 μm and molybdenum as ferromolybdenum <40 μm. Samples were pressed at a pressure of 600 MPa and sintered under the following conditions: a) for 80 min at 875°C, heating rate 35°C min^{-1}, b) during heating to a temperature of 1120°C, zero holding time, heating rate 17°C min^{-1}, c) for 1 h at 875°C + 2 h at 1120°C in a retort furnace in cracked ammonia (see the data in Figs. 5.31 and 5.32).

Table 6.4 gives the mechanical properties of sintered samples for all three conditions.

In this case, particular attention was paid to the impact of the addition of 0.5% Mo on the mechanical properties of manganese steel with a manganese content of 4.5% sintered at three different conditions. The reason for the differential impact of both types of iron powder on the properties of manganese steel was previously explained.

After sintering at 875°C the studied properties of steels with the addition of molybdenum are lower than those of pure manganese steel due to insufficiently high temperature for the diffusion of molybdenum into the iron matrix. The hardness of all steels shows that even at this low sintering temperature the iron matrix was alloyed with manganese and partially with carbon and only to a small extent with molybdenum. The density of all samples slightly increased only after sintering during heating to a temperature of 1120°C. The tensile strength of the steel with added molybdenum in relation to the type of iron powder increased by 184 MPa (H) and 303 MPa (R) respectively, which exceeded the values obtained for the manganese steel. A further increase in the yield strength of the steel with the molybdenum addition was achieved by stepwise sintering and with longer sintering time. It can be seen that a relatively significant increase in yield strength was already recorded during heating to sintering temperature. For the manganese steel one should take into account the partial loss of manganese by sublimation during holding at 1120°C. The mechanical properties of steels prepared on the basis of the powder R are significantly lower than those prepared on the basis of powder H.

These data show how increasing sintering temperature and sintering time increase gradually the strengthening effect of molybdenum on the strength properties of manganese steel. Consequently, the addition of molybdenum to the manganese steel is very efficient [104].

The effect of sintering time on the microhardness of Fe–0.32C-4Mn and Fe–4Mn–0.5Mo–0.32C steel microstructures sintered for 3, 10, 30 and 60 min at 1100°C and 1200°C was also studied. Measurement of the microhardness of microstructural constituents of these steels was carried out on cylindrical samples (diameter 10 mm, 10 mm length). Manganese was added in the form of FeMnC and molybdenum as ferromolybdenum, both with the particle size <40 μm. Samples were pressed at a pressure of 600 MPa and sintered at 1120°C and 1200°C in cracked ammonia at a dew point of –30°C. The microhardness values of the microstructural constituents of the investigated steels are shown in Table 6.5, where the samples A and C were prepared on the basis of iron powder H, B and D on the basis of iron powder RZ.

Based on these microhardness values, it can be concluded that sintering for 3 min already resulted in alloying of the matrix as a

Table 6.5. Microhardness (HV 0.01) values of microstructure constituents of Fe–4Mn–0.32C and Fe–4Mn–0.5Mo–0.32C steel specimens prepared on the basis of Hametag (H) and RZ (R) iron powders. Sintering for 3 (A), 10 (B), 30 (C) and 60 (D) min at 1100°C and 1200°C in cracked ammonia, dew point –30°C. The Fe–4Mn–0.5Mo–0.32C steel specimens are denoted by C and D [102,103]

Steel - designation/ H or R iron powder	Sintering temperature [°C]	Sintering time [min]			
		3	10	30	60
		Microhardness HV 0.01			
A/H	1100	116–129	136–181	143–160	122–151
B/R		105–110	143–160	128–148	126–151
C/H (Mo)		116–138	160–181	126–170	151–206
D/R (Mo)		105–116	143–160	136–143	116–143
A/H	1200	105–122	170–221	115–143	128–143
B/R		110–128	160–193	110–128	118–139
C/H (Mo)		126–160	151–206	125–163	128–151
D/R (Mo)		118–1149	136–173	116–138	105–122

*Microhardness values – 25 measurements for each stage

result of the condensation of vapour on the surface of iron powder particles in the compact. Another phenomenon is that the relatively highest values were measured after sintering for 10 min, reflecting a longer period of sublimation of manganese from ferromanganese with its partial diffusion inside the particles. After sintering for 30 min the microhardness values were lower due to more extensive alloying of the interior of the iron powder particles by diffusion of manganese from the surface layers thus reducing the Mn concentration gradient. After sintering for 60 min the microhardness values were the lowest due to reversed partial sublimation of manganese from the samples into the atmosphere and partial decarburisation. The effect of the addition of molybdenum (samples C and D) is expressed in all cases by higher microhardness values than samples without molybdenum (A, B). The effect of sintering temperature is especially marked for the microhardness values after 10 min sintering. In all cases, the microhardness values of the samples prepared on the basis of iron powder H were higher compared to the samples prepared on the basis of powder R.

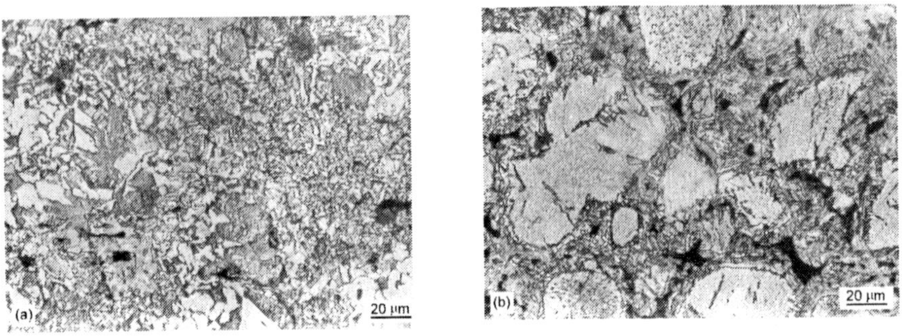

Fig. 6.9. Microstructures of Fe–4Mn–0.32C steels sintered for 10 min at 1200°C: (a) prepared on the basis of H iron powder (A), (b) prepared on the basis of R iron powder (B). Nital etched [103–105].

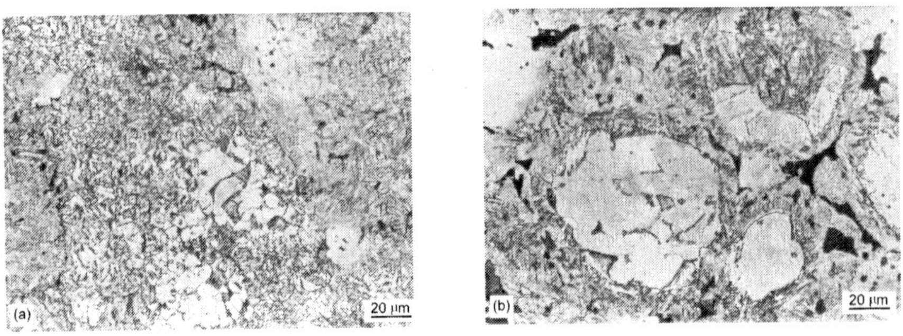

Fig. 6.10. Microstructures of Fe–4Mn–0.5Mo–0.32C steels sintered for 10 min at 1200°C: (a) prepared on the basis of H iron powder (C), (b) prepared on the basis of R iron powder (D). Nital etched [103–105].

Figures 6.9 and 6.10 show the microstructure of the samples tested after 10 min sintering at 1200°C. The microhardness values are given in Table 6.5.

6.1.4. Effect of sintering temperature on properties of Fe–Mn–C, Fe–Cr–C and Fe–Mo–C steels

ASC100.29 iron powder, graphite, elemental Mo powder <32 μm and electrolytic manganese <5 μm were used for preparation of the compacts. Electrolytic Cr <63 μm and aluminothermic Cr <10 μm were used for the Cr steels. The test bars from the powder mixtures were compacted at 600 MPa and sintered in a laboratory pusher furnace in technical purity hydrogen in steel boxes under a Al_2O_3 + 5% graphite getter.

The tensile strength and impact energy values of these steels in dependence on sintering temperature are shown in Fig. 6.11.

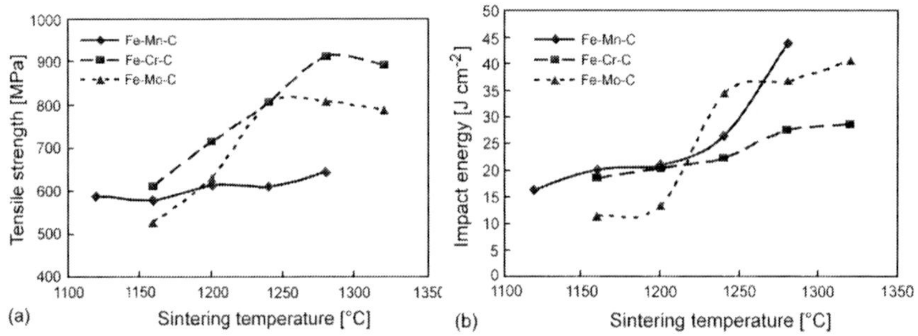

Fig. 6.11. Mechanical properties of sintered Fe–2Mn–0.3C, Fe–1Cr–0.7C and Fe–1.5Mo–0.7C steels as a function of sintering temperature in the range of 1120 to 1300°C during 60 min: (a) – tensile strength, (b) impact energy. Fe–Mn–C, 90 min, H_2/getter [53].

To explain the course of the ultimate strength and impact energy of the sintered steels alloyed with Mn, Cr and Mo, it is necessary to notice specifically the Mn steel compared to the other steels. The tensile strength of the Mn steel practically does not change from the temperature of 1120°C. On the other hand, the impact energy increases sharply from about 1200°C. This is due to two factors, which are analyzed in section 4.3. The particle size of the electrolytic manganese carrier <5 μm is inappropriate. Manganese rapidly sublimates from such particles already during the early stages of heating in which the temperature is not yet sufficient for the diffusion of manganese from the surface of the particles into their core and thus part of the Mn vapour escapes into the sintering atmosphere. The second factor is the holding time and high sintering temperature. Under such conditions, manganese sublimates at high rate back from the alloyed matrix into the atmosphere. Thus, in the final state the sample certainly contained a significantly lower manganese content than what was supplied to the system. This is reflected in increasing values of impact energy.

6.2. Effect of liquid phase forming elements on properties of Fe–Mn–C steels

6.2.1. Effect of copper

The addition of copper to the manganese alloyed steel was used because of its strenghtening effect on the mechanical properties of the mixed Cu-containing powders. Another reason is that copper forms in a mixed system a liquid phase at 1085°C. Based on knowledge of the

Table 6.6. Density (ρ) and mechanical properties of Fe–4.5Mn–0.45C and Fe–4.5Mn–1Cu–0.45C steels. 600 MPa, 3 h, 1120°C, Mn and C as FeMnC [28,121–123]

Steel Fe–	ρ [g cm^{-3}]	Rm [MPa]	A [%]	HV10	Sintering conditions
4.5Mn–0.45C	7.0	785	1.9	186	3 h, 1120°C
4.5Mn–1Cu–0.45C	6.9	728	2.3	168	
4.5Mn–0.45C	7.0	767	2.3	217	1 h, 1050°C + 2 h, 1120°C
4.5Mn–1Cu–0.45C	6.9	750	3.0	197	

*In the sintered condition the samples contained 4.27 – 4.37% Mn and 0.15– 0.21% C

impact of copper on the properties of sintered steels alloyed with Ni, Cu and Mo, it was assumed that the copper liquid phase accelerates and makes more uniform the alloying of the sintered Mn steels.

Samples of Fe–4.5Mn–C and Fe–4.5Mn–1Cu–C were prepared from Hametag iron powder, manganese in the form FeMnC and electrolytic copper. Samples were pressed at a pressure of 600 MPa and sintered: a) for 3 h at 1120°C, b) short-term induction sintering for 1 and 3 min at temperatures between 1200 and 1300°C in cracked ammonia at a dew point of –30°C and forged. Specially prepared samples of the same composition were produced by pouring the powder mixture freely into a metal box of dimensions 10 × 50 × 100 mm and sintered at 1120°C for 1 and 3 h. Thus, the sintered plates were freely forged with a reduction of 50% (upset-forging) after heating for 10 min at 1100°C. Standard test bars were manufactured by machining from forged plates.

The mechanical properties of the sintered Fe–Mn–C and Fe–Mn-Cu–C samples are shown in Table 6.6.

6.2.1.1. Mechanical properties

The density and mechanical properties in dependence on copper addition and sintering conditions are listed in Table 6.6.

The tensile strength of the Mn steel with a copper addition was by 57 MPa lower and hardness by 18 HV 10 lower compared with the values for the manganese steel without copper addition.

Table 6.7 gives the density and mechanical properties of short-time induction sintered and forged Fe–Mn–C and Fe–Mn–Cu–C steels.

Table 6.7. Density (ρ) and mechanical properties of Fe–4.5Mn–0.45C (MnC) and Fe–4.5–Mn–1Cu–0.45C (MnCuC) steels in the sintered and forged conditions. Mn and C as FeMnC [1,49,121–124]

Stage	Sintering conditions [°C/min]	ρ [g cm^{-3}]	Rm [MPa]		A [%]	HV 30	
			MnC	MnCuC	MnCuC	MnC	MnCuC
Sintered	1200/1	6.6	445	413	0	169	153
	1200/3	6.6	542	475	0	172	151
	1300/1	6.5	565	518	1.1	155	155
	1300/3	6. 6	589	561	2.0	155	145
Forged	1200/1	7.6	622	576	1.0	248	296
	1200/3	7.6	737	710	1.0	257	329
	1300/1	7.6	749	741	1.9	239	300
	1300/3	7.6	805	737	1.3	230	231

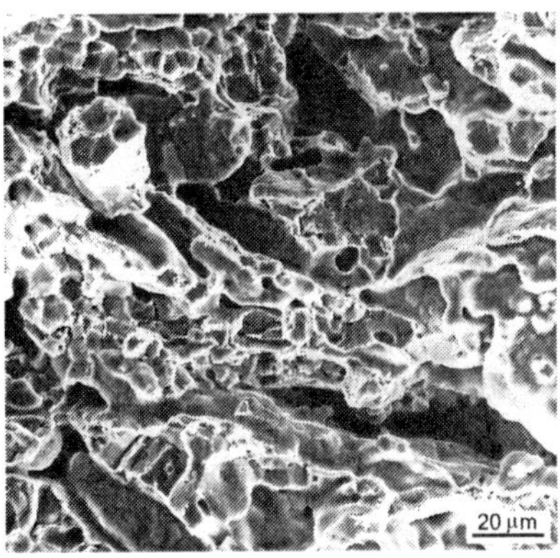

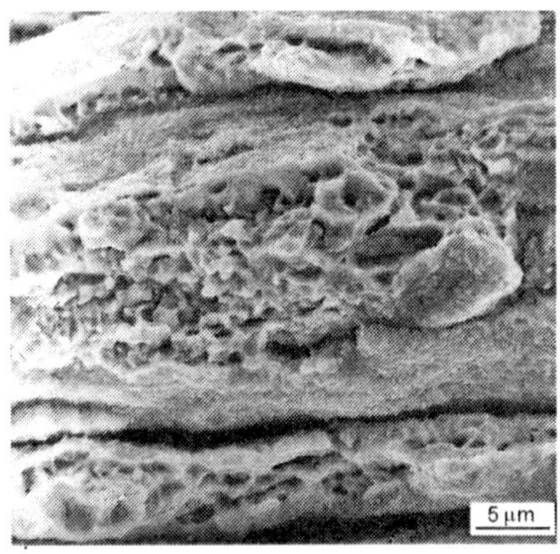

Fig. 6.12 (left). Fracture of Fe–4.5Mn–0.45C steel sintered for 3 h at 1120°C. SEM.
Fig. 6.13 (right). Detailed view of small dimple features in fracture of steel shown in Fig. 6.12. SEM.

The hardness of the samples with the addition of copper in the forged state is significantly higher than that of the samples without added copper.

Copper did not increase the mechanical properties of the sintered manganese steel nor the tensile strength after short-time induction sintering or forging with the exception of hardness.

Figure 6.12 shows the characteristic fractures of Fe–4.5Mn–0.45C steel and Fig. 6.13 in more detail the view on the fracture surface of the steel shown in Fig. 6.12.

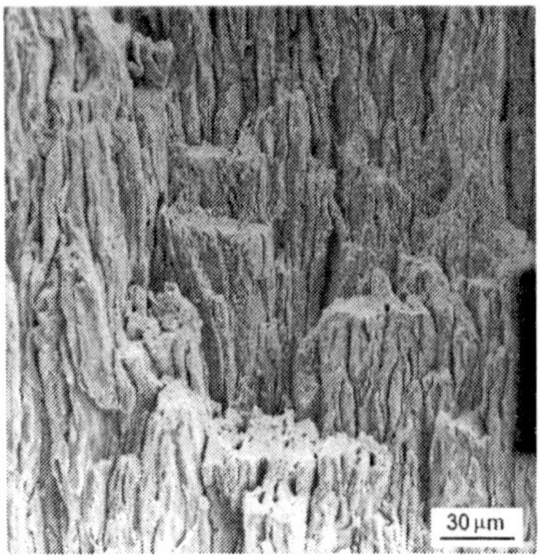

Fig. 6.14. Fracture of Fe–4.5Mn–1Cu–0.45C steel sintered for 3 h at 1120°C. SEM.

The fracture of FeMnC steel is mixed, ductile with dimples and transcrystalline with a small amount of intercrystalline fracture. The fracture of FeMnCuC steel is shown in Fig. 6.14 and is atypical. Longitudinal columns on the fracture surface show that the primary iron powder was Hametag iron powder. This is probably the specific case of ductile (lamellar tearing) failure propagating in cascades across the fracture surface in the direction of tensile load. This fracture is not described in the classification of fractures of conventional steels [124].

6.2.1.2. Effect of copper on mechanical properties of loose bulk sintered and upset-forged manganese steel

The mechanical properties of the manganese steel samples processes in this fashion with and without additional 1% of copper are shown

Table 6.8. Density (ρ) and mechanical properties of Fe–4.5Mn–0.45C a Fe–4.5Mn–1Cu–0.45C steel prepared by upset-forging of sintered freely poured powder mixtures for 1 and 3 h at 1120°C [121,125]

Steel	ρ [g cm^{-3}]	Rm [MPa]	A [%]	HV 10	Sintering
Fe–4.5–0.45C	7.15	566	2.4	144	1 h 1120°C
Fe–4.5–1Cu–0.45C	7.26	513	0.8	187	
Fe–4.5–0.45C	7.15	603	2.9	253	3 h 1120°C
Fe–4.5–1Cu–0.45C	7.15	356	0.5	263	

*As sintered carbon content was after sintering for 1 h 0.35%, after sintering for 3 h for FeMnC 0.29% and 0.18% for FeMnCuC steel

in Table 6.8. Samples for tensile testing were prepared by machining of upset forged samples.

Sintering for 3 h increased the hardness of both types of steels. Even in this case the steel with the addition of copper had a lower tensile strength than the samples without additives and even forged sintered bulk powders and to a greater extent after sintering for 3 h.

The results in Tables 6.6, 6.7 and 6.8 clearly show that the addition of copper to the manganese steel reduces its strength properties.

6.2.2. Effect of tin

To test the effect of tin on the properties of manganese steel, samples were prepared on the basis of Hametag iron powder. Manganese and

Table 6.9. Density (ρ) and mechanical properties of sintered Fe–(2.5–5.5)Mn–xC and Fe–(2.5–5.5)Mn–1Sn–xC steels; Hametag iron powder, 600 MPa, 3 h, 1120°C, cracked ammonia, dew point –30°C [68,125]

Steel Fe–	ρ [g cm^{-3}]	Rm [MPa]	A [%]	HV 10	*K* [J]
2.5Mn–0.25C	7.10	245	3.2	120	17.5
2.5Mn–1Sn–0.25C	7.10	308	3.9	107	20.4
3.5Mn–0.35C	7.07	418	3.1	189	14.8
3.5Mn–1Sn–0.35C	6.95	399	3.5	172	11.8
4.5Mn–0.45C	7.05	632	1.6	190	10.0
4.5Mn–1Sn–0.45C	6.92	599	1.9	192	8.0
5.5Mn–0.55C	6.97	713	1.3	231	9.9
5.5Mn–1Sn–0.55C	6.90	560	2.0	232	6.0

carbon were added as FeMnC <40 μm and tin in an amount of 1% (melting point of tin is 232°C). Samples were pressed at a pressure of 200 and 600 MPa. Table 6.9 gives the characteristics of the samples pressed at a pressure 600 MPa.

The Fe–2.5Mn–0.25 C steel with the addition of 1% tin reached a slightly higher tensile strength of 308 MPa and 20.4 J impact energy compared with the manganese steel with the same manganese content. In steels with higher manganese content the tin addition reduces the tensile strength and impact energy. The relative impact of the addition of tin on the ultimate strength of the Mn steel is shown

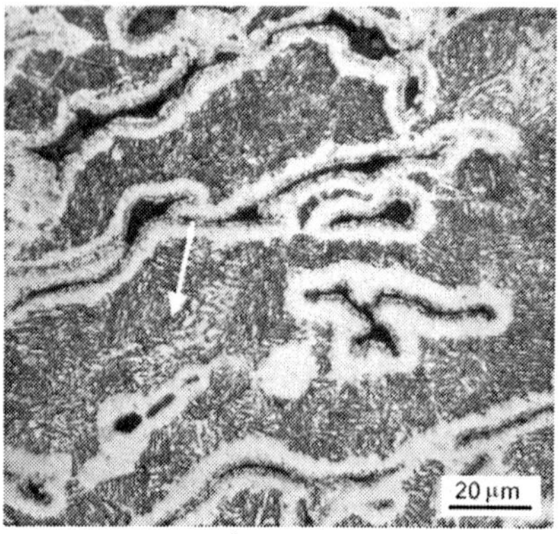

Fig. 6.15. Microstructure of the Fe–4.5Mn–0.05P–0.45C steel. 600 MPa, sintering 3 h at 1120°C. Nital etched.

Fig. 6.15 according to the values listed in Table 6.9. The lowest values of impact energy were obtained for the Fe–4.5Mn–1Sn–C and Fe–5.5Mn–1Sn–C steels and at the same manganese content the highest hardness values were attained.

According to these results, the addition of tin to the manganese steel caused a reduction in the tensile strength and impact energy of the studied Mn steels.

6.2.3 Effect of phosphorus

Phosphorus is a common alloying elements in the mixed Fe-P–(C) systems in which it contributes to increase the strength and toughness properties. Phosphorus in these steels particularly strengthens the interparticle necks. This increases the strength and to less extensive alloying of the internal part of the iron powder particles with phosphorus and by this the toughness properties increase. Phosphorus is a ferrite stabilising element, thereby reducing the solubility of carbon in the iron matrix. As a result of the formation of a liquid phase at 1050°C phosphorus causes spheroidisation of the pores. This reduces the proportion of open pores in the sintered compacts.

Given the known properties of phosphorus-alloyed sintered steels, the effect of phosphorus on the properties of the sintered manganese steel was also studied. Test samples were prepared on the basis of Hametag iron powder <0.16 mm with the addition of 4.5% Mn in the form of FeMnC <40 mm. Phosphorus in the amount of 0.05% was added in the form of red phosphorus and the amount of 0.5% as ferrophosphorus. Samples for tensile and impact tests were pressed at 600 MPa and sintered for 3 h at 1120°C and in the latter case for 1 h at 1050°C + 2 h at 1120°C in technical purity hydrogen. Some of the

Table 6.10. Density (ρ) and mechanical properties of Fe–4.5Mn–0.45C (A), Fe–4.5Mn–0.05P–0.45C (B) a Fe–4.5Mn–0.5P–0.45C (C) steels. 600 MPa, 0.05% P as red phosphorus, 0.5% as ferrophosphorus [125,126]

Steel	ρ [g cm^{-3}]	Rm [MPa]	A [%]	HV 10	K [J]	Sintering [°C/h]
A	7.0	797	1.9	193	8.9	
B	6.83	342	0.9	152	1.6	1120/3
C	6.51	brittle		262	n. d.	
A	6.98	787	3.3	217	9.1	
B	6.78	215	1.7	191	2.2	1050/1 + 1120/2
C	6.50	193	0.7	235	1.7	
B	6.86	266	0.0	152	2.2	1120/1
C	6.57	208	0.0	148	1.0	

samples after heating for 10 min at 1100°C were upset-forged with 50% reduction and after forging annealed for 60 min at 650°C. Test bars from the upset-forged samples were manufactured by machining.

The properties of the studied sintered manganese steels with the addition of phosphorus are shown in Table 6.10.

The carbon and manganese contents in the sintered steels A were 0.20% C and 4.4% Mn, in steel B 0.6% C and 4.5% Mn, and in steel C 0.17% C and 4.3% Mn, respectively. Acid insoluble matter in steel C 0.43% was introduced into the system by ferrophosporus

The addition of two quantities of phosphorus to manganese steel sintering condition caused a reduction in density of the samples, and the addition of 0.5% P up to P ~0.3 g cm^{-3}. Samples with 0.5% P additions after sintering for 3 h at 1120°C has burst into the mounting clamps of the machine. Therefore, additional samples were sintered with phosphorus stepped, as shown in the table. Samples with 0.5% addition of phosphorus sintered for 1 h showed the fragility of such as after sintering for 3 h. Part of the samples sintered under the terms of Table 6.10 was forged and their properties are listed in Table 6.11. Test bars for tensile testing and impact energy of the forged samples were produced by machining. The properties of the manganese steel with the addition of phosphorus step-sintered and then upset forged are shown in Table 6.12. As regards the carbon and manganese content of the sintered steels with the addition of phosphorus, it

Table 6.11. Density (ρ) and mechanical properties of upset forged and annealed manganese and manganese–phosphorus steels. Upset-forging by reduction of 50 %, annealing 1 h at 650°C [125,126]

<table>
<tr><th>Steel
Fe–</th><th>ρ
[g cm^{-3}]</th><th>Rm
[MPa]</th><th>A
[%]</th><th>HV 10</th><th>Sintering
[°C/h]</th></tr>
<tr><td>4.5Mn–0.45C</td><td>7.75</td><td>812</td><td>3.6</td><td>206</td><td>1120</td></tr>
<tr><td>4.5Mn–0.05P–0.45C</td><td>7.67</td><td>419</td><td>2.8</td><td>221</td><td rowspan="2">1050/1+
1120/2</td></tr>
<tr><td>4.5Mn–0.5P–0.45C</td><td>7.60</td><td>587</td><td>2.5</td><td>190</td></tr>
</table>

Table 6.12. Density (ρ) and mechanical properties of Fe–4.5Mn–0.45C, Fe–4.5Mn–0.05P–0.45C a Fe–4.5Mn–0.5P–0.45C manganese steels compacted at a pressure of 600 MPa and sintered for 1 h at 1050°C + 2 h at 1120°C [123,125]

Steel Fe–	ρ [g cm^{-3}]	Rm [MPa]	A [%]	HV 10
4.5Mn–0.45C	7.75	645	3.6	206
4.5Mn–0.05P–0.45C	7.76	419	2.8	221
4.5Mn–0.5P–0.45C	7.60	587	2.7	190

should be noted that in these cases only the carbon content decreased whereas the manganese content remained practically unchanged.

The upset-forged samples with the addition of phosphorus reached relatively high values of tensile strength and elongation compared to sintering without the addition of phosphorus.

The above results show that phosphorus significantly reduces the properties of the sintered manganese steels.

The microstructures of samples with the addition of phosphorus are fundamentally different from the manganese steel with no added phosphorus. Such a structure is shown in Fig. 6.16 for the steel with the addition of 0.05% P. On the surface of all iron particles the open pores of the compact showed a white layer of uniform thickness of ~5–6 μm. A detailed view of the microstructure inside the grains of the white layer is shown in Fig. 6.16. The microstructure of samples with the addition of 0.5% phosphorus is shown in Fig. 6.17. It is a coarse-grained ferritic structure with rounded pores which is due to the stabilizing effect of phosphorus in ferrite and spheroidisation of the pores.

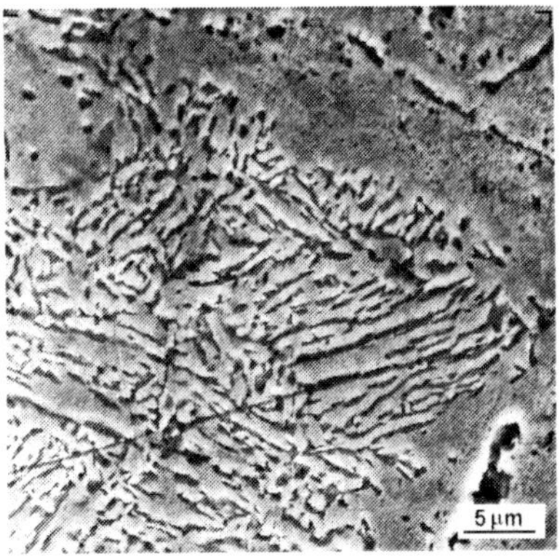

Fig. 6.16. Detailed view of the microstructure of the interior of individual grains under the white layer formed by fine pearlite indicated by the arrow in Fig. 6.15, SEM.

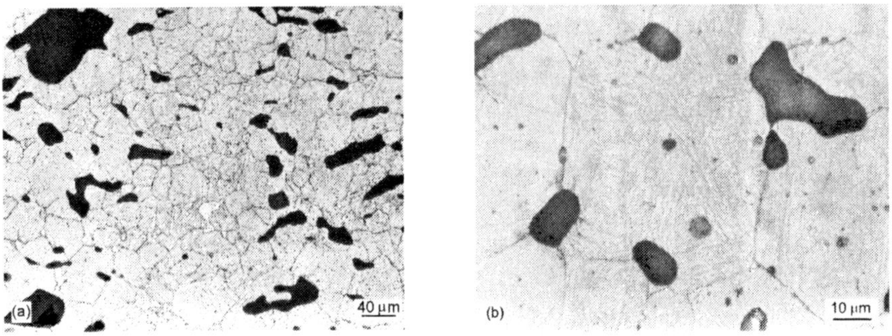

Fig. 6.17. Ferritic microstructure of Fe–4.5Mn–0.5P–0.45C steel at two magnifications. Sintering 3 h at 1120°C. Nital etched.

In comparison with the fractures of the samples with the addition of phosphorus, fractures of the samples of Fe–0.45C–4.5Mn steel is shown in Fig. 6.12. Fracture of the samples with the addition of 0.05% P after the tensile test is shown in Fig. 6.18. It is a flat transcrystalline fracture showing the original characteristics of the Hametag powder particles. This fracture also shows the white layer shown in Fig. 6.15 at the bottom of the picture. A view inside the grains at a higher magnification is in Fig. 6.19. Intercrystalline fracture of samples with 0.5% P, which is reflected in the tensile test as brittle, is shown in Fig. 6.20 where one can see the white layer and spheroidised pores.

Based on these results, it can be concluded that phosphorus has a significant deteriorating effect on the mechanical properties of the manganese steel also on the character of structures and fractures.

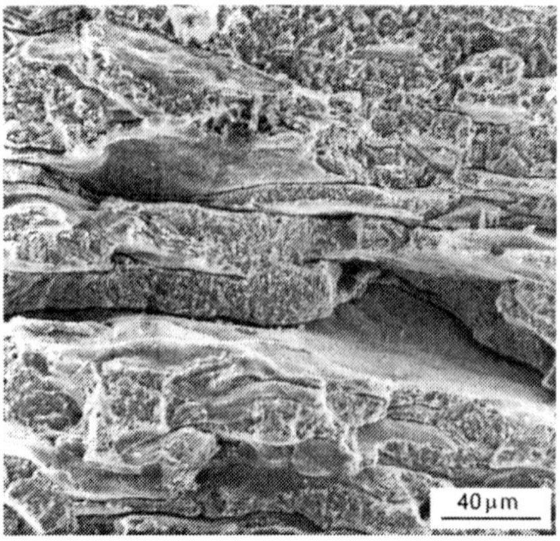

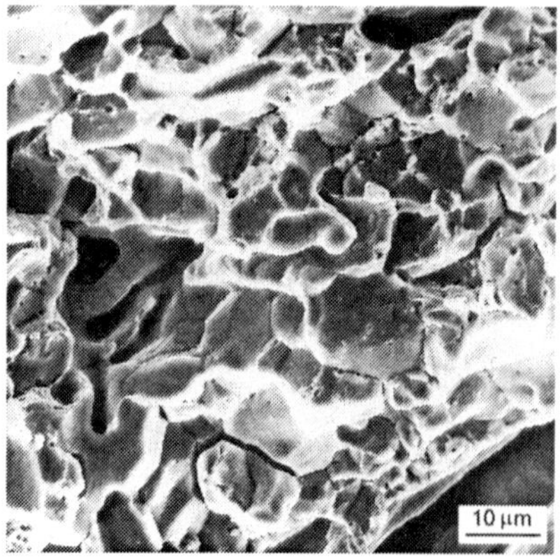

Fig. 6.18 (left). Fracture of Fe–4.5Mn–0.45C steel with addition of 0.05% P. 600 MPa, Hametag iron powder. SEM.
Fig. 6.19 (right). Fracture of Fe–4.5Mn–0.45C steel with addition of 0.05% P at higher magnification. 600 MPa, Hametag iron powder. SEM.

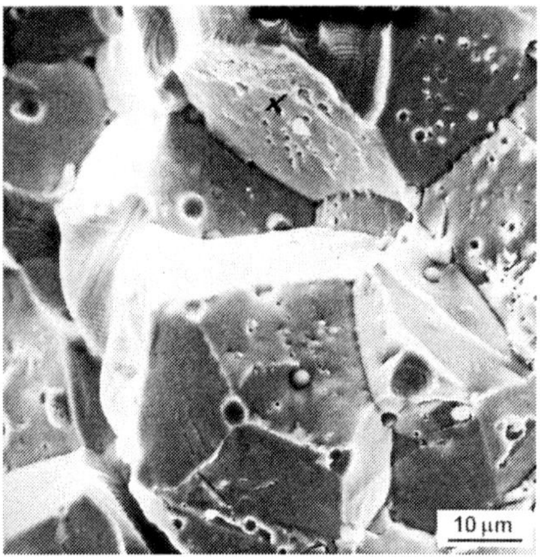

Fig. 6.20. Intercrystalline fracture of Fe–4.5Mn–0.5P–0.45C steel. White layer indicated by x, sintering for 3 h at 1120°C, SEM.

6.3. Effect of boron

Boron belongs to the elements forming the liquid phase below the common sintering temperature and, therefore, can be used as an enhancing element aimed for increasing the density and mechanical properties of sintered components. Boron forms a eutectic (Fe–3.74B) at 1174°C and has a unipolar solubility with iron. In cases when the prealloyed matrix is used instead of plain iron powder, boron diffuses into the matrix. By this way the extent of densification increases together with the matrix hardness and reduction of the final fraction of the eutectic in the microstructure (bipolar solubility) compared with the Fe–B system.

The addition of boron to the sintered material has two roles, namely strengthen the matrix and, under certain conditions, produce on the surface of the parts hard boride layers by boriding without halides, as will be seen below. Therefore, both roles of boron can be used for the systems prepared on the basis of the mixed and prealloyed powders. Research on the impact of the two roles of boron was also carried out on systems without manganese, especially based on prealloyed powders with chromium to allow comparison of the results and gain insight into the preferential effect of boron on some of the studied material systems.

Compared with the classic boriding of ferrous materials, in this work in the part dealing with boriding the activator was manganese due to its high vapour pressure or boriding was carried out without manganese in all cases without halides.

6.3.1. Effect of boron addition on properties of Fe–Mn–C steel

The specimens of Fe–(2,3,4)Mn were prepared on the basis of sponge iron powder NC100.24. Medium-carbon ferromanganese (81.2% Mn, 1.5% C) milled in air with <40 μm was used as a manganese carrier. Boron in the amount of 0.1, 0.2 and 0.4 mass % as ferroboron (Fe18B) <40 μm, 0.4% of graphite and of 0.8% of HW wax as a lubricant were admixed to the powders used. The tensile strength bars from the powder mixes were compacted at 600 MPa and sintered uniformly for 60 min at 1150 and 1200°C in hydrogen in a laboratory pusher furnace.

6.3.1.1. Properties of Fe–Mn–(B)–C steels

Density and mechanical properties. The sintered density as a function of manganese and boron additions, and of sintering conditions, is shown in Fig. 6.21. The green density of the samples was ~6.9 g cm^{-3}. Higher density was achieved after sintering at higher tenmperature. A boron addition of 0.4% to the Fe–2Mn–0.4C steel increased the density to 7.06 g cm^{-3} after sintering at 1150°C and to to 7.13 g cm^{-3} after sintering at 1200°C.

The tensile strength values of the investigated Fe–Mn and Fe–Mn–B steels as a function of manganese and boron addition and sintering temperature are shown in Fig. 6.22. The tensile strength of Fe–2Mn steel was in the range 485 to 518 MPa and that of Fe–4Mn steel in the range 562 to 586 MPa in dependence on sintering temperature. Slightly higher tensile strength values were obtained by sintering at a lower temperature.

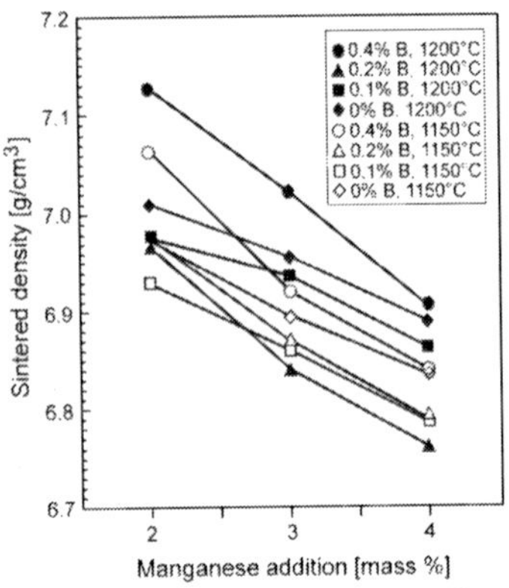

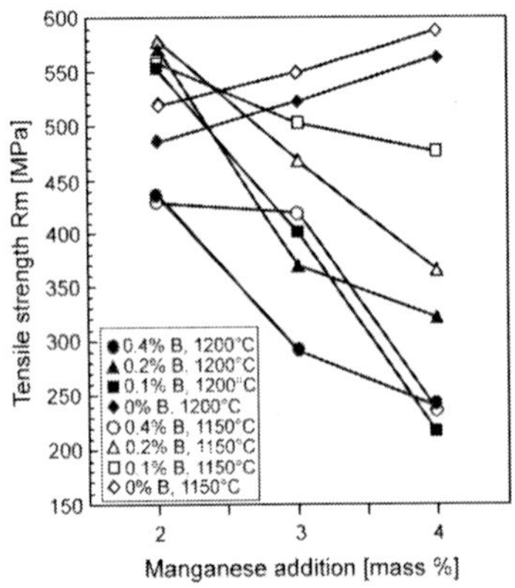

Fig. 6.21 (left). Sintered density of Fe–Mn–C and Fe–Mn–B–C steels as function of manganese and boron addition and of sintering temperature. 600 MPa, 1 h, 1150 and 1200°C, hydrogen [127].

Fig. 6.22 (right). Tensile strength values of Fe–Mn steels in dependence on manganese and boron addition and sintering temperature. 600 MPa, 1 h, 1150 and 1200°C, hydrogen [127].

The addition of 0.1 and 0.2% B increased the tensile strength to 551–572 MPa for Fe–2Mn steel. All boron additions caused a decrease of tensile strength, especially with 0.4% B for Fe–4Mn steel to 234 MPa. Sintering at 1150°C showed to be more effective for Fe–2Mn–0.2B–0.4C steel than at 1200°C.

The hardness values increased with increasing manganese and boron addition, as shown in Fig. 6.23. The hardness of the manganese steel without boron addition was in the range 124–149 HV 10 (Fe–2Mn) to 203–238 HV 10 (Fe–4Mn) in dependence on sintering temperature. The boron addition resulted in a significant increase of hardness, especially for Fe–4Mn steel, up to 384 HV 10 after sintering at 1150°C.

Microstructure and fracture. The microstructure of the Fe–2Mn–0.4C steel is heterogeneous and formed by ferrite, pearlite and bainite, as shown in Fig. 6.24(a). With increasing Mn content the microstructure is formed only by bainite, as shown in Fig. 6.24(b).

The microstructures of all Fe–Mn steels sintered with assistance of the boron liquid phase are more homogeneous with some differences in their character in dependence on boron addition, especially of those sintered at 1200°C, compared to the boron-free steels. No eutectic was formed at the grain boundaries for the steels with 0.1% boron addition only. The microstructure with coarse grains and eutectics was observed for the Fe–Mn steels with 0.4% boron addition only.

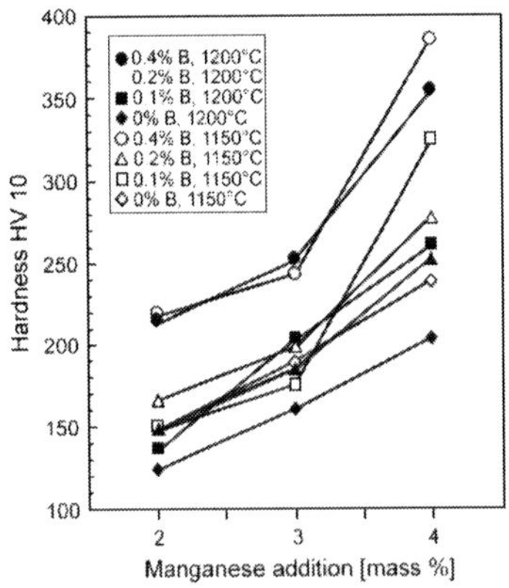

Fig. 6.23. Hardness of Fe–Mn steels in dependence on manganese and boron addition and on sintering temperature. 600 MPa, 1 h, 1150 and 1200°C, hydrogen [127].

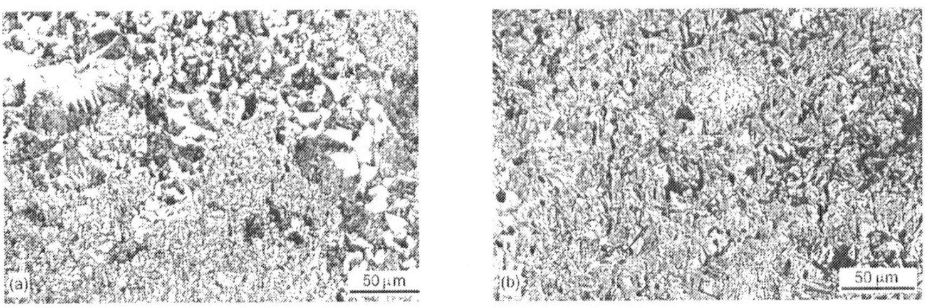

Fig. 6.24. Microstructure of manganese steel sintered at 1200°C. (a) – Fe–2Mn–0.4C, (b) Fe–4Mn–0.4C steel. Nital etched [128].

The pores were rounded, speroidised and isolated mainly at the grain boundaries. A eutectic network formed in the microstructure; its portion increased at higher manganese content as shown in Figs. 6.25(a),(b). According to the analysed microstructures, boron diffused into the Mn-alloyed matrix grains already at a sintering temperature of 1150°C as compared with the eutectic temperature of 1174°C for the Fe–B system.

The fracture of the Fe–Mn–C steels was characterised by transgranular ductile mode with dimple features and line facets as shown in Fig.6.26. On the fracture of the steels with boron addition are present also cleavage facets as shown in Fig.6.27. The fracture of Fe–Mn steel specimens with 0.4% boron addition occurred mainly through a eutectic phase and coarse grain boundaries. The fracture of the specimen with cleavage facets in integranular mode is shown in Fig.6.28.

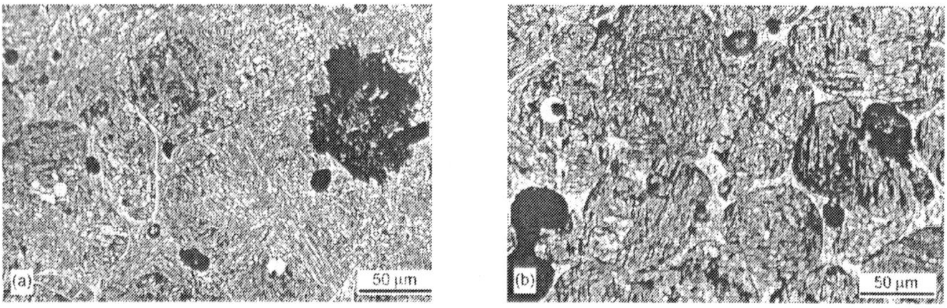

Fig. 6.25. Microstructure of manganese steel with addition of boron sintered at 1200°C. (a) – Fe–2Mn–0.4B–0.4C, (b) – Fe–4Mn–0.4B–0.4C steel. Nital etched [128].

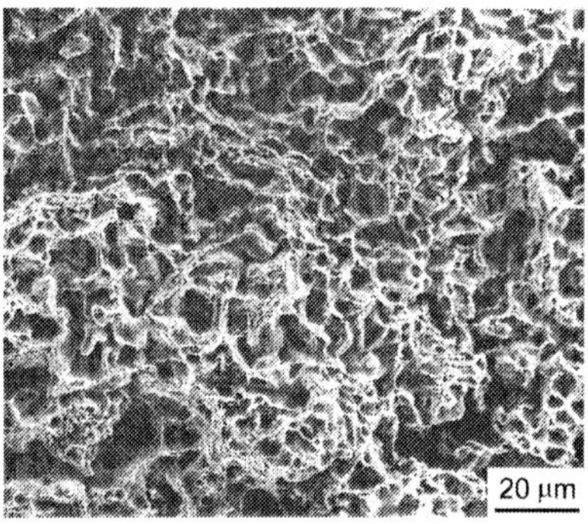

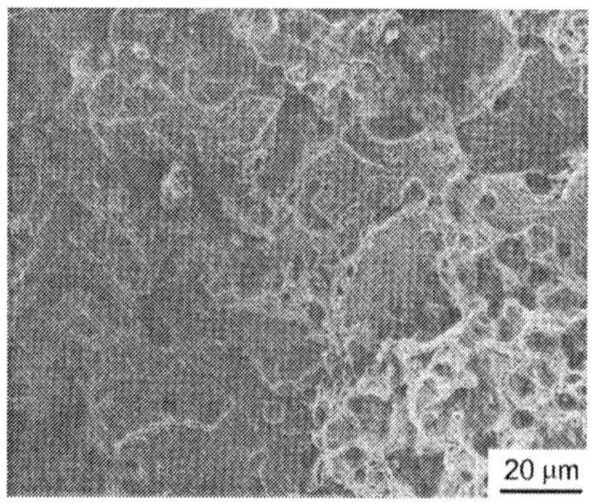

Fig. 6.26 (left). Fracture surface of Fe–2Mn–0.4C steel specimen sintered at 1200°C; Rm = 485 MPa, 124 HV 10 [128] SEM.
Fig. 6.27 (right). Fracture surface of Fe–2Mn–0.2B–0.4C steel specimen sintered at 1200°C. Rm = 571 MPa, 149 HV 10. SEM [128].

According to the presented results boron addition caused the increase in density of the specimens, more homogeneous microstructure compared to that without boron. 0.1 and 0.2% of boron increased the tensile strength of Fe–2Mn steel up to 557 and 578 MPa in relation to sintering temperature, and an increase of hardness up to 384 HV 10 for Fe–4Mn–0.4B steel.

6.3.2. Effect of boron addition on tribological properties of Fe–Mn–C steels

The specimens for tribological tests were prepared from the same powder mixes based on NC100.24 iron powder with the addition of boron in the amount of 0.1, 0.2 and 0.4% with the mean particle size <20 μm. Manganese was added as medium-carbon ferromanganese (81.2% Mn, 1.5% C).

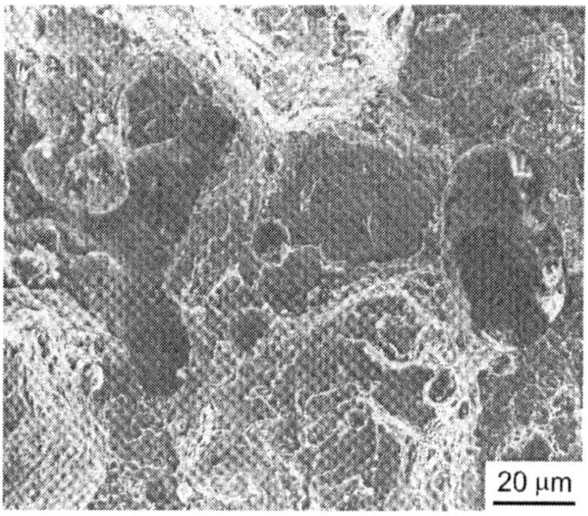

Fig. 6.28. Fracture surface of Fe–2Mn–0.4B–0.4C steel specimen sintered at 1200°C. Rm = 433 MPa, 213 HV 10. SEM [128].

The impact energy bars (10 × 10 × 55 mm) were compacted from the powder mixes at 600 MPa and sintered for 60 min at 1150°C in hydrogen in a pusher furnace as in the previous case. The test specimens for tribological tests were the halves of the broken impact energy test bars. The tribological tests were performed on both sides of two specimens perpendicular to the compacting direction from each tested steel.

The dry-run tribological (sliding) properties were tested by the disc–block method shown in Fig. 5.50. The tests were conducted under the following conditions: the radial loading force 100N, a steel disc (ASTM D2) with a diameter of 50 mm (thickness 10 mm) heat treated to 53–54 HRC with a freshly ground surface (Ra = 0.15 μm), sliding speed v = 0.5 m s^{-1}, sliding distance 1000 m, number of rotations 197 n min^{-1}. To determined the dry friction behaviour of the tested steels, one specimen from each steel was exposed to the seizure test.

The effect of manganese and boron additions on the sintered density of the tested steels is shown in Fig. 6.29. The density decreased with manganese and boron additions compared to that of Fe–Mn–C steel. The dependence of hardness as a mean value of 12 measurements on manganese and boron addition is shown in Fig. 6.30. The hardness of Fe–Mn–C steel increased from 154 (Fe–2Mn) to 242 HV 10 (Fe–4Mn). A boron addition of 0.4% caused a large increase in the hardness of Fe–4Mn steel up to 440 HV 10. The similar dependences for both properties of the same steels sintered at 1200°C are shown in Figs. 6.21 and 6.23. The increase of hardness caused by boron proved that boron at a sintering temperature of

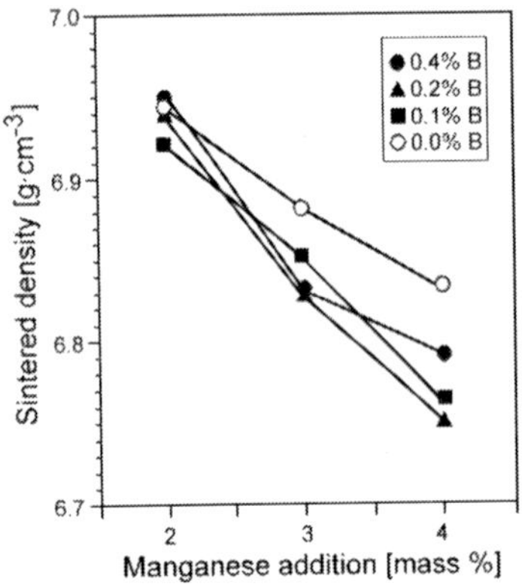

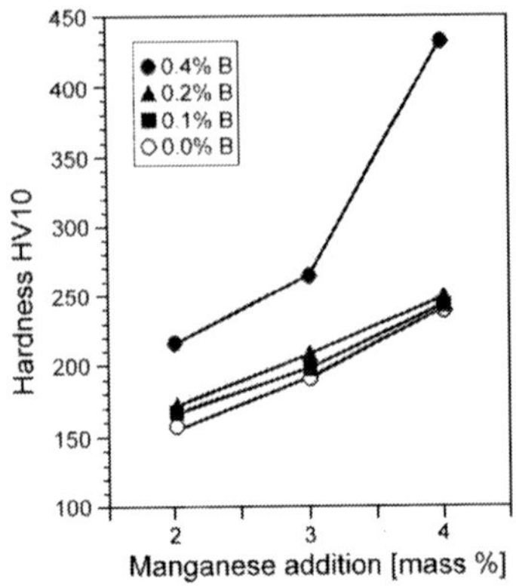

Fig. 6.29 (left). Dependence of sintered density of Fe–Mn and of Fe–Mn–B steels on manganese and boron addition. 600 MPa, 60 min, 1150°C, hydrogen [129].
Fig. 6.30 (right). Dependence of hardness of sintered Fe–Mn and Fe–Mn–B steels on manganese and boron addition. 600 MPa, 60 min, 1150°C, hydrogen [129].

1150°C diffused into the matrix alloyed with manganese in spite of the fact that the eutectic temperature of the Fe–B system is 1174°C.

The friction coefficient of the investigated steels in dependence on the manganese and boron addition is shown in Fig. 6.31. Boron in general caused an increase of the friction coefficient compared to that of Fe–Mn steel. The highest friction coefficient was recorded for the steels with 0.2 and 0.4% B and the lowest one for the Fe–3Mn (0.31) steel and boron-containing steels, especially those with 0.4% B (0.38).

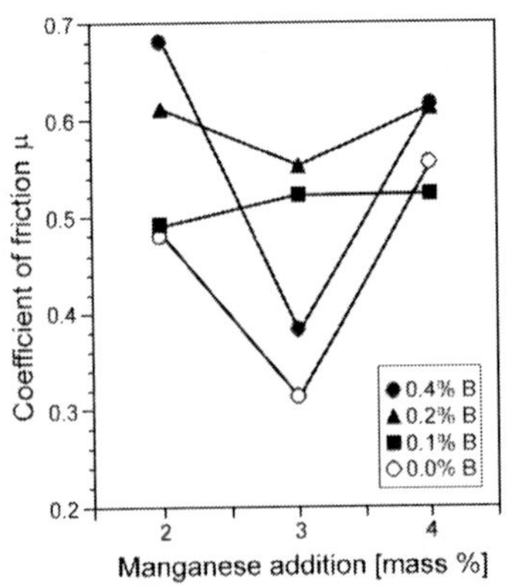

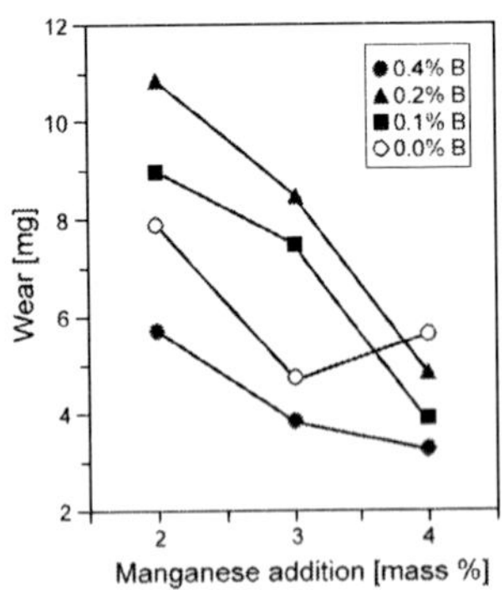

Fig. 6.31 (left). Friction coefficient of Fe–Mn and of Fe–Mn–B steels as function of manganese and boron addition. 600 MPa, 60 min, 1150°C [129].
Fig. 6.32 (right). The dependence of wear of Fe–Mn and of Fe–Mn–B steels on manganese and boron addition. 600 MPa, 60 min, 1150°C [129].

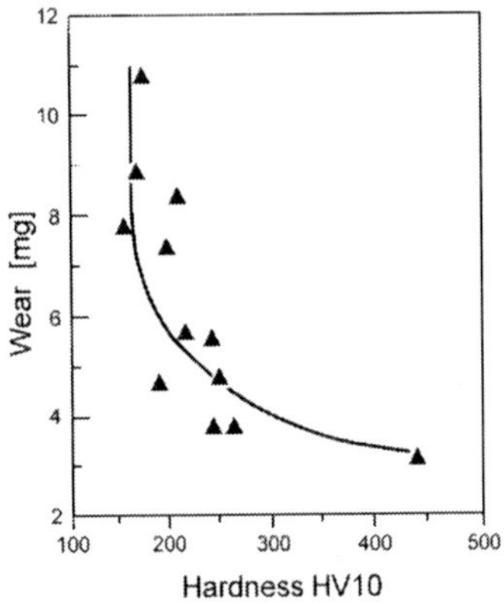

Fig. 6.33. Relationship between dry wear and hardness of Fe–Mn and Fe–Mn–B steels [129].

The dependence of the wear values of the Fe–Mn and Fe–Mn–B steels on manganese addition is shown in Fig. 6.32. According to this diagram the wear decreased with increasing manganese addition. A boron addition of 0.1 and 0.2% caused an increase of wear for Fe–2Mn and Fe–3Mn steels compared to the Fe–Mn boronless steel. The lowest wear was exhibited by all Fe–Mn steels with the boron addition of 0.4%.

The relation between the wear and hardness of the investigated steels is shown in Fig. 6.33. The wear of these steels decreased to ~4 mg relatively steeply up to the hardness of ~250 HV 10 and slowly to ~3 mg at 440 HV 10.

Seizure test. The hardness values of the investigated steels, the friction coefficient, the temperature in the contact zone and the load at which the seizure test was finished are given in Table 6.13.

In this case, the relatively highest friction coefficient value (0.49) was obtained for the boron-free samples independently of the

Table 6.13. Hardness (HV 10), friction coefficient (μ), temperature (T) and the highest testing load (F) at which seizure of the investigated steels occurred in dependence on manganese and boron addition in mass % [129,130]

Mn [%]	2				3				4			
B [%]	0.0	0.1	0.2	0.4	0.0	0.1	0.2	0.4	0.0	0.1	0.2	0.4
HV 10	154	167	172	215	191	197	207	264	242	244	250	440
μ	0.49	0.45	0.45	0.42	0.49	0.45	0.45	0.42	0.49	0.53	0.42	0.45
T [°C]	155	230	230	235	160	173	170	230	170	153	275	105
F [N]	1100	*1200	*1300	*1300	1100	1200	1300	1400	1200	1100	*1500	*1300

*without seizure

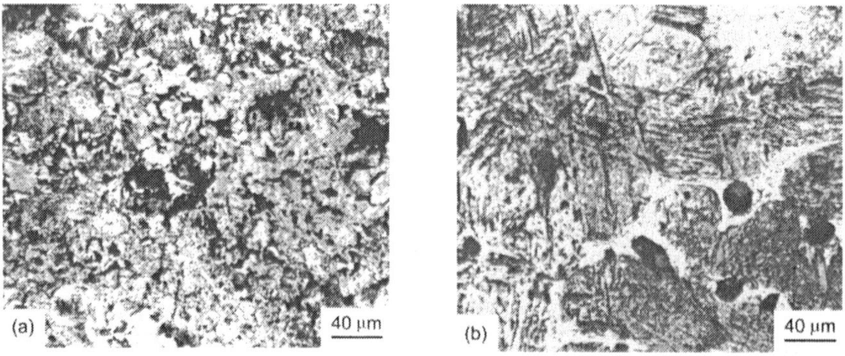

Fig. 6.34. Microstructure of: (a) Fe–3Mn–0.4C steel, (b) Fe–4Mn–0.4B–0.4C steel. 60 min, 1150°C. Nital etched [129].

manganese addition at a relatively low temperature in the range 0.42 to 0.45, except for the Fe–4Mn–0.1B steel. The most uniform course of the friction moment and by this way of the friction coefficient in this test was recorded for the Fe–2Mn–0.2B steel.

The seizure tests of boron-free samples were finished at the relatively lowest temperature with increasing tendency with increasing manganese addition.

Figure 6.34 shows the characteristic microstructures of the investigated steels responsible for their tribological characteristics. The microstructure shown in Fig. 6.34(a) was without ferrite grains. As shown in Fig. 6.34(b), the eutectics Fe–B–(Mn,C) was observed at the grain boundaries with spheroidized pores and martensite in Fe–4Mn–0.4B–0.4C steel.

The effect of boron addition on the properties of Fe–Mn–C steel expressed in:

–highest hardness attained for Fe–4Mn–C steel with 0.4% B

–an increase of the friction coefficient

–the lowest wear exhibited of the Fe–Mn–C steels with 0.4% B

–high wear resistance of the steels with boron addition at the seizure tests at the load of 1300 N.

6.3.3. Effect of boron addition on mixed low-alloyed Fe–Mn–Cr–Mo–C steels

The test samples (10 mm diameter, 10 mm long) of a Fe–0.7% Mn–1.65% Cr–0.2% Mo–0.7% graphite mixture were prepared. The tensile strength bars were prepared on the basis of water-atomised iron powder ASC100.29 (A) without and with the addition of 0.2% B (B). Chromium was added in form of a master alloy (Fe–33Cr–5C–7.6 (Mn,Mo,Ni)) with a melting point of 1130–1140°C, manganese as electrolytic manganese, molybdenum as ferromolybdenum and boron as ferroboron (Fe18B), all with particle size <40 μm.

The samples compacted at 500 MPa were sintered for 60 min at 1120, 1150 and 1200°C in vacuum (10^{-3} mm Hg), in cracked ammonia and in a $95N_2/5H_2$ atmosphere in a laboratory pusher furnace.

The effect of sintering conditions on the density of samples of both types of powder mixtures is shown in Fig. 6.35. The density of both types of steels after sintering in cracked ammonia was lower than in the green state. The density of samples sintered with the addition of boron in vacuum increased with increasing temperature. This was due to both boron and the liquid phase of Fe–Cr–C master alloy. The density as in the steel A samples, *i.e.* 7.36 g cm^{-3} was also recorded for the steel B samples sintered at 1200°C in the $95N_2/5H_2$ atmosphere. The hardness of the investigated steels depending on the sintering conditions is shown in Fig. 6.36. The steel A samples without boron reached the highest hardness of 191 HV 10 after sintering at 1200°C. In contrast, the hardness of the samples B with boron reached 288 HV 10 after vacuum sintering at 1150°C.

Figure 6.37 shows the dependence of the microhardness of the euetctics for the steel C. This is a mixed Fe-0.2B system without the addition of carbon prepared under the same conditions as the steels A and B.

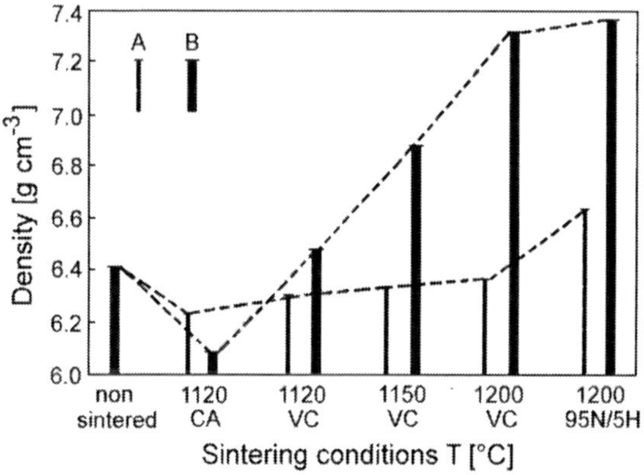

Fig. 6.35. Sintered density of steels A and B in dependence on sintering conditions in cracked ammonia (CA), in vacuum (VC) and in $95N_2/5H_2$ atmosphere for 60 min at 1120°C [130].

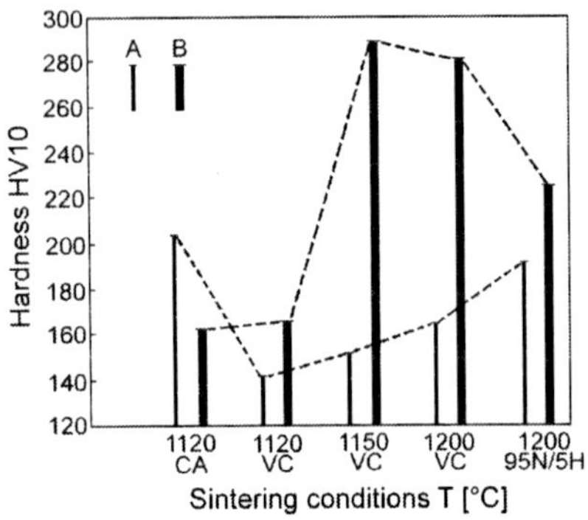

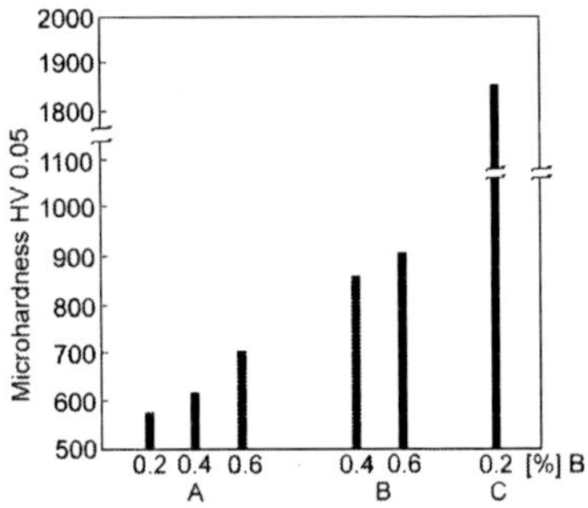

Fig. 6.36 (left). Hardness of steels A and B in dependence on sintering conditions in cracked ammonia (CA), in vacuum (VC) and in $95N_2/5H_2$ atmosphere for 60 min at 1120°C [130].

Fig. 6.37 (right). Microhardness HV 0.05 of the eutectics in A, B and C steels in dependence on boron addition. Sintering 60 min at 1120°C; D – Fe–0.2B, E – prealloyed Fe–0.7Mn–1Cr–0.2Mo powder [130].

The highest microhardness of the eutectic was recorded for the steel C, up to 1850 HV 0.05. In the steels A and B, the effect of boron addition was also evident. X-ray analysis in the eutectic of steel C detected the presence of Mn, Cr, Mo and Fe.

The results show that the boron addition had a significant effect on all the investigated properties of the manganese- and chromium-alloyed steels.

6.3.4. Effect of boron addition on properties of Fe–Cr–Mo–V–(Mn) sintered steels

This part of investigation was focused on the effect of boron addition on the properties of steels based on three grades of water-atomised and vacuum-annealed prealloyed powders with Cr as the main alloying element. The compacts were prepared from the following powders:

- Fe–1Cr–0.3Mo–0.3V (KIP 103V) coded here K1
- Fe–3Cr–0.3Mo–0.3V (KIP 30CRV) coded here K2
- Fe–1Cr–0.7Mn–0.3Mo (KIP 4100V) coded here K3

all with particle size <160 μm.

Graphite in the amount of 0.2% and 0.2% B as ferroboron <40 μm and 0.8% zinc stearate as a lubricant were admixed to the powders. The tensile strength bars were compacted at 690 MPa and sintered at 1120°C in cracked ammonia in a retort furnace, and at 1200°C and 1280°C in hydrogen in a laboratory pusher furnace for sintering time being uniformly 60 min.

Table 6.14. Starting chemical composition in mass [%] of water-atomised and vacuum-annealed prealloyed powders K1, K2 and K3

Powder	C	Mn	Cr	Mo	V	O
K1	0.01	0.08	1.07	0.32	0.29	0.15
K2	0.01	0.09	3.02	0.29	0.23	0.14
K3	0.01	0.62	0.93	0.27	–	0.19

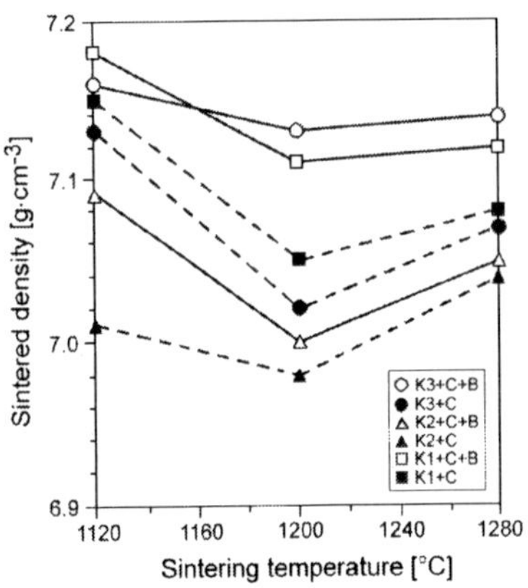

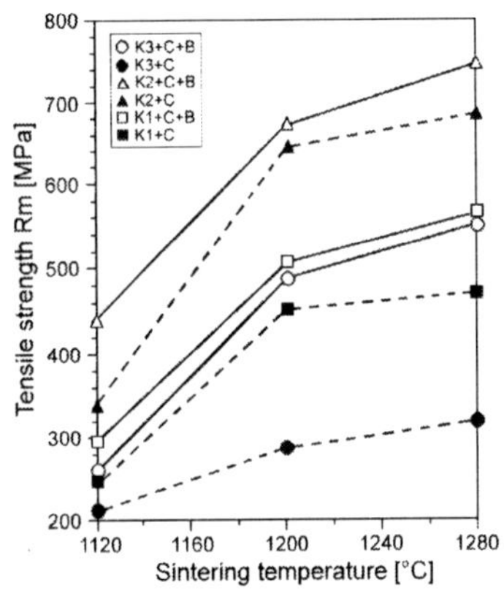

Fig. 6.38 (left). Sintered density of K1, K2 and K3 steels with 0.2% graphite addition without and with 0.2% B in dependence on sintering temperature. 690 MPa, hydrogen, sintering time 60 min [131–133].

Fig. 6.39 (right). Tensile strength of K1, K2 and K3 steels with 0.2% graphite addition without and with 0.2% B in dependence on sintering temperature. 690 MPa, hydrogen, sintering time 60 min [131–133].

The chemical composition of these base powders is given in Table 6.14.

The effect of sintering temperature on the as sintered density of the tested steels is shown in Fig. 6.38. The green density of the compacts was ~7.0 g cm^{-3}. As follows from the figure the highest increase in density was attained after sintering at 1120°C, except for the K2 steel. The highest density of 7.18 g cm^{-3} was attained by the K3-B steel and slightly lower by the K1-B steel. A marked decrease of density was observed after sintering at 1200°C followed by a small increase for the K1 and K2 steels without and with boron addition.

Contrary to density, the lowest strength was attained for all the steels after sintering at 1120°C, as shown in Fig. 6.39. The relatively highest increase in tensile strength was obtained after sintering at 1200°C. The K2-B steel after sintering at 1280°C has a tensile

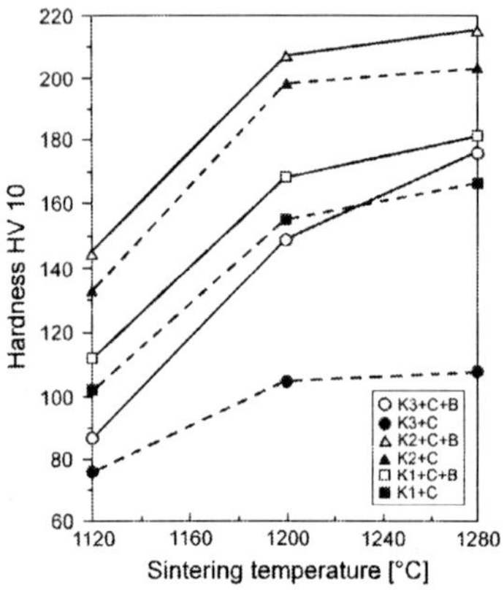

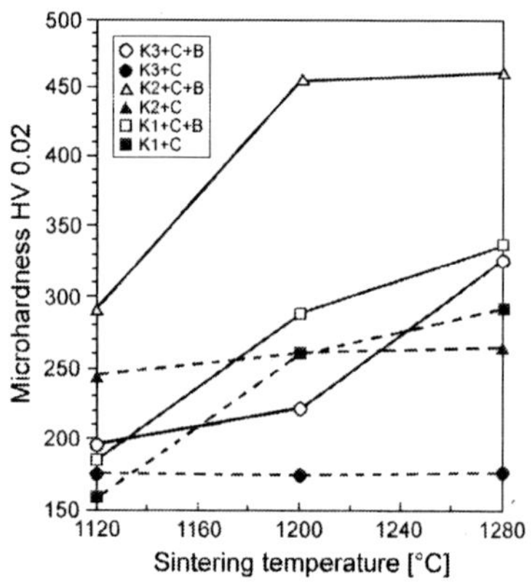

Fig. 6.40 (left). Hardness of K1, K2 and K3 steels with 0.2% graphite without and with 0.2% B addition in dependence on sintering temperature. 690 MPa, hydrogen, sintering time 60 min [131–134].

Fig. 6.41 (right). Miocrohardness of K1, K2 and K3 steel structures with 0.2% graphite without and with 0.2% B addition in dependence on sintering temperature. 690 MPa, hydrogen, sintering time 60 min [131–134].

strength of 754 MPa. The lowest strength was recorded for the K3 steel after sintering at all three temperatures. It is assumed that this was caused by demanganisation of this steel with increasing sintering temperature.

The addition of 0.2% boron caused a significant increase in the yield strength of the steels K1 and K3 to 560 and 573 MPa sintered at 1280°C at a low carbon concentration. A similar course of tensile strength, depending on the sintering temperature, has hardness, Fig. 6.40.

Figure 6.41 shows the microhardness of microstructural components in dependence on sintering temperature. According to the course of the dependences, the microstructure of steel K3 is ferritic. The addition of boron caused a significant increase in the microhardness of the structural components of all three steels, mostly for steel K2, *i.e.* containing 3% Cr, which was composed mainly of lower bainite [133]. The microstructures of these steels with the addition of 0.2% B did not contain eutectics. The test results of the effect of the addition of 0.4% B on the properties of the microstructure of prealloyed Fe–0.7Mn–1Cr–0.2Mo steel after sintering in an atmosphere containing nitrogen ($75N_2/25H_2$, dew point of –34°C) show that sintering in the atmosphere containing a high proportion of nitrogen can lead to a large increase in the density of the material system and strengthen the matrix also without formation of the eutectic compared with sintering in atmospheres with a low nitrogen content [135,137,138].

6.3.5. Manganese-assisted pack boriding of sintered steels

Boriding is a chemical-thermal process for creating hard layers with a low friction coefficient on the surface of parts produced from ferrous materials with full density and sintered steels. Various boriding processes, boriding agents and the results with the characteristics of the hard layers were described in, for example, [139–142].

As will be seen below, boriding of the sintered steels was in this case studied in two processes, namely two-stage and single-stage, without the use of halides as activators because the halides are always linked with environmental and corrosion problems. The single-stage process was realised taking into account the results achieved in the two-stage process.

6.3.6. Sintering and pack boriding – two-stage process

Two-stage means that sintering occurred at higher temperature and pack boriding at lower temperature. Pack boriding, comparable to pack cementation, is the most widely used boriding process for wrought ferrous and nonferrous parts because of its relative easy handling, safety and the potential to change the composition of the powder mix compared to the other boriding processes. The process involves packing of the annealed, cleaned, smooth parts into a boriding powder mix in a steel container, placing it into a furnace, and heating up to the process temperature.

Boride layer thickness from ~0.025 to 0.125 mm is specified for abrasive wear applications, and layers of the thickness less than ~0.025 mm are specified for adhesive wear. The disadvantages of boriding treatments of wrought steel parts are that the processes are inflexible and labor-intensive, making the process less cost effective [141]. They cause enviromental problems due to the use of halides as activators and, therefore, these processes are not used in PM structural parts production.

The results attained at boriding of steel specimens sintered at 1120°C followed by pack boriding at 1050°C will now be discussed. The investigation of such a boriding process for PM parts as was conducted is undoubtedly better for the furnace and also for the powder parts. From this reason, boriding was carried using a powder mixture consisting of ferroboron, high-carbon ferromanganese as an activator, and alumina as a diluent preventig caking of the boriding agent. This process was investigated with the presumption of its possible application for PM parts.

The following section gives the data also on boriding of the steels with no manganese because they provide information for other alloyed systems, especially those alloyed with Cr which are the subject of special attention for application in PM production. For the investigation of this process, the standard tensile test bars 5 mm thick were prepared from the following powders:

- Fe – (water-atomised iron powder W–V–16) coded here A,
- Fe–2% Ni–0.5% Mo (prealloyed Astaloy A powder) coded here B,
- Fe–1.75% Ni–1.5% Cu–0.5% Mo (diffusion prealloyed powder Ultrapac A), coded here C.

The powder mixtures with the addition of 0.3% graphite and 0.5% zinc stearate as a lubricant were compacted at 600 MPa and sintered for 1 h at 1120°C to prepare the experimental samples. The sintered samples were packed into the boriding mix in a steel box, covered with a steel plate and treated in flowing cracked ammonia (dew point –30°C, 12 l min^{-1}) for 2 h at 1050°C. Both sintering and boriding processes were carried out in a laboratory retort furnace.

Some of the specimens were austenitized for 0.5 h at 850°C in nitrogen, quenched in oil, and tempered in air for 0.5 h at 200°C. The specimens were tested as sintered, as sintered and heat-treated, as pack-borided, and as pack-borided and heat-treated.

6.3.6.1. Properties of two-stage borided steels

Density. Densities of the green, sintered and borided specimens are shown in Fig. 6.42. The sintered density increased by 1.0 to 2.0% compared to the green density. The boriding caused an additional increase in density by 1.7 to 3.6% compared to the sintered values. This proved that even this relatively low-temperature treatment (1050°C) increased the density indicating an unexpected post-sintering enhancement effect during the boriding process. Higher density levels were obtained with alloyed steels and particularly with the FeNiCuMo steel.

Tensile strength and hardness. The tensile strength values of the steels in all tested states are shown in Fig. 6.43. Boriding of the sintered steels increased the tensile strength from 365 to 640 MPa, *i.e.* tensile strength increased by 50, 53 and 40% for the steels A, B and C, respectively, compared to the sintered ones. The UTS values of the heat-treated borided steels A, B and C, respectively, reached

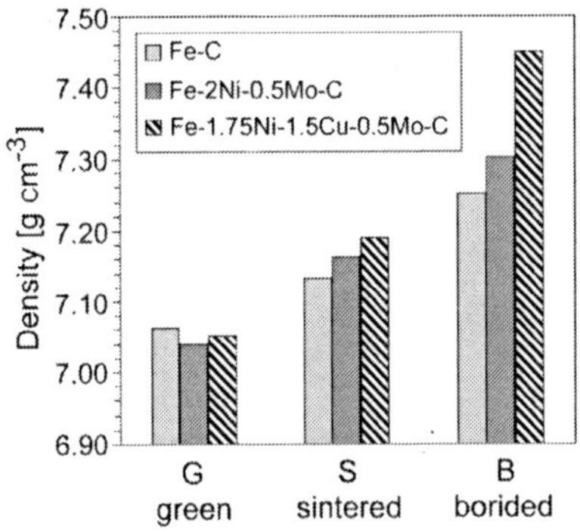

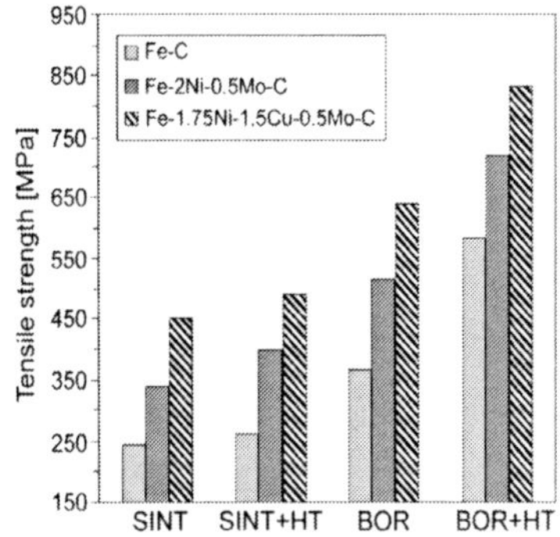

Fig. 6.42 (left). Density of green, sintered, and borided steel specimens Fe–0.3C, Fe–2Ni–0.5Mo–0.3C and of Fe–1.75Ni–1.5Cu–0.5Mo–0.3C; boriding for 2 h at 1050°C [143].
Fig. 6.43 (right). Tensile strength of sintered (SINT), sintered and heat-treated (SINT+HT), borided (BOR), and borided and heat-treated (BOR+HT) steels. A (0.12% C, 0.81% Mn), B (0.10% C, 0.84% Mn) and C (0.14% C, 1.12% Mn) [143,144].

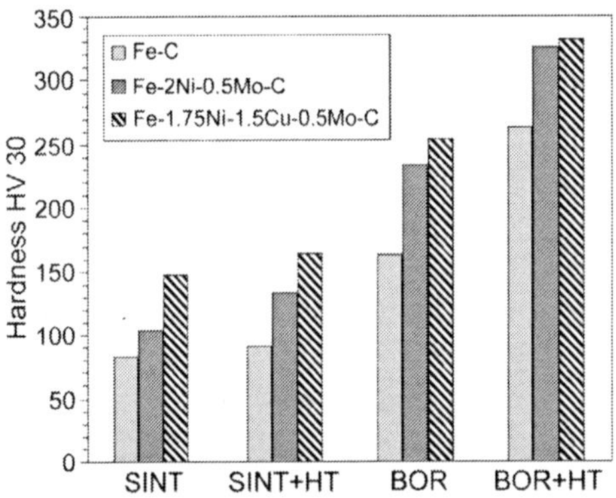

Fig. 6.44. Hardness HV 30 of the steels as in Fig. 6.43 [144].

582 to 835 MPa, *i.e.* increase of 124, 81 and 70% compared to the non-borided steels.

The sintered carbon in the specimens ranged from 0.1 to 0.14%. During boriding, the manganese content in the borided specimens increased to 0.81–1.12% by diffusion from the boriding mixture in the direction to the specimen interior.

The corresponding hardness values of the investigated steels are shown in Fig. 6.44. The hardness of the borided specimens increased by 96, 125 and 72% for the steels A, B and C, respectively, compared to the sintered ones. The hardness of the sintered specimens, similarly as tensile strength, was not affected by heat treatment. With the same carbon content, the heat treatment of the borided specimens increased

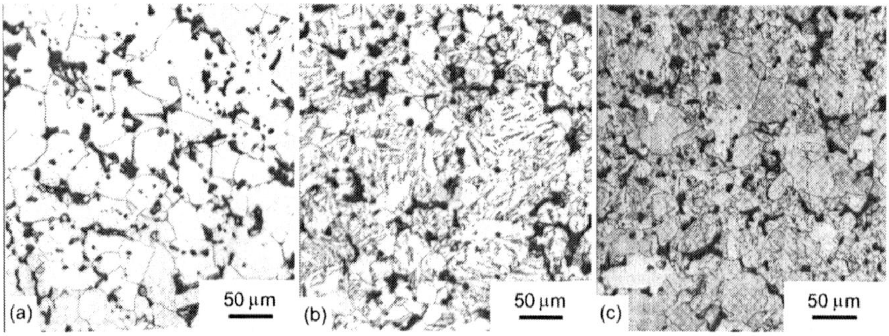

Fig. 6.45. Microstructures of steels sintered in cracked ammonia for 1 h at 1120°C. (a) Fe–0.3C (A), (b) Fe–2Ni–0.5Mo–0.3C (B), (c) Fe–1.75Ni–1.5Cu–0.5Mo–0.3C (C). Nital etched [143].

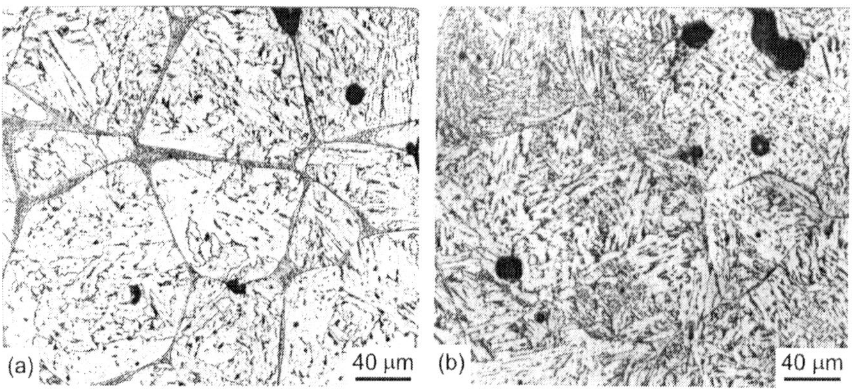

Fig. 6.46. Core microstructure of the steels: (a) Fe–2Ni–0.5Mo–0.3C (B), (b) Fe–1.75Ni–1.5Cu–0.5Mo–0.3C (C) pack borided 2 h at 1050°C. Nital etched [144].

their hardness by 190, 145 and 103% when compared to the borided ones for the A, B and C steels, respectively.

Microstructure. Figure 6.45 shows the microstructures of the sintered A, B and C steels, respectively, formed by ferrite, ferrite-pearlite and ferrite–coarse bainite. The mean microhardness values of these steels were 122, 151 and 138 HV 0.05. The microstructures of the borided specimens are shown in Fig. 6.46.

A white boride surface layer was formed on all specimens. The thickness of the white layer depended on the composition of the steel. The eutectic-like surface layer formed on steel C was the thickest, as shown in Fig. 6.47. The formation of such a layer on the FeNiCuMoC steel sample at 1050°C resulted from the presence of copper in the steel which gives a low-melting eutectic with boron at 1013°C [145].

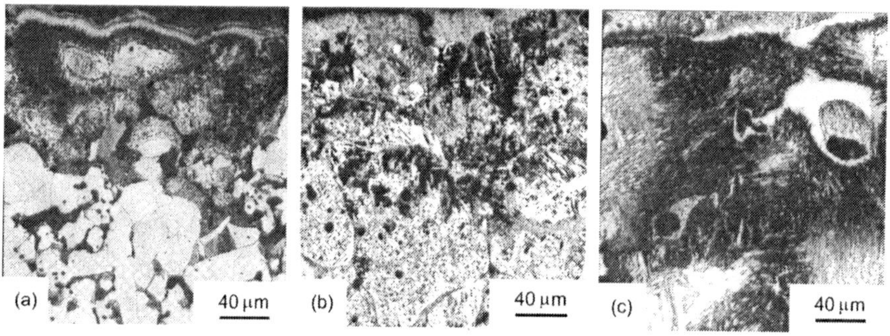

Fig. 6.47. Microstructures of surface boride layers with subsurface zone formed in specimens borided for 2 h at 1050°C. (a) Fe–0.3C, (b) Fe–2Ni–0.5Mo–0.3C (B), (c) Fe–1.75Ni–1.5Cu–0.5Mo–0.3C (C). Nital etched [143].

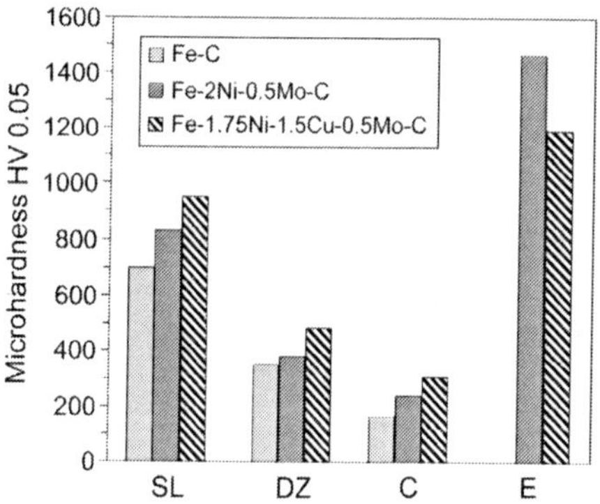

Fig. 6.48. Microhardness (HV 0.05) of borided steels. Sl – surface layer, DZ – diffusion zone, C – core, E – eutectics [143].

Microhardness. Boriding caused increased the microhardness of the microstructure constituents in the specimen cros-section, contrary to that of the wrougth steel specimen on which only a surface boride layer formed. The microhardness of the surface layer, diffusion zone, core grains and of the eutectics is given in Fig. 6.48.

The microhardness as a mean value of 20 measurements in the surface layer was in the range of 699 to 963 HV 0.05, depending on the chemical composition. The microhardness of the white surface layer in alloy C was the highest one.

The structure of the surface layer was eutectic-like. Figure 6.49 shows the microstructures of the surface layer, partly with the diffusion zone and the microstructure in the core in the borided state for the FeNiCuMoC sample and after sintering only. The difference in the character of the sintered and borided specimen core microstructures is shown in Figs. 6.49(c),(d). The changes in Mn, Ni and Cu contents, measured in 6 points, from the surface

Table 6.15. Microanalysis of pack-borided FeNiCuMoC steel sample conducted at 6 points shown in Fig. 6.49. EDX analysis, SEM [145,146] El.– element

El.	No. of analysed point					
	*1	*2	**3	4	5	+6
	Content mass [%]					
Mn	16.13	17.75	11.58	4.65	0.00	0.00
Ni	1.82	0.53	1.55	1.70	0.00	0.00
Cu	2.31	1.59	3.19	2.10	1.04	1.26

*Boride layer, ** white lyer, +core of the specimen

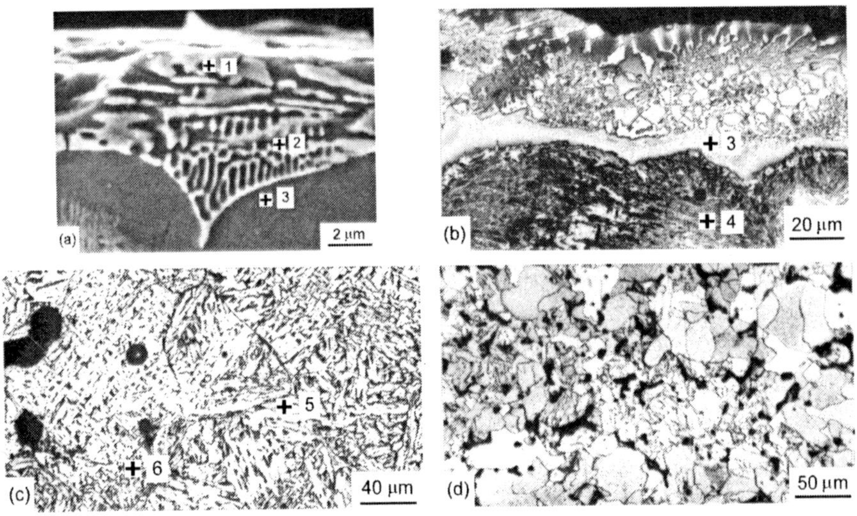

Fig. 6.49. Microstructure of FeNiCuMoC (C) steel specimen pack-borided in the mixture containing Mn. (a), (b) – surface complex borided layer, (c) – microstructure of the borided core specimen, (d) – sintered microstructure. Nital etched [145,146].

layer up to the core of the specimen caused by boriding are given in Table 6.15. The high manganese content in the surface layer (16.13%) was recorded with decreasing tendency of Mn concentration to the core (0%). In the case of Ni and Cu there was unexpected reverse diffusion from the core into the surface layer documented by their contents measured at points 5 and 6. The lowest Cu content was recorded in the eutectic-like subsurface layer with the lowest microhardness of ~400 HV 0.05 (point 3, 3.18% Cu).

Sinter boriding – single-stage process. This process was performed in two stages differing in the composition of the boriding mixtures. Manganese as an activator of the process was used in the first stage, and in the second stage the boriding mixture consisted only of boron and alumina. The single-stage pack boriding process was applied only during sintering. The samples were prepared from the powders and then processed as in the previous case for A, B and C steels.

The green compacts were packed with the boriding mixture in a steel box, covered with a steel plate. Sinter boriding was carried out in a laboratory pusher furnace for 1 h at 1120°C in cracked ammonia (dew point –30°C). As a reference, green compacts were sintered without boriding.

Sinter boriding with assistance of manganese. Boriding powder mix consisted as in previous cases of ferroboron, ferromanganese as carrier of manganese and alumina as diluent.

Density values of the processed specimens are given in Table 6.16 [149,150]. Pack sinter boriding caused a marked increase of density of the tested steels in comparison with the green and sintered densities. The alloyed steels, especially the copper-alloyed steel C, showed greater densification after sinter boriding than did the Fe–C steel.

The microstructure of sintered specimens was ferritic–pearlitic depending on the base composition of the matrix. Sinter boriding

Table 6.16. Density and dimensional changes of samples A, B and C. Sinter pack-borided for 1 h at 1120°C with assistance of manganese. 600 MPa, cracked ammonia [148,149]

Steel	Density [g cm^{-3}]			Shrinkage length [%] +
	Green	Sintered	Sint. bor.	
Fe–0.3C (A)	7.00	7.03	7.13	0.33
Fe–2Ni– 0.5Mo–0.3C (B)	6.91	7.01	7.16	0.63
Fe–1.75Ni–1.5Cu–0.5Mo–0.3C (C)	6.96	7.04	7.21	0.45

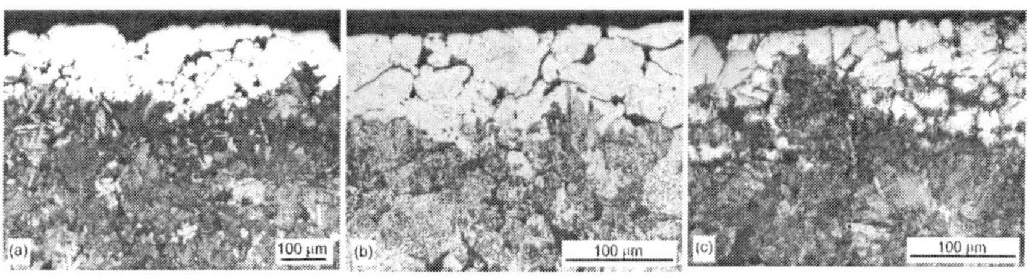

Fig. 6.50. Boride layers on PM steel samples pack-sinter borided 1 h at 1120°C in cracked ammonia. Steel: (a) – A, (b) – B, (c) – C as in Table 6.16. Nital etched [149].

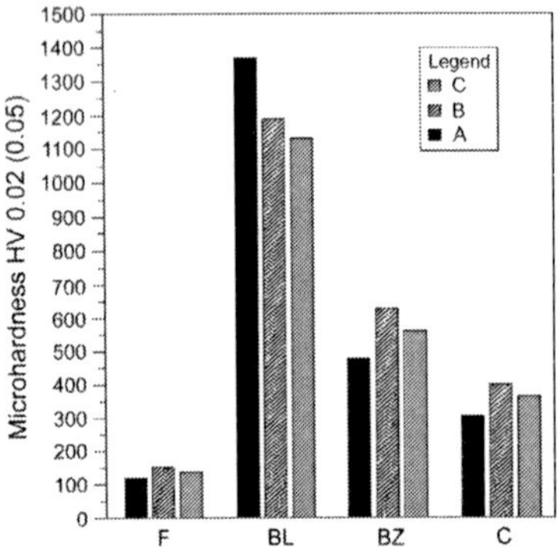

Fig. 6.51. Microhardness of ferrite in sintered reference samples (F) microstructure and of PM steels sinter pack-borided 1 h at 1120°C. BL – boride surface layer, BZ – diffusion zone, C – core grains. Legend: A, B, C as in Table 6.16, 15–60 parallel indents for each value [149,150].

resulted in formation of white surface boride layers on all steel samples, as shown in Fig. 6.50.

The microhardness values of ferrite in sintered microstructure, boride layers and in the core regions are shown in Fig. 6.51. The highest microhardness of 1370 HV 0.05 was measured on the finest boride layer in material A. There was also a marked increase of the core microhardness compared to the sintered state. This is probably due to modest alloying of the matrix with manganese and boron of the alloyed steels compared with the Fe–C steel.

The results indicate that the sinter boriding process with assistance of manganese performed during sintering can be realised only without halides. The process resulted in formation of hard layers with a thickness of 40 to 180 μm on the surface of the tested Fe–C and Fe–Ni–Mo–Cu steels with the microhardness of 1100 to 1370 HV 0.05.

Sinter boriding without assistance of manganese. These experiments were conducted on a boriding mixture consisting of ferroboron and alumina with the aim to verify if the use of manganese as a boriding activator is necessary. The first experimental series was carried out as a single-stage process – sinter boriding for 1 h at 1120°C in cracked ammonia. The second series was used for comparison as two-stage boriding – sintering of the samples for 1 h at 1120°C followed by boriding for 1 and 2 h at 1050°C. Both processes used cracked ammonia.

The following powders were used for the preparation of the samples for the first series:

- Fe – (ASC100.29 water-atomised iron powder) coded here A,
- Fe–2% Ni–0.5% Mo (Astaloy A prealloyed powder) coded here B,
- Fe–1.75% Ni–1.5% Cu–0.5% Mo (Distaloy SA, diffusion partly prealloyed powder) coded here C,

all with addition of 0.3% graphite and of 0.8% HW wax as a lubricant.

The cylindrical samples (10 mm diameter, 10 mm long) were compacted at 600 MPa and sinter-borided for 1 h at 1120°C.

The following powders were used for the preparation of the samples for the second series:

–Fe–1% Cr–0.3% Mo–0.3% V (103 V, water-atomised and vacuum annealed powder), coded here K1,

–Fe–3% Cr–0.3% Mo–0.3% V (30CRV, water-atomised and vacuum anneaed powder), coded here K2.

Graphite in the amount of 0.2, 0.5 and 0.9% and 0.8% HW wax as a lubricant were admixed to the powders. The powder mixtures were then compacted at 700 MPa to prepare the impact toughness and bending strength bars.

Sinter boriding – 1th boriding series. Figure 6.52 shows the microstructures of Fe–C specimens sinter-borided in the boriding mix without manganese for 1 h at 1120°C in cracked ammonia. The uniform thickness of the surface boride layer with a sharp transition into the interior of the sample was the characteristic change of the core microstructure in this process, Figs. 6.52(a),(b). The uniform surface boride layers were also formed on the FeNiMoC and FeNiCuMoC steel samples shown in Figs. 6.53 and 6.54. A marked distinction in the character of the boride layers was observed when compared with, for example, Fig. 6.49.

The highest microhardness value was measured in the surface layer on the sample of A steel as shown in Table 6.17.

Two stage boriding – 2nd series of boriding without assistance of manganese. Characteristic microstructures of surface boride layers

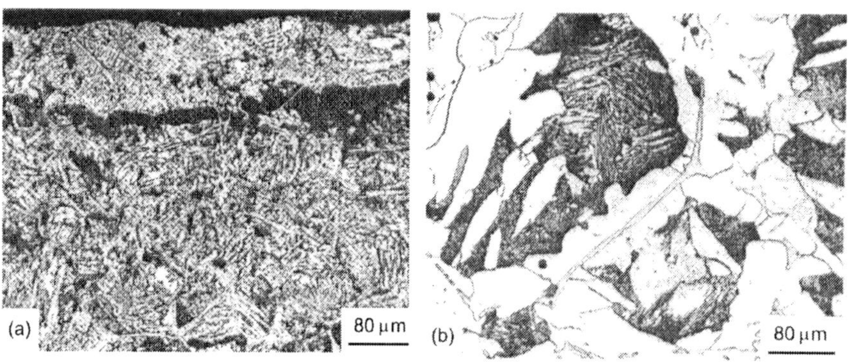

Fig. 6.52. Microstructure of: (a) surface boride layer, (b) core in Fe–0.3C specimen sinter-borided for 1 h at 1120°C in cracked ammonia in a boriding mix without manganese. Nital etched [151].

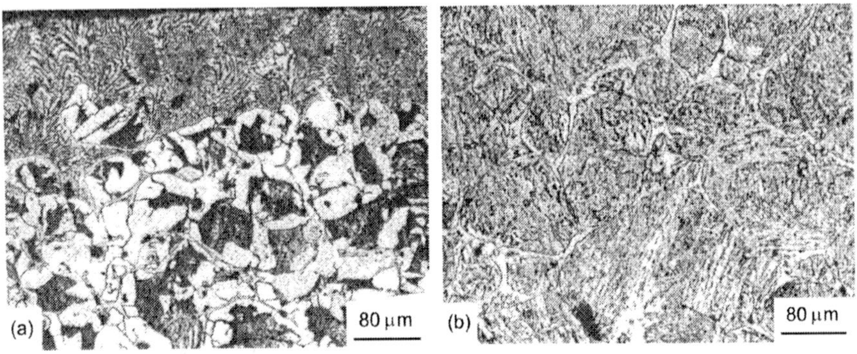

Fig. 6.53. Microstructure of: (a) surface boride layer, (b) core in Fe–2Ni–0.5Mo–0.3C specimen sinter-borided for 1 h at 1120°C in cracked amonia packed in a boriding mix without manganese. Nital etched [151].

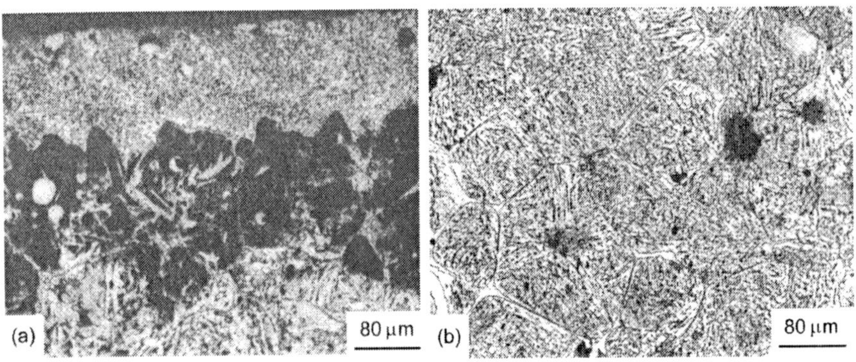

Fig. 6.54. Microstructure of: (a) surface boride layer, (b) core in Fe–1.75Ni–0.1.5Cu–5Mo–0.3C steel specimen sinter-borided for 1 h at 1120C in cracked in a boriding mix without manganese. Nital etched [151].

Table 6.17. Highest microhardness values of ferrite in as sintered steel specimens, of the borided surface, of diffusion layers and of the matrix core grains sinter-borided 1 h at 1120°C in cracked ammonia without manganese in a boriding mix [148,151]

Steel	Microhardness HV 0.05			
	Ferrite – as sintered	Boride layer	Diffusion zone	Core
Fe–0.3C	125	1280	451	305
Fe–2Ni– 0.5Mo–0.3C	168	1185	611	378
Fe–1.75Ni–1.5Cu–0.5Mo–0.3C	154	1124	533	355

Table 6.18. Microhardness HV 0.05 of the surface boride layer and of the core of the K1 and K2 specimens with various graphite addition pack-borided 1 and 2 h at 1050°C. K1 and K2 steel composition see in Fig. 6.55 [151]

Treatment [°C/h]	HV 0.05 value	Steel					
		K1 + 0.5C		K2 + 0.2C		K2 + 0.9C	
		Layer	Core	Layer	Core	Layer	Core
1050/1	Mean	1588	459	1719	320	–	–
	Range	1329–2125	276–534	1426–2148	188–441	–	–
1050/2	Mean	–	–	–	–	1772	1910
	Range	–	–	–	–	1586–2191	1509–2274

on K1 and K2 steels, both with addition of 0.2 and 0.5% graphite borided 1 h at 1050° C are shown in Fig.6.55.

The white layers were formed in the surface material area of the specimens. The change in the feature of the surface boride layer and of the subsurface zone in comparison with Figs. 6.52–6.54 was affected by the difference in the base chemical composition of the tested steels.

Microhardness values of the surface boride layers and of the core of the specimens K1 and K2 with various graphite additions after two-stage boriding without assistance of manganese are listed in Table 6.18.

The effect of the base composition of the steels, alloyed with carbide- and nitride-forming elements, such Cr, Mo and V, was manifested in their relatively high microhardness values of the surface boride layers and of the specimen cores. Higher graphite additions resulted in higher microhardness values of the boride layer as well of the core specimens.

The thermodynamics of the boriding of the tested steels without manganese addition was more beneficial for the formation of an uniform surface boride layers on the specimens, compared with manganese-assisted boriding. The results proved that the boriding process proceeded in this case in the presence of a gas phase facilitating the transport of boron from the solid powder mix on the metal surface and then into the core of the sample. It follows from it, that activator in this case could be only hydrogen and nitrogen in the atmosphere used. It is a unique situation for powder metallurgy.

From the thermodynamic point of view the investigated model system contains 3 phases:

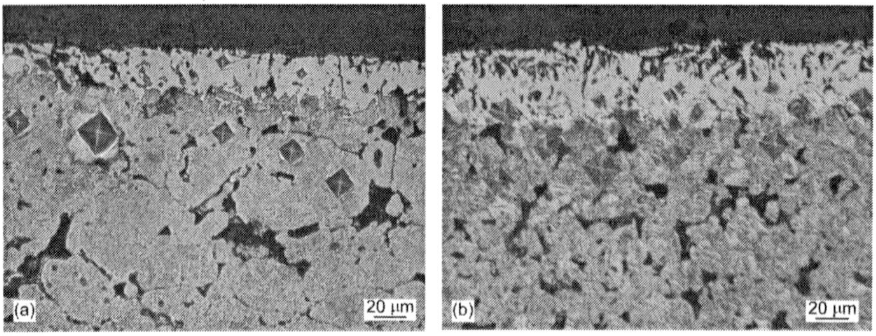

Fig. 6.55. Microstructure of surface boride layers formed at boriding 1 h at 1050°C on specimens: (a) K1 (Fe–1Cr–0.3Mo–0.3V–0.5C), (b) K2 (Fe–3Cr–0.3Mo–0.3V–0.2C) steel. Etching in (H_2O, HF, HNO_3) solution [151].

1. gas phase – 'sintering' atmosphere and all potential gas products originating during the process,

2. solid phase – base metal phase and all products originating during the process which become a part of the base metal phase,

3. solid phase – powder boriding mix, which is the source for B (and Mn when Mn is used as an activator).

The following results can be concluded from the thermodynamic (TD) analysis of the Fe–B–H–N–(Mn–O) system, considering the mentioned phases:

–transfer of boron from the powder mix containing ferroboron as a boron carrier in the TD model process occurs through the compound $B_3H_3N_{3(g)}$,

–presence of oxygen in the system does not affect the composition of the metal phase, but can affect the transfer mechanism of boron from the boriding mix into the metal phase,

–transfer of manganese from the powder mix, if used, is from the thermodynamic point of view an independent process and is realised through the manganese vapour causing only alloying of the matrix including the surface layer with decreasing tendency in concentration from the surface layer to the interior of the sample,

–oxygen in the system from the point of view of the manganese transfer forms MnO in the metal phase if the gas phase contains >0.7% O_2.

These experiments carried out and the TD analysis proved that the manganese in the boriding powder mix does not fulfil the role of an activator.

Only hydrogen and nitrogen in the atmosphere can be used as an activator in the boriding process and the effect of their mutual portion on the mechanism and kinetics of the reactions can affect the properties of borided materials. These results, relating to the transfer of boron through a gas activator ($BHN_{(g)}$) in the pack boriding process are particularized that the boron can be transferred through the compound $BN_{(g)}$ [151,152].

The conclusion considering the transfer of B through the compound $B_3H_3N_{3(g)}$ is universally valid not only for PM materials but also for wrought steels. The H–N containing atmospheres as the base sintering atmospheres in powder metallurgy, facilitate the further development and possible realisation of this process in production of PM parts [151].

7

Properties of sintered and powder forged steels based on prealloyed powders

Using manganese as alloy element in sintered steels, also in combination with Cr, is an attractive alternative to the more common but expensive alloy additives, especially Ni, Cu and Mo, also in high concentrations. The main peculiarities of Mn that should be considered is its high affinity for oxygen which is slightly higher than that of Cr – and much higher than that of Cu, Ni and Mo. Chapters 2 and 3 refuted in detail concerns regarding the oxidation of manganese in the samples during sintering in the atmosphere which does not meet thermodynamic requirements on the purity of the sintering atmosphere. The high vapour pressure of manganese and hence its sublimation in the form of vapour ensure successful alloying and sintering of parts from powders containing manganese. Of the prealloyed powders, discussed in this chapter, at least two of them contain manganese in an amount of ~0.7%. Thus, we can compare the properties of steels based on prealloyed Cr–Mo and Ni–Cu–Mo powders with those that also contain manganese, taking into account the variations in chemical composition and purity and processing conditions.

7.1. Properties of sintered and powder forged steels based on prealloyed Fe–Cr–Mn–Mo–(V) powders with high oxygen content

At the time of research, these sintered steels, based on prealloyed powders alloyed with chromium and manganese and some with vanadium, *i.e.* three elements with a high affinity for oxygen, could not be obtained from producers of iron powders. Vanadium has a high affinity for nitrogen and with carbon forms carbides V_4C_3 and nitrides with nitrogen. A small amount of vanadium in the steel softens the grains, pearlite and martensite. Vanadium has the strongest effect by increasing the ratio of yield to tensile strength with a slight reduction in toughness. The manganese content in conventional mild steel is in the range 0 to 0.25% and 0.15 to 0.25% V. The advantage of chromium prealloyed powders being the main alloying element is its low hardening effect on ferrite without carbon allowing good compressibility of such powders.

Therefore, for this research it was necessary to ordered powders as specified by investigators which could be obtained at that time only in the condition after dispersion without subsequent annealing, which are specified below. Also, at that time only one box with charcoal was placed in industrial furnaces to prevent decarburisation of parts for a certain number of boxes of parts. When sintering compacts, with the properties and sintering conditions presented below, no measures were taken against decarburisation of the compacts during sintering in the laboratory pusher furnace.

Investigations were therefore conducted on the water-atomised prealloyed following powders without preliminary annealing with the following chemical composition:

A – 0.06% C, 0.35% Mn, 2.95% Cr, 0.35% Mo, 0.2%, 1% O, 0.35% acid insoluble matter,

B – 0.01% C, 0.30% Mn, 1.1% Cr, 0.28% Mo, 0.25% V 0.95% O, 2.9% acid insoluble matter,

C – 0.01% C, 0.7% Mn, 1.1% Cr, 0.3% Mo, 0.9% O, 2.6% acid insoluble matter.

The high content of oxygen and acid insoluble matter was characteristic of the powders in the as supplied state.

To prepare the powder samples with the particle size <0.25 mm, graphite was added in an amount of 0.4% (index 1) and 0.8% (index 2) and 0.5% zinc stearate as a lubricant. The powder mixtures were pressed to standard tensile strength bars with the density of 6.0, 6.5 and 6.8 g cm^{-3} and prismatic rods with the density of 6.5 g cm^{-3} for forging.

All compacts were sintered for 30 min at 1200°C in cracked ammonia in the laboratory pusher furnace. Samples were forged to bars measuring 10 × 10 × 80 mm after heating for 5 min at 1050°C in an atmosphere $Ar/5H_2$. From the forging temperatures the bars were cooled in air without subsequent tempering. The forged bars were used to manufacture, by turning, the threaded rods 6 mm in diameter for tensile tests.

7.1.1. Mechanical properties

The density and mechanical properties of the studied Fe–Cr–Mn–Mo–V (A) steel with the addition of graphite (A1, A2) in the sintered state are shown in Table 7.1.

The mechanical properties are affected by density. Higher values of yield and tensile strength were recorded for steel A1 with a density of 6.85 g cm^{-3} with the addition of 0.4% graphite compared to steel A2 with a density of 6.82 g cm^{-3} with the addition of 0.8% graphite. According to fracture analysis, these differences are caused by a greater fraction of intergranular facets on the A2 steel fracture surface as compared with steel A1.

The chemical composition of the samples after forging from 1050°C, which corresponds to the sintering conditions, is presented in Table 7.2.

These powders are characterised by a high content of AIM (acid insoluble matter) and oxygen and different contents of manganese and chromium. High decarburisation of the sintered samples was due to sintering conditions, which did not prevent their decarburisation. Edges of the samples were ferritic. At sintering in cracked ammonia the carbon in the compacts also contributed to the reduction of surface oxides on the powder particles. A negative aspect is the high oxygen content and also the high content of acid insoluble matter in this state. Given that the powders B and C contain the same content of manganese, the high content of acid insoluble matter can not be

Table 7.1. Density and mechanical properties of Fe–Cr–Mn–Mo–V steel in the sintered state with graphite addition. A1 – 0.4%, A2 – 0.8%, 30 min, 1200°C, cracked ammonia [153]

Steel	Density [g cm^{-3}]	Rp0.2 [MPa]	Rm [MPa]	A [%]	Hardness HV 30
A1	6.04	212	221	0.20	110
	6.54	380	386	0.24	138
	6.85	526	533	0.32	142
A2	6.05	169	173	0.10	123
	6.56	388	395	0.20	146
	6.82	447	454	0.00	178

Table 7.2. Concentration of carbon, oxygen, nitrogen and of acid insoluble matter (AIM) in Fe–Mn–Cr–Mo–(V) steels in the forged state. Sintering at 1200°C for 30 min in cracked ammonia [153]

Steel	Concentration [mass %]				Graphite [mass %]
	C	O_2	N_2	AIM	
A1	0.06	0.33	0.029	0.20	0.4
A2	0.19	0.29	0.012	0.15	0.8
B1	0.07	0.19	0.022	2.03	0.4
B2	0.32	0.12	0.011	1.28	0.8
C1	0.05	0.18	0.014	1.95	0.4
C2	0.22	0.15	0.012	1.06	0.8

Remark: AIM – acid insoluble matter. Powders: A – Fe–0.35Mn–2.95Cr–0.35Mo–0.2V–1.0O, 3% acid insoluble matter, B – Fe–0.30Mn–1.1Cr–0.2Mo–0.2V–0.9O, 9% acid insoluble matter, C – Fe–0.7Mn–1.1Cr–0.3Mo–0.9O, 2.6% acid insoluble matter.

Table 7.3. Density and mechanical properties of Fe–Mn–Cr–Mo–(V) steels in the forged state [153]

Steel	Density [$g\ cm^{-3}$]	Rp0.2 [MPa]	Rm [MPa]	A [%]	Hardness HV 30
A1	7.62	595	612	0.0	252
A2	7.60	821	640	0.7	261
B1	7.66	392	443	12.5	214
B2	7.68	555	729	6.1	218
C1	7.64	303	446	12.5	177
C2	7.68	445	646	8.5	200

attributed to the impact of manganese, whose concentration in the powder C was higher.

The density and mechanical properties of Fe–Mn–Cr–Mo–(V) steels in the forged state are listed in Table 7.3.

Hardmetal indexable inserts were used for the production of tensile bars from the forgings because of their high hardness. Higher values of tensile strength were recorded for the steel with the addition of 0.8% graphite compared with samples with the initial addition of 0.4% graphite. On the basis of these mechanical properties it is not possible specifically to evaluate the effect of vanadium in the steels B and C. All values of the mechanical properties showed a negative effect of high concentrations of oxygen and impurities (AIM), as evidenced by their values in Table 7.2. These results showed that prealloyed powders, alloyed with chromium and manganese in particular, require a significantly higher chemical purity than those that were used in this work.

7.1.2. Properties of powder forged steels based on industrially produced chromium- and manganese-prealloyed powders of high oxygen content

Investigations were carried out on Fe–1Cr–0.7Mn–0.3Mo water-atomised prealloyed powders from three manufacturers, with the compositions as supplied presented in Table 7.4.

The powders coded A, C, D and E were produced according to the specification of the authors of this work. The oxygen content of all powders was in the range 0.41–1.03%. For this reason the powders were reduction annealed prior to further application.

Table 7.4. Initial chemical composition of water-atomised Fe–Cr–Mn prealloyed powders as supplied by three producers [154,155]

Alloy element	C	Mn	Cr	Mo	Si	P	S	O
Code	Chemical composition [mass %]							
A	0.15	0.70	1.00	0.31	0.02	0.013	0.01	0.52
B	0.12	0.84	0.88	–	0.03	0.011	0.01	0.41
C	0.03	0.70	1.03	0.53	0.03	0.020	0.02	1.03
*D	0.03	0.70	0.98	0.35	0.08	0.020	0.02	0.78
*E	0.08	0.20	2.90	0.30	0.05	0.020	0.02	0.60

Remark: Powders *D and *E with 0.25% V. The powders B and D from one producer and C and E from another producer.

7.1.2.1. Reduction annealing in gas atmosphere

To reduce the oxygen concentration, the powders were reduction annealed at 900–150°C for 45 min: a) in cracked ammonia (dew point –48°C) and b) in hydrogen + 2% natural gas (dew point –48°C). By annealing at 1150°C the oxygen concentration decreased to 0.16%, but the powder was cacked. Annealing of the powder at 1100°C in hydrogen with the addition of natural gas caused the decrease of the oxygen content up to ~0.12%. The microstructure of the sample (10 mm diameter, 10 mm height), compacted at 600 MPa from the powder annealed under b) conditions and sintered at 1300°C for 30 min in cracked ammonia was without inclusions (unetched).

7.1.2.2. High temperature sintering

The samples (10 mm diameter, 10 mm height) from the powder A (0.52% O as supplied) with addition of graphite in the amount of 0.2, 0.4, 1.0 and 1.4% were compacted at 200 and 600 MPa and sintered for 30 min at 1200, 1250 and 1300°C in cracked ammonia under getter (alumina + 3% graphite). The carbon and oxygen contents obtained in the as sintered samples are listed in Table 7.5.

Table 7.5. As sintered carbon (C_c) and oxygen (O) content [mass %] of sintered Fe–0.7Mn–1Cr–0.3Mo (0.52% O) steel based on prealloyed powder in dependence on graphite addition, compacting pressure and sintering temperature [154,155]

Graphite [%]	Compacting pressure [MPa]	Sintering [°C/min]					
		1200/30		1250/30		1300/30	
		O	C	O	C	O	C
0.2	200	0.20	0.18	0.18	0.12	0.17	0.08
	600	0.20	0.18	0.12	0.10	0.16	0.10
0.6	200	0.20	0.48	0.17	0.32	0.07	0.24
	600	0.20	0.52	0.13	0.52	0.08	0.41
1.0	200	0.10	0.78	0.11	0.72	0.08	0.51
	600	0.11	0.80	0.08	0.80	0.06	0.65
1.4	200	0.09	0.91	0.09	0.72	0.06	0.80
	600	0.11	0.90	0.07	0.08	0.07	0.78

The increasing sintering temperature resulted in the decrease of the oxygen and carbon contents in the samples. The relatively lowest decrease in the carbon content was observed after sintering at 1200°C at a constant oxygen content of 0.2% and 0.6% of graphite addition. The 0.6% graphite addition caused the decrease of oxygen content in the samples from 0.52% in the starting powder to 0.06% after sintering at 1300°C. The 1.4% graphite addition was not more effective in comparison with 1.0% graphite. The effect of compacting pressure – porosity – in the decrease of the carbon content compared with the oxygen content was higher. To achieve a low oxygen content in the specimens based on the powder with a high starting oxygen content high sintering temperature is required and the graphite addition should be at least ~0.6%.

Table 7.6. As supplied chemical composition of K1, K2 and K3 prealloyed powders [mass %]

Powder grade	Base elements						
	C	Cr	Mn	Mo	V	N	O
	Chemical composition [mas %]						
K1 (103V)	0.01	1.07	0.08	0.32	0.29	–	0.15
K2 (30CRV)	0.01	3.02	0.09	0.29	0.23	0.002	0.24
K3 (4100V)	0.01	1.13	0.62	0.27	–	0.001	0.19

Remark: Concentration of N and O determined later by analysis.

7.2. Properties of powder forged steels based on commercial prealloyed Fe–Cr–Mn–Mo–V powders

This part of the study describes the properties of steels prepared on the basis of the following Fe–Cr–Mn–Mo–V prealloyed powders, supplied with a relatively low oxygen content:

Fe–1Cr–0.3Mo–0.3V (KIP 103V) – coded here K1,
Fe–3Cr–0.3Mo–0.3V (KIP 30CRV) – coded here K2,
Fe–1Cr–0.7Mn–0.3Mo (KIP 4100) – coded here K3.

The chemical composition of these prealloyed powders as supplied after vacuum annealing is presented in Table 7.6.

The powders were admixed with 0.2, 0.5 and 0.9% of graphite and with 0.7% of zinc stearate as a lubricant, respectively.

Tensile strength and impact energy bars (10 × 10 × 55 mm) were compacted at 690 MPa and sintered for 60 min at 1120°C in a retort furnace in cracked ammonia in boxes under getter (alumina + 5% graphite).

7.2.1. Density and mechanical properties

7.2.1.1. Density

The green and sintered density of the compacts in dependence on graphite addition is shown in Fig. 7.1. The green density of K1, K2 and K3 compacts decreased with increasing graphite addition with small differences in the decreasing values with the increase of the graphite addition in the range of ~6.9 to 7.0 g cm^{-3}. With the increase of the graphite addition, the as sintered density was also influenced by the composition of the powders. The highest values of the as sintered density were recorded for steel K3 with 0.62% Mn and no vanadium, *i.e.* from 7.12 to 7.02 g cm^{-3}. The lowest density was obtained for the samples prepared on the basis of the K2 powder.

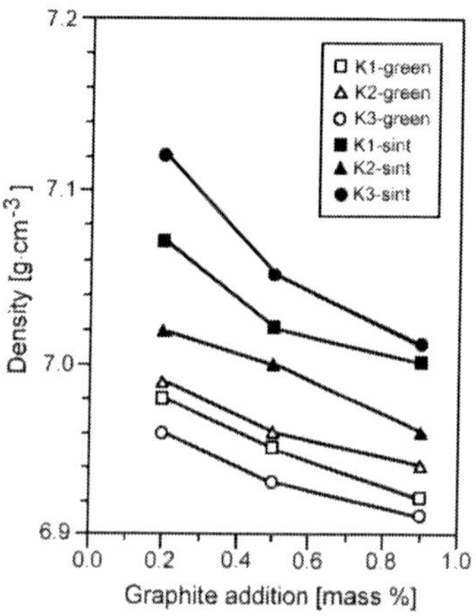

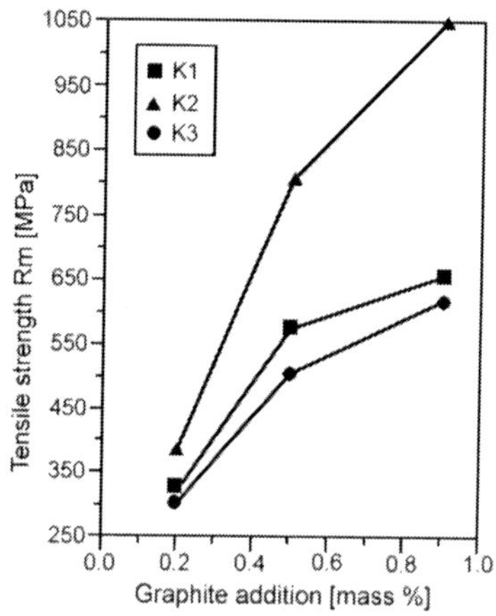

Fig. 7.1 (left). Dependence of green and as sintered density of compacts based on K1, K2 and K3 powders on graphite addition; 690 MPa, 60 min, 1120°C, cracked ammonia [156–158].

Fig. 7.2 (right). Dependence of tensile strength of sintered compacts based on K1, K2, and K3 powders on graphite addition; 690 MPa, 60 min, 1120°C, cracked ammonia [156–158].

7.2.1.2. Mechanical properties

The concentrations of the combined carbon in the samples of the examined sintered steels are given in Table 7.7.

The relationship between the strength of the investigated steels and the addition of graphite shows Fig. 7.2. These combined carbon concentrations of the studied steels suggest that sintering under a getter (alumina graphite + 5% graphite) did not prevent decarburisation of the samples, which was also reflected in their lower mechanical properties. The highest values of ultimate tensile strength, 1046 MPa, was recorded for K2 steel with 3% of chromium

Table 7.7. Combined carbon concentrations of K1, K2 and K3 steels in relation to the amount of the graphite addition. 690 MPa, 60 min, 1120°C, cracked ammonia

Steel	Graphite addition [mass %]		
	0.2	0.5	0.9
	Concentration of as sintered carbon in the compacts [mass %]		
K1 (103V) – 1Cr	0.19	0.32	0.60
K2 (30CRV) – 3Cr	0.15	0.28	0.58
K3 (4100V) – 1Cr–0.7Mn	0.14	0.30	0.59

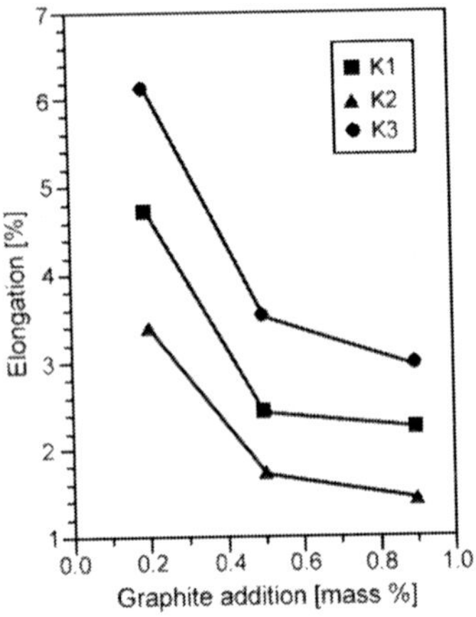

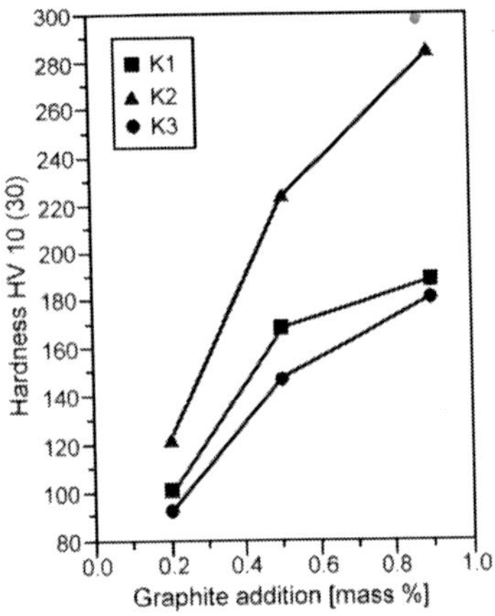

Fig. 7.3 (left). Dependence of K1, K2 and K3 steel elongation on graphite addition. 690 MPa, 60 min, 1120°C, cracked ammonia [156–158].
Fig. 7.4 (right). Dependence of hardness of K1, K2 and K3 steels on graphite addition. 690 MPa, 60 min, 1120°C, cracked ammonia [156–158].

and 3% and 0.58% of carbon, and 804 MPa for a carbon content of 0.28%. The K1 and K3 steels with the same carbon tensile strength reached 600 MPa and 640 MPa and 500 MPa and 555 MPa, respectively. In this case, the lowest tensile strength value was shown by the K3 steel with 0.62% Mn.

The dependence of elongation of these steels on graphite addition is shown in Fig. 7.3. The highest ductility was obtained for the steel K3, in the range 6.1–2.9%, and the lowest for the steel K2 in the range 3.2–1.4% with a declining trend with increasing addition of graphite.

The dependence of hardness on graphite addition is shown in Fig. 7.4. The hardness values, depending on the addition of graphite, exhibit the same course as the tensile strength. The K1 and K2 steels attained the hardness values in the range of 92 to 188 HV 10 and the K2 steel of 122 to 285 HV 30.

The dependence of impact energy on the addition of graphite to the studied steels, determined on unnotched specimens, is shown in Fig. 7.5. The form of these curves is approximately the same as of the elongation curves. The highest values of impact energy were obtained for K3 steel, in the range 42.5 to 25 J, and the lowest for steel K2 in the range 25 to 20 J, depending on the increasing addition of carbon.

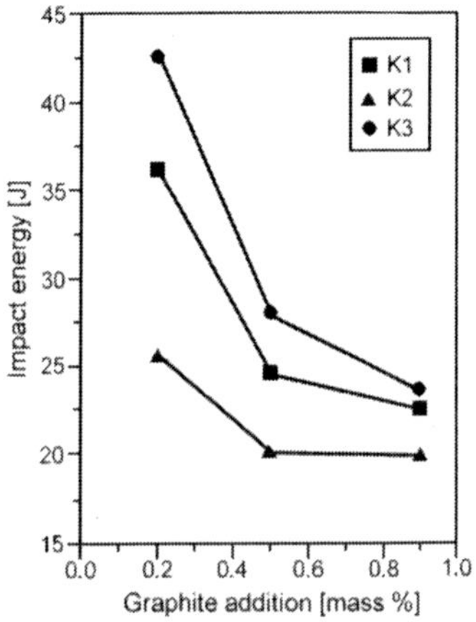

Fig. 7.5. Dependence of impact energy (unnotched samples) values of the K1, K2 and K3 steels on graphite addition. 690 MPa, 60 min, 1120°C, cracked ammonia [156–158].

The results show that all three investigated steels were characterised by higher sintering activity resulting in higher density in comparison with the green densities. The tensile strength and hardness significantly reflect the strengthening effect of the chromium concentration connected with the effect of carbon and the differences in the effect of Cr concentration of 1% and 3%. These effects were reflected in lower values of elongation and impact energy because it can be seen that the values of these properties of the steel with 3% Cr are the lowest. On the other hand, K3 steel containing 0.62% Mn and 1% Cr and no vanadium steels showed the lowest tensile strength and hardness values as compared with the K1 and K2 steel but had the highest elongation and impact energy values. The difference in ductility values between the K2 and K3 steels (0.15% C) is 17 J in the values of impact energy.

7.3. Development and manufacturing of rolling bearing rings by powder forging of steels based on prealloyed Fe–Ni–Mo and Fe–Cr–Mn–Mo powders

Research of the production of rings of rolling bearing rings by powder forging (PF) of sintered steels was a part of an industrial project. Its main objective was to reduce production costs for the production of roller bearing rings compared with their production from a conventional bearing steel whilst maintaining their lifetime and dynamic durability at least equivalent to the standards for bearings.

Only a small amount of information was available before the start of the project which indicated that the bearing rings with the outer diameter up to 180 mm made by powder forging, based on a prealloyed Fe–Ni–Mo powder (grade Ancorsteel FD4600) were installed in engines for heavy trucks for transcontinental transport and were characterised as 'heavy duty parts' [159–161]. The rolling bearing rings are axisymetrical in the form and, therefore, highly suitable for powder forging, in particular with regard to the fact that production of rolling bearings from wrought bearing steel has the character of mass production.

In the project, the properties of steels were studied on three types of prealloyed powders in the sintered state, and after forging and heat treatment under different processing conditions, initially in laboratory conditions. The resultant mechanical and toughness properties and the values of the contact fatigue resistance of the studied steels were then taken into accunt at forging inner and outer rolling bearing rings, which represent the main types of bearings.

Part of the project was also the development of experimental equipment for the production of chromium and manganese prealloyed powders by atomisation with water and oil + nitrogen, and with water + inhibitors, and an automatic forging line with a capacity of 600 pcs/h and then an automatic forging line for industrial applications at 1200 pcs/h.

The chemical composition of the prealloyed powders used in the development and manufacture of rolling bearing rings and, for comparison, of wrought bearing steels, are given in Table 7.8.

Table 7.8. Chemical composition of prealloyed powders used for investigation of PF bearing rings and of wrought bearing steels [mass %] [114]

Powder	C	Mn	Ni	Mo	Cr	Si	S+P+Cu (max)	O	AIM
1 – Astaloy A	<0.10	0.20	1.90	0.56	–	–	–	0.13	0.19
2 – KIP 4100	0.15	0.70	–	0.31	0.99	–	–	0.54	1.54
3 – Sumiron 4100S	0.014	0.74	–	0.27	1.00	–	–	0.07	n.d.
(a) Wrought bearing steel SAE 52100, DIN 105Cr6	0.95–1.1	0.25–0.40	0.30 (max)	–	0.9–1.1	0.50–0.30	0.82 (max)	–	–
(b) Wrought bearing steel ASTM A485, DIN 100Cr–6Mn, SKF 3	0.9–1.1	0.9–1.2	0.30 (max)	–	0.8–1.2	0.40–0.70	0.82 (max)	–	–

Remark: AIM – acid insoluble matter. Prealloyed Sumiron 4100S powder has not been available for more than 20 years now.

Prealloyed Astaloy A powder was selected for investigation on the basis on the previously cited work, despite the fact that a wrought bearing steel of such a composition does not exist. Prealloyed powders KIP 4100 and 4100S Sumiron were chosen because they were alloyed with chromium and manganese and powder Sumiron 4100 had the lowest oxygen content in the supplied state. The S designation in trade mark of Sumiron 4100S powder means that this powder contained ~0.05% S as a machining aid. No other Fe–Cr–Mn prealloyed powders were available at the time of the project.

The powders for the preparation of the compacts were mixed with 0.8% of zinc stearate as a lubricant and with different amounts of graphite (0.3–1.0%). The samples for the investigation of the mechanical properties (preforgings 10 mm × 10 mm × 80 mm) were manufactured under the following conditions: powder mixtures were compacted to a density of ~6.5 g cm^{-3}, followed by sintering at 1120°C for 1 h (powder Astaloy A) and at 1250°C for 0.5 h (powders KIP 4100 and Sumiron 4100S) in cracked ammonia. The sintered specimens (preforgings) were coated by a water–graphite emulsion followed by heating in a retort furnace in a $97N_2/3H_2$ atmosphere at the temperatures of 880 to 1150°C for 10 min. The powder forging of the rings was initially done manually in a closed die in a crank press (1600 kN) – air cooling, and later on a semi-automatic forging line.

7.3.1. Mechanical properties of powder forged steels based on prealloyed powders

7.3.1.1. Mechanical properties of PF steel based on Astaloy A powder

Rectangular samples (10 × 12 × 80 mm) from powder Astaloy A (0.52% O) with admixed 0.4 and 0.8% graphite and a lubricant were compacted to the density of 6.5 g cm^{-3} and sintered at 1250 and 1300°C for 30 min in a laboratory pusher furnace in cracked ammonia under getter. The sintered samples were powder forged from the temperature of 1100°C followed by air cooling without tempering. The carbon and oxygen content and the mechanical properties of powder forged and heat-treated steel samples based on Astaloy A powder are given in Table 7.9 and carbon and oxygen content and the mechanical properties of steels based on KIP 4100 powder are given in Table 7.10.

Heat treatment increased the yield and tensile strength with a ductility reduction. At a small increase of the density of the forged specimens, all the mechanical properties increased. Specimens with

Table 7.9. Mechanical properties of as forged and as heat-treated Fe–2Ni–0.5Mo steel based on Fe–2Ni–0.5Mo powder (Astaloy A). Forged density 7.78 g cm^{-3}, C_c = 0.47% [154–156]

Density [g cm^{-3}]	Re [MPa]	Rm [MPa]	A [%]	Z [%]	HB	KC [J]	*Rp0.2 [MPa]	*Rm [MPa]	*A [%]
7.66	637	862	7.1	11.4	148	37	–	–	–
7.72	657	925	7.9	14.0	159	39	–	–	–
7.73	664	930	10.3	25.5	211	67	1105	1267	5.7

Remark: * Quenching: 850°C, 30 min, N_2, oil; 200°C, 30 min, air; KC – impact energy

Table 7.10. Carbon (Cc) and oxygen (O) content and mechanical properties of powder forged steel based on KIP 4100 powder (Fe–0.7Mn–1Cr–0.30Mo–0.54O) in dependence on 0.4 and 1.0% graphite addition and on sintering temperature; 30 min, 1250°C and 1300°C, cracked ammonia, forged density 7.78 g cm^{-3} [155,162,163]

Graphite [%]	C_c [%]	O [%]	Rp0.2 [MPa]	Rm [MPa]	A [%]	Z [%]	Sintering [°C]
0.4	0.04	0.16	413	512	17.0	22.1	1250
1.0	0.35	0.23	1199	1266	2.2	10.0	1250
0.4	0.19	0.09	546	700	8.6	18.5	1300
1.0	0.32	0.18	1411	1545	2.6	9.8	1300

the density of 7.83 g cm^{-3} shows the highest tensile strength of 930 MPa, elongation 10.3%, and impact energy 67 J.

7.3.1.2. Mechanical properties of PF steel based on KIP 4100 powder

Graphite in the amount of 0.4 and 1% and a lubricant were added to the powder for the preparation of the samples. The samples, compacted to a density of 6.5 g cm^{-3} were sintered at two temperatures in cracked ammonia under a getter. Forging was carried out after heating the sintered specimens for 10 min at 1100°C in nitrogen, and after forging the specimens were cooled in air. Table 7.10 shows the content of combined carbon and oxygen after sintering and the mechanical properties.

The marked decrease of carbon content after sintering was caused by the insufficient effect of the getter. The oxygen content in the forged samples was affected by the sintering temperature and by the amount of graphite addition to the powders. The lowest tensile strength of 512 MPa and the highest elongation and reduction of area values were recorded for the steel with the lowest carbon content (0.04%) after sintering at 1250°C and similarly after sintering at 1300°C. The as forged tensile strength of 1266 MPa after sintering at 1250°C and of 1545 MPa after sintering at 1300°C was reached for both steels with the combined carbon content of 0.35% and 0.32% respectively (1% graphite). Sintering of the samples in cracked ammonia had a detrimental effect reflected in the high degree of decarburisation of the specimens leading to large differences in the mechanical properties.

The experimental results show that the Fe–Cr–Mn prealloyed powders, including powders with a high starting oxygen content, showed after forging the mechanical properties which would satisfy the requirements on the manufacture of PF rings of the rolling bearings. The both values of the strength properties in the forged and heat-treated conditions are close to the values of wrought bearing steel.

7.3.1.3. Mechanical properties of PF steel based on Sumiron 4100S powder

This section describes the mechanical and toughness properties of powder forged steels prepared on the basis of Sumiron 4100S powder (Fe–1Cr–0.74Mn–0.1Mo–0.07O) in dependence on the graphite addition and forging temperature. The properties of the steel based on this powder, taking into account also a very low oxygen content of this powder in the as supplied state, were investigated with the aim of obtaining the values of the mechanical and toughness properties which would enable these steels to be used for the manufacture of rings of the rolling bearings and for comparing them with the properties of the steels based on the powders Astaloy A and KIP 4100. Natural graphite in the amount of 0.4%, 0.6% and 0.8% and 0.8% of zinc stearate as a lubricant were added to the steels coded as S4, S6 and S8 powder steels.

Samples (rectangular bars) 8 × 15 × 78 mm in size were compacted to the density of 6.5 g cm^{-3} and sintered at 1250°C for 30 min in cracked ammonia under a getter (alumina + 2% graphite) in a laboratory pusher furnace. The sintered samples were coated with a water–graphite emulsion. The samples were heated in a retort furnace in a $97N_2/3H_2$ atmosphere at the temperature of 800, 900, 1050 and 1150°C for 10 min and manually forged in a crank press (capacity 1600 kN) to the bars 10 × 10 × 80 mm in size for the determination of the mechanical and toughness properties as well as of carbon, oxygen and nitrogen content.

The specimens for the determination of the following properties were prepared from PF bars by machining without prior annealing:

– tensile test bars 6 mm in diameter with thread heads M10,

– fracture toughness testing 10 × 10 × 55 mm in size with an U3 notch,

– fracture toughness testing 10 × 5 × 55 mm in size with an U2 notch 3 mm deep were manufactured by machining. In the notch root a crack 0.22 mm wide and 2 mm deep was formed by electrical discharge machining after heat treatment.

Heat treatment (through hardening) was carried out by quenching of the forged samples from the austenitisation temperature of 800, 830, 860 and 890°C (holding time 30 min in nitrogen) in oil 50°C followed by tempering at 165°C for 60 min in air.

Table 7.11. Carbon and oxygen contents and mechanical and toughness properties of powder forged and heat-treated steel based on Sumiron 4100S powder as a function of graphite addition and forging temperature. Sintering 30 min at 1250°C [164,165]

a) Forging temperature 1050°C

Graphite [%]	C_c [%]	O [%]	Re [MPa]	Rm [MPa]	A [%]	*HRC	KCU3 [J]	*K_{IQ} [MPa·m$^{1/2}$]
0.4	0.34	0.08	685	841	11.0	51	37	57
0.6	0.45	0.05	752	925	9.50	58	26	39
0.8	0.62	0.04	900	1103	7.00	62	23	27

b) Forging temperature 1150°C

Graphite [%]	C_c [%]	O [%]	Re [MPa]	Rm [MPa]	A [%]	*HRC	KCU3 [J]	*K_{IQ} [MPa·m$^{1/2}$]
0.4	0.32	0.03	680	838	14.5	52	45	n. d.
0.6	0.44	0.03	752	923	11.0	59	29	n. d.
0.8	0.63	0.02	900	1100	9.5	62	24	47

Remark: As forged density 7.83 g cm^{-3}; heat treatment: 830°C, 30 min, N_2, oil, 165°C, 1h, air; *as heat-treated

Table 7.12. Powder forged combined carbon (C_c), oxygen (O) and nitrogen (N) concentration (mass %) of S4, S6 and S8 steels in dependence on forging temperature [163]

Steel	Element	Forging temperature [°C]			
		800	900	1050	1150
		As sintered contents [mass %]			
S4	C	0.39	0.36	0.34	0.32
	O	0.09	0.08	0.07	0.03
	N	0.02	0.02	0.01	0.01
S6	C	0.50	0.45	0.44	0.44
	O	0.10	0.09	0.05	0.05
	A	0.02	0.02	0.01	0.01
S8	C	0.71	0.67	0.63	0.63
	O	0.11	0.08	0.03	0.02
	N	0.02	0.01	0.01	0.01

Powder forged basic properties. The carbon and oxygen content and the mechanical and toughness properties of PF and heat-treated steel based on Sumiron 4100S powder as a function of the graphite addition and forging temperature are listed in Table 7.11.

Table 7.12 gives the carbon, oxygen and nitrogen concentrations in dependence on forging temperature.

The listed concentrations of C, O and N of these PF steels as a matter of fact correspond approximately to the as sintered ones. The short holding time at the forging temperature caused only a small loss in the carbon content compared to the as sintered one. This is also valid for the oxygen and nitrogen content. It means a progressive decrease of the oxygen and nitrogen content in the samples occurred during sintering of S8 steel, as confirmed by the values listed for the forging temperature of 1150°C (0.02% O, 0.01% N), Table 7.11.

Yield and tensile strength. The density of the forgings was in the range from 7.71 to 7.83 g cm^{-3}. As shown in Fig. 7.6, the highest yield strength values were obtained after forging at 900°C affected by higher carbon concentration. The samples S8 forged at 1050°C attained the highest tensile strength of 1120 MPa shown in Fig. 7.7.

The hardness values were in the range of 301 to 349 HV 10 for S4 steel, of 311 to 367 for S6 steel and of 345 to 381 HV 30 for S8 steel.

Elongation, reduction of area and impact toughness. The effect of forging temperature on elongation and reduction of area of the investigated steels is shown in Figs. 7.8 and 7.9.

In contrast to the yield strength, tensile strength and hardness, and the elongation and reduction of area values markedly increased with

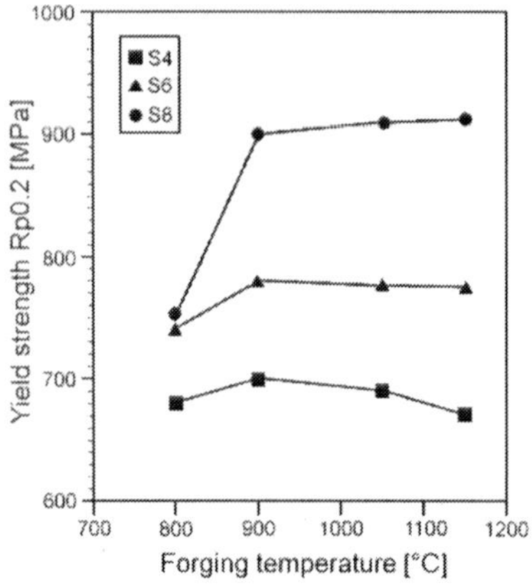

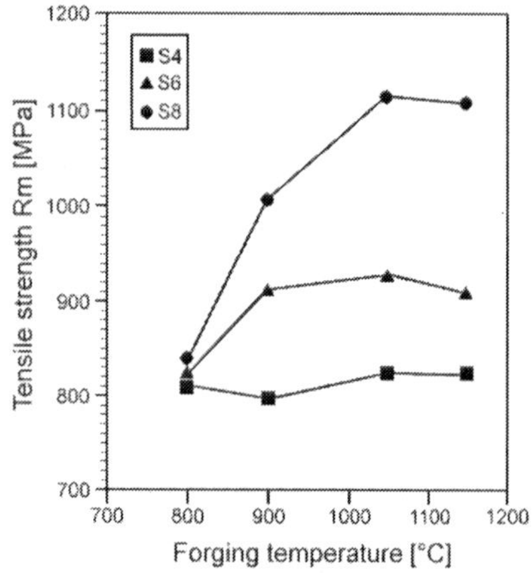

Fig. 7.6 (left). The dependence of yield strength of PF S4, S6 and S8 steels on forging temperature [164–167].

Fig. 7.7 (right). The dependence of tensile strength of PF S4, S6 and S8 steels on forging temperature [164–167].

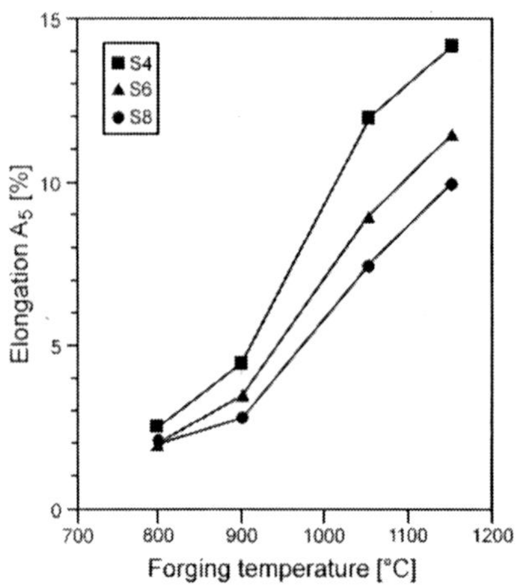

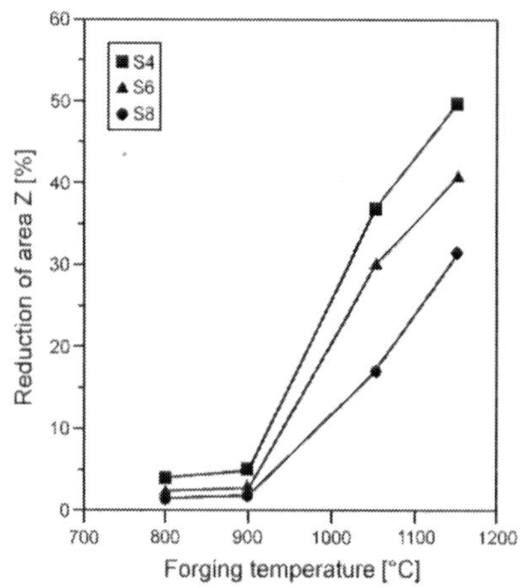

Fig. 7.8 (left). The dependence of elongation of PF S4, S6 and S8 steels on forging temperature [164–167].

Fig. 7.9 (right). The dependence of reduction of area of PF S4, S6 and S8 steels on forging temperature [164–167].

increasing forging temperature to 1050°C and 1150°C. The highest elongation of 14% was recorded for the S4 steel forged at 1050°C.

The notch impact toughness values of the investigated steels in dependence on forging temperature, shown in Fig. 7.10, present the same tendency as the elongation and reduction of area values in dependence on forging temperature. The highest values of impact energy were attained for the S4 steel forged at 1150°C. The impact energy values decreased with increasing C content in the samples.

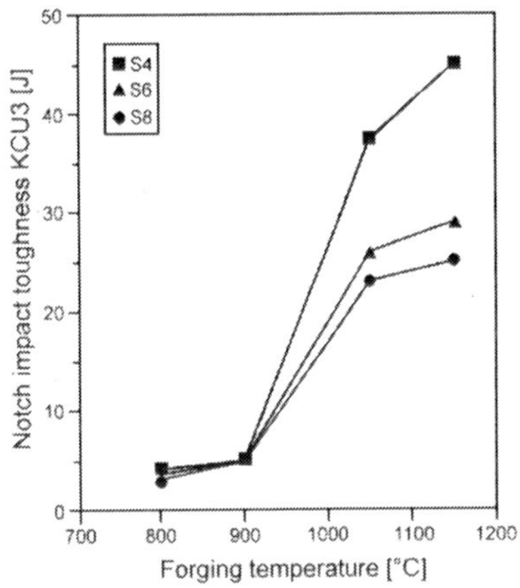

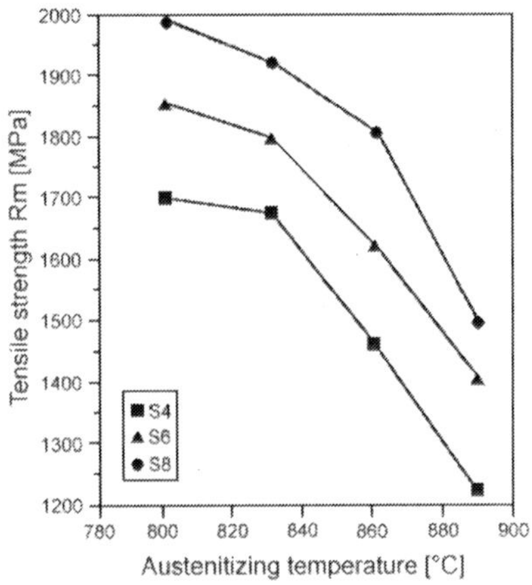

Fig. 7.10 (left). The dependence of notch impact toughness (KCU3) values of PF S4, S6 and S8 steels on forging temperature [164–167].

Fig. 7.11 (right). The dependence of as heat-treated tensile strength values of PF S4, S6 and S8 steels on austenitizing temperature [164–167].

As heat-treated tensile strength and hardness. The PF specimens were heat-treated at 1050°C. The dependence of the tensile strength of the PF heat-treated specimens of the S4, S6 and S8 steels on austenitisation temperature is shown in Fig. 7.11.

The highest tensile strength values in the range of 1700 to 1988 MPa obtained the samples quenched from 800°C in dependence on carbon content. A relative uniform drop in tensile strength at an increase of the quenching temperature from 830 to 890°C was 20–26% (1500 to 1201 MPa). The dependence of as heat-treated hardness of these steels on austenitisation temperature is shown in Fig. 7.12. The highest hardness was obtained by quenching from 820°C. This temperature corresponds with the optimum quenching temperature applied for wrought bearing steel. Increasing quenching temperature did not affect the hardness of the samples. The S8 steel reached the hardness of 62 HRC (0.63% C_c). A relatively high hardness of 58 HRC was also obtained for S6 steel. The hardness of 62 HRC is required for the rolling bearing rings made from wrought bearing steel.

Fracture toughness. Fracture toughness of the structural steels is an important property mainly for dynamically highly loaded parts and, therefore, the fracture toughness is a characteristic property of the wrought bearing steels requiring high contact fatigue endurance. The fracture toughness values of the investigated PF (S4, S6, S8)

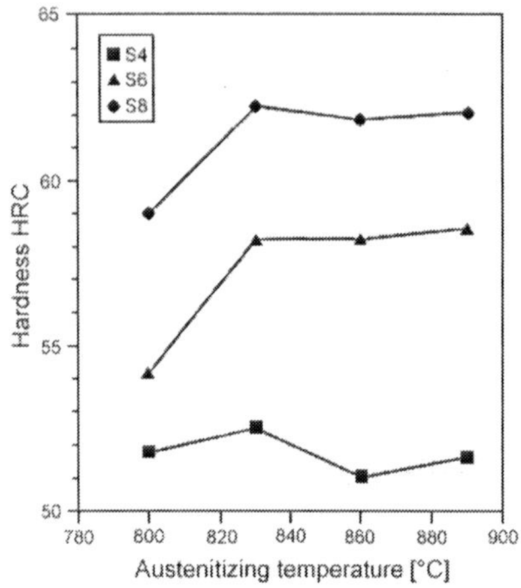

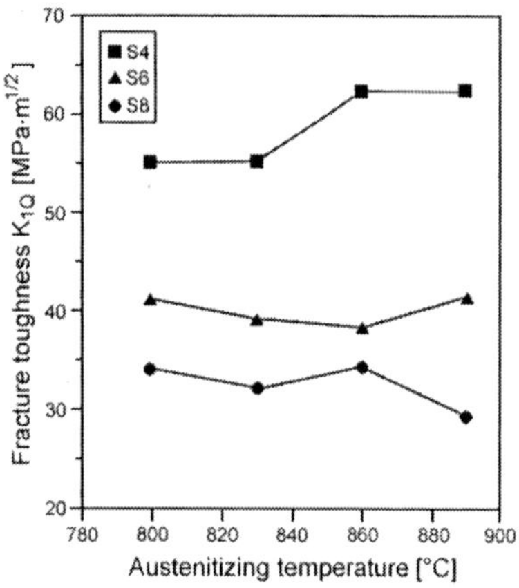

Fig. 7.12 (left). The dependence of as heat-treated hardness HRC values of PF S4, S6 and S8 steels on austenitizing temperature [164–167].
Fig. 7.13. (right). The fracture toughness of PF S4, S6 and S8 steels in dependence on austenitizing temperature [164–167].

steels in dependence on austenitizing temperature are shown in Fig. 7.13.

The fracture toughness values of the PF steels at austenitizing temperature of 1050°C, as shown in Fig. 7.13, depended predominantly on the carbon content without a marked effect of the quenching temperature. The increase in carbon content from 0.39% (S4) to 0.71% (S8) caused the decrease in fracture toughness values by ~3%. The minimum fracture toughness value of 21 MPa · m$^{1/2}$ is required for wrought bearing steel. The highest fracture toughness of 62 MPa · m$^{1/2}$ was reached by the S4 steel forged at 1050°C.

In contrast to the values obtained after forging at 1050°C, the S8 samples forged at 1150°C after quenching from 830°C and tempering showed the fracture toughness of 47 MPa · m$^{1/2}$.

Microstructure. The microstructure of the PF S4 samples was highly homogeneous characterised by fine ferrite–pearlite grains, as shown in Fig. 7.14(a). The microstructure of the as heat-treated S4 steel samples was formed by martensite and retained austenite, Fig. 7.14(b). The martensite in the microstructure of the as heat-treated S8 sample was finer, as shown in Fig. 7.14(c). The PF steels with such characteristics can fulfil high requirements for the production of rolling bearing rings. This is also confirmed by the result which shows that the structures of all three steels with different carbon contents did not contain any non-metallic inclusions when examined

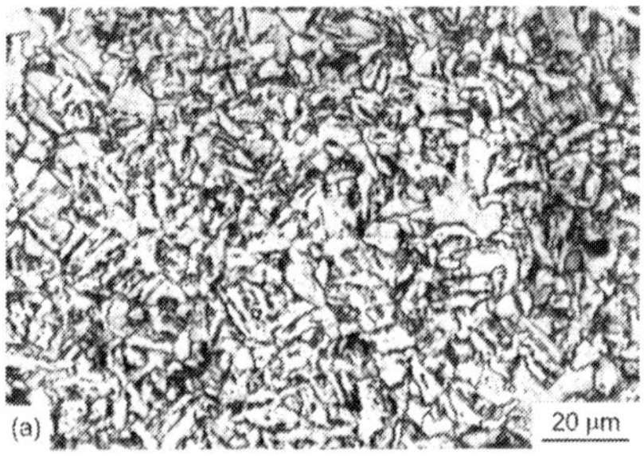

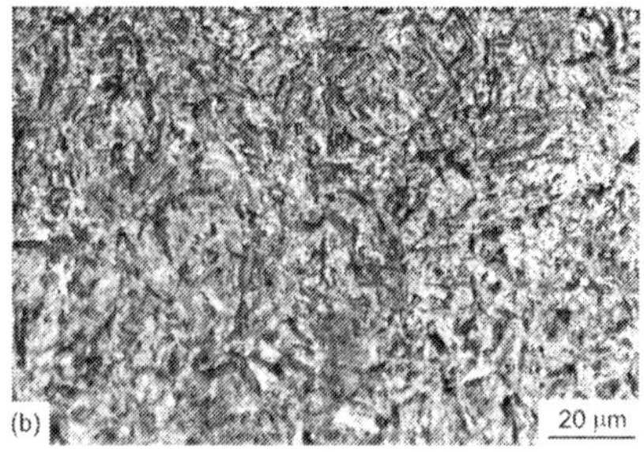

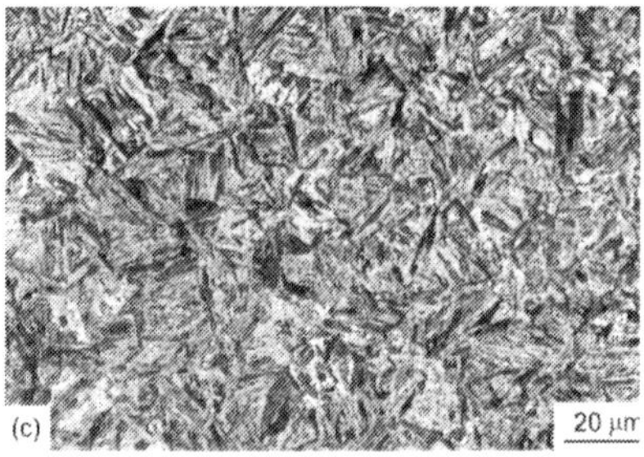

Fig. 7.14 (a). Microstructure of PF S4 steel sample, (b) microstructure of as heat-treated S4 steel sample. Forging temperature 1050°C, quenching temperature 830°C, (c) microstructure of as heat-treated S8 steel sample. Forging temperature 1050°C, quenching temperature 830°C. Nital etched.

by optical metallography. This is in agreement with the low oxygen content in the PF samples based on Sumiron 4100S powder given in Table 7.12.

7.3.2. Contact fatigue endurance of powder forged steels based on prealloyed powders

In addition to porosity, the contact fatigue resistance, to which all rolling bearings are subjected, of the PF steels is also influenced by all other mechanical and toughness properties and structural characteristics in the heat-treated state. For this reason, the contact fatigue resistance of the PF steel is given by the standards for wrought bearing steel which is the condition for application of these materials in the manufacture of the rolling bearing rings. This characteristic is the most important property of the steels not only for bearings but also for gears and all other moving parts working under contact fatigue loading.

Initially, bars with a diameter of 4 mm and 80 mm long were produced from the forged samples based on the powders Astaloy A (A), KIP 4100 (K) a Sumiron 4100S (S) and, for comparison, from wrought bearing steel (SAE 52100). The test bars were case-hardened under industrial conditions to a hardness of 60 HRC and then ground to the surface roughness required for the raceways of the rings of the rolling bearings. The contact fatigue test of the bars

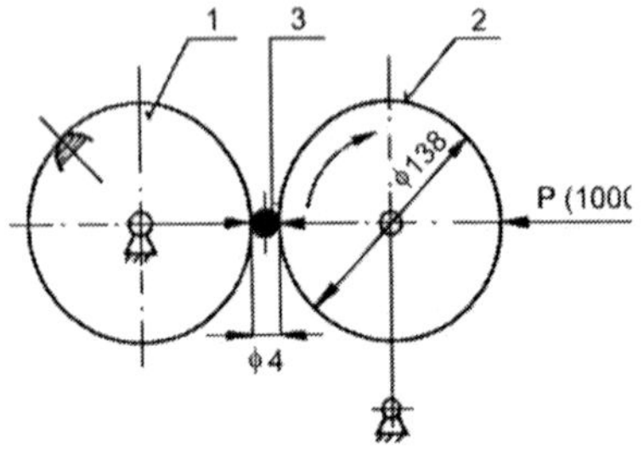

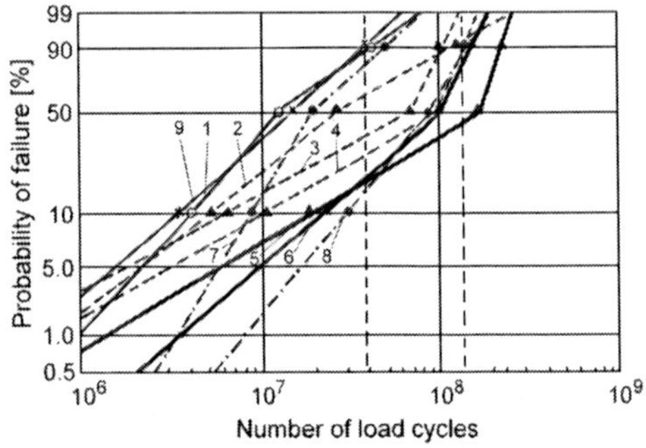

Fig. 7.15 (left). Scheme of rolling contact fatigue endurance test equipment – RMAT method. 1 – supporting disc 2 – pressing disc, 3 – sample [168].
Fig. 7.16 (right). Weibull probability to failure of PF case-hardened samples prepared from prealloyed Fe–2Ni–0.5Mo (A), Fe–Mn–Cr (K, S) and of wrought bearing steel. 1 – Astaloy A, 2 to 6 Sumiron 4100S, 7 and 8 KIP 4100 (K), 9 – wrought bearing steel (AISI 52100, 100Cr6) [165,169].

were carried out by the RMAT method shown in Fig. 7.15. Rolling contact fatigue endurance tests were carried out at 494 rev s^{-1} at a Hertzian pressure of $5 \cdot 10^3$ MPa. 10 to 15 traces were made on each sample until pitting occurred.

The results of the tests are shown in Fig. 7.16 as the relation between the Weibull probability to failure and the number of cycles applied up to pitting formation. Pitting formation on the sample was demonstrated by vibration of the test equipment and at this moment the testing on this trace was stopped followed by the testing on a new trace. The test results of the samples of all three steels prepared from the A, K and S powders are also presented in Table 7.13.

The contact fatigue endurance of powder forged Fe–2Ni–0.5Mo (Astaloy A) steel (curve 1) was on the level of wrought bearing steel (curve 9). The rolling contact fatigue endurance of PF steel samples prepared from both Fe–0.7Mn–1Cr–0.2Mo powders was universally

Table 7.13. Weibull probability to failure and loading cycles for A1, K3, S5 and S6 steels with designation of curve number shown in Fig. 7.16

Weibull probability	Prealloyed powders			
	A1	K3	S5	S6
	Number of cycles			
L_{10}	$2.0 \cdot 10^6$	$8 \cdot 10^6$	$2.0 \cdot 10^7$	$2.2 \cdot 10^7$
L_{50}	$1.1 \cdot 10^7$	$7 \cdot 10^7$	$1.8 \cdot 10^8$	$1.1 \cdot 10^8$
L_{90}	$3.0 \cdot 10^7$	$1 \cdot 10^8$	$1.5 \cdot 10^8$	$2.4 \cdot 10^8$

higher than that of the samples prepared from Astaloy A powder. It is necessary to note that the starting oxygen content of powder A was 0.13%, of powder K 0.54% with 1.54% of AIM and of powder S 0.07%, Table 7.8.

Table 7.13 lists the Weibull probability values to failure L_{10}, L_{50} and L_{90} and loading cycles for A1, K3, S5 and S6 steels according to Fig. 7.16.

The Weibull probability to failure L_{10}, L_{50} and L_{90} (10, 50 and 90%) from the group of equal samples giving an equal or better fatigue endurance was calculated on the basis of the results obtained by testing of the mentioned steels until pitting occurred.

The measured values of the PF Fe–0.7Mn–1Cr–0.2Mo steel (K) compared with the data given in Ref. 170, were on the same level or higher than those for the vacuum-degassed wrought bearing steel. The highest values of L_{50} and L_{90} were recorded for the samples based on the S powder (curves 5 and 6), namely of 2.4×10^8 cycles taking into account also the lower density of the S samples of 7.75 g cm^{-3}. Lower L_{90} values were recorded for the steel based on the K powder (curve 8). The contact fatigue endurance of the steel based on the A powder was on the level of that manufactured from wrought bearing steel – curve 1. This shows that the rolling contact fatigue endurance of the powder forged steels based mainly on the Fe–Cr–Mn prealloyed powders was more influenced by the character of the microstructure in dependence on alloying and heat treatment and that the 'full' – total density of the samples in some small range was not a primary condition for their high contact fatigue endurance [154,155].

The results confirmed the possibility of using chromium- and manganese-prealloyed powders for production of highly loaded parts subjected to rolling contact fatigue, especially of rolling bearing rings instead of those based on the Ni-alloyed powder. It means the Fe–Mn–Cr–Mo PF steels exhibited higher resistance to the initiation of the failure formation and mainly to the failure propagation than the PF steel based on Astaloy A powder and as the samples prepared from wrought bearing steel.

In terms of the oxygen content in prealloyed powders listed in Table 7.8, the difference in the initial oxygen content values between the powder K (0.52%) and powder S (0.07%) can be regarded as the reason why the samples produced from powder K showed a greater scatter of the values of Weibull probability to fracture and also lower values of L_{90} than the samples from powder S.

The results show that industrial production of the rolling bearing rings by forging of powder steels should be based mainly on Fe–Cr–Mn–Mo prealloyed powders with an extra low starting oxygen content with minimum of impurities.

7.3.3. Processing conditions and base characteristics of powder forged rolling bearing rings and their properties

On the basis of previous results on the contact fatigue resistance of samples of powder PF Sumiron 4100S for the production of rings, tests were also carried out for comparison on the samples of the Astaloy A and KIP 4100 powders. The properties of these powders, depending on the carbon content in relation to the forged and heat-treated conditions, were used a a basis for assuming that the bearing rings from these powders would reach the minimum value of lifetime required by the standards for bearings from wrought bearing steel. This created material conditions for the application of these prealloyed powders for the manufacture of rolling bearing rings by forging to 'full' density, which is always the condition for dynamically loaded components. It was also necessary to determine the optimum density of sintered blanks and forging conditions and thereby the deformation force to achieve the highest density of the forgings.

7.3.3.1. Effect of density of sintered blanks on deformation force

The values of deformation force in forging blanks were used to determine their optimum density in the sintered state. For this purpose, sleeves were pressed powder from Astaloy A powder with the addition of 0.3% graphite in four tool dimensions, diameter 50 mm/diameter 34 mm to diameter 92 mm/diameter 70 mm, which corresponded to the dimensional range of the selected PM forged rings with the addition of material for mechanical finishing operations. For these tests, sleeve-shaped samples were pressed to a density of 5.8, 6.2 and 6.9 g cm^{-3} and sintered for 1 h at 1120°C in cracked ammonia. Blanks were graphitized in a 20% aqueous suspension of graphite. The samples were forged after heating at 1050°C in an atmosphere $97N_2/3H_2$ in a 1600 kN forging press. The mass of all sintered blanks was the same, 106 g. Deformation force was measured by the strain gauge method with recording of a graph.

A forged sleeve is shown in Fig. 7.17 and the effect of the density of sintered blanks on the deformation force is in Fig. 7.18.

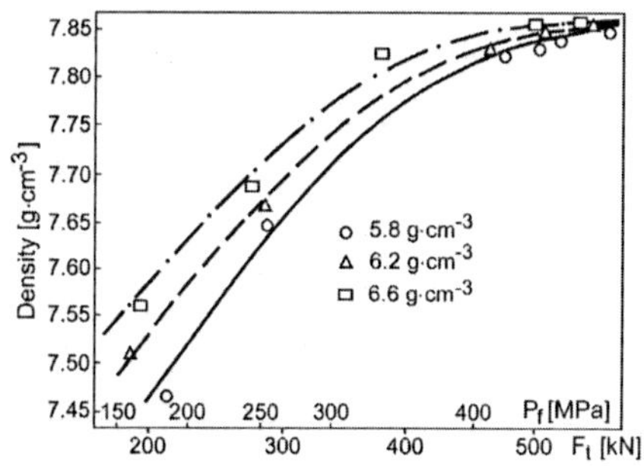

Fig. 7.17 (left). A sleeve (forging) produced by forging of samples with different density in the sintered condition in the tests of determination of the deformation force to obtain the maximum density of the forging. Forging, diameter 50 mm/34 mm, 106 g.

Fig. 7.18 (right). Effect of the density of PM forgings of sleeves on deformation force F_t and forging force P_t in dependence on the density of the sintered blank, 1050°C, Ar, forging, diameter 50 mm/diameter 34 mm [171].

The density of the blanks had no effect on achieving 99% density of the forgings of the sleeves. Based on these results, the blanks in all cases were pressed to a density of ~6.5 g cm^{-3}. To achieve 'full' density, the deformation force was increased by 25 to 50% as compared with the force required to achieve about 99% relative density of the PF sleeves.

To achieve the high toughness of PF steel rings with the aim to minimize subsurface porosity, blanks of different sizes were tested. In the case of the blanks corresponding to the forging tool dimensions (outer or inner diameter) compaction took place only by axial forging force. Such conditions, with the height reduced by 20 to 30%, resulted in satisfactory geometric shapes of all areas of forgings, but with a relatively high subsurface porosity.

Forging of the blanks with a smaller outer or inner diameter resulted in a combination of axial and lateral spreading of the material in the tool. Lateral spreading of material to the outer diameter of the tool resulted in incomplete formation of the inner surface of the ring resulting in the formation of a fold in the material shown in Fig. 7.19 for the outer and inner rings of a tapered roller bearing.

Lateral spread of material in the tool axis in the direction of reducing the internal diameter of the forged blanks by ~6% while maintaining the mass of the blanks, resulted in the lowest subsurface porosity of the rings, with no collapse of the outer surfaces of the forging. Under such forging conditions, subsurface porosity of the rings was detected up to 0.2 mm from the surface. This will minimize

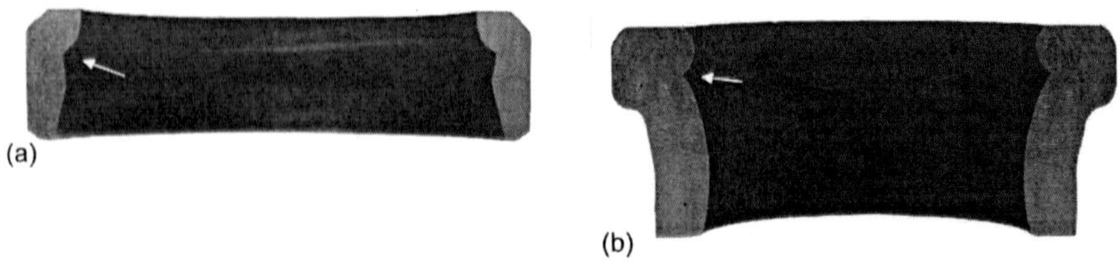

Fig. 7.19. Folds of the material on the inner surface of the forging; a) outer, b) inner ring of a tapered bearing.

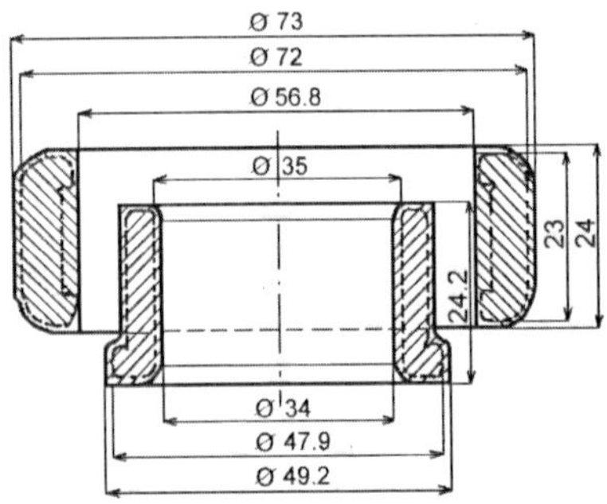

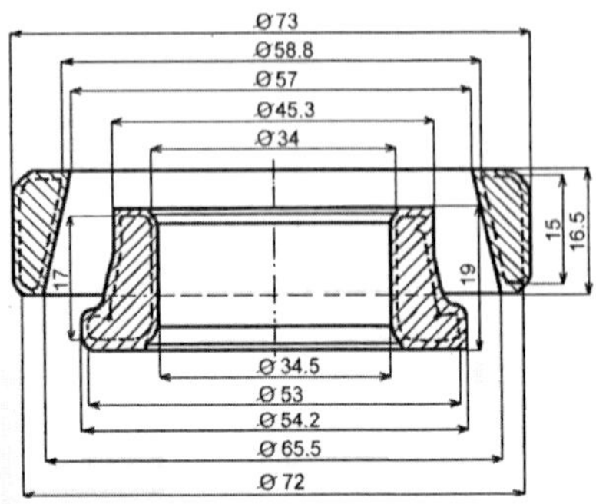

Fig. 7.20 (left). Dimensions of the outer and inner rings of the PM roller bearing in the forged state (outer contours) in the finished state (inner contours – dashed line) [172–174].
Fig. 7.21 (right). Dimensions of the outer and inner rings of the PM tapered bearing in the forged state and in the finished state (lines as in Fig. 7.20) [172–174].

material allowances for finishing turning and grinding. These inner and outer functional surfaces of the finished rings are therefore pore-free.

Figures 7.20 and 7.21 show the dimensions of the finished outer and inner rings of cylindrical roller and tapered roller bearings and forgings with allowances for finishing turning and grinding, so that the subsurface layer of the finished rings was pore-free. The total material allowances for turning and grinding operations were within the range of 0.6 to 1 mm for both diameters and heights of the individual rings.

7.3.4. Examination of powder forged bearing rings properties

The rings for the lifetime and durability testing were prepared by sintering the blanks with a density of 6.5 g cm^{-3} for 30 min at 1120°C in cracked ammonia in an industrial furnace and forged in a 1600 kN crank press after heating for 10 min at 1050°C in a $97N_2/3H_2$

atmosphere. The blanks were soaked in a graphite emulsion.

17 rings were forged and examined:

10 outer rings – outer diameter from 65 to 120 mm, mass of PF rings was 144 to 655 g,

4 inner rings – outer diameter 40 to 60 mm, mass of PF rings was 96 to 284 g,

3 thrust bearing rings – outer diameter 52 to 67 mm, mass of PF rings was 68 to 158 g.

The ratio of the mass of the powder used for the manufacture of the rings to the mass of the classic bearing steel for all the tested rings was 23.6–71.6%.

As also shown in Figs. 7.20 and 7.21, the total use of the starting material for manufacturing of PF bearing rings in relation to the net mass of the finished rings was in the range 51 to 80% and by classical technology – bars, in the range 19 to 51% [175].

Figure 7.22 shows the set of all the tested PF rings and Fig. 7.23 the rolling bearings assembled with one outer or inner PF ring and

Fig. 7.22. A set of 7 tested PF ball, cylindrical, tapered roller and thrust bearing rings [176,177].

Fig. 7.23. Rolling bearings assembled with one outer or inner PF ring with corresponding inner or outer wrought bearing steel ring and individual thrust bearing rings [176–178].

with a second wrought bearing steel ring. The compound bearings were tested for the lifetime and durability of PF rings.

7.3.4.1. Durability and failure of powder forged rolling bearing rings

The dynamic characteristics of rolling bearings, such as contact fatigue endurance, were tested on standardized testing stations fitted with 20 assembled bearings until failure of the fifth bearing at the normalized load. Failure was manifested by vibrations of the testing station. Time to failure of the fifth bearing is evaluated as the bearing lifetime in hours and as the basic dynamic load rating (BLDR) of the test series and of failed single rings (C_i/C_{cat}) within the corresponding tested series of PF case-hardened rolling bearing rings. C_i is valid for the PF ring and C_{cat} is the catalogue value for the given bearing.

The basic dynamic load rating (BLDR) was calculated according to equation:

$$L_{10} = (10^6/60\ n) \times (C/P)^k,$$

where L_{10} is the basic (catalogue) durability in (10^6 cycles), n – is the frequency of revolution (4200 min^{-1}), C – is the basic dynamic load rating (N), P – is the equivalent dynamic load rating (N), and $k = 3$ for ball bearings (point contact) and 10/3 for tapered and cylindrical rolling bearings (line contact). The used ratio ranged from 2.28 to 4.41, *i.e.* the stress above the stresses which occur during common operational exploitation of the rolling bearings.

The real dynamic load rating of an individual damaged bearing C_i and of the tested series value was calculated on the basis of the measured L_{10} value. The ratio C_i/C_{cat} (in %) defines the basic dynamic load rating (BLDR) by means of which the properties of all tested bearing rings with PF rings are evaluated.

Table 7.14 gives the corresponding basic verified data about the durability (BLDR) of the 30207/1 (tapered bearing) and 6208/1 (ball bearing) rings with line and point rolling contact applications with some differences in density, manufactured under approximately comparable conditions of the corresponding whole series.

The minimum BLDR catalogue value for wrought steel bearings is 125%. The outer ball bearing ring (6208 bearing) made from Astaloy A powder achieved the BLDR of 150% and 163% of the test series (point contact) and those of the outer tapered rolling bearing 30207

Table 7.14. Reproducible results of the basic dynamic load rating (BLDR) of some test series of PF failed rings (C_i/C_{cat}) within corresponding tested series of PF case-hardened rolling bearing rings. 6028 – ball bearing, point contact, 30207 – tapered bearing, line contact [176,179]

Bearing type	Density [g cm^{-3}]	Material	HRC	C_i/C_{cat} (individual rings) [%]					BDLR [%]
6208/1	7.84	A4	61	70	101	141	>206	>206	150
6208/1	7.77–7.81	S3	60–63	199	no failure up to 206				256
6308/1	7.82	A4	62–63	99	178	187	188	189	163
30207/1	7.83	A4	62–64	165	199	201	>201	>201	188
30207/1	7.70–7.83	A4	59–61	71	122	133	>151	>151	133
30207/2	7.78–7.81	K3	58–64	117	165	174	186	212	169
30207/2	7.76–7.83	A3	61–63	105	111	131	136	140	124

Remark: /1 – outer ring, /2 inner ring; A – Astaloy A powder, S – Sumiron 4100S powder, K – KIP 4100 powder; 3,4 – graphite addition in tenths of a percent

achieved BLDR of 133% and 188% (line contact). The inner 30207 tapered bearing rings attained the BLDR values of 124%. According to these results, the BLDR values of the PF bearing rings made from Astaloy A powder fulfilled the requirements for durability of rolling bearings. The PF inner rings of 30207 bearing manufactured from KIP 4100 powder achieved the BLDR about 169%.

The PF rings prepared on the basis of Sumiron 4100S powder achieved the highest BLDR value of 256% when the test was interrupted due to test capacity reasons. No PF ring failed during the test time also with the lower carbon content in the ring as in wrought bearing steel, as listed in Table 7.8. This was manifested in high homogeneity of the microstructure, Fig. 7.15.

The durability (contact fatigue life) of the bearing rings based on Fe–1Cr–0.7Mn–0.2Mo powder (Sumiron 4100S) with low oxygen content and high chemical purity substantially exceeded the durability of wrought bearing rings. Contact fatigue lifetime (h) improved by 200% and 500%.

To verify the durability of PF bearing rings, 10 000 ball bearings 6029 were produced, which were assembled with both PF rings and under inspection fitted to technical equipment, with special additional indication PM. No damage was found in any of the bearings even after a long time. This confirmed the reproducibility of the properties of powder steels and the production conditions and durability of the PF bearing rings.

Failure of PF races. About 300 damaged PF bearing rings prepared under different conditions which were damaged by spalling (pitting) were examined. Two groups of the failure reasons followed from these analyses. The first group of the causes for pitting formation were the powder metallurgy reasons namely pores and free (unforged) powder particle surfaces in the ring material which were sporadically found. This led to the localized intercrystalline fracture of the ring broken additionally by impact [179]. The second group, which is known from wrought bearing steels rings analyses, but caused by other factors are the white etching areas (WEA) – butterflies.

The defects initiating crack formation at the rolling contact fatigue test are situated at a certain depth below the surface coinciding with the zone of maximum stresses and never on the surface of the raceway. The mean depth of pitting appearing on metallographical cross-sections of more than 100 tapered and ball bearing failed rings was measured and the values are given in Table 7.15.

White etching areas. The white etching areas form basically in the rings made from wrought bearing steel where the formation of butterflies is associated with non-metallic inclusions. The formation of butterflies in PF rings was associated with the pores which acted as internal notches due to the same fatigue contact loading. They were formed in two forms, namely as 'white bands' forming an angle of 40° to 50° with the raceway tangent without clear presence of pores as shown in Fig. 7.24. The butterflies formed at a pore are shown in Fig.7.25. The microcracks were found in all butterflies (transformed areas of the martensitic structure) shown in Fig. 7.26 and Fig. 7.27.

Table 7.15. Measured (Z_m) and calculated (Z_c) mean depth of failure arise areas from raceway surface with calculated shear stress τ_{max}. 30207 tapered roller bearing, 6209 ball bearing, 1 – outer ring, 2 – inner ring [179]

Bearing ring type	Z_m [μm]	Z_c [μm]	τ_{max} [MPa]
30207/1	.72	70	325
30207/2	94	80	353
6209/2	240	300	653

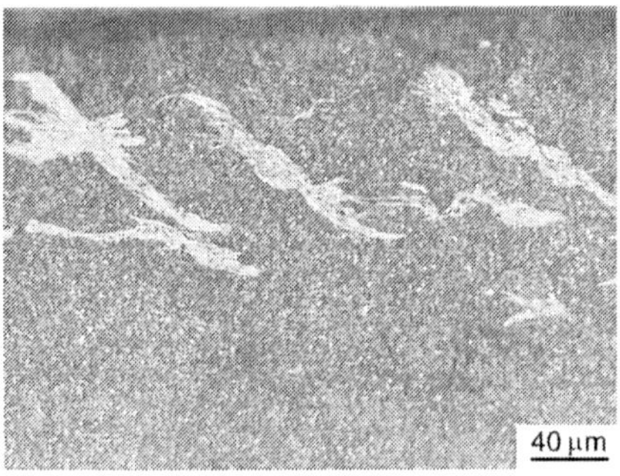

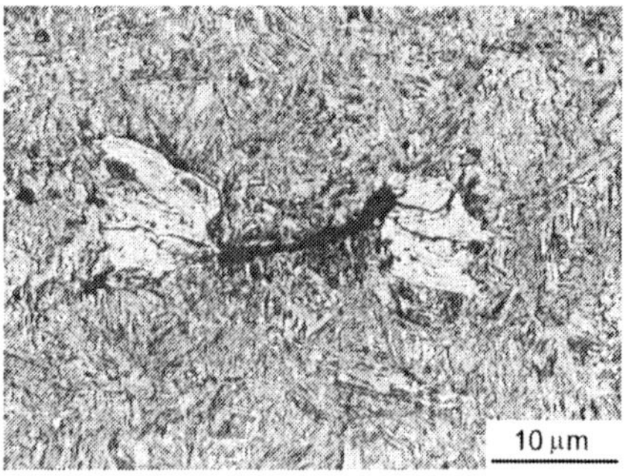

Fig. 7.24 (left). White bands in the microstructure in an inner tapered roller bearing ring (30207/2). Astaloy A powder, ρ_0 = 7.84 g cm^{-3}, 61 HRC, C_i = 123%. Nital etched [179–182].
Fig. 7.25 (right). Butterflies formed at the pore in an inner ball bearing ring (6209/2). Astaloy A powder, ρ_0 = 7.85 g cm^{-3}, 60 HRC, C_i = 127%. Nital etched [179–182].

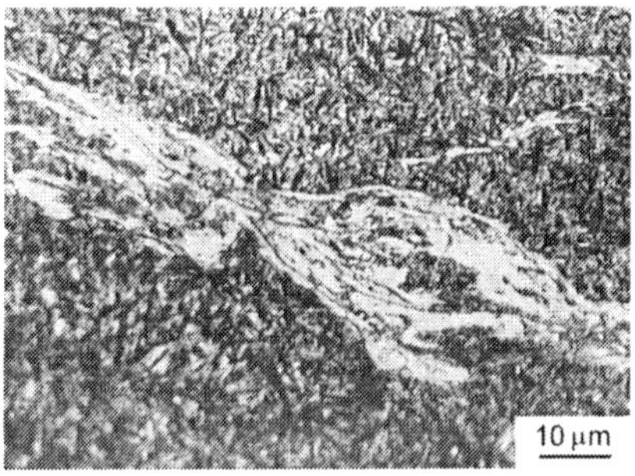

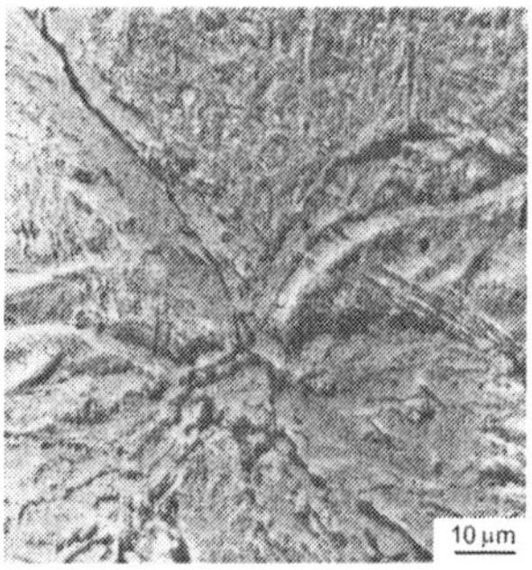

Fig. 7.26 (left). Microcracks in the butterfly in an inner tapered roller bearing ring (30207/2). Astaloy A powder, ρ_s = 7.85 g cm^{-3}, 60 HRC, C_i = 127%. Nital etched [179–182].
Fig. 7.27 (right). Microcracks in the butterfly spreading in different directions from its centre. Astaloy A powder, ρ_s = 7.83 g cm^{-3}, 61 HRC, C_i = 132%. Nital etched [179–182].

The microhardness values of white etching areas, measured on metallographic cross-sections reached 1470 HV 0.05. According to a number of results, white etching areas were formed by oversaturated ferrite as a consequence of carbide dissolution and by this way the decay of martensite [183].

7.3.5. Experimental equipment for atomisation of prealloyed Fe–Cr–Mn–Mo powders with low oxygen content

The results of the lifetime and durability tests of the PF forged rolling bearing rings showed that the rings made on the basis of prealloyed

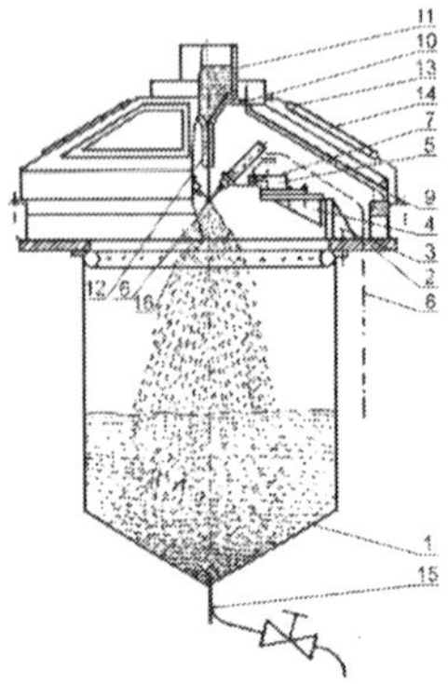

Fig. 7.28. Simplified section through experimental equipment for producing metallic powders by atomisation with water and with oil–nitrogen. 1 – cylindrical vessel, 6 – atomising nozzles, 8 – water or supply by the pressure pump, 11 – graphite crucible with molten metal [184].

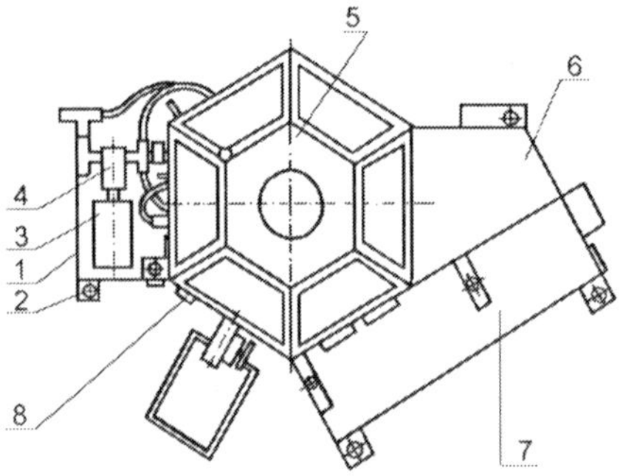

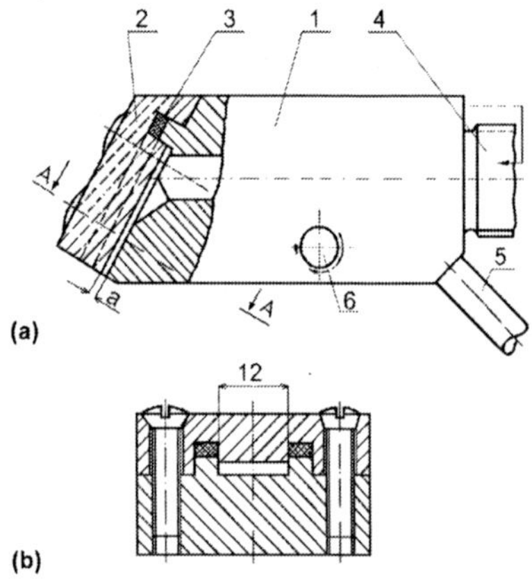

Fig. 7.29 (left). Floor plan of equipment according to Fig. 7.28 [184].
Fig. 7.30 (right). Water or liquid gas nozzle for atomising molten metals to powder [184]. 1 – body, 2 – front, 3 – seal, 4 – pressure water or oil, 5 – pressure nitrogen, 6 – adjustment of the position, *a* – thickness of pressure medium jet (12 × *a*).

powder Fe–0.7Mn–1Cr–0.2Mo with the oxygen concentration <0.10%, show several times greater life and durability than the rings produced from wrought bearing steel as well as the rings made from Astaloy A powder.

Experimental atomisation equipment whose scheme is shown in Fig. 7.28 and the floor plan in Fig. 7.29 was developed to ensure efficient production of atomised prealloyed powders alloyed with manganese and chromium and with a low oxygen content.

Atomisation equipment was up to half filled with water for spraying water or oil for atomisation with oil and nitrogen. In the latter case, before atomisation the fuel tank was filled with nitrogen to avoid possible ignition of fuel vapour spraying by atomised metal particles.

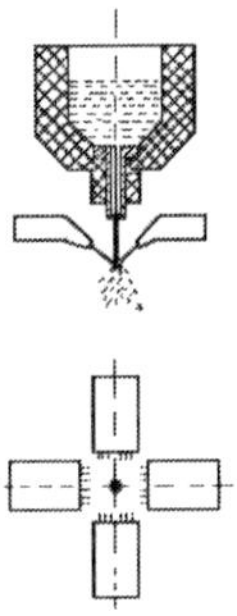

Fig. 7.31. Diagram showing distribution of four flat water or oil–nitrogen nozzles for producing metallic powders by atomisation [184].

The device was equipped with four flat nozzles shown in Fig. 7.30(a) with the section through the nozzle A–A in Fig. 7.30(b). This was a nozzle that sprayed water or oil–nitrogen medium. Arrangement of spray nozzles is shown in Fig.7.31. The nozzle dimensions were 8×1 mm. The thickness of the jet nozzle and, therefore, of the jet of the atomisation medium can be increased by increasing the seal thickness. The atomisation liquid medium in an amount of 120 l min^{-1} was supplied to the nozzles by a gear pump at a pressure of 15 MPa. In the case of atomisation with oil + nitrogen, nitrogen was supplied from gas cylinders.

7.3.5.1. Atomisation of Fe–0.7Mn–1Cr–0.2Mo powder with water or oil + nitrogen

5–6 kg metal charges were prepared from deep drawing sheet steel, high-carbon ferromanganese, ferrochromium and ferromolybdenum. Deoxidation wass carried out by adding 20 g of aluminum.

The charges were melted in a medium-frequency crucible furnace with a frequency of 4000 Hz, which was built above the crucible with an exit nozzle. The flow time of the given quantity of molten metal with a temperature of 1650°C was about 1 min. The powders were annealed after atomisation.

The oxygen content of the water-atomised FeMnCrMo powders was in the range 0.55–0.63%, *i.e.* on the level of the powders discussed in section 7.12.

The oxygen content of the powders atomised with oil + nitrogen was <0.1%, which corresponds to the oxygen content of Sumiron 4100S powder.

The powder forged steel from this powder reached a yield strength of 900 MPa, tensile strength of 1100 MPa, elongation of 15% and reduction in area of 50%. In the heat-treated condition, the steel with the combined carbon content of 0.62% reached a tensile strength of 1900 MPa and a hardness of 62 HRC. These mechanical properties are similar to those shown in Figs. 7.6, 7.7, 7.11 and 7.12.

7.3.5.2. Production of Fe–0.7Mn–1Cr–0.2Mo powders by water atomisation with inhibitors

The atomisation tests with the aim to produce the powders with the lowest oxygen content in this case were carried out also with water + 17 inhibitors of different chemical composition (5 g of each inhibitor in 10 l of water) and for comparison again with water. The results are presented in Table 7.19.

At plain water atomisation, the lowest oxygen content in the powder was in the range 0.71–0.46%, *i.e.* on the same level as in the previous case.

At atomisation with water with addition of the most active inhibitors, the powders reached the oxygen content in the range of 0.012–0.036%, *i.e.* on the level reached in oil–nitrogen atomisation and on the level of Sumiron 4100S powder.

These experiments proved that the Fe–Cr–Mn–Mo prealloyed powders with a low oxygen can be produced also by atomisation with water with the addition of special inhibitors.

Table 7.19. Chemical composition of the Fe–0.7Mn–1Cr–0.2Mo powders atomised by water and by water with inorganic inhibitors and annealed at 1000°C for 30 min in hydrogen (d. p. –30°C) [154,155]

*No. of melt	Agent	Content [mass %]					
		C	Cr	Mn	Mo	Si	O
20	Water+1	0.09	0.88	0.68	0.15	0.12	0.012
16	Water+2	0.14	1.83	0.85	n.d.	0.02	0.014
17	Water+3	0.18	1.69	0.71	n.d.	0.06	0.017
24	Water+4	0.12	0.74	0.57	0.15	0.06	0.021
25	Water+5	0.33	0.86	0.76	0.12	0.11	0.036
27	Water	0.13	0.88	0.58	0.15	0.07	0.464
26	Water	0.12	0.91	0.57	0.16	0.16	0.447
8	Water	0.22	1.55	1.33	n.d.	0.06	0.710

Remark: *internal numbers, 1 – 5 different inhibitors

8

Processing and properties of hybrid Fe–(Cr)–xMn–(Mo)–(V) steels

This chapter describes the methods for preparation of hybrid sintered steels on the basis of prealloyed Fe–Cr–Mo–(V) and Fe–(0.85,1.5)Mo powders with a manganese addition and their mechanical, toughness, structural and fracture characteristics in the as sintered and as sinter hardened conditions.

8.1. Properties of hybrid steels based on prealloyed Fe–Cr–Mo–V powders

In this case, the powders are also alloyed with vanadium in addition to chromium and manganese. From the thermodynamic point of view, these elements are characterised by high affinity for oxygen.

The samples were prepared form the prealloyed water-atomised vacuum-annealed powders:

– Fe–1Cr–0.3Mo–0.3V (KIP 103V), coded here K1,

– Fe–3Cr–0.3Mo–0.3V (KIP 30CRV), coded here K2,

The chemical composition of the K1 powder was: 0.01C, 0.08Mn, 1.07Cr, 0.32Mn, 0.29V, 0.15O, and of the K2 powder: 0.01C, 0.09Mn, 3.02Cr, 0.29Mo, 0.23V, 0.24O, both in mass %. The apparent density of the powders was 2.8 g cm^{-3}.

Manganese in the amount of 0–3.0% in the form of high-carbon ferromanganese (6.7% C, 0.2% Si, 0.2% P, 76% Mn, balance Fe, milled in air, mean particle size 20 μm), natural graphite in the amount of 0–0.24% considering the carbon content in ferromanganese used and 0.8% HW wax as a lubricant were added to the powders K1 and K2 were added.

The investigated material systems are coded here: K1, K2 – 0% Mn, K11, K21 – 1% Mn, K12, K22 – 1.5% Mn, K13, K23 – 2% Mn, K14, K24 – 3% Mn (all with 0.24% C).

The standard tensile, transverse rupture and impact energy test specimens were compacted at 690 MPa and sintered: a) at 1120°C for 60 min in cracked ammonia (dew point –33°C, flow rate 7 l min^{-1}) in a retort laboratory furnace under a getter (alumina + 5% graphite) in a double steel box, b) at 1200°C for 60 min in hydrogen in a laboratory pusher furnace also under a getter (alumina + 5% graphite) in a double steel box.

8.1.1. Mechanical and toughness properties and microstructure

The chemical composition and density of the sintered samples are given prior to the evaluation of the mechanical and toughness properties of the investigated steels in dependence on the processing conditions.

As sintered chemical composition. The carbon content in the K1 group sintered samples was in the range 0.21 to 0.23% and in the K2 group 0.23–0.24% (1120°C), *i.e.* similar to the values prior to sintering (0.24% C). A small manganese content loss in the samples was observed due to the known reverse sublimation of manganese (demanganisation). A higher manganese loss at a higher sintering temperature was determined. An increase in the manganese content for the K1 and K2 steel specimens (without Mn addition) to 0.13–0.21% compared with the starting Mn content in the powder (0.08–0.09%) was caused by the Mn vapour escaping from the specimens with the higher Mn content followed by their condensation on the surfaces of the specimens with lower or no Mn content.

Density. The as sintered density of the tensile test specimens based on the powder K1 was in the range 7.06 to 7.04 g cm^{-3} and the density of the specimens based on the powder K2 was in the range 7.01 to 6.95 g cm^{-3}. The manganese addition caused a decrease of the sintered density of the steels based on the powder K1 by 0.3 to 1.7% and of the steels based on the powder K2 by 0.6 to 1.3% compared with the density of the steels without manganese addition.

8.1.1.1. Mechanical properties

Tensile strength (Rm). The dependence of tensile strength on manganese content for the investigated steels is shown in Fig. 8.1. The effect of higher sintering temperature demonstrated on all tested mechanical properties, namely on those based on the K2 powder was investigated. The manganese addition to the powder K1 after sintering at 1120°C caused a linear increase of tensile strength from 291 MPa (0% Mn) to 594 MPa (3% Mn), *i.e.* by 100 MPa/1% Mn. In the case of the K2 powder the addition of only 1% Mn increased the tensile strength from 395 MPa (0% Mn) to 652 MPa (1% Mn), *i.e.* an increase by 257 MPa/1% Mn without any further increase of tensile strength with increasing Mn addition. The highest tensile strength of 661 MPa (density 7.11 g cm^{-3}) in steel K22 (1.5% Mn) and of 929 MPa (density 7.05 g cm^{-3}) without the effect of 2 and 3Mn additions after sintering at 1200°C for 1 h was recordd.

The relatively highest tensile strength values after sintering at both temperatures were obtained with the total Cr+Mn content of about 4–4.5% at the carbon content in the range 0.21 to 0.24%.

Hardness. The dependence of hardness on manganese addition of all the investigated steel specimens is shown in Fig. 8.2. There are some differences in hardness values depending on the test bar type, and they are greater for the samples based on the powder K2 due to demanganizing of the test sample surfaces. The highest hardness of 185 HV 10 in the steel K1 group and of 240 HV 10 in the steel

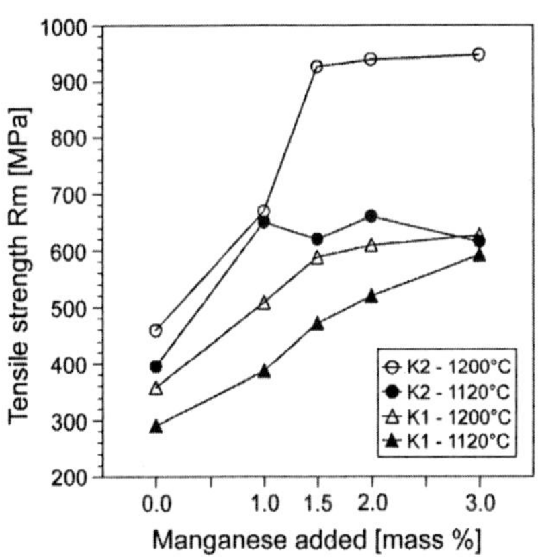

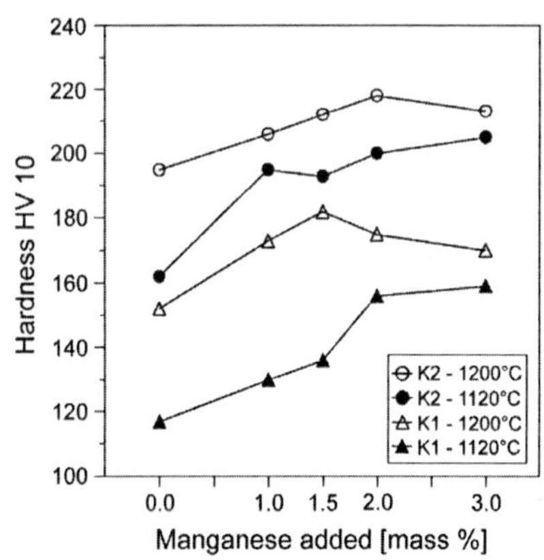

Fig. 8.1 (left). Dependence of tensile strength of steels based on the powders K1 and K2 sintered at 1120°C and 1200°C for 60 min on manganese addition [185,186].
Fig. 8.2 (right). Dependence of surface hardness HV 10 (tensile specimens) of the steels as in Fig. 8.1 [185,186].

group K2 with addition of 3% Mn was attained. The hardness of 173 HV 10 in steel K12 and of 255 HV 10 in steel K22 was attained after sintering at 1200°C.

Transverse rupture strength (TRS). The transverse rupture strength of the investigated steels in dependence on manganese addition is shown in Fig. 8.3. The course of TRS curves is partly similar to the curves for Rm. TRS increased with increasing manganese content. The steels based on the powder K1 sintered at 1120°C had the TRS in the range 729 to 1095 MPa and the steels based on the powder K2 in the range 721 to 1052 MPa. Sintering at 1200°C caused an increase in TRS for K13 and K14 steel by ~120 MPa (maximum 1207 MPa) and for K22–K24 steels by ~235 MPa (maximum 1297 MPa) compared to those sintered at 1120°C. The addition of 1.5% Mn to the powder K2 caused after sintering 1200°C an increase of TRS by ~300 MPa compared to the material without Mn addition.

The sintered density of the transverse rupture test specimens based on the powder K1 was in the range 7.00 to 6.88 g cm^{-3} and density of those based on the powder K2 was in the range 6.92 to 6.85 g cm^{-3}.

Impact energy. The dependence of impact energy (KC) of the investigated steels on the manganese addition and on sintering temperature was recorded as in previous cases, *i.e.* from 16.6 to 11.8 J for the K1 steel group and from 12.4 to 8.5 J for the K2 steel group, Fig. 8.4. Sintering at 1200°C caused a marked increase of KC values for K2 steel group to 18.0–11.2 J. Impact energy of K1 steel

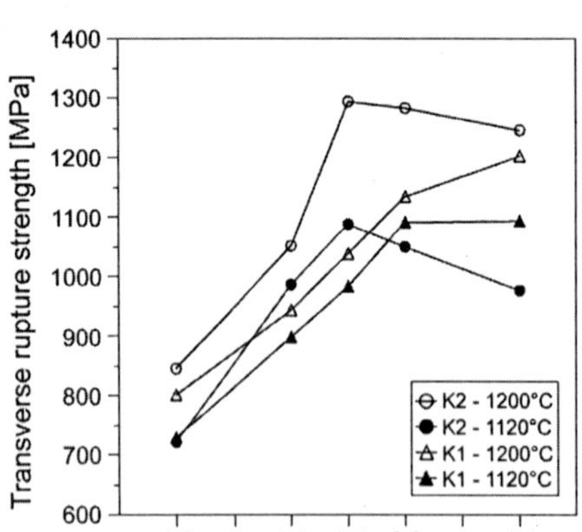

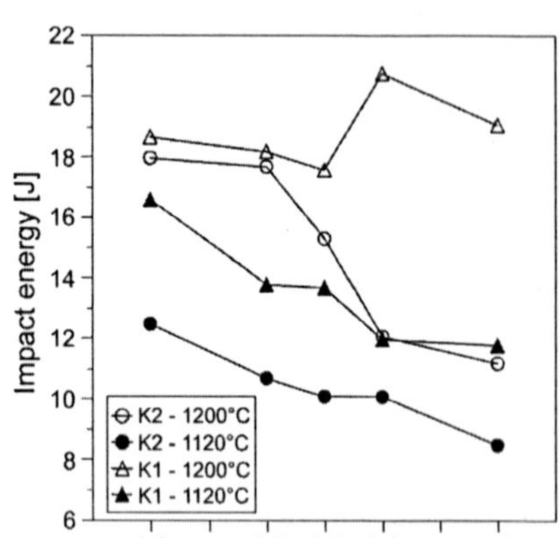

Fig. 8.3 (left). Dependence of TRS on manganese addition and on sintering temperature of the steels based on the K1 and K2 powders [185,186].

Fig. 8.4 (right). Dependence of impact energy (KC) on manganese addition and on sintering temperature of the steels based on the K1 and K2 powders [185,186].

group sintered at 1200°C reached average values of 18.9 J (maximum of 20.8 J at 2% Mn) without any dependence on manganese addition.

8.1.1.2. Microstructure

No Mn oxide networks were observed on the unetched metallographic cross-sections. The microstructure of the samples based on the powder K1 without manganese addition was ferritic–pearlitic with the ferrite microhardness of 194 HV 0.02 and with pearlite microhardness of 272 HV 0.02. A substantial alloying of the matrix by manganese was achieved by addition of 1% Mn. A bainitic homogeneous microstructure with the microhardness of 405 HV 0.02 was formed. The microstructure shown in Fig. 8.5(a) of the sample K1 with 3% Mn was homogeneous with the microhardness of 410 HV 0.02.

The white grains with the microhardness of 247 HV 0.02 were observed only sporadically, *i.e.* a substantially lower portion as compared with the plain iron matrix alloyed with manganese. The microstructure of the samples based on the powder K2 without manganese addition was also ferritic–pearlitic with the microhardness of ferrite grains of 281 HV 0.02. The bainitic microstructure with the microhardness of 405 HV 0.02 is shown in Fig. 8.6.

The microstructure analysis of the steel K21 (1Mn) showed that the matrix of the starting powders is more advantageous for alloying with manganese than the matrix of the plain iron powder. It was also the proof of the manganese vapour formation during heating and its condensation in the compact.

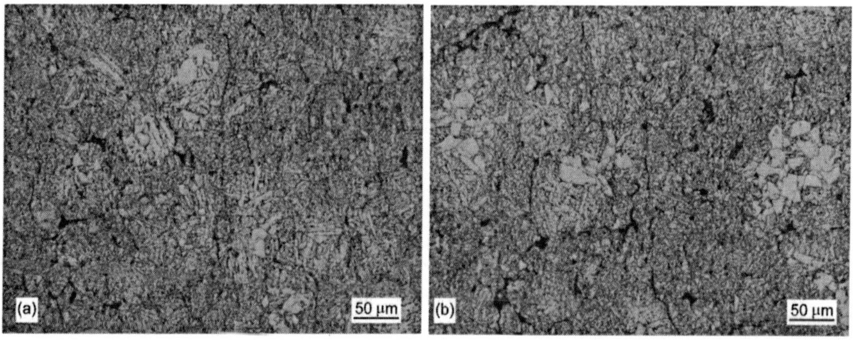

Fig. 8.5. Microstructure of the Fe–1Cr–0.3Mo–0.3V–1.0Mn–0.23C steel sintered at 1120°C. (a) – K12 (1.5Mn); Rm = 473 MPa, TRS = 965 MPa, KC = 12.8 J, 136 HV 10, (b) – K13 (2Mn, 1200°C), *Rm* = 611 MPa, TRS = 1137 MPa, KC = 20.8 J, 175 HV 10. Nital etched.

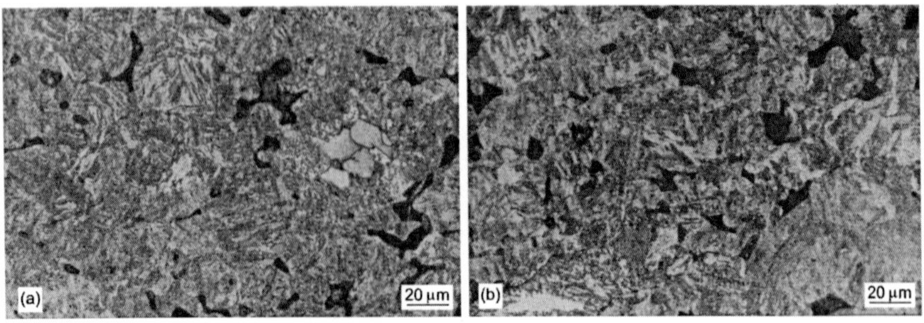

Fig. 8.6. Microstructure of the Fe–3Cr–0.3Mo–0.3V–1.0Mn–0.24C steel sintered at 1200°C. (a) specimen K21 (1.5Mn); *Rm* = 670 MPa, TRS = 1053 MPa, KC = 17.7 J and 206 HV 10, (b) specimen K22 (1.5Mn); *Rm* = 929 MPa, TRS = 1297 MPa, KC = 15.3 J and 212 HV 10. Nital etched.

8.1.2.1. Fractures of Fe–Cr–Mo–V–Mn–V steels

(a) Fractures of K1 steels without Mn addition. The fractures of steels based on the K1 and K2 prealloyed powders without Mn addition had ductile dimple morphology as shown in Fig. 8.7 and Fig. 8.8. The load-bearing cross-section of the sample K2 is larger compared to K1 with much finer dimples. This corresponded to the difference in tensile strength values, Fig. 8.1.

(b) Fractures of K1 steels with Mn addition. Another character of the fractures was exhibited by the K1 steels with manganese addition. The deciding feature was that the fracture in which 'inclusions' form is not known in wrought steels as well in sintered Fe–Mn steels.

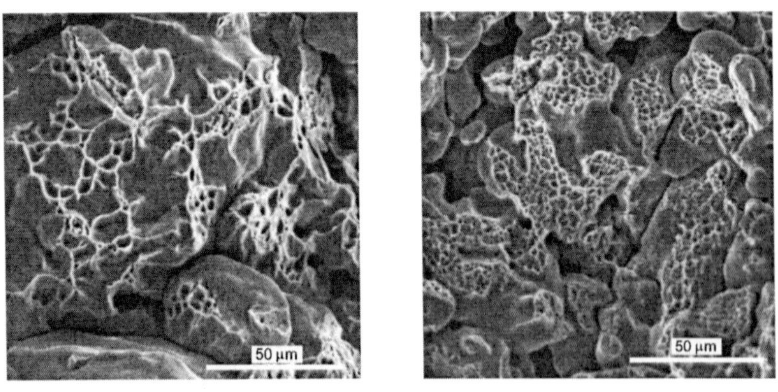

Fig. 8.7 (left). Fracture of a tensile test sample of K1 steel without Mn addition. 1200°C, 60 min, H_2. SEM [187,188].
Fig. 8.8 (right). Fracture of a tensile test sample of K2 steel without Mn addition. 1200°C, 60 min, H_2. SEM [187,188].

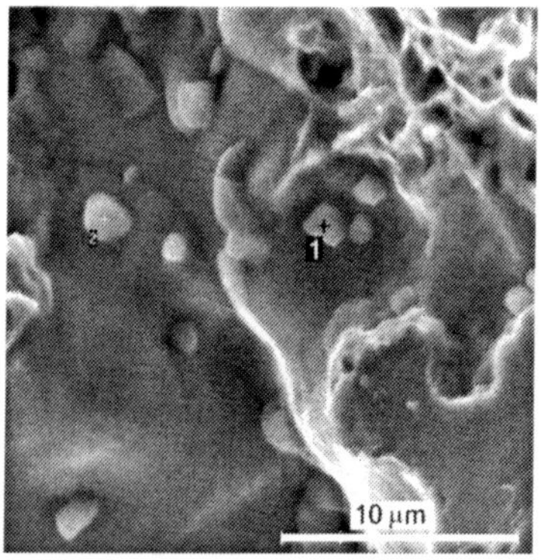

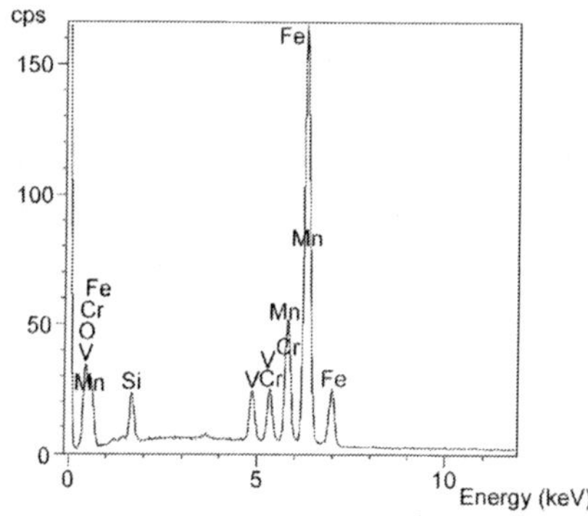

Fig. 8.9 (left). Fracture surface of K14 (3Mn) steel. 1120°C, 60 min, cracked ammonia, (original magnification ×10000). SEM.
Fig. 8.10 (right). Spectrum of the 'inclusion' 1 shown in Fig. 8.9.

The fracture shown in Fig. 8.9 through a large pore of the K14 steel. According to the spectrum shown in Fig. 8.10, the chemical composition of these inclusions confirms that they were formed by Fe, Mn, Cr and V, *i.e.* by the base alloying elements of the K1 steel.

Since these inclusions are formed by metallic elements, they can be referred to as 'metallic' inclusions (further also inclusions). It is possible to presume that small ferromanganese particles before total sublimation of manganese from them and dissolution of the remaining iron in the matrix acted as nuclei for the formation of these inclusions.

Figure 8.11 shows the detailed view of the fracture surfaces in the samples of K14 (3Mn) steels with a large number of 'inclusions' in comparison with the previous case. The inclusions are approximately spherical. The spectrum shown in Fig. 8.12 also shows the peak of the oxygen content and the concentrations of other elements in the analysed inclusions in comparison with the previous case, Fig. 8.10.

(c) *Fractures of the steels based on K2 (3Cr) powder*. Figure 8.13 shows the fracture surface of the K24 (3Mn) sample sintered for 60 min at 1200°C in hydrogen, and Fig. 8.14 is the spectrum of one of the 'inclusions'.

In this case, the fracture can be characterised as more uniform with uniform dimples containing spherical 'metallic' inclusions which were formed, in addition to iron, by the largest amount of chromium, a smaller amount of manganese and an even smaller amount of vanadium and oxygen reflected in a small peak. As shown in Fig. 8.15, such inclusions were also formed on the free fracture surfaces through a pore of the sintered specimens similarly as shown

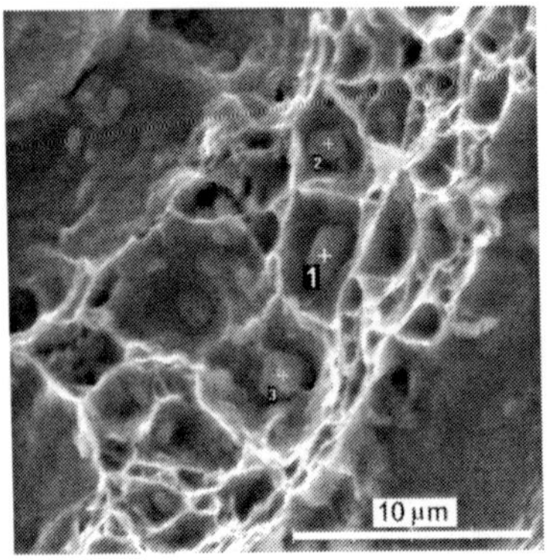

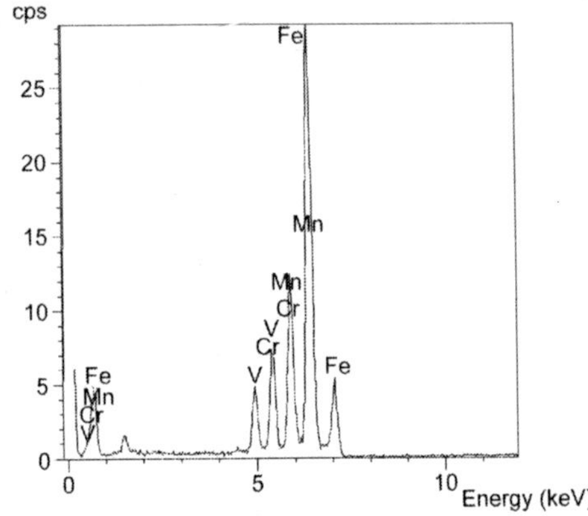

Fig. 8.11 (left). Fracture surface of K14 (3Mn) steel. 1120°C, 60 min, cracked ammonia (original magnification ×10000). SEM.
Fig. 8.12 (right). Spectrum of 'inclusion' 1 shown in Fig. 8.11.

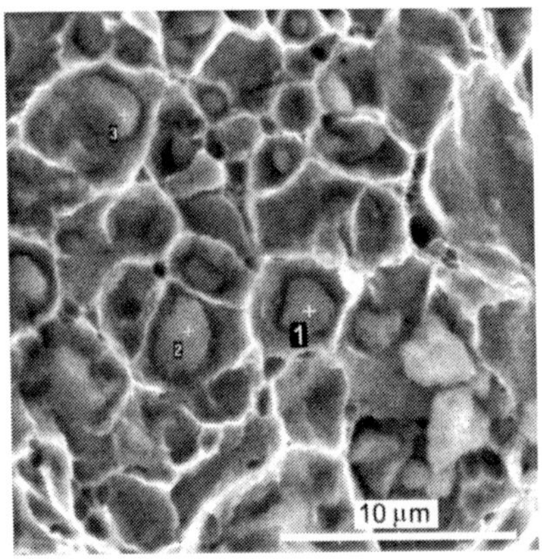

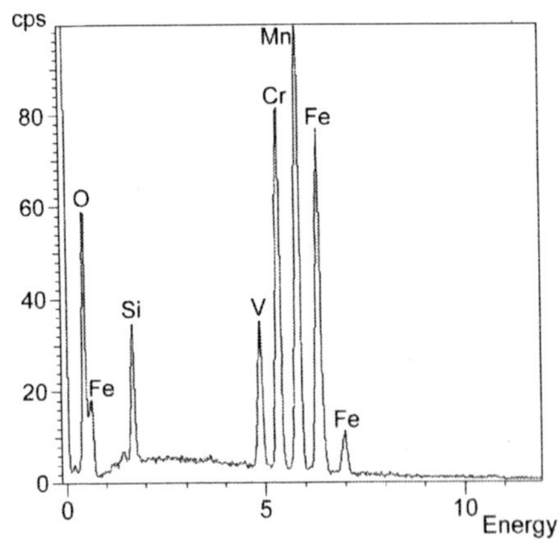

Fig. 8.13 (left). Fracture of a tensile test sample K2 (3Mn). 1200° C, 60 min, H_2, (original magnification ×10000). SEM [187].
Fig. 8.14 (right). Spectrum of inclusion 1 shown in Fig. 8.13 [187].

in Fig. 8.9, but they are smaller than, for example, in Fig. 8.13. The spectrum of the inclusion 3, shown in Fig. 8.16, confirms its basic equal composition as in the previous case, *i.e.* alloying with Fe, Mn, Cr and V. The distribution of the inclusions corresponded to the possible distribution of ferromanganese particles in a green compact. In this case, they were sintered together with the base prealloyed powder particles of the K2 steel.

Figure 8.17 shows the fracture of K24 steel with spherical inclusions and the spectrum of one of the inclusions.

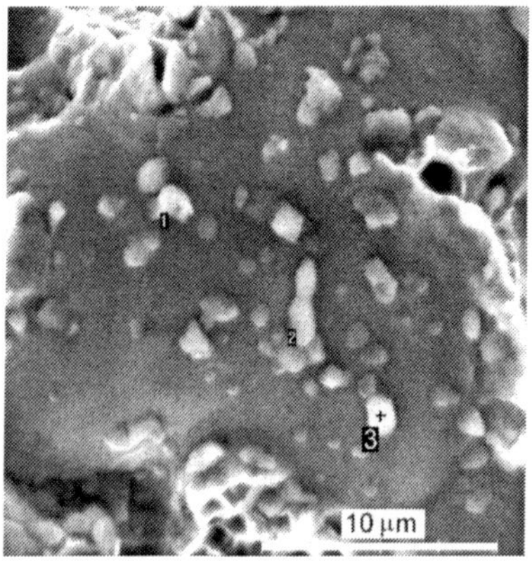

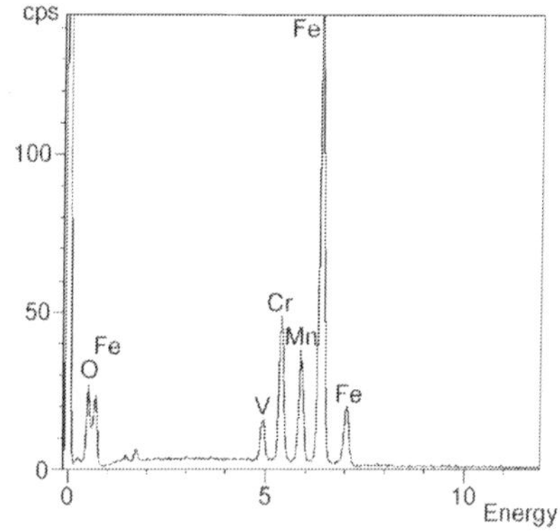

Fig. 8.15 (left). Fracture through a pore – free non-failed surface of the steel K2 with 3Mn addition (original magnification x 10000). SEM [187].
Fig. 8.16 (right) Spectrum of inclusion 3 shown in Fig.8.15.

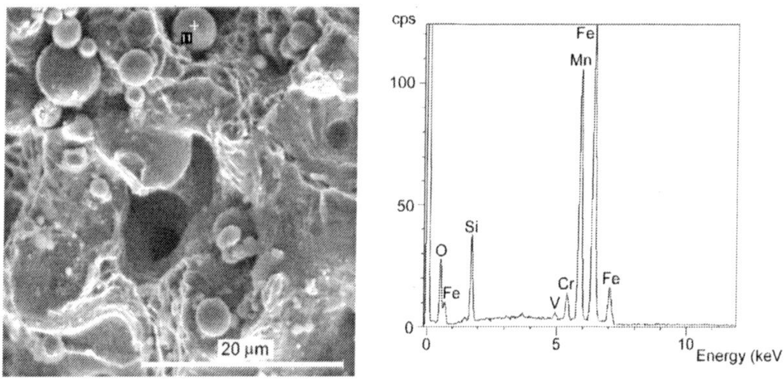

Fig. 8.17 (a) Fracture of K2 (3Mn) with spherical inclusions in ball form formed in K24 steel (3Mn). 1200°C, 60 min, H_2 (original magnification – ×10000). SEM. (b) Spectrum of the inclusion 11.

The fracture of K24 (3Mn) steel shown in Fig. 8.17 does not exhibit the characteristic dimple morphology as in the previous cases. The spherical inclusions of different sizes formed the clusters. This was caused by insufficient mixing of the ferromanganese powder with the base K2 powder.

The size of the dimples on the fracture surfaces of the K2 steel with Mn addition was more uniform with spherical inclusions of approximately the same size as on the fracture surfaces of the K1 steel. The size of these inclusions measured on the photographs was in the range ~1–3 μm. According to the analysis of the spectra, the 'metallic' inclusions on the fracture surfaces of the K2 steel contained higher concentrations of Mn, Cr and V in comparison with those analysed on the steel K1 fracture surfaces.

(d) Effect of manganese addition on the composition of the 'metallic' inclusions in fractures of 3Cr and 1Cr steels with different manganese addition. In addition to the previously mentioned spectra of the analysed 'metallic' inclusions, these inclusions were also analysed on the fracture surfaces of other samples of the steels K2 and K1, sintered for 1 h at 1200°C in hydrogen. The concentrations of the individual elements in the inclusions are presented in Table 8.1.

The results show that all the analysed 'metallic' inclusions were characterised by the highest manganese concentration. The highest chromium concentration was recorded on the inclusions on the fracture surfaces of the steel K21. The data show that all the analysed 'metallic' inclusions exhibited the highest Mn concentration in spite of the fact that the tested steels contained different Mn contents. The 'inclusions' on the fracture surface of this steel contained also the highest Cr concentration which may be attributed to diffusion interaction of 3Cr+1Mn. On the other hand, in the case of K14 steel with 1% Cr this may be attributed to the diffusion interaction of 1Cr+3Mn.

The vanadium concentration in the analysed 'metallic' inclusions in the dimples on the fracture surfaces of the steels K21, and in particular, K14 was several times higher than in the initial condition

Table 8.1. Concentrations of the metallic elements of three 'metallic' inclusions in fractures of K21 and K13 steels and of four in fractures of K14 steel (EDX, LINK analyzer)

Fe–Cr–xMn–0.3Mo–0.3V steel	FeK	MnK	CrK	VK	MoL	SiK
K21 – 3Cr–1Mn	35.1	23.6	5.1	8.7	1.4	1.5
	29.2	31.2	24.8	6.3	1.7	6.8
	39.1	25.1	28.9	5.7	0.6	0.6
K13 – 1Cr–2Mn	32.4	65.3	0.7	0.1	0.5	1.0
	51.7	41.2	1.0	0.3	1.0	4.8
	54.1	38.4	1.0	0.2	1.7	4.7
K14 –1Cr–3Mn	56.0	18.5	9.3	12.9	0.4	2.9
	62.3	15.4	9.3	12.0	0.2	0.7
	30.0	25.3	29.5	14.5	0.1	0.6
	10.7	40.3	19.2	19.6	4.4	5.8
Mean	44.6	29.8	14.4	7.4	1.1	2.7

(0.3%). It may be seen that the low concentration, similar to that in the initial condition, was recorded only for the inclusions in the K13 (2Mn) steel. The analysed appearance of the fracture surfaces of, in particular, the steel K2 shows that the highest values of tensile and bending strength were obtained as a result of adding 1.5% Mn, as shown in Fig. 8.1 and 8.2.

The results of these analyses and of the previously mentioned spectra of the individual inclusions on the fracture surfaces of the steels Fe–(1.3)Cr–xMn–0.5Mo–0.3V, sintered at temperatures of 1120°C and 1200°C, show that the nuclei for the formation of these inclusions were ferromanganese particles from which manganese did not sublimate completely, *i.e.*, already during preheating of the compacts, as indicated by the average manganese content of 29.8% in the inclusions. This means that no inclusions could be formed by the rest of one particle of high-carbon manganese which contained 76% Mn.

Manganese reacted by diffusion already during preheating in particular with vanadium from the base matrix, and the rest of the ferromanganese particles acted as nuclei for the formation of the spherical inclusions. This diffusion-active nucleus (Fe, Mo, V) reacted by diffusion with other elements from the matrix (Fe, Cr, Mo, Si). Silicon was introduced into the materials by ferromanganese particles which contained only 0.2% Si. The concentrations of the individual elements were influenced by the chromium content of the starting powder (1% Cr, 3% Cr) and by the manganese addition. (The dimples on the fracture surfaces of the samples prepared under the same conditions using the prealloyed powder Fe–3%Cr–0.5%Mo (Astaloy CrM) did not contain the previously mentioned and analysed inclusions).

8.2. Processing and properties of hybrid Fe–Cr–Mo–xMn–C steels

8.2.1 Properties of Fe–3Cr–0.5Mo steel with manganese addition

The specimens were prepared from Fe–3% Cr–0.5% Mo prealloyed powder (Astaloy CrM), coded here H. The manganese addition in the amount of 1, 2 and 3 mass % in the form of medium-carbon ferromanganese (80% Mn, 1.1% C, 0.98% Si, particle size <45 μm) milled in a ball mill in air, natural graphite in the amount of

Table 8.2. Sintered density (ρ_s, g cm^{-3}), carbon (C_c) and Mn content of hybrid H1, H2 and K2 steel samples. *Mn added, C_c and Mn in mass %

Steel	H1, 1250°C/30 min				H2, 1180°C/40 min				K2, 1200°C/60 min				
*Mn	0	1	2	3	0	1	2	3	0	1	1.5	2	3
ρ_s	6.98	7.06	7.01	6.97	7.05	7.04	7.01	6.98	7.01	7.03	7.04	7.02	6.99
C_c	0.21	0.21	0.22	0.22	0.20	0.21	0.22	0.23	0.20	0.22	0.23	0.20	0.20
Mn	0.10	0.89	1.93	2.71	0.25	0.95	1.86	2.80	0.17	0.87	1.24	1.76	2.75

0.24% and 0.8% HW as a lubricant were added to the base powders.

The standard tensile, bend (transverse rupture) strength and impact energy test bars were compacted at 600 MPa. The samples were sintered at: a) 1250°C for 30 min in a laboratory tube furnace in a steel box in cracked ammonia (dew point –30°C), coded here H1, b) 1180°C for 40 min in an industrial pusher furnace equipped with a delubing zone in a steel box in $70N_2/30H_2$ atmosphere (dew point of –50°C), coded here H2.

8.2.1.1. Base characteristics and mechanical properties

Density and chemical composition. The as sintered base characteristics of hybrid Fe–Cr–Mo–Mn steel (3Cr) samples sintered under laboratory and industrial conditions are listed in Table 8.2. The properties of K2 steel are presented for comparison with those based on the H powder.

The manganese and carbon contents in specimens sintered under laboratory as well as under industrial conditions in the atmospheres with the dew point of –30°C and –50°C were on the starting levels without a marked loss of carbon and manganese. In box sintering of many compacts, whether in laboratory or industrial conditions, the compacts were under a 'cloud' of manganese vapour enveloping every compact, *i.e.* the total batch of the specimens, even those with a lower or no addition of manganese, as was previously explained. This proves manganese contents in 0Mn samples (0.10%, 0.25%, 0.17%).

8.2.1.2. Mechanical properties

The dependence of the mechanical properties of the H1, H2 and K2 steels on manganese addition is shown in Figs. 8.18–8.21. The highest UTS of 940 MPa and TRS of 1300 MPa was obtained for the K2 steel with 1.5% Mn. The UTS of 856 MPa and TRS of 1397 MPa was recorded for the steel H1 with 2% Mn. The steel H2 with

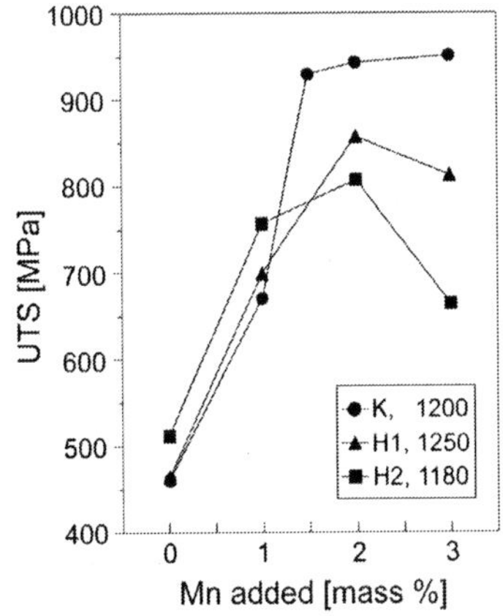

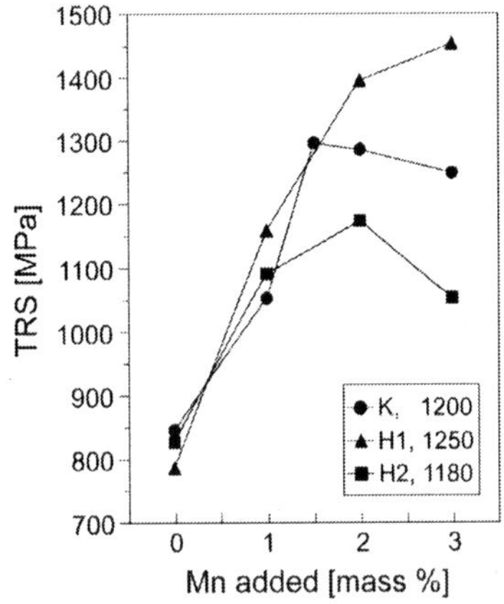

Fig. 8.18 (left). Dependence of tensile strength (UTS) of H1 (1250°C), H2 (1180°C) and K (K2) (1200°C) steels on manganese addition [189].
Fig. 8.19 (right). Dependence of transverse rupture strength (TRS) of H1 (1250°C), H2 (1180°C) and of K (K2) (1200°C) steels on manganese addition [189].

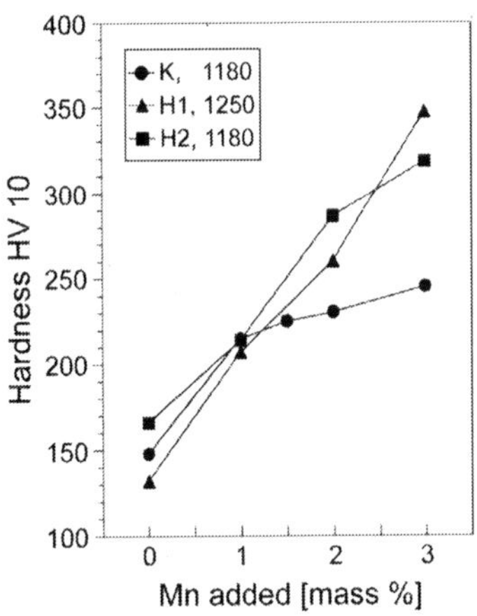

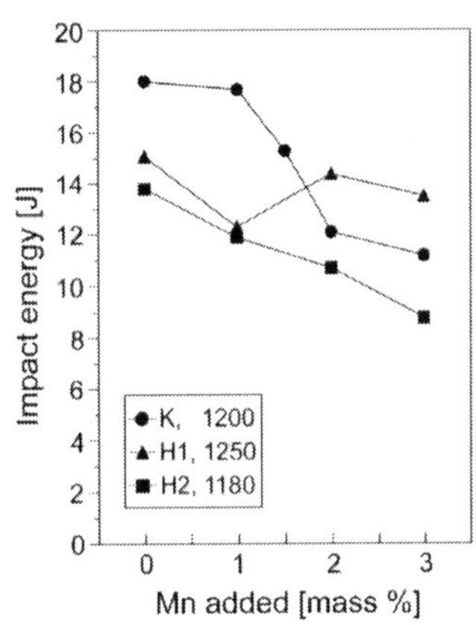

Fig. 8.20 (left). Hardness HV 10 of H1 (1250°C) and H2 (1180°C) steels in dependence on manganese addition [189].
Fig. 8.21 (right). Impact energy of H1 (1250°C) and H2 (1180°C) steels in dependence on manganese addition [189].

the same Mn addition showed a tensile strength of 806 MPa and TRS of 1175 MPa. Hardness increased and impact energy decreased with increasing manganese addition.

The hardness of the H1 and H2 steels increased with increasing manganese addition up to 350 HV 10 with no difference between them, whereas the K2 steel showed the highest hardness of

240 HV 10 with 3% Mn. The impact energy of the H1 steel in dependence on Mn addition was in the range 15 to 14 J. The impact energy of H2 steel decreased from 14 to 8.5 J with increasing Mn addition. The K2 steel showed the highest impact energy value of ~18 J with 1% Mn.

8.2.1.3. Microstructure and fracture

Microstructure. The characteristic microstructures of the H1 and H2 steels with the highest and the lowest tensile strength are shown in Fig. 8.22. The microstructures exhibit relatively high homogeneity, as shown by optical micrographs.

The microhardness measurements of the microconstituents in the microstructures of the H1 and H2 steels confirmed the increasing values with increasing Mn addition at total lower values afer sintering under industrial conditions compared to the values after laboratory sintering listed in Table 8.3.

According to the microhardness values, the formation of martensite was detected in individual microvolumes of the microstructures of the steels with 1% Mn at slow furnace cooling rates in the

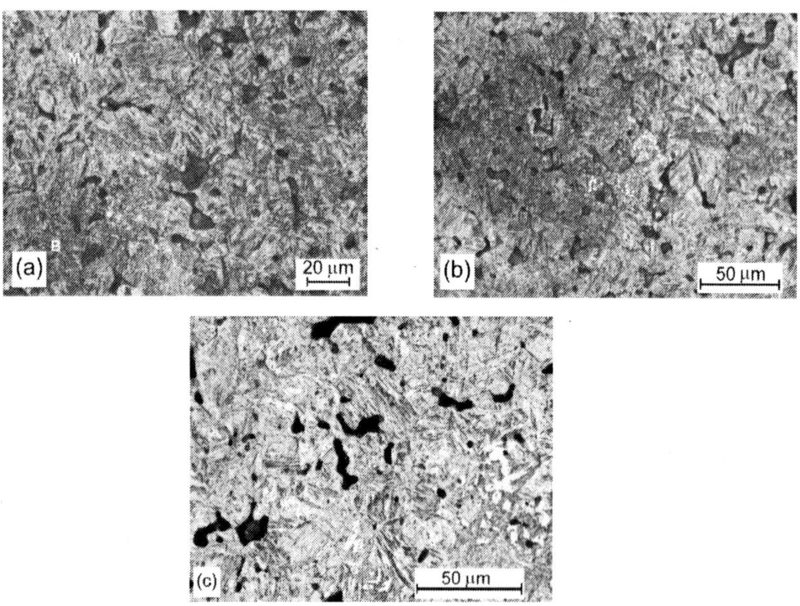

Fig. 8.22. Optical micrographs of the cores of the steel tensile strength samples: (a) H1 (2 Mn, Rm = 856 MPa), (b) H1 (3 Mn, Rm = 811 MPa), (c) H2 (2 Mn, Rm = 806 MPa). Nital etched.

Table 8.3. Microhardness (HV 0.025) values of the microconstituents in the microstructures of the H1 and H2 steels. *Mn added

Steel	H1, 1250°C/30 min				H2, 1180°C/40 min			
*Mn [%]	0	1	2	3	0	1	2	3
HV 0.025	152–294	220–532	270–702	329–819	185–327	240–492	255–534	281–577

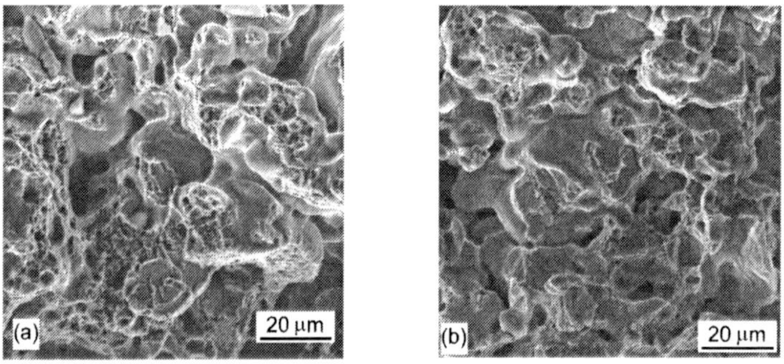

Fig. 8.23. Fracture surfaces of: (a) H1 (2Mn) steel (Rm = 856 MPa), (b) H2 (1.5Mn) steel (Rm = 806 MPa) – the highest tensile strength values. SEM.

range of 492 to 819 HV 0.025. Higher microhardness values in the microstructure of the H1 and H2 steels with carbon contents of only 0.24% C is to be noted. The real 'homogeneity' of the microstructures of these steels was characterised by fine pearlite, bainite and martensite in dependence on manganese addition, as follows from the microhardness values.

Fracture. Figure 8.23 shows the fractures of the steels H1 and H2 with the highest tensile strength values. They are characterised by the ductile dimple morphology and by individual small transgranular cleavage facets without inclusions in the dimples as in fractures of the K1 and K2 steels.

The presented results confirm that the sintered steels with high mechanical and toughness properties can be produced on the basis of hybrid powders, using prealloyed Fe–3Cr–0.5Mo powder with admixed manganese in the amount of 1 to 3% in the form of medium-carbon ferromanganese sintered under laboratory conditions at 1250°C as well under industrial sintering at 1180°C.

8.3. Effect of manganese carrier on properties of hybrid Fe–Cr–Mo–xMn steels

The following powders for the preparation of the samples for investigation were used:

(*i*) prealloyed Fe–3Cr–0.5Mo (Astaloy CrM),

(*ii*) manganese carriers:

– electrolytic Mn (supplied, particle size <20 μm, 0.37O) – EMn,

– medium-carbon ferromanganese (80% Mn, particle size <45 μm, 0.67O) – FeMn,

– high-carbon ferromanganese (76% Mn, particle size <45 μm, 0.77O) – FeMnC,

(both ferromanganese grades were milled in a laboratory ball mill in air).

Manganese in the amount of 1, 2 and 3% in the form of the mentioned manganese carriers and of 0.8% HW wax as a lubricant were admixed to the Fe–3Cr–0.5Mo powder. The starting carbon content in all mixtures was 0.24% corresponding to the carbon brought to the system by high-carbon ferromanganese for the 3% Mn addition. Natural graphite in the amount to 0.24% C was added to the compositions without the addition of manganese, including those alloyed with FeMn.

The standard tensile, transverse rupture and impact energy test bars were compacted to a density of 7.0 g cm^{-3} and sintered at:

– 1250°C for 30 min in cracked ammonia (dew point –30°C) in a laboratory tube furnace,

– 1180°C for 40 min in a $70N_2/30H_2$ atmosphere (dew point –50°C)

Table 8.4. Sintered density (ρ_s, g cm^{-3}), and carbon (C_c) and manganese (Mn) content of hybrid Fe–3Cr–0.5Mo–(0,1,2,3)Mn–C steels sintered at 1180°C and at 1250°C. All C_c, Mn values in mass %, *Mn added

Sint. temp. [°C]		Manganese carrier									
			EMn			FeMn			FeMnC		
	*Mn	0	1	2	3	1	2	3	1	2	3
1180	ρ_s	7.05	7.04	7.00	6.97	7.04	7.01	6.98	7.04	7.03	7.05
1250		6.98	7.07	7.03	6.98	7.06	7.01	6.97	7.04	7.02	6.97
1180	C_c	0.20	0.20	0.23	0.25	0.21	0.23	0.24	0.21	0.24	0.24
1250		0.21	0.22	0.23	0.26	0.24	0.22	0.26	0.25	0.26	0.27
1180	Mn	0.09	0.88	1.94	2.98	0.98	1.96	2.91	0.92	1.88	2.95
1250		0.09	0.88	1.87	2.89	0.94	1.88	2.86	0.87	1.92	2.92

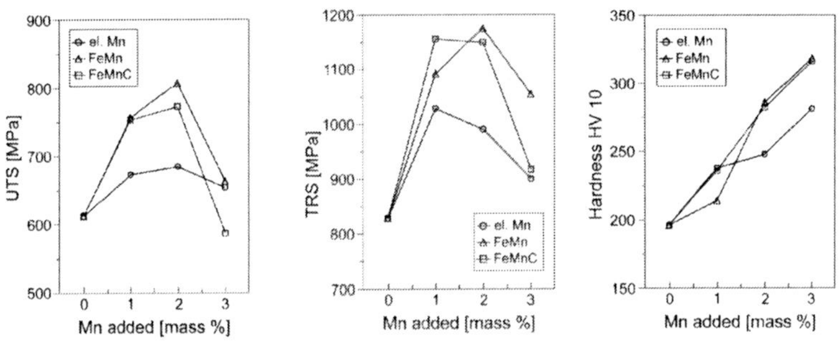

Fig. 8.24. The dependence of tensile strength, transverse rupture strength and hardness of hybrid Fe–3Cr–0.5Mo steels on manganese addition and manganese carrier grade [190].

in an industrial pusher furnace in a separate steel box placed between the boxes with other parts.

Density (ρ_s), *carbon and manganese content and dimensional changes.* Table 8.4 gives the as sintered density and carbon and manganese contents for the tested steel samples.

The sintered densities of the samples regardless of the manganese carrier grade were similar to the green density (7.0 g cm^{-3}). The carbon content in all steels was on the level of the starting value ~0.24% and the manganese content corresponded also to the added amount. The loss of C and of Mn according to the given values was not determined despite the fact that sintering was carried out in the atmospheres not fulfilling the thermodynamic requirements for their purity. A minor increase in the manganese content of the samples without manganese addition due to combined sintering of all samples was caused by manganese vapour in which the sintered compacts were held.

8.3.1. Mechanical properties

8.3.1.1. Sintering at 1180°C

The dependences of tensile strength, transverse rupture strength and hardness on the manganese addition and on the manganese carrier grade are shown in Fig. 8.24. The tensile and transverse rupture strength increased after adding 1 and 2% Mn. Hardness increased uniformly with increasing manganese addition.

The highest tensile strength of 800 MPa and transverse rupture strength of 1175 MPa were recorded for the manganese addition in the form of medium-carbon ferromanganese (FeMn), *i.e.* an increase by 100 MPa/1% Mn for UTS and by 150 MPa/1% Mn for TRS caused by the 1% Mn addition. The lowest values were recorded when electrolytic manganese was used as the manganese carrier. Hardness was increasing with increasing manganese addition up to ~300 HV 10 without a significant effect of the manganese carrier grade.

8.3.1.2. Sintering at 1250 °C

Mechanical properties. The dependences of tensile and transverse rupture strength and hardness on the manganese addition and manganese carrier grade are shown in Fig. 8.25.

Regarding the manganese carrier grade, the same effect as in the previous case was observed. The lowest tensile strength and transverse rupture strength values with the addition of manganese in the form of electrolytic manganese and the highest ones again with medium ferromanganese were achieved. In this case, the Fe–3Cr–0.5Mo–2Mn–0.22C steel has a tensile strength of 850 MPa and the Fe–3Cr–0.5Mo–3Mn–0.26C and the transverse rupture strength of 1455 MPa. Hardness increased with increasing manganese addition up to ~330 HV 10 without any significant effect of the manganese carrier grade. All the mechanical properties of the investigated steels were higher after sintering at 1250°C compared to those shown in Fig. 8.24.

Impact energy. Table 8.5 gives the impact energy values of Fe–3Cr–0.5Mo–xMn–C samples based on the CrM powder sintered under industrial and laboratory conditions.

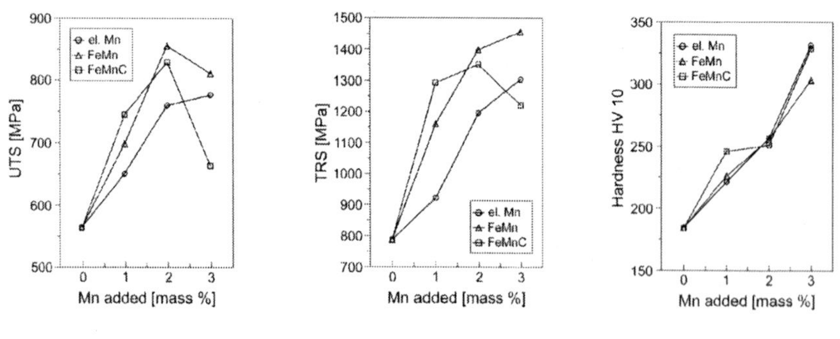

Fig. 8.25. Dependence of tensile strength, transverse rupture strength and hardness of hybrid Fe–3Cr–0.5Mo–xMn–0.24C steels on manganese addition and on manganese carrier grade. Sintering at 1250°C for 30 min. Density and carbon contents in Table 8.3 [190].

Table 8.5. Impact energy values (unnotched, KC) of Fe–3Cr–(1,2,3)Mn–0.5Mo–0.24C steel samples sintered at 1180°C and 1250°C. *Mn added, S. t. – sintering temperature

S. t. [°C]		Manganese carrier									
			EMn			FeMn			FeMnC		
	*Mn [%]	0	1	2	3	1	2	3	1	2	3
1180	KC [J]	13.8	11.9	9.1	8.7	12.9	10.7	8.8	12.3	11.0	7.9
1250		15.1	13.3	9.4	8.4	10.3	14.4	13.5	13.5	14.7	12.7

Higher impact energy values were recorded after sintering at 1250°C than at 1180°C. Marked differences in impact energy values were determined after industrial sintering compared to laboratory sintering. Higher values were attained when manganese was added as both ferromanganese carriers compared to those when manganese was added as EMn. In all cases the highest values were attained with 2% Mn addition.

8.3.1.3. Microstructure and fracture

Figure 8.26 shows the microstructure of the Fe–3Cr–0.5Mo–0.21C_c and Fe–3Cr–2Mn–0.5Mo–0.22C_c steels based on the CrM powder. Figure 8.26(a) shows the microstructure of the samples without Mn addition and Fig.8.26(b) with addition of 2Mn in the form of FeMn which resulted in the highest tensile strength in dependence on manganese carrier grade. The significant difference in the microstructures is shown between the specimens unalloyed with manganese and those alloyed with 2% Mn.

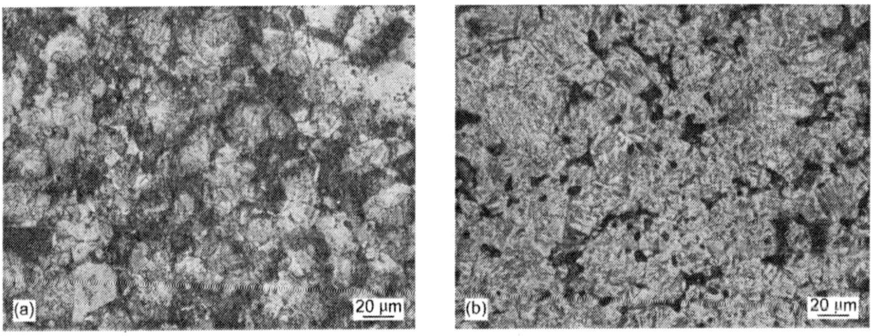

Fig. 8.26. Microstructures of Fe–3Cr–xMn–0.5Mo–xC steels sintered at 1250°C. (a) – without Mn addition, (b) – with 2% Mn addition, Mn carrier FeMn. Nital etched.

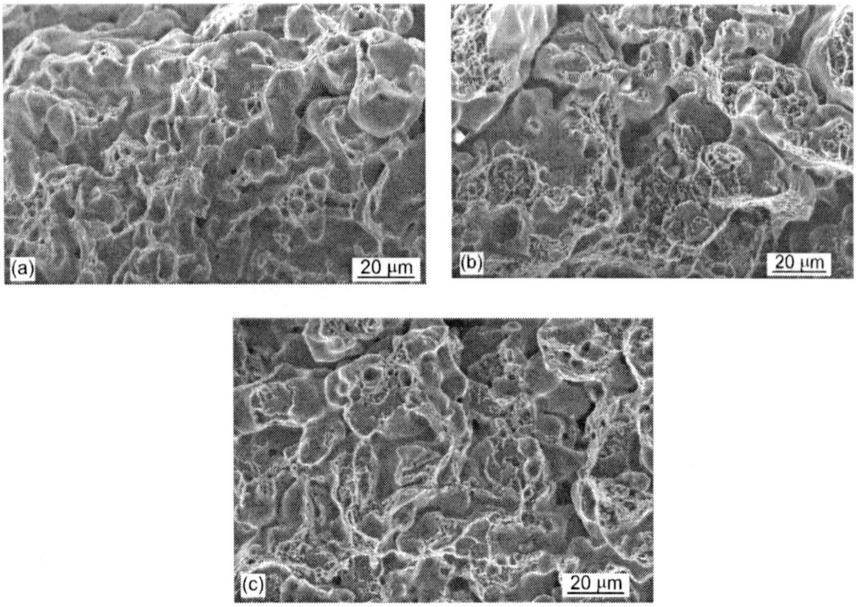

Fig. 8.27. Fractures of Fe–3Cr–xMn–0.5Mo–C steels sintered at 1250°C. (a) – without Mn addition, (b) – 2% Mn addition, Mn carrier – FeMn, (c) – 2% Mn, Mn carrier – FeMnC. SEM.

Figure 8.27 shows the fractures of the tensile strength samples of Fe–3Cr–xMn–0.5Mo–C steels sintered at 1250°C.

The fractures shown can be characterised by ductile failure with dimple morphology. The difference between them is only in the size of the dimples and on the load-bearing cross–section. It seems that the load bearing cross-section in the Mn-alloyed samples is larger and the dimples are finer when compared with Fig. 8.27(a) (0Mn) (the highest tensile strength of 850 MPa). Individual transgranular brittle facets were observed on the fracture of the samples alloyed with high-carbon ferromanganese, Fig. 8.27(c).

The tensile and transverse rupture strength values proved that the highest values were attained when Mn was added in the form of medium-carbon ferromanganese and the lowest ones when electrolytic manganese was used as the Mn carrier.

8.4. Sintered and sinter hardened hybrid steels

According to current knowledge, sinter hardening requires the powders possessing sufficient hardenability to form in the microstructure martensite, or possibly, martensite–bainite. The common sinter

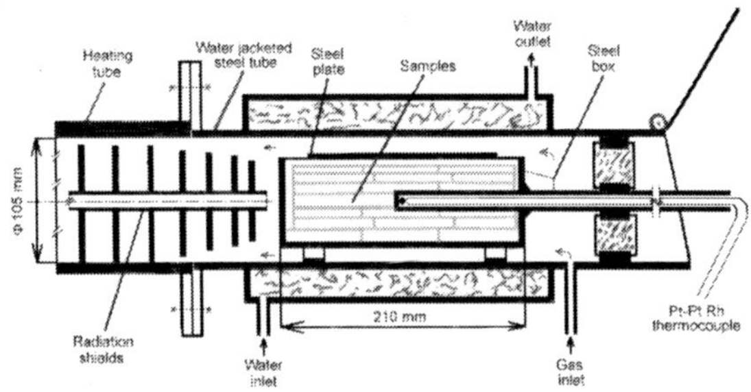

Fig. 8.28. Steel box in the sinter hardening position with the arrangement for the continuous measurement of the sintering temperature of the sample batch [191].

hardening steels consist of standard powder mixtures, or hybrid steels with an addition of 2% Cu and/or (1, 2)% Ni in spite of the problem with non-martensitic Ni-rich areas in the sinter hardened microstructure.

Regarding the main physical–metallurgical characteristics of Mn and Cr, concerning their hardenability, they fulfil the conditions for application for the preparation of hybrid sinter hardened steels based on Fe–Cr–Mo–Mn powders. For this reason they were studied in the present work with the aim to investigate the effect of manganese addition and of manganese carrier grades on the mechanical properties and structure of sintered and sinter hardened hybrid Fe–3Cr–0.5Mo–xMn–0.24C steel after sintering in low purity dissociated ammonia without addition of any other hardenability raising elements.

The samples for testing the tensile, transverse rupture and impact energy were prepared under the conditions described in Section 8.3. Sintering of all three types of the samples of the total weight of ~1200 g together in the steel box shown in Fig. 8.28 with the arrangement for the continuous measurement of the temperature in the centre of the batch, was carried out for 30 min at 1250°C in dissociated ammonia (dew point –30°C, 5 l/min) in a laboratory tube furnace at a heating rate of 0.13°C s^{-1} and furnace cooling at 0.06°C s^{-1}. To prevent the samples from sticking together, they were dusted with alumina in the steel box. The samples were dewaxed during preheating by holding for 30 min at 600°C.

Table 8.6. Sintered density (ρ_s), combined carbon (C_c), manganese (Mn) content (mass %) and linear dimensional change ($\Delta l/l$) of hybrid Fe–3Cr–0.5Mo–xMn–0.24C steel sintered for 30 min at 1250°C

Mn carrier			EMn			FeMn			FeMnC	
Mn added [%]	0	1	2	3	1	2	3	1	2	3
ρ_s [g cm^{-3}]	6.98	7.07	7.03	6.98	7.06	7.01	6.97	7.04	7.02	6.97
C_c [%]	0.21	0.22	0.23	0.23	0.24	0.22	0.23	0.22	0.23	0.23
Mn [%]	0.10	0.88	1.97	2.87	0.94	1.88	2.86	0.087	1.92	2.92

Note: The shavings for chemical analysis were prepared by the milling of 3 tensile test bars

8.4.1. As sintered properties

8.4.1.1. Basic characteristics

The density values and carbon and manganese contents and dimensional changes of the investigated steel samples measured on tensile strength bars are listed in Table 8.6.

The analysed carbon and manganese contents of all steels showed in relation to their starting levels only relatively small ± differences, *i.e.* on the level of the starting ones, without some marked effect of the manganese carrier grade. In contrast, in the samples without Mn addition, the manganese content increased to a mean value of 0.10%. As shown in Table 8.5, the carbon content of all the steels with small differences was on the level of the starting content of 0.24% C. No decarburizing or demanganizing of the sintered samples took place. The linear length changes of the tensile strength bars were in the range of +0.06 to –0.3% with a tendency to zero change at 3% Mn.

8.4.1.2. Mechanical properties

The dependence of the tensile and bending strength and hardness of tested specimens on the manganese addition and manganese carrier grade is shown in Fig. 8.29.

The highest tensile strength of 856 MPa and transverse rupture strength of 1397 MPa were obtained with 2% Mn addition (Mn as FeMn). The 1Mn steel reached the tensile strength of 745 MPa, transverse rupture strength of 1292 MPa at the hardness of 210 HV 10. The effect of the manganese carrier grade on the strength properties was recorded. The lowest values of the tested properties attained the steels with manganese addition in form of electrolytic manganese. The use of FeMnC compared to FeMn as the manganese

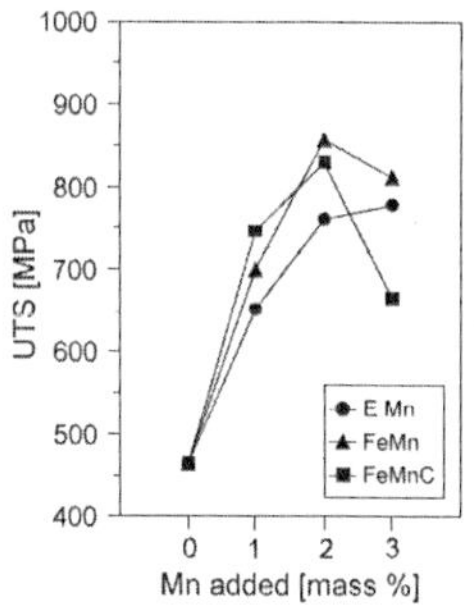

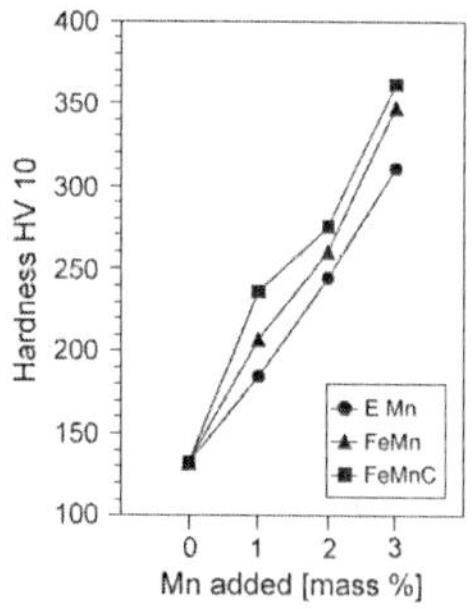

Fig. 8.29. Tensile strength (UTS) and apparent hardness of sintered hybrid Fe–3Cr–0.5Mo–xMn–0.24C steels in dependence on manganese addition and on manganese carrier grade. ($Rp_{0.2}$: 0Mn = 375 MPa, 1Mn (Mn as EMn) = 602 MPa, 1Mn (Mn as FeMn) = 657 MPa, 1Mn (Mn as FeMnC) = 601 MPa, 2Mn (Mn as FeMn) = 747 MPa) [191,192].

Table 8.7. Impact energy KC (unnotched) values of sintered hybrid Fe–3Cr–0.5Mo–xMn–0.24C steels in dependence on manganese addition and on manganese carrier grade. Other as in Table 8.5

Mn carrier		No	EMn			FeMn			FeMnC		
Mn added [%]		0	1	2	3	1	2	3	1	2	3
KC [J]	mean	15.1	13.3	9.4	8.4	12.3	14.4	13.5	13.5	14.7	12.7
	range	13.5–17.1	12.1–14.5	9.0–10.2	7.8–9.2	10.6–14.0	12.5–15.9	12.1–14.5	12.5–14.1	12.7–16.0	10.6–14.1

carrier resulted in a marked decrease of the strength properties of 3Mn steels. Hardness, measured on the metallographic cross-sections, increased linearly with increasing manganese content by ~117 HV 10/1% Mn without any marked effect of the manganese carrier grade.

The impact energy values of the investigated steels are listed in Table 8.7. Evidently, relatively small differences in impact strength in dependence on manganese content and manganese carrier were observed compared to the strength properties and hardness. The highest impact strength was shown by the 2Mn steels alloyed with FeMn and FeMnC and, on the other hand, the lowest ones by the 3Mn steels. It means the behaviour of impact energy values of the steels alloyed with FeMn and FeMnC corresponded to that of tensile strength. Considering the knowledge that the impact strength of the wrought steels containing >4.5% Mn decreased, it can be supposed that the same effect also applies to the impact strength as well as for tensile strength of the tested hybrid 3Mn steels. Special attention

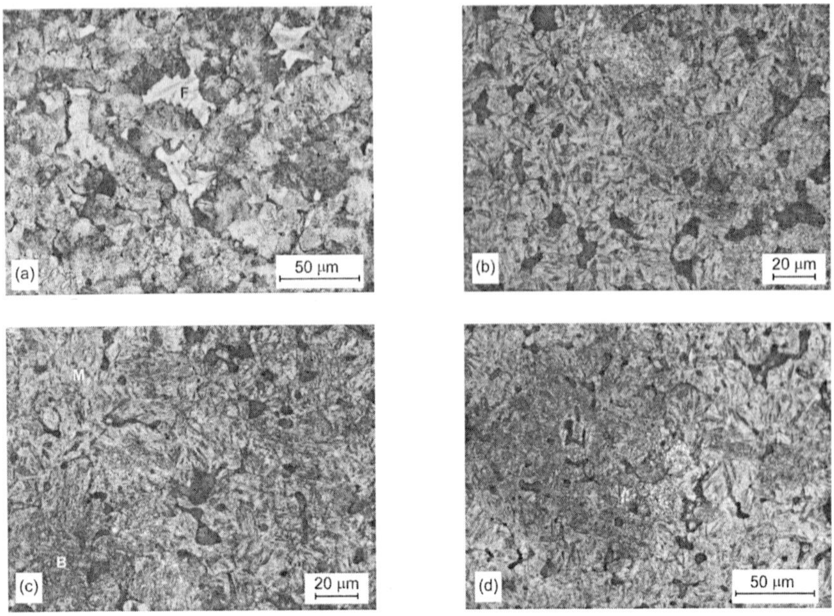

Fig. 8.30. Microstructures in the core of sintered hybrid Fe–3Cr–0.5Mo–(0,1,2,3) Mn–xC_c steel samples. Mn added as FeMn. (a) 0Mn: 0.10Mn–0.21C_c, *Rm* = 464 MPa, 132 HV 10; scattered ferrite, fine pearlite, upper bainite hardly discernible, (b) 1Mn: 1.03Mn–0.24 C, *Rm* = 699 MPa, 207 HV10; fine pearlite, upper and lower bainite, martensite, (c) 2Mn: 2.13Mn–0.22C_c, *Rm* = 856 MPa, 260 HV 10; scattered fine pearlite, upper and lower bainite, martensite, (d) 3Mn: 3.30Mn–0.26C_c *Rm* = 811 MPa, 347 HV 10; lower bainite and upper bainite, martensite. Nital etched.

should be paid to the impact energy of the steels alloyed with EMn which only decreased with increasing manganese content.

8.4.1.3. Microstructure and microhardness

The characteristic microstructures of steels with the highest tensile strength are shown in Fig. 8.30. They exhibited microstrcture heterogeneity, also characteristic for the hybrid steels, but lower than that known for mixed alloy steels.

Microhardness measurements of microconstituents, listed in Table 8.7 as minimal and maximal values, are used to characterise the nature of the phases in each steel in greater detail than is possible by examining the micrographs.

The microhardness values indicate that the microstructures of all materials are heterogeneous (contain phases ranging from ferrite up to martensite), somewhat depending on manganese addition and, to a minor degree, on the manganese carrier (diffusion interface prevailing between the phases). The microstructure heterogeneity was affected also by the structural and substructural heterogeneity of the base CrM particles which is best observable in the 0Mn steel, Fig. 8.29(a). There was a small number of white ferritic grains, brightly grey largest areas corresponding to the cores of the starting CrM powder particles and darker areas surrounding them with the microhardness up to 294 HV 0.025. The martensite isles with the microhardness of 509 HV 0.025 and higher were alredy observed in steels with 1% Mn.

The microhardness of martensite increased up to 937 HV 0.025 with increasing Mn content in the tested steels. The same is also valid for the minimum microhardness values increased up to 422 HV 0.025, corresponding to ferrite, fine pearlite and bainite in relation to the manganese content.

The hardness values up to 68 HRC were determined, which were also measured in some range and were affected by the manganese addition and the manganese carrier grade.

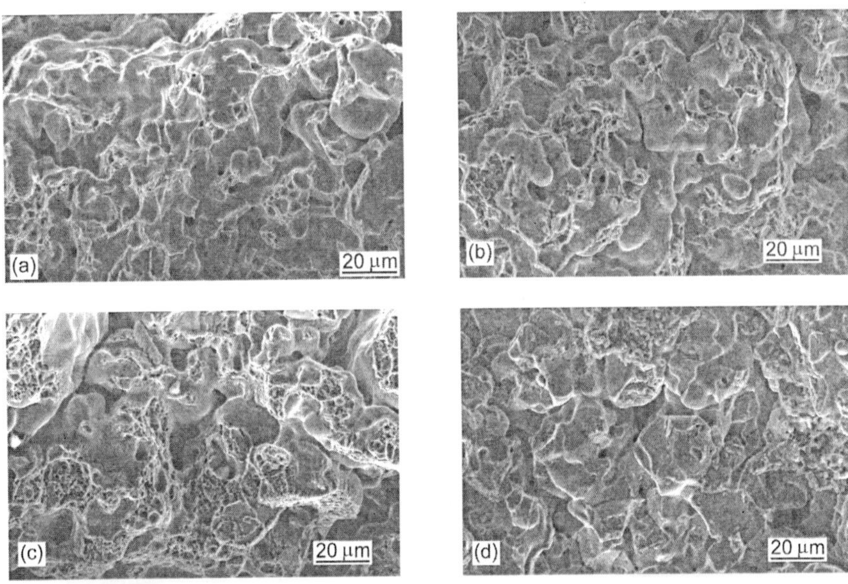

Fig. 8.31. Fracture surfaces of sintered hybrid Fe–3Cr–0.5Mo–(0,1,2,3)Mn–C steel samples with the highest tensile strength value obtained. Mn added as FeMn. (a) 0Mn: 0.10Mn–0.21C_c, Rm = 464 MPa, 132 HV 10, (b) 1Mn: 1.03Mn–0.24 C_c, Rm = 699 MPa, 207 HV10, (c) 2Mn: 2.13Mn–0.22C_c, Rm = 856 MPa, 260 HV 10, (d) 3Mn: 3.30Mn–0.26C_c, Rm = 811 MPa, 347 HV 10. SEM.

8.4.1.4. Fracture

Figure 8.31 shows the fracture surfaces of tensile strength bars of steels of maximum tensile strength (Mn as FeMn). The fractures of steels without and with 1 and 2% Mn exhibited prevailing ductile dimple morphology. The fracture of 3Mn steel had a mixed character, *i.e.* mainly intercrystalline and partly transcrystalline cleavage fractures and dimples in a minor fraction which demonstrated a decrease of the strength properties. The non-metallic inclusions were not found in dimples on the fracture surfaces. This confirmed the high-purity of the matrix powder particles and of the manganese carrier powder.

8.4.2. As sinter hardened properties

The position of the steel box during sinter hardening is shown in Fig. 8.28. The samples for sinter hardening after sintering were furnace cooled to 900°C, and then the box with the samples was pushed into the water-jacketed exit zone. The cooling rate in the temperature range 900 to 400°C was ~0.3°C s^{-1}.

8.4.2.1. Mechanical properties

The investigated steels by sinter hardening obtained higher values of tensile and bending strength and hardness, compared to those of the sintered ones shown in Fig. 8.32. The same tendency for both tensile and transverse rupture strength in relation to the manganese addition and manganese carrier grade as in Fig. 8.28 was observed as in the

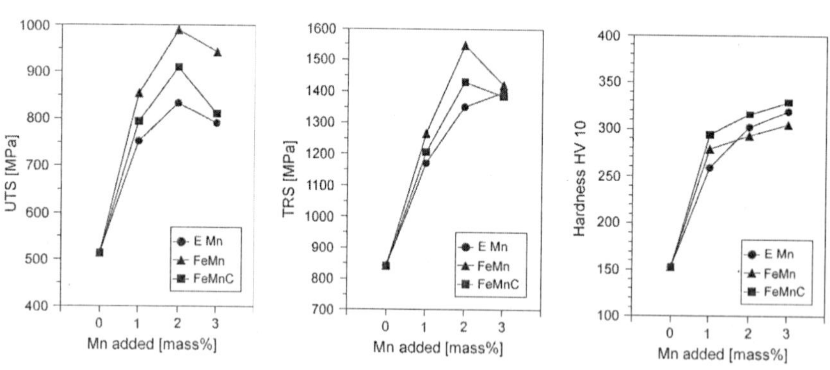

Fig. 8.32. Tensile strength (UTS), transverse rupture strength (TRS) and apparent hardness of sinter-hardened hybrid Fe–3Cr–0.5Mo–xMn–0.24C steels in dependence on manganese addition and on manganese carrier grade. Yield strength ($Rp_{0.1}$) of the steel: 0Mn = 340 MPa, 1Mn (Mn as EMn) = 736 MPa, 1 Mn (Mn as FeMn) = 618 MPa, 2Mn (Mn as FeMn) = 947 MPa, 1Mn (Mn as FeMnC) = 661 MPa [191–193].

Table 8.8. Microhardness (HV 0.025) and corresponding (micro) HRC values of microconstituents in the microstructures of sintered hybrid Fe–3Cr–0.5Mo–xMn–0.24C steels

Mn carrier	No		EMn			FeMn			FeMnC	
Mn added [%]	0	1	2	3	1	2	3	1	2	3
HV 0.025	152–294	153–509	255–687	313–785	220–532	270–702	329–819	352–557	381–719	422–937
HRC	(6)–29	(7)–51	23–59	31–63	15–51	26–60	25–65	27–53	39–61	43–68

Note: a) Microhardness of the starting CrM particles of different size was in the range of 77 to 103 HV 0.025. Cementite carbide in pearlite is $(Fe,Mn,Cr)_3C$. b) Measured HV 0.025 values of the phases: ferrite – 152–180, fine pearlite – 200–290, upper bainite – 310–380, lower bainite – 420–500, martensite – 550–937.

Table 8.9. Impact energy values (KC) (unnotched) of sinter hardened hybrid Fe–3Cr–0.5Mo–xMn–0.24C steel samples in dependence on manganese addition and manganese carrier grade

Mn carrier		No	EM			FeMn			FeMnC		
Mn added [%]		0	1	2	3	1	2	3	1	2	3
KC [J]	mean	16.7	13.3	10.5	8.4	13.8	14.8	12.5	10.6	12.2	9.5
	range	13.7–18.8	12.9–14.4	8.3–11.7	7.9–9.3	12.9–15.3	12.1–16.3	10.8–14.5	9.7–12.1	10.3–15.2	8.1–12.1

sintered state. (The results correlate with the low carbon content and the measured cooling rate which in terms of common sinter hardening process is regarded as low). In spite of this the mechanical properties obtained at lower carbon contents and lower cooling rate were high. The highest hardness of martensite was achieved with 0.5–0.6% C. The recommended cooling rates for sinter hardening usually are ~1.0–2.5°C s^{-1}. The highest tensile strength of 989 MPa and the bending strength of 1548 MPa were obtained for the 2% Mn steel (Mn as FeMn). In this way an increase by 238 MPa/1% Mn in tensile strength and by 353 MPa/1% Mn in transverse rupture strength was obtained. The 1Mn steel reached the tensile strength of 854 MPa, transverse rupture strength of 1265 MPa and the hardness of 294 HV 10. The decrease of strength and impact properties was recorded for the 3% Mn in the sintered state. A further increase of manganese to 2 and 3% caused only a relatively small linear increase in hardness of 293 to 316 HV 10 at 2% Mn and of 305 to 329 HV 10 at 3% Mn.

Table 8.10. Microhardness (HV 0.025) and corresponding (micro) HRC values of microconstituents forming the microstructures of hybrid sinter hardened Fe–3Cr–0.5Mo–xMn–0.24C steels. *Mn added

Mn carrier	No	EM			FeMn			FeMnC		
*Mn [%]	0	1	2	3	1	2	3	1	2	3
HV 0.025	136–483	272–629	311–687	460–705	260–608	340–652	460–683	265–669	391–724	500–775
HRC	(5)–48	26–57	31–60	46–60	24–56	34–58	46–59	25–59	40–63	49–63

Note: Measured HV 0.025 value for the phases: ferrite –136–170, fine pearlite 220–270, upper bainite – 300–380, lower bainite – 410–490, martensite – 520–775.

The sinter hardening conditions used did not affect to any significant extent the impact energy values of the investigated steel samples listed in Table 8.8, as compared to the sintered ones listed in Table 8.9. The highest sinter hardened KC value was determined also in the 2Mn steel samples. An increase of impact energy was obtained also by the 0Mn steel. On the other hand, a relatively marked decrease in impact energy values compared to the as sintered ones was recorded for the steels with Mn addition as FeMnC.

The microhardness range values of microconstituents formed as sinter hardened microstructures of tested steels, given in Table 8.10, characterise the phases in more detail than in the case of the sintered microstructures. An increase of the microhardness of softer phases in the range of 272 to 500 HV 0.025 and more uniform values of martensite in the range 629 to 775 HV 0.025 in the microstructures of all steels was recorded. It can be deduced from this that the as sinter hardened microstructures exhibited lower heterogeneity compared to the as sintered ones. All 1Mn steels also contained fine pearlite and martensite with the microhardness of 608 to 669 HV 0.025. This proves the effect of the higher cooling rate on the phase transformation of austenite of the tested hybrid Fe–Cr–Mo–Mn–C steels at a low carbon content of 0.24% C. Greater uniformity of microstructures, now containing predominantly lower bainite and martensite with a minimum of fine pearlite compared to the as sintered state, should be noted.

The strength and toughness properties and characteristics of the structural microconstituents of Fe–3C–0.5Mo–xMn–0.24C steels confirm that chromium and manganese with 0.5% Mo in the hybrid sintered steels fulfil the requirements on processing by the sinter hardening process also for low carbon contents (0.24% C) in contrast to other elements of the alloyed systems used for sinter hardening.

8.5. Properties of hybrid Fe–Cr–0.5Mo–xMn–C steels sintered under industrial conditions

The standard tensile, transverse rupture (bend) strength and impact energy test bars from the powder mixtures were prepared under the same conditions as in the previous cases. The effect of the manganese additions of 1, 2, 3% and of manganese carrier grades with the uniform starting carbon content of 0.24% C on the properties of all steels was investigated.

The samples (total mass ~1200 g) placed in a steel box (330 × 330 × 80 mm) between the boxes with the sintered structural parts of other composition, were sintered alone for 40 min at 1180°C in a $75N_2/30H_2$ atmosphere with an inlet dew point of –50°C in an industrial pusher furnace at a heating rate of ~15°C/min, and cooling rate of ~10°C/min, both calculated on the basis of time and length of the corresponding furnace zones. Schematic of the furnace with the temperature profile is shown in Fig. 8.33.

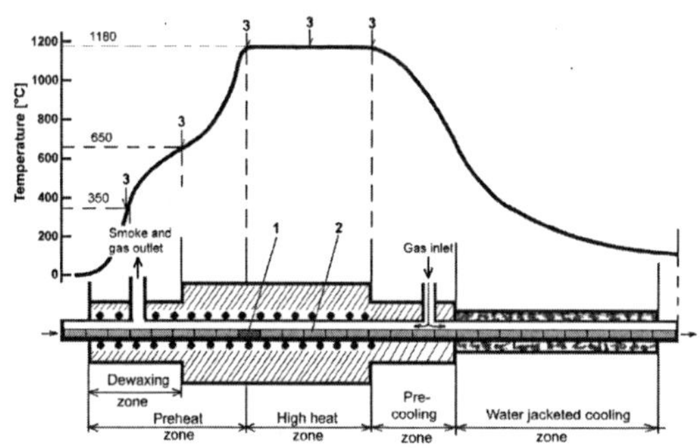

Fig. 8.33. Schematic of the pusher sintering furnace used and the corresponding temperature profile [194].

8.5.1. Basic characteristics

The density values and carbon and manganese contents of the investigated steel sample bars are presented in Table 8.10.

The analysed carbon and manganese contents of all steels showed, in relation to their starting levels, only relatively very small differences, *i.e.* neither decarburizing nor demanganizing without some effect of the manganese carrier grade was determined.

Table 8.11. Sintered density (ρ_s) (tensile bars), combined carbon (C_c) and manganese content (mass %) of hybrid Fe–3Cr–0.5Mo–(0,1,2,3)Mn–0.24C steels. *Mn added

Mn carrier	No.	EMn			FMn			FMnC		
*Mn [%]	0	1	2	3	1	2	3	1	2	3
ρ_s [g cm^{-3}]	7.05	7.04	7.00	6.97	7.04	7.01	6.98	7.04	7.03	7.02
C_c [%]	0.20	0.20	0.23	0.25	0.21	0.23	0.24	0.21	0.24	0.24
Mn [%]	0.25	0.88	1.94	2.86	0.95	1.96	2.93	0.94	1.95	2.92

Note: Carbon and manganese contents were determined from three tensile bar shavings.

In contrast, in the samples without Mn addition, an increase of manganese content to a mean value of 0.25% with the effect of manganese carrier grade was recorded.

8.5.2. Mechanical properties

Figures 8.34–8.37 show the dependence of tensile strength, bend strength, hardness and of impact energy values on manganese addition and manganese carrier grade.

The highest tensile strength of 806 MPa and bend strength of 1175 MPa were obtained with 2% Mn addition (Mn as FeMn). By this means an increase in tensile strength by 131 MPa/1Mn and in bend strength by 174 MPa/1Mn in relation to the tensile strength of 0Mn steel were achieved. The relative highest increase of both strength

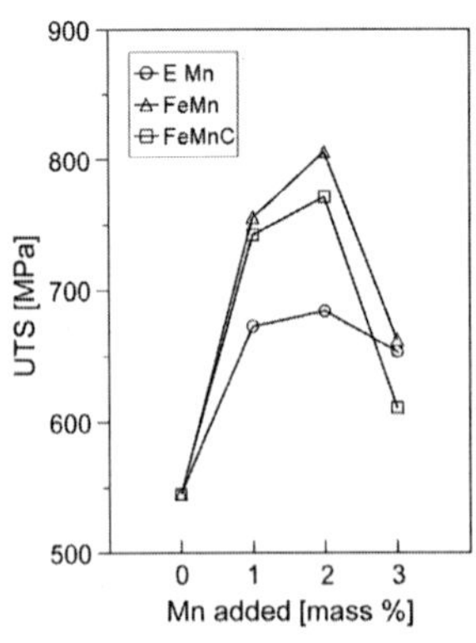

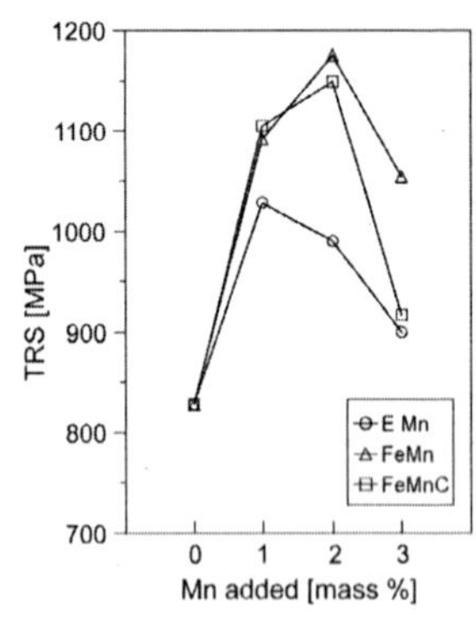

Fig. 8.34 (left). Tensile strength (UTS) of hybrid Fe–3Cr–xMn–0.5Mo–0.24C steels in dependence on manganese addition and on manganese carrier grade [194].
Fig. 8.35 (right). Transverse rupture strength (TRS) of hybrid Fe–3Cr–xMn–0.5Mo–0.24C steels in dependence on manganese addition and on manganese carrier grade [194].

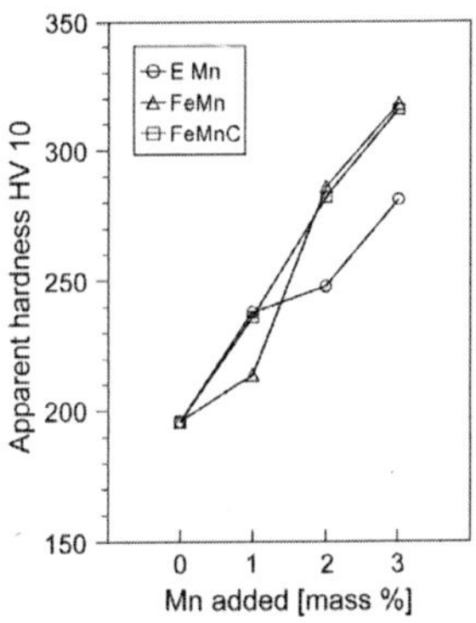

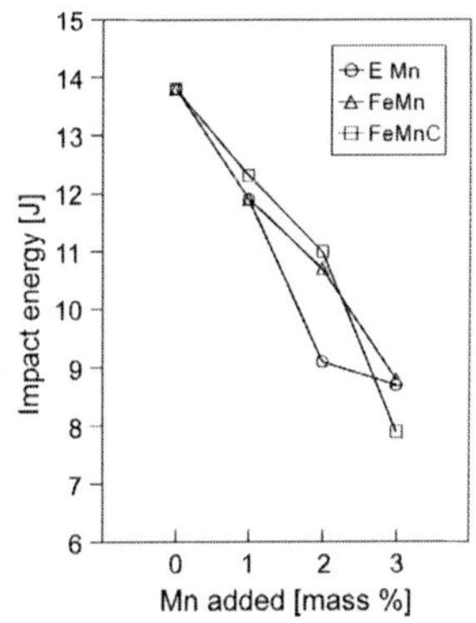

Fig. 8.36 (left). Apparent hardness (HV 10) of hybrid Fe–3Cr–xMn–0.5Mo–0.24C steels in dependence on manganese addition and on manganese carrier grade [194]. **Fig. 8.37** (right). Impact energy of hybrid Fe–3Cr–xMn–0.5Mo–0.24C steels in dependence on manganese addition and on manganese carrier grade [194].

properties was obtained with 1Mn addition, as demonstrated by the tensile strength value of 745 MPa and the bend strength of 1150 MPa (both Mn as FeMnC). A marked decrease of both strength properties was obtained for the 3Mn containing steels, especially those alloyed with manganese in the form of both ferromanganese grades. The effect of the manganese carrier grade was observed mainly on the strength properties of the steels with manganese contents above 1%. The lowest values of tensile and bend strengths were attained in the steels with manganese addition in the form of electrolytic manganese.

Hardness up to 318 HV 10 increased linearly with increasing manganese addition, with smaller values at 2 and 3Mn added as electrolytic manganese. Evidently, the impact energy values of the investigated steels decreased with increasing manganese content in the range of 13.8 J for 0Mn steel to 7.9 J for 3Mn steel without some marked effect of the manganese carrier grade.

8.5.3. Microstructure and fracture

8.5.3.1. Microstructure

Figure 8.38 shows the characteristic microstructures of the steels with the highest tensile strength (Mn added as FeMn). These steels exhibited a relatively homogeneous microstructure, characteristic of hybrid manganese-containing steels, and a significantly higher homogeneity than that of other mixed alloy PM steels.

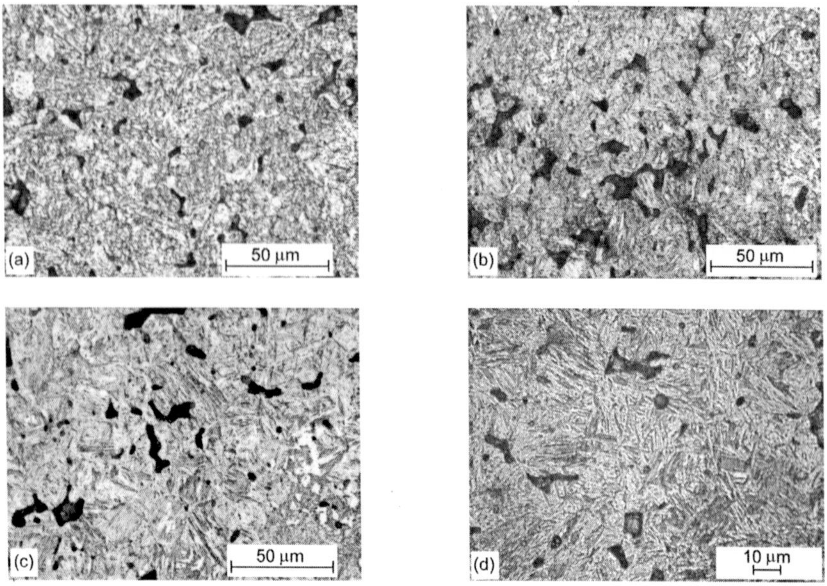

Fig. 8.38. Optical micrographs in core of hybrid Fe–3Cr–0.5Mo–xMn–C steel samples with the highest strength properties. Mn added as FeMn. (a) 0Mn (0.25Mn–0.20C_c, UTS = 506 MPa), (b) 1Mn (1.14Mn–0.21C_c, UTS = 756 MPa), (c) 2Mn (2.06Mn–0.23C_c, UTS = 806 MPa), (d) 3Mn (3.11Mn–0.24C_c, UTS = 663 MPa), Nital etched [194]. Note: (d) – higher magnification to show the microstructure in more detail.

The microhardness values of the microconstituents are shown in Fig. 8.39 as minimal (grey) and maximal values (black). These microhardness data were used to characterise the phases in each steel in more detail than is possible from the micrographs in Fig. 8.38. The microhardness values confirm that the microstructures contained fine pearlite, upper and lower bainite and martensite. The manganisation of the manganese-free steel to 0.25% Mn follows from Fig. 8.39 according to the determined microhardness values in the range of 185 to 327 HV 0.25. This was the result of the effect of manganese vapour formed by sublimation from the manganese carrier in the compact during sintering.

(Microhardness of the starting CrM powder particles of different size was in the range from 77 to 103 HV 0.025. Measured HV 0.025 values of the phases were: fine pearlite – ~185–281, upper bainite – ~312–343, lower bainite – ~459–492, martensite (M) – 503–659).

As shown in Fig. 8.39, the microhardness of the martensite islands increased in the range from 503 to 659 HV 0.025 with an increase in the Mn content. This also applies to the minimum microhardness values which increased from 228 to 343 HV 0.025, corresponding to

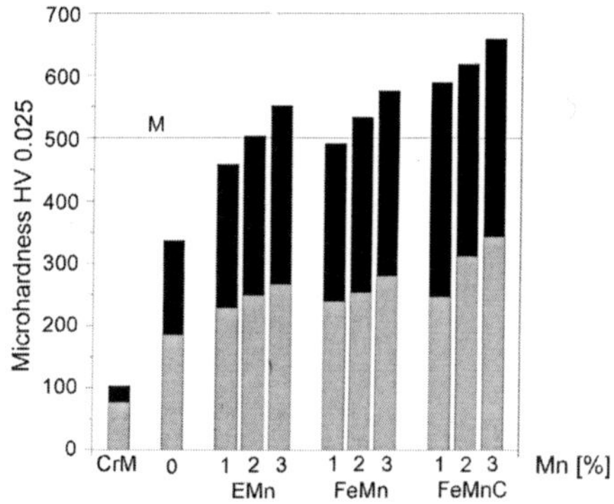

Fig. 8.39. Microhardness (HV 0.025) values of microconstituents in microstructures of hybrid Fe–3Cr–0.5Mo–xMn–0.24C steels (50 indentations in each microstructure). CrM–base prealloyed powder (0Mn, 0C).

fine pearlite and upper bainite, depending on the manganese content. The effect of the manganese carrier grade is demonstrated by the microhardness ranges. The martensite islands (590 HV 0.025) were observed already in steel with 1% Mn to which manganese was added as FeMnC. Increasing minimal and maximal microhardness values with increasing manganese addition reflect the increasing proportion of harder phases in the microstructures.

It can be assumed that some optimum ratio of softer phases (pearlite, bainite) and martensite at 2% Mn resulted in the highest strength properties, Figs. 8.34 and 8.35. In contrast, in the 3Mn steels a relatively high proportion of brittle martensite to that of bainite resulted in a marked decrease in the strength properties. The increasing microhardness of all phases with increasing manganese additions was reflected in the continuous increase of apparent hardness and decrease of impact energy, Fig. 8.37. The effect of the manganese carrier on the microstructure of the steels, confirmed by the microhardness values, is reflected in the strength properties.

8.5.3.2. Fracture

Figure 8.40 shows representative fracture surfaces of 1Mn, 2Mn and 3Mn steel specimens of maximum tensile strength (manganese added as FeMn) and of 3Mn steel (manganese added as FeMnC). The fracture of the 1Mn steel exhibited a ductile dimpled rupture morphology and small transgranular cleavage facets, Fig. 8.40(a). The fracture surface of the manganese-free steel exhibited a smaller fraction of ductile dimpled fracture and is comparable with that of the 1Mn steel. The fracture surface of 2Mn steel exhibited a mixed morphology, Fig. 8.40(b) with primarily large transgranular ductile rupture and a small intergranular cleavage facets. In contrast, the fracture surface of

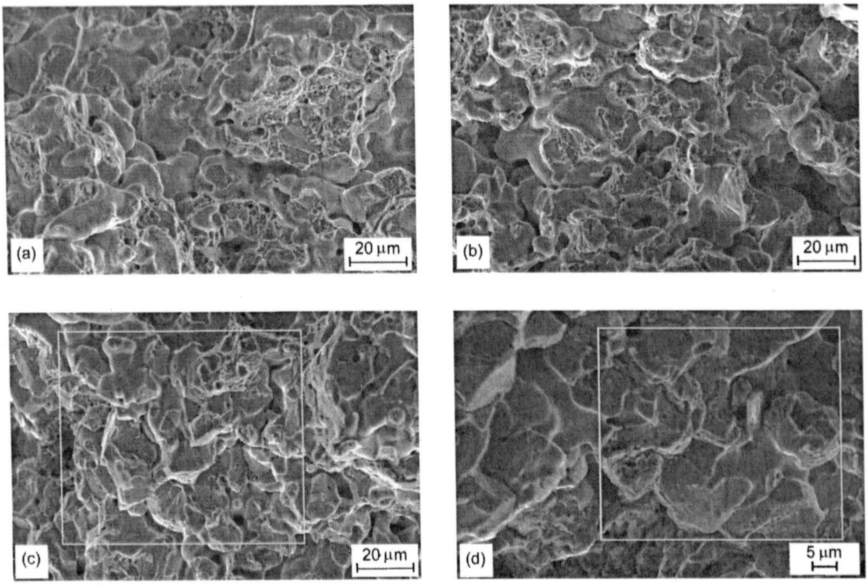

Fig. 8.40. Fracture surfaces of hybrid Fe–3% Cr–0.5% Mo–(1,2,3)% Mn steel tensile bars. (a), (b), (c) (Mn added as FeMn), (d) 3Mn steel (Mn added as FeMnC). (a) 1Mn (1.14% Mn–0.2% C_c UTS = 756 MPa, 214 HV 10), (b) 2Mn (2.06% Mn–0.23C_c, UTS = 806 MPa, 286 HV 10), (c) 3Mn (3.11% Mn–0.24% C_c, UTS = 663 MPa, 318 HV 10), (d) 3Mn (3.05% Mn–0.24% C_c, UTS = 586 MPa, 316 HV 10). SEM/SEI. Note: (d) – higher magnification of details from Fig. 8.40(c) [194].

3Mn steel showed mainly transgranular and intergranular cleavage, smooth intergranular decohesion and a small number of dimpled facets, Fig. 8.40(c). The fracture shown in Fig. 8.40(d) (manganese added as FeMnC) is characterised primarily by intergranular smooth decohesion facets and, to a minor extent, by transgranular cleavage facets and scattered dimples. No non-metallic inclusions were found in the dimples on the fracture surfaces.

8.5.4. Effect of industrial sintering conditions on properties and microstructures of hybrid Fe–3Cr–0.5Mo–Mn–C steels

The results obtained using hybrid powders, *i.e.* mixtures of the prealloyed Fe–3Cr–0.5Mo powder (CrM) and three additives containing manganese and low carbon content prove that manganese, due to its high vapour pressure, affected effectively this material system also under industrial sintering conditions.

As a result, a small amount of manganese was lost by oxidation to MnO which has no adverse effect on the surface concentration

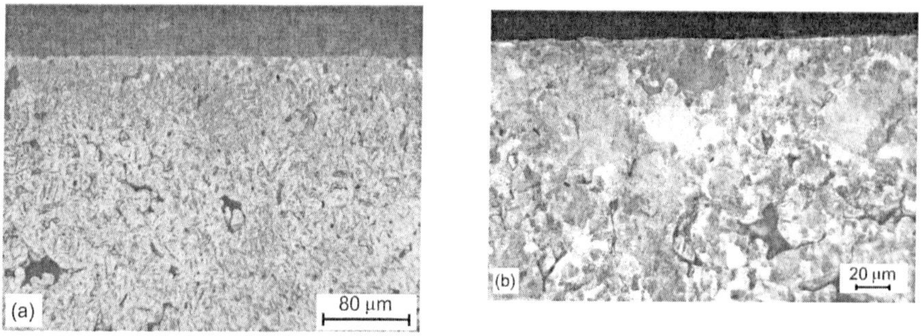

Fig. 8.41. Representative microstructures of surface of: (a) hybrid Fe–3Cr–0.5Mo–3Mn–0.24C steel tensile bars (Mn as FeMnC), (b) Fe–3Mn–0.5C steel roller part (120 g). Tensile bars and roller sintered independently in the pusher furnace for 40 min at 1180°C in $70N_2/30H_2$ atmosphere, dew point of –50°C. Nital etched [194].

of manganese on the specimens containing manganese sintered in one box. The microstructures of the surface layers of hybrid Fe–Cr–0.5Mo–Mn–C steel and of a Fe–Mn–C steel also confirm the absence of surface demanganizing, Fig. 8.41.

Industrial sintering conditions in continuous furnaces differ from laboratory sintering conditions. During industrial sintering, boxes containing the specimens/parts move through the furnace from the entrance to the exit and the atmosphere is in a counter flow direction, *i.e.* it moves in a direction opposite to that of the movement of the boxes, Fig. 8.33. Normally, with decreasing temperature, the atmosphere flowing through the sintering and preheating zones is increasing contaminated by H_2O and CO_2 – the products of prior reduction processes.

It is necessary to pay special attention to the atmosphere in the dewaxing zone, considering strict thermodynamic requirements for oxygen partial pressure in the sintering atmosphere for manganese steels. As the parts in this zone heat up, the lubricant melts, evaporates and breaks down into complex smaller molecules (for example, methane, ethane, butane, propane, carbon monoxide, carbon dioxide, water). This decomposition starts as low as 150°C, peaks near 450°C, and is usually completed at 550°C. The atmosphere plays a role in polymer burnout since it can provide reaction species to help break apart the polymer molecules.

The dew point of the sintering atmosphere at the boundary dewaxing–preheat zone at 650°C should be <–90°C what was not fulfilled based on the thermodynamics. However, this criterion was not fulfilled and the specimens in the box were exposed for 60 min to the effect of an oxidizing atmosphere (manganese starts to oxidise in air at 400°C).

These results prove that there was no oxidation of manganese carrier particles in the compacts in this zone and that the oxide films on the surface of the CrM powder particles in compacts, if they existed, were removed (reduced).

Inasmuch as at least 3Cr steel exhibits a protective action against oxidation, particularly at low temperatures, a coherent film of Cr_2O_3 usually forms as a subscale or spinel such as $FeCr_2O_4$ at the metal/scale interface can be formed. The surface of Fe–3Cr–0.5Mo particles may be covered with this oxide or spinel.

This phenomenon indicates that components alloyed with manganese (mixed or hybrid) can be sintered under industrial conditions in boxes in continuously operating furnaces equipped with a dewaxing zone and in an atmosphere which does not meet the requirements for the purity of the atmosphere, based on Ellingham–Richardson diagram. This fact also facilitates the practical sintering of steels containing elements such as chromium and manganese to increase strength and hardenability.

The diffusion of manganese into the Fe–Cr–Mo matrix is controlled by several factors as lattice defects, shape of the starting prealloyed powder particles, particle size distribution and work hardening cold-compacted contact areas between particles increasing the dislocation density.

The result of the complex interactions between the manganese–diffusion mechanism and the characteristics of powder particles is the formation of heterogeneous sintered microstructures.

A clear demonstration of this are the differences in microhardness values of individual steel microstructures, Fig. 8.39, notwithstanding the observation that the microstructures appear to be relatively homogeneous as shown in Fig. 8.38. This proves that the heterogeneity of the microstructure caused by these factors is also retained, to some extent, in austenite. This means that individual microvolumes in the microstructure, not equal in size, were alloyed by variable concentrations of manganese and carbon.

The consequence of this is that the products of the austenite transformation in different grains, and even in the same grain–microvolumes, can differ. In this case, the manganese alloying effect due to chemical heterogeneity resulted in heterogeneous microstructures of the sintered steels. This resulted in the transformation of austenite in microstructures (with several phases corresponding to austenite grain–microvolumes with manganese and carbon concentrations), namely, fine pearlite, bainite and martensite with an initial content of 0.24% C.

The formation of martensite in these steels is the consequence of the diffusionless $\gamma \rightarrow \alpha$ transformation of austenite to martensite at slow cooling rates, similar to that in air in wrought steels containing 5–6% Mn. This also occurs during continuous industrial sintering, similar to martensite transformation of iron–carbon steels. The microstructure of this transformation is a oversaturated α-solid solution with the concentration corresponding to the γ-solution, *i.e.* manganese martensite without the formation of the tetragonal lattice. The formation of martensite causes a self-phase strengthening. Furthermore, as a rule, in wrought steels with ≥1.5% Cr, the martensite is formed when the manganese content reaches ~2.5%. This explains the formation of martensite in the tested steels with a 1% Mn addition. Similarly, bainite forms in wrought iron–manganese–carbon steels containing ~2% Mn and with chromium, since we can assume that an even lower Mn content is needed for the formation of bainite.

These factors support the diffusionless transformation in hybrid sintered Fe–3Cr–0.5Mo–xMn–0.24C in grain-microvolumes containing a minimum 2.5% Mn to martensite, and to bainite in steels containing <2% Mn, Fig. 8.38.

The microhardness measurements show convicingly that in sintered multi-phase structures it is necessary to study the bainite in more detail. Bainite, which is the main constituent forming the shown microstructures with the impact on mechanical properties, is classified under isothermal transformation conditions as upper bainite (slow cooling rates) and lower bainite (lath bainite, high cooling rates). The transformation of austenite in wrought steels containing carbon and some alloying elements progresses in a complex manner under continuous cooling conditions and, therefore, the phases are hardly to define.

This is the case in PM because the continuous cooling is intrinsic to sintering performed in the laboratory, under industrial sintering conditions, and in sinter hardening process. The microstructures are heterogeneous and formed by grain-microvolumes with various manganese and carbon concentrations as documented by the relatively wide range of bainite microhardness values. This applies to all sintered steels, whether mixed, hybrid, or prealloyed.

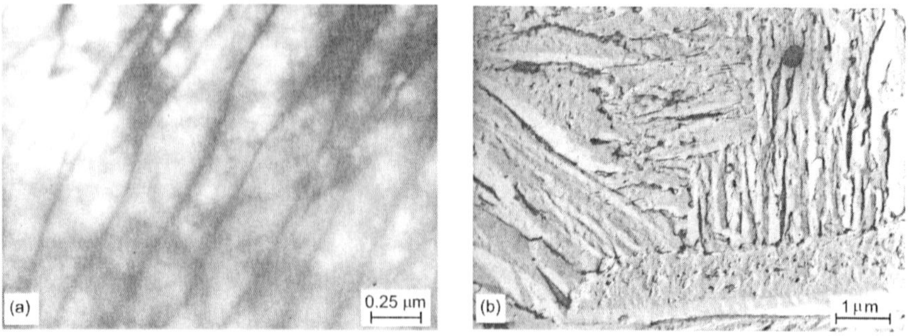

Fig. 8.42. Bainite structures of sintered mixed Fe–1.4Mn–0.5Cr–0.5Mo–0.6Cu–0.07C steel, porosity 7% (sintered density 7.28 g cm^{-3}). (a) thin foil – bainite lath ferrite (original magnification ×44000), (b) carbon replica shadowed with Au (original magnification ×14000). TEM.

Based on the characteristics of the austenite transformation to bainite under continuous cooling, it can be assumed that bainite forms with various morphologies.

The characteristic feature of granular bainite formation is the existence of dispersed martensitic–austenite aggregates (islands of martensite and austenite (M–A) in the highly dislocated ferritic matrix. The characteristic bainite structures of a mixed sintered steel cooled from 900°C in nitrogen are shown in Fig. 8.42.

It can be assumed that marked decrease in the strength of 3Mn steels under slow cooling rates was caused by a larger fraction of martensite and of granular bainite in the microstructures. Lower bainite can be excluded because the carbides that are formed within its laths are more effective barriers to crack propagation than are carbides localized at the interfaces in upper bainite. As shown in Fig.8.40(c), the fracture of 3Mn steel (alloying via FeMn) is mixed and is formed by transgranular and intergranular facets, and to a lesser extent by a ductile dimpled morphology. The transgranular microcracks are localized in the large martensitic plates, and intergranular rupture is the result of boundary brittleness in martensitic structures caused by acicular martensite and/or by some other factors. The fracture surface of a 3Mn steel (alloying with FeMn) shown in Fig. 8.40(d) is mixed, formed mostly by smooth intergranular ruptures – grain boundary decohesion along the former austenitic boundaries, due to embrittlement. This can be also caused by zones with different structures (ferrite/austenite and/or martensite layers) in austenite matrices in the vicinity of grain boundaries, as can be the case with granular bainite.

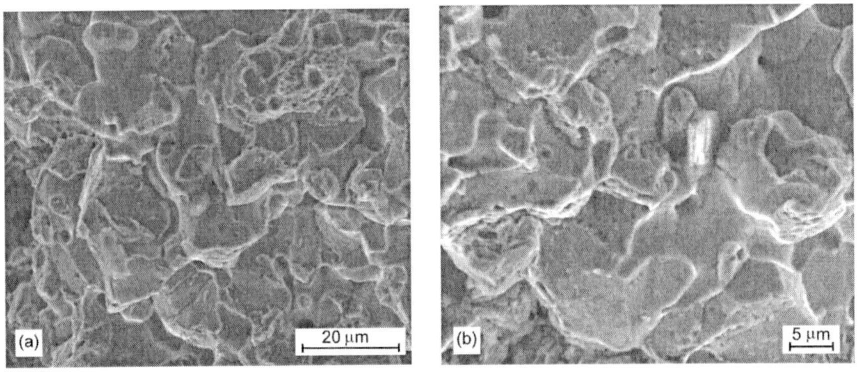

Fig. 8.43. Details of the fractures of 3Mn steels – (a) Mn added as FeMn, (b) Mn added as FeMnC. SEM.

The proportion of intergranular zones may be large. Figure 8.43 shows the surfaces of 3Mn steels alloyed via FeMn and FeMnC at higher magnification. Inasmuch as an incident of the formation of granular bainite is the coarsening of the primary grains, this region becomes very brittle and hard contributing to the rupture by grain boundary decohesion. In this case of alloying with FeMnC of lower chemical purity than with FeMn it might be considered that eventual impurity segregation at the grain boundaries increases their brittleness, causing a larger decrease of the strength.

The formation of a relatively large fraction of martensite and granular bainite in the Fe–3Cr–0.5Mo–3Mn steel sintered under normal sintering conditions is predicted from a physical–metallurgical perspective and the decrease in strength limits its utility.

The type of manganese carrier resulted in a range of strength levels of the steel, without marked effects on hardness and impact energy. A similar effect of the same manganese carrier on the strength of sintered iron–manganese–carbon steels in relation to the iron powder grades was recorded. Electrolytic manganese has a high-purity whereas both ferromanganese grades are ternary, multicomponent high manganese–iron–carbon alloys. These alloys differ in chemical activity because of differences in composition; electrolytic manganese exhibits the highest manganese sublimation rate. A difference between electrolytic manganese and the ferromanganese grades used in this study also exists with respect to the particle size distribution; electrolytic manganese had a particle size <20 µm and for ferro-manganese grades it was <45 µm. The large range in the particle size distribution is reflected in the prolonged sublimation time of the manganese and condensation during the preheat stage (larger particles). Manganese diffusion alloying from the surface layer into the interior of the base particles in the compact is advantageous compared to with fine-particle electrolytic manganese (sublimation for a short time at lower temperatures). The adverse effect of high–carbon ferromanganese on the 2Mn and the 3Mn steels probably reflects the higher fraction of a complex iron–manganese carbide $(Mn,Fe)_7C_3$ of high hardness in the microstructures. The complex concurrent diffusion of three elements from one source (carbon, manganese, silicon), and the lower chemical purity compared with medium-carbon ferromanganese are other factors influencing the role of the manganese carrier. Based on the presented results, commercial medium-carbon ferromanganese was lowest in cost and optimum manganese carrier for alloying of mixed or of the sintered steels based on prealloyed powders.

On the basis of the previous results it is possible to state conclusively that the properties of the hybrid system Fe–3% Cr–0.5% Mo–(1–3%)Mn–0.24% C reflect successful sintering in an industrial pusher furnace at 1180°C in the $70N_2/30H_2$ atmosphere, not withstanding the high affinity of manganese and chromium for oxygen.

The highest tensile strength (806 MPa) and bend strength (1175 MPa) was achieved when 2% Mn was added in the form of medium-carbon ferromanganese.

The effect of the ferromanganese grade was observed primarily on strength.

Demanganizing or decarburizing of the sintered hybrid iron–chromium–molybdenum –manganese steel compared to the starting values in relation to the mechanical properties practically did not occur.

Heterogeneous microstructures, formed as a result of slow cooling rates, consisted of martensite, lower and upper bainite, and fine pearlite. The proportion of these microconstituents depended on the amount of manganese added and on the microconcentrations of manganese and carbon in grain-microvolumes formed in heterogeneous austenite. Martensite forms with slow cooling in the microvolumes in steels containing 4–5% Mn.

Different ferromanganese grades affected mostly strength; there was a significant decrease in these properties in 3Mn steels. The combination of a 3% Cr steel with 3% Mn adversely affected the strength. The lowest strengths were achieved when manganese was added in the form of electrolytic manganese.

The results confirm the synergic hardening effect of chromium, manganese and molybdenum also at a low carbon level on the mechanical properties of Fe–3% Cr–0.5% Mo–(1,2,3)% Mn–0.24% C steels sintered under standard industrial conditions.

The hybrid material system can be attractive in the production of cost-effective high-performance PM parts, without addition of other elements to enhance hardenability, and as a substitution for nickel.

8.6. Properties of hybrid Fe–(0.85, 1.5)Mo–Mn–C steels

Molybdenum has high affinity for carbon and strongly improves the strength mainly at higher temperatures. Molybdenum in PM belongs to the alloying elements with low affinity for oxygen. Sintered molybdenum alloy steels can be prepared as mixed based on plain iron powder and hybrid based on prealloyed powders both with the addition of manganese and carbon. Molybdenum in the amount of (0.5–1)% is a common alloying element in some sintered high strength steels.

In this study, the properties of hybrid Fe–(0.85,1.5)Mo–xMn–xC steels and for comparison the mixed Mo-alloyed steels based on plain iron powder sintered under different conditions are presented.

8.6.1. Properties of Fe–0.85Mo–xMn–0.3C steels

The samples were prepared as mixed based on plain iron powder and as hybrid based on prealloyed Fe–Mo powder with addition of manganese. For the preparation of the samples the following powders were used:

- prealloyed Astaloy 85Mo powder (Ast85Mo),
- elemental molybdenum,
- ASC100.29 iron powder,
- manganese in amount of (1,2,3%) in the form of medium-carbon ferromanganese,
- natural graphite in amount of 0.5% or 0.3% and 0.8% HW wax as lubricant.

The tensile strength, transverse rupture strength and impact energy bars from both powder mixtures were compacted to the density of 7.0 g cm^{-3} and sintered for 60 min at 1200°C in dissociated ammonia (dew point –30 °C) in a tube furnace.

Table 8.12. Green (ρ_g) and sintered density (ρ_s) and Mn and C_c contents for Fe–0.85Mo–(1,2,3)Mn–0.3C prepared as mixed (premix) and as hybrid based on Fe–0.85Mo–prealloyed powder (prealloy) steels and green density (ρ_g). Sintering at 1200°C for 60 min, *Mn added

	Fe–0.85Mo (prealloy)				Fe–0.85Mo (premix)			
*Mn [%]	0 Mn	1 Mn	2 Mn	3 Mn	0 Mn	1 Mn	2 Mn	3 Mn
ρ_g [g cm^{-3}]	6.91	6.89	6.86	6.81	6.94	6.94	6.91	6.89
ρ_s [g cm^{-3}]	7.02	7.00	6.94	6.86	7.08	7.06	6.98	6.91
Mn [%]	0.12	0.90	1.95	2.86	0.09	0.97	1.95	2.92
C_c [%]	0.22	0.17	0.23	0.22	0.22	0.21	0.23	0.23

8.6.1.1. Sintered density and manganese and carbon contents

The results are listed in Table 8.11. A small increase in the as sintered density was recorded when compared to the green density. Also minimal differences were observed in density between the hybrid and premixed compacts. The manganese content in the 0Mn specimen was 0.12% and carbon content 0.24%. The as sintered manganese and carbon contents in the manganese-containing steels were on the level of the starting ones. No demanganizing or decarburizing was recorded after sintering in the atmosphere with the dew point of –30°C.

8.6.1.2. Mechanical properties

Figures 8.44–8.47 show the dependences of the tensile strength,

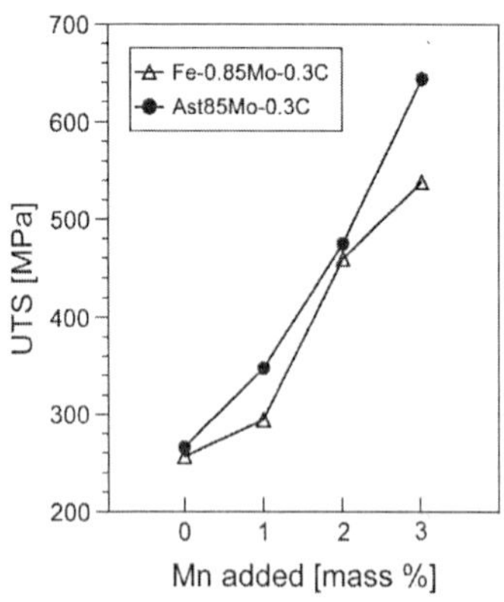

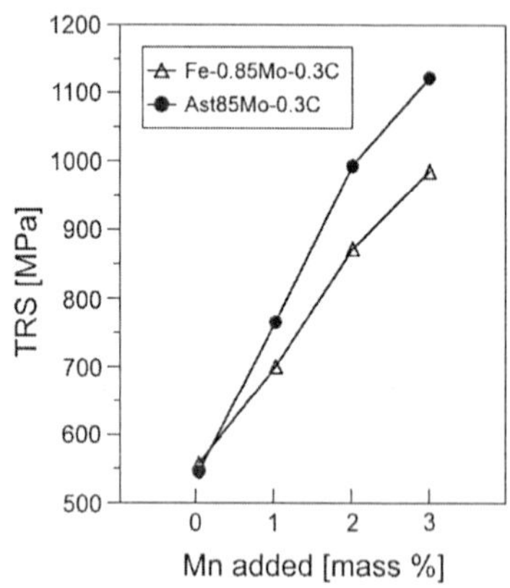

Fig. 8.44 (left). Dependence of tensile strength of hybrid and mixed Fe–0.85Mo–0.3C steels on manganese addition. Sintering at 1200°C [195].
Fig. 8.45 (right). Dependence of transverse rupture strength of hybrid and mixed Fe–0.85Mo–3C steels on manganese addition. Sintering at 1200°C [195].

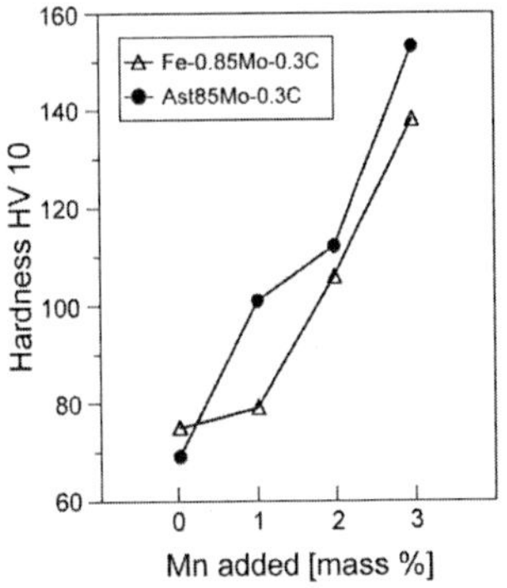

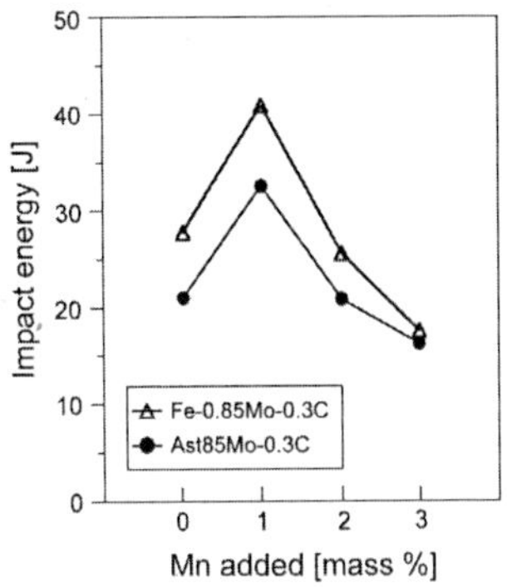

Fig. 8.46 (left). Dependence of hardness of hybrid and mixed Fe–0.85Mo–0.3C steels on manganese addition. Sintering at 1200°C [195].
Fig. 8.47 (right). Dependence of impact energy of hybrid and mixed Fe–0.85Mo–0.3C steels on manganese addition. Sintering at 1200°C [195].

transverse rupture strength, elongation, impact energy and hardness values on manganese addition for both mixed and hybrid steels. The mechanical properties of the mixed steels, except impact energy, are lower compared to those based on the prealloyed powder.

The hybrid Fe–0.85Mo–3Mn–0.3C steel reached the tensile strength of 645 MPa and the mixed one of 539 MPa. Manganese addition caused an increase in UTS of hybrid steel by 126 MPa/1% Mn. The TRS of hybrid steel attained 1121 MPa with 3% Mn. Impact energy of mixed steels was higher compared to the hybrid steel. Impact energy increased in both cases with 1% Mn addition compared to the steel without Mn addition, *i.e.*, by 13.1 J for mixed steel. Hardness similarly as tensile strength was increasing with increasing manganese addition linearly up to 153 HV 10.

Fig. 8.48. Microstructure of: (a) Fe–3Mn–0.3C, (b) hybrid Fe–0.85Mo–3Mn–0.3C steel. Sintering at 1200°C for 3 min. Nital etched.

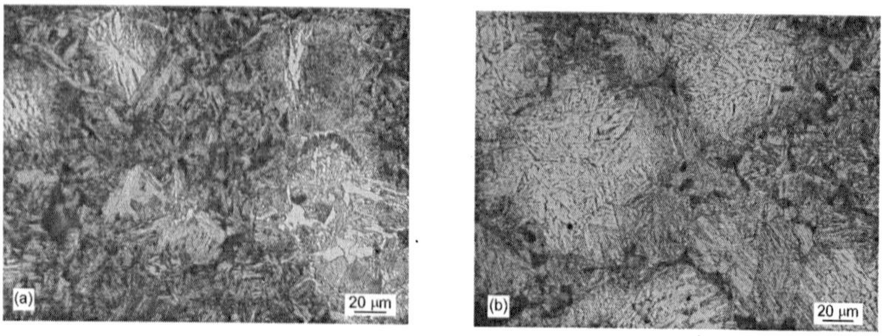

Fig. 8.49. Microstructure of: (a) Fe–3Mn–0.3C, (b) hybrid Fe–0.85Mo–3Mn–0.3C steel. Sintering at 1200°C for 60 min. Nital etched.

8.6.1.3. Microstructure

Figures 8.48 and 8.49 show the microstructures of mixed Fe–3Mn–0.3C and of hybrid Fe–0.85Mo–3Mn–0.3C steels sintered for 3 and 60 min at 1200°C. In both cases the surfaces of the based powder particles after sintering for 3 min were alloyed with manganese. Thicker layers were formed on the base particles of the hybrid steels. The cores of the prealloyed powders were alloyed with manganese homogeneously with a portion of bainite compared to those prepared as mixed. The microstructure after sintering for 60 min of both systems was formed by bainite.

8.6.2. Properties of Fe–0.85Mo–Mn–0.5C steel

The specimens were prepared under the conditions as in section 8.6.1 which were in this case sintered at 1150°C and 1200°C in cracked ammonia with a dew point of –30°C.

Table 8.13. Green (ρ_g), sintered density (ρ_s) of Fe–0.85Mo–(1,2,3)Mn–0.5C steel specimens prepared as mixed (premix) and as hybrid based on Fe–0.85Mo prealloyed powder (prealloy) after sintering at 1150°C and 1200°C

1150°C, 60 min	Fe–0.85Mo (prealloy)				Fe–0.85Mo (premix)			
	0 Mn	1 Mn	2 Mn	3 Mn	0 Mn	1 Mn	2 Mn	3 Mn
ρ_g [g cm^{-3}]	6.89	6.85	6.86	6.82	6.94	6.91	6.89	6.89
ρ_s [g cm^{-3}]	6.92	6.95	6.90	6.80	7.00	6.99	6.93	6.87
1200°C, 60 min	**Fe–0.85Mo (prealloy)**				**Fe–0.85Mo (premix)**			
	0 Mn	1 Mn	2 Mn	3 Mn	0 Mn	1 Mn	2 Mn	3 Mn
ρ_g [g cm^{-3}]	6.89	6.87	6.86	6.82	6.94	6.92	6.87	6.87
ρ_s [g cm^{-3}]	6.99	6.97	6.93	6.82	7.02	7.00	6.92	6.87

8.6.2.1. Density

Table 8.13 gives the green and sintered density of both material systems sintered at 1150°C and 1200°C. A small increase in sintered density was recorded compared to green density. Minimal differences were observed in density between the specimens based on the prealloyed powder compared to those based on ASC100.29 powder.

8.6.2.2. Mechanical properties after sintering at 1150°C

Figures 8.50–8.53 show the mechanical properties of hybrid and mixed Fe–0.85Mo–xMn–0.5C steels in dependence on manganese addition.

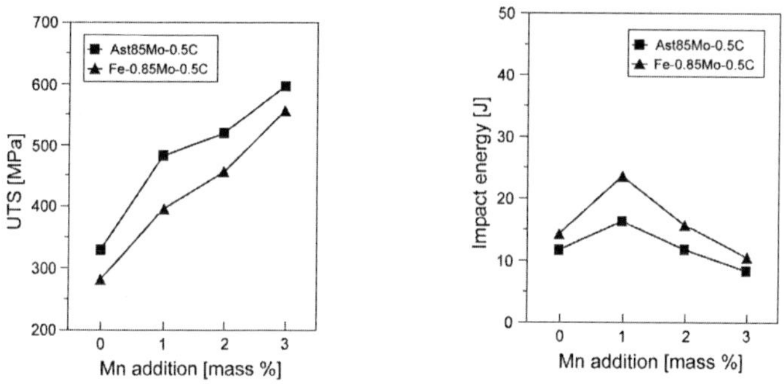

Fig. 8.50. Dependence of tensile strength and impact energy values of hybrid Fe–0.85Mo–0.5C and mixed Fe–0.85Mo–0.5C steels sintered at 1150°C for 60 min in cracked ammonia in dependence on manganese addition [196].

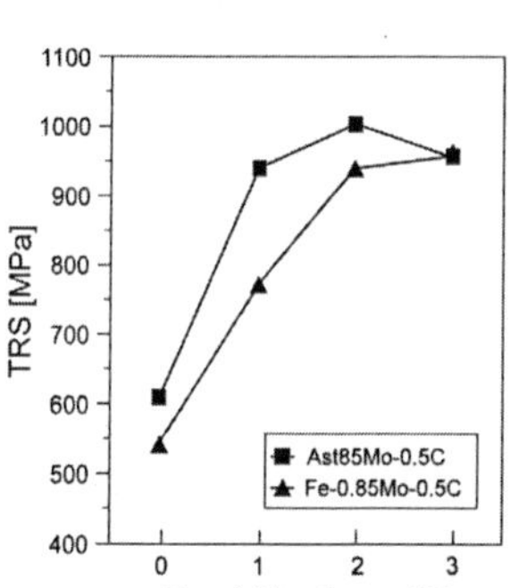

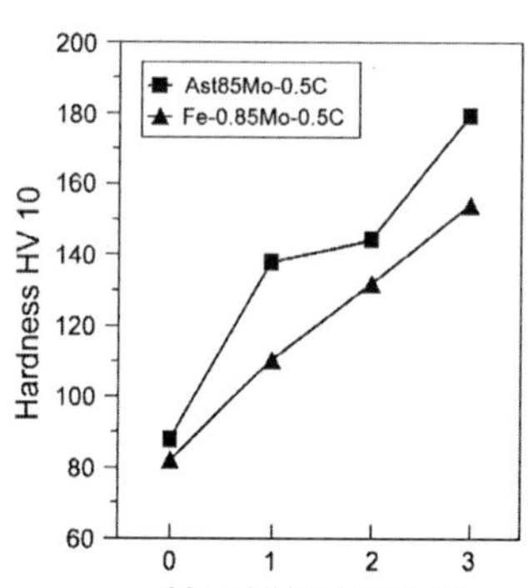

Fig. 8.51 (left). Dependence of transverse rupture strength of hybrid Fe–0.85Mo–0.5C and mixed Fe–0.85Mo–0.5C steels sintered at 1150°C for 60 min in cracked ammonia in dependence on manganese addition [196].
Fig. 8.52. (right). Dependence of hardness of hybrid Fe–0.85Mo–0.5C and mixed Fe–0.85Mo–0.5C steels sintered at 1150°C for 60 min in cracked ammonia on manganese addition [196].

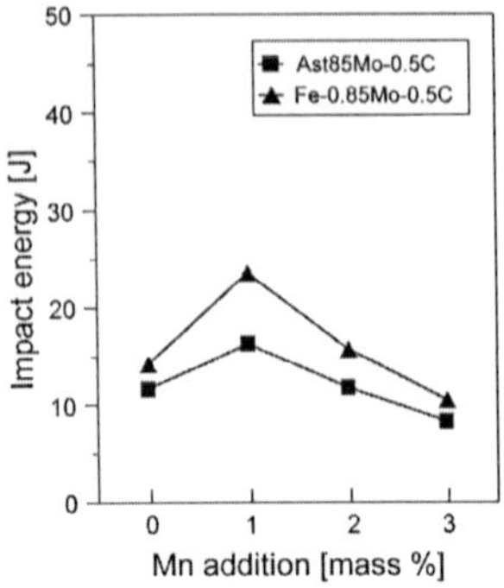

Fig. 8.53. Dependence of impact energy of hybrid Fe–0.85Mo–0.5C and mixed Fe–0.85Mo–0.5C steels sintered at 1150°C for 60 min in cracked ammonia on manganese addition [196].

The mechanical properties of mixed steels, except impact energy, are lower compared to the hybrid steels ones. Whereas the sintering in dissociated ammonia was carried out, it is possible to expect a lower combined carbon content in the materials than added 0.5% graphite.

The hybrid Fe–0.85Mo–3Mn–0.5C steel reached the tensile strength of 591 MPa and the mixed of 563 MPa. The manganese addition to the hybrid system increased the tensile strength by 97 MPa/1% Mn. The highest transverse rupture strength was 970 MPa with 2% Mn. The impact energy of the mixed system was higher than that of the hybrid system. Impact energy increased in both steel systems with 1% Mn addition to 24 J and 15 J, respectively. Hardness increased linearly with increasing manganese addition.

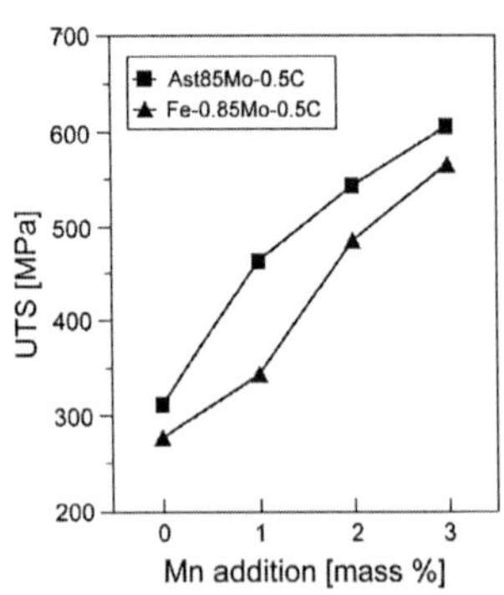

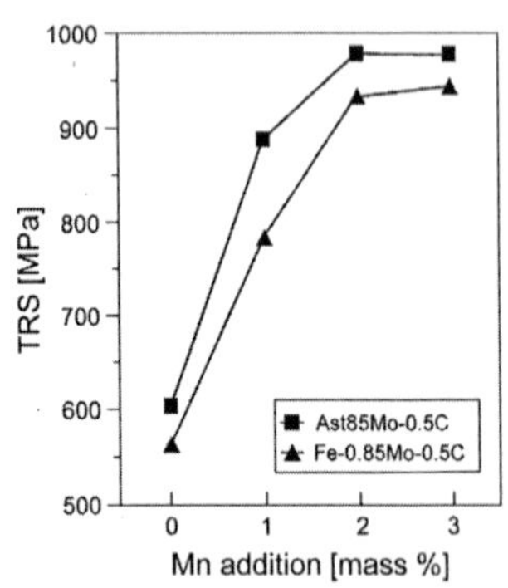

Fig. 8.54 (left). Dependence of tensile strength of hybrid and mixed Fe–0.85Mo–0.5C steels on manganese addition. Sintering at 1200°C for 60 min in cracked ammonia [196].

Fig. 8.55 (right). Dependence of transverse rupture strength of hybrid and mixed Fe–0.85Mo–0.5C steels on manganese addition. Sintering at 1200°C for 60 min in cracked ammonia [196].

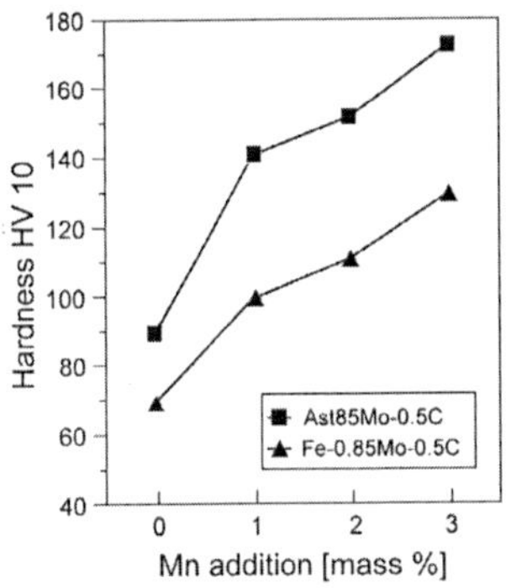

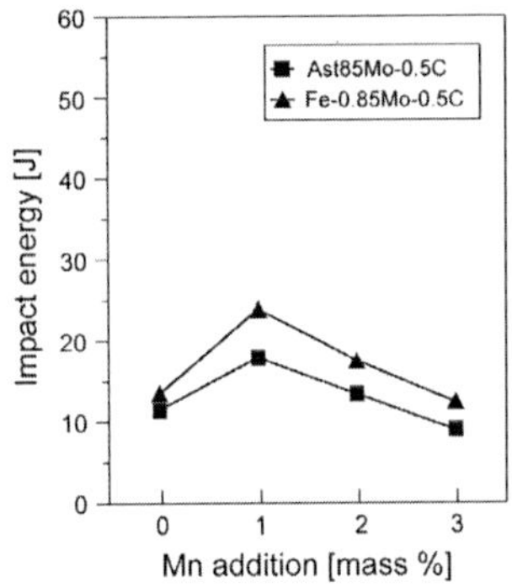

Fig. 8.56 (left). Dependence of hardness of hybrid and mixed Fe–0.85Mo–0.5C steels on manganese addition. Sintering at 1200°C for 60 min in cracked ammonia [196].
Fig. 8.57 (right). Dependence of impact energy of hybrid and mixed Fe–0.85Mo–0.5C steels on manganese addition. Sintering at 1200°C for 60 min in cracked ammonia [196].

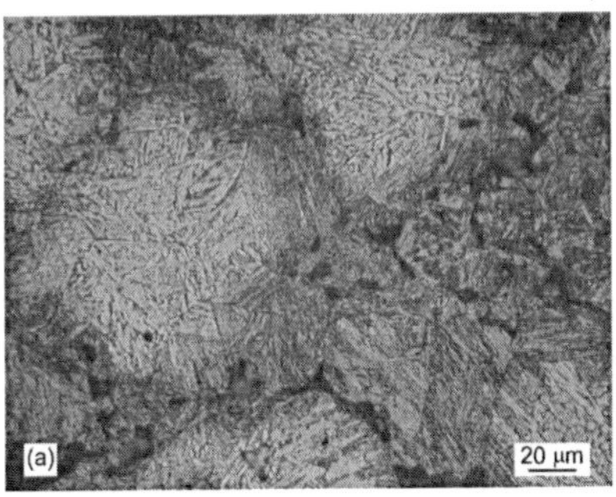

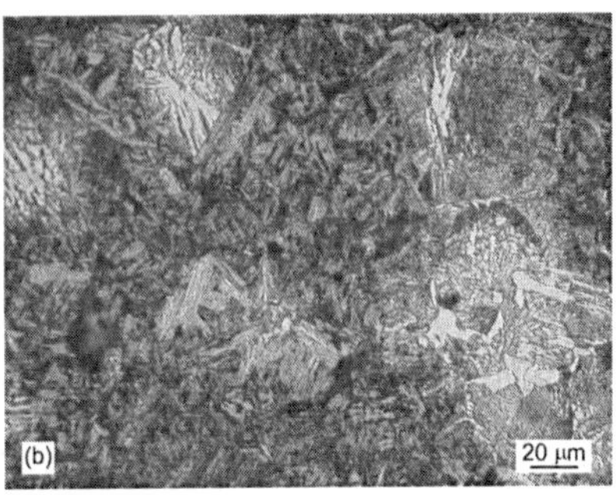

Fig. 8.58. Microstructure of: (a) hybrid Fe–0.85Mo–3Mn–0.5C, (b) Fe–3Mn–0.5C. Sintering at 1200°C for 60 min in cracked ammonia. Nital etched.

8.6.2.3. Mechanical properties after sintering at 1200°C

Figures 8.54–8.57 show the dependences of tensile strength, transverse rupture strength, hardness and of impact energy on manganese addition of hybrid and mixed steels.

The differences between hybrid and mixed steels were on the same level as after sintering at 1150°C. The tensile strength of the 3Mn steel was 593 MPa. Sintering at 1200°C did not affect the mechanical properties of these steels.

The microstructure character of both material systems contributes to the evaluation of mechanical properties. Figure 8.58 shows the microstructure of the Fe–0.85Mo–3Mn–0.5C and Fe–3Mn–0.5C steels sintered at 1200°C for 60 min. The microstructure of the hybrid steel was more homogeneous compared to Fe–Mn–C steel. The larger differences in the alloying of matrices with manganese were observed in the microstructures of the mixed steels compared to the hybrid steels. It means that the Fe–Mo–prealloyed matrix is more advantageous for alloying with manganese in the gas (Mn vapour)–solid (base powder) phase system.

8.6.3. Properties of Fe–0.85Mo–xMn–0.6C steel

0.6% of graphite was added to the samples prepared on the basis of the Fe–0.85Mo prealloyed powder used for the determination of the mechanical properties. The bars compacted at 600 MPa were sintered at 1120°C and 1200°C for 60 min in a tube furnace in cracked ammonia with a dew point of –30°C.

8.6.3.1. Mechanical properties, microstructure and fracture

Figure 8.59 shows the tensile strength, hardness and impact energy of hybrid Fe–0.85Mo–xMn–0.6C steel sintered at 1120 and 1200°C in dependence on manganese addition.

Sintering at 1200°C resulted in only slightly higher mechanical properties in relation to the manganese addition in comparison with the values after sintering at 1120°C. The tensile strength and hardness increase linearly with increasing manganese addition. A higher tensile strength, *i.e.* 760 MPa, and a hardness of 200 HV10 were recorded for the Fe–0.8Mo–3Mn–0.6 C steel in comparison with the steels with the addition of graphite of 0.3 and 0.5%, respectively. The highest impact energy (24 J) was recorded for the steels with 1% Mn.

Figure 8.60 shows the microstructures of the steels with different manganese content sintered at 1120°C and 1200°C. As can be seen,

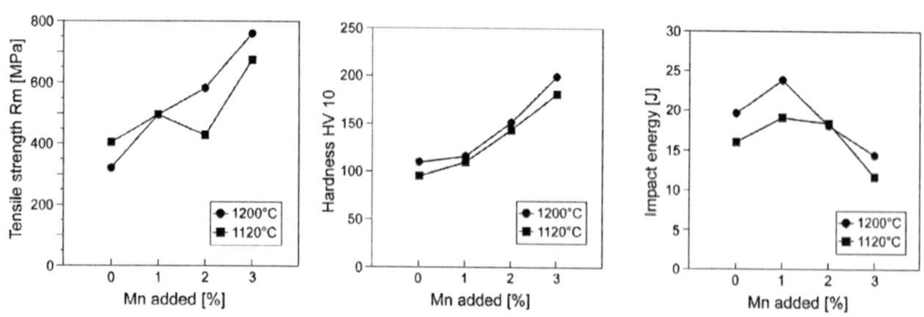

Fig. 8.59. Dependence of mechanical properties of hybrid Fe–0.85Mo–Mn–0.6C steel on manganese addition in the form of FeMn. Sintering for 60 min at 1120°C and 1200°C [197].

Fig. 8.60. Microstructure of hybrid Fe–0.85Mo–3Mn–0.6C steel, (a) sintered at 1120°C for 60 min, (b) sintered at 1200°C for 60 min. Nital etched.

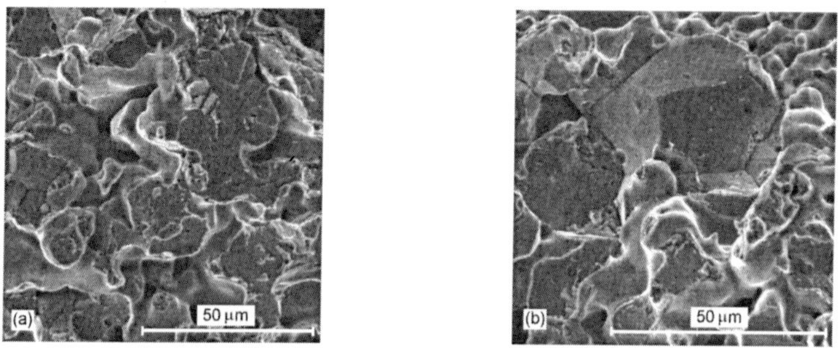

Fig. 8.61. Fracture surfaces of hybrid Fe–0.85Mo–3Mn–0.6C steel sintered at: (a) 1120°C for 60 min, (b) 1200°C for 60 min. SEM.

the homogeneity of the structure was higher after sintering at the higher temperature.

Figure 8.61 shows the fracture surfaces of hybrid Fe–0.85Mo–3Mn–0.6C steel sintered at two temperatures. The fracture of the sample sintered at 1120°C is mostly brittle transcrystalline. On the other hand, the fracture of the sample sintered at 1200°C is transcrystalline and intercrystalline, as clearly indicated at a higher magnifications in Fig. 8.61(b). This was caused by the higher graphite addition, *i.e.* 0.6% in comparison with the steels with 0.5 or 0.3% graphite addition.

8.6.4. Properties of industrially sintered mixed and hybrid Fe–0.85Mo–xMn–C steels

The test bars were produced from ASC100.29 iron powder and prealloyed Fe–0.85Mo powder (Ast85Mo). Manganese was added in the amount of 1, 2 and 3% as FeMn and graphite in the amount of 0.3% and 0.6%. The specimens from the powder mixtures were

Table 8.14. Green (ρ_g) density, sintered density (ρ_s) and manganese and carbon content of Fe–0.85Mo–(1,2,3)Mn–0.3C tensile strength bars prepared as mixed (premix) and as hybrid (prealloy). Industrial sintering at 1180°C for 40 min in a $70N_2/30H_2$ atmosphere. *Mn added

	Ast85Mo (prealloy)				Fe–0.85Mo (premix)			
*Mn [%]	0 Mn	1 Mn	2 Mn	3 Mn	0 Mn	1 Mn	2 Mn	3 Mn
ρ_g [g cm^{-3}]	6.92	6.91	6.87	6.85	6.96	6.95	6.93	6.92
ρ_s [g cm^{-3}]	7.00	6.96	6.90	6.84	7.02	7.03	6.96	6.90
Mn [%]	0.11	0.97	1.90	2.86	0.08	1.07	2.12	2.86
C_c [%]	0.29	0.28	0.26	0.27	0.27	0.26	0.27	0.28

compacted at 600 MPa and sintered in an industrial pusher furnace at 1180°C for 40 min in a $70N_2/30H_2$ atmosphere with a dew point of –50°C at a cooling rate of ~10°C/min.

Table 8.14 gives the green and sintered density values and as sintered manganese and carbon content values.

After industrial sintering compared to the laboratory sintering, Table 8.12, in terms of the effect of manganese and carbon contents on the mechanical properties of the tested steels it can be stated that no decarburizing or demanganizing of the samples occurred in spite of different dew points of sintering atmospheres used (–30 or –50°C).

8.6.4.1. Mechanical properties

Figures 8.62 and 8.63 show the mechanical and toughness properties of the tested steels in dependence on manganese addition. The effect of the preparation method of the base powder mix was determined.

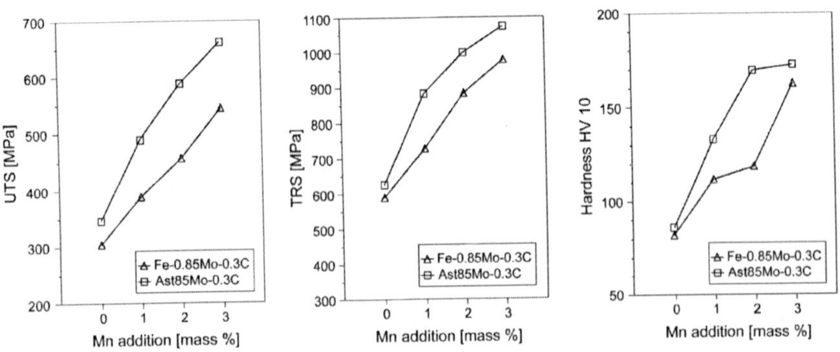

Fig. 8.62. Tensile strength, transverse rupture strength and hardness of prealloy and premix Fe–0.85Mo–0.6C steels industrially sintered at 1180°C for 40 min in $70N_2/30H_2$ atmosphere (dew point –50°C) in dependence on manganese addition [198].

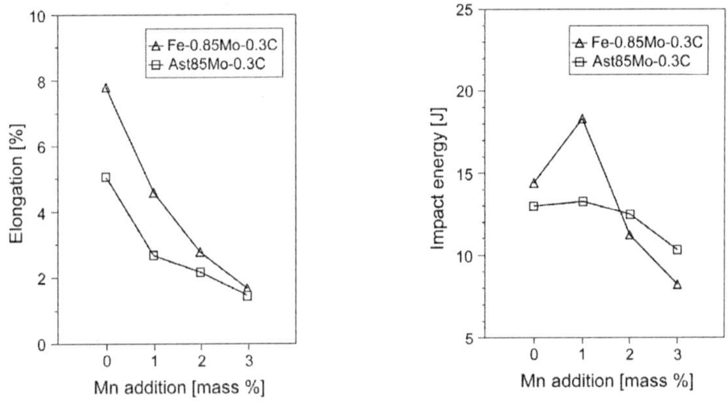

Fig. 8.63. Elongation and impact energy of hybrid and premix Fe–0.85Mo–0.6C steels industrially sintered at 1180°C for 40 min in a $70N_2/30H_2$ atmosphere (dew point –50°C) in dependence on manganese addition [198].

The mechanical properties, such as tensile strength, transverse rupture strength and hardness of the hybrid prealloyed materials were higher than of the mixed ones. The differences in the properties, caused by higher hardenability of the Mo-prealloyed powder compared to plain iron powder increased with increasing manganese addition. Higher activity of the base prealloy Ast85Mo powder for alloying by Mn–vapour compared to the premix based on ASC100.29 powder without manganese addition (0Mn) manifested itself in higher tensile and transverse rupture strength.

The tensile strength of the hybrid steel increased by 107 MPa/1% Mn and of the premix steel by 80 MPa/1% Mn. The prealloy Fe–0.85Mo–3Mn–0.6C material sintered at 1120°C in a $70H_2/30N_2$ atmosphere had a tensile strength of 460 MPa, transverse rupture strength of 750 MPa and elongation <1%. Contrary to the strength properties, the premix system attained higher elongation decreasing with increasing manganese addition to an equal value at 3% Mn. The main difference in elongation between both systems was already evident in the manganese-less state.

8.6.5. Properties of sintered hybrid Fe–(1.5, 0.85)Mo–Mn–0.5C steels

This study decribes the effect of manganese addition on the properties of sintered hybrid Fe–(0.85,1.5Mo)–0.5C steels prepared on the basis of prealloyed powders after sintering at 1120°C.

Samples were prepared from the following powders:

– prealloyed Fe–1.5Mo (Astaloy Mo, AstMo),
– prealloyed Fe–0.85Mo (Astaloy 85Mo, Ast85Mo),
– high-carbon ferromanganese (76% Mn, 6% C, balance Fe), milled in air, particle size <40 μm,
– natural graphite CR 12.

Manganese as a high-carbon ferromanganese powder (FeMnC) was added in the amount of 0, 1, 2, 3 or 4% to the Ast85Mo and AstMo powders (hybrid systems), all with 0.5% C coming from high-carbon ferromanganese completed to 0.5% C to the mixtures with lower manganese additions with graphite and with 0.8% HW lubricant.

8.6.5.1. Properties of the investigated steels

The density and the carbon and manganese content of the steels determined on tensile strength bars are listed in Table 8.15. The density of the steels decreased with increasing manganese addition as known from the Fe–Mn–C systems considering the lower density of manganese. The carbon and manganese contents are on the level of the added contents. It means that no demanganisation occurred. A small loss of carbon in relation to the starting 0.5% was determined.

Mechanical properties. The mechanical properties of Fe–(0.85,1.5)Mo–0.5C steels with manganese addition are shown in Fig. 8.64.

In Fe–1.5Mo–0.5C steel, the tensile strength, transverse rupture strength and impact energy values increased with increasing manganese addition up to 3% Mn. The addition of 4% Mn caused a drastic decrease of the tensile and transverse rupture strength of Fe–1.5Mo–C steel. It means that in the Fe–1.5Mo steel, based on prealloyed powder with 0.5% C, the manganese addition was

Table 8.15. Sintered density (ρ_s) and carbon (C_c) and manganese content (Mn) of Fe–(1.5,0.85)Mo–xMn–0.5C steels. Compacting pressure 600 MPa, sintering at 1120°C for 1 h, dissociated ammonia; Mn and C_c in mass %

Fe–1.5Mo–xMn–0.5C	0Mn	1Mn	2Mn	3Mn	4Mn
ρ_s [g cm^{-3}]	6.89	6.89	6.83	6.80	6.73
C_c [%]	0.44	0.40	0.43	0.43	0.47
Mn [%]	0.05	1.04	2.04	2.70	4.27
Fe–0.85Mo–xMn–0.5C	0Mn	1Mn	2Mn	3Mn	
ρ_s [g cm^{-3}]	6.99	6.97	6.93	6.82	none added
C_c [%]	0.41	0.41	0.38	0.40	
Mn [%]	0.12	1.22	2.32	3.25	

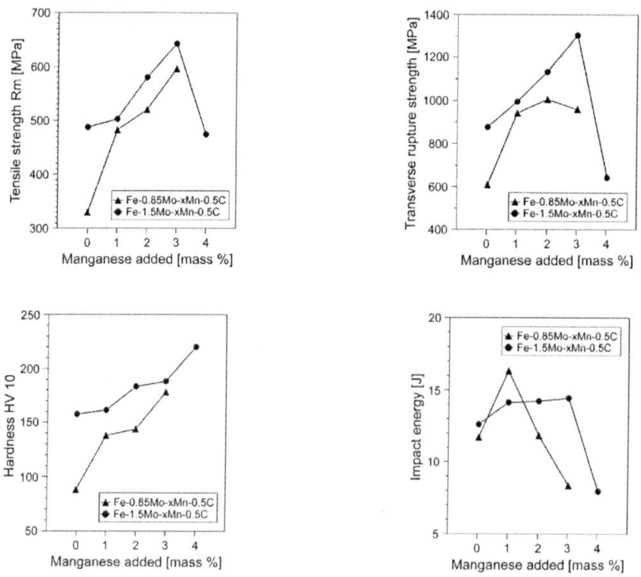

Fig. 8.64. The dependence of tensile strength, transverse rupture strength, hardness and impact energy (unnotched specimen) of Fe–(1.5,0.85)Mo–0.5C steels on manganese addition. Sintering at 1120°C for 60 min, dissociated ammonia [199].

restricted to 3%. The Fe–1.5Mo–0.5C steel with 3% Mn attained the tensile strength of 643 MPa, *i.e.* an increase by 52MPa/1% Mn, and the transverse rupture strength of 1303 MPa with 3% Mn, *i.e.* an increase by 142 MPa/1% Mn. The manganese addition in the amount of 1 to 4% increased the hardness from 158 to 221 HV10. The impact energy increased with increasing Mn content up to 3%. The highest impact energy value of 16.5 J was recorded for the Fe–0.85Mo–C steel, *i.e.* impact energy of this steel increased with addition of 1% Mn by 27% to 15 J compared to that without Mn addition despite of higher strength properties and hardness caused by this Mn addition. Hardness was increasing with increasing manganese to 180 HV 10.

All properties of Fe–0.85Mo–xMn–0.5C steel were practically lower compared to the Fe–1.5Mo–xMn–0.5C steels. It can be from the results deduced that for achievement the optimum combination of properties of sintered Fe–(0.85,1.5)Mo–xMn–0.5C steels the C and Mn content and sintering conditions should be mutually adapted.

Microstructure. The microstructure of the Fe–1.5Mo–1Mn–0.5C steel, Fig. 8.65(a), is homogeneous bainitic (Rm = 488 MPa) and that of the Fe–1.5Mo–3Mn–C steel, Fig. 8.65(b), is formed with finer bainite (Rm = 503 MPa). The microstructure of the

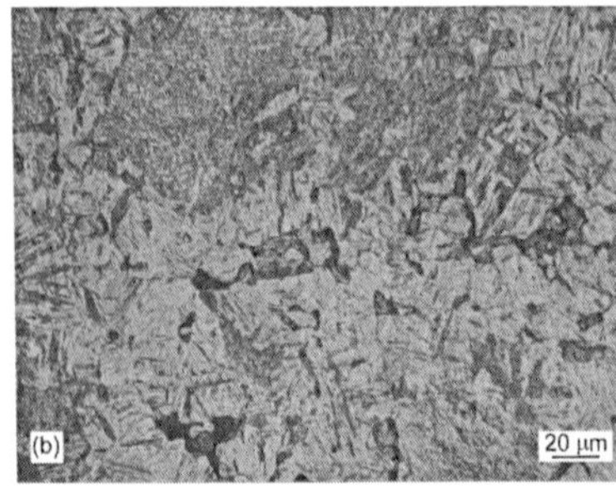

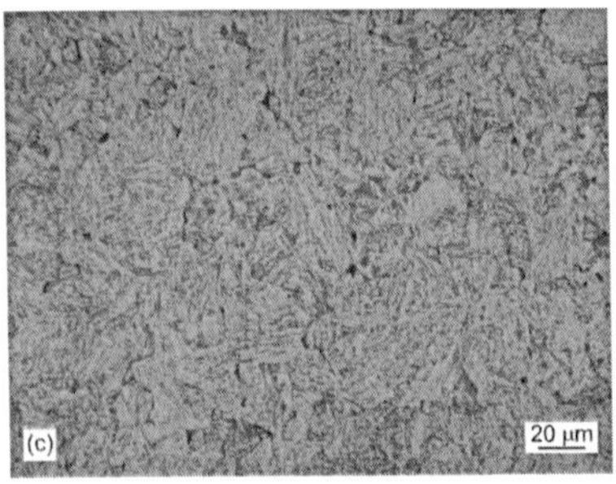

Fig. 8.65. Microstructure of hybrid Fe–1.5Mo–(1,3,0)Mn–0.5C steels sintered at 1120°C for 60 min. (a) – Fe–1.5Mo–1Mn–0.5C, (b) – Fe–1.5Mo–3Mn –0.5C, (c) – Fe–1.5Mo–0Mn–0.5C. Nital etched.

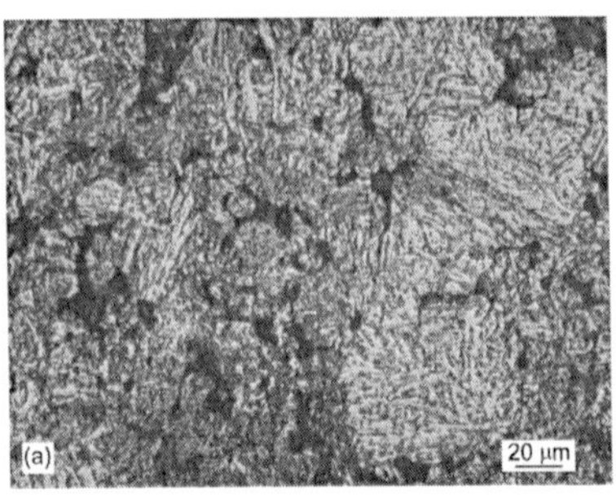

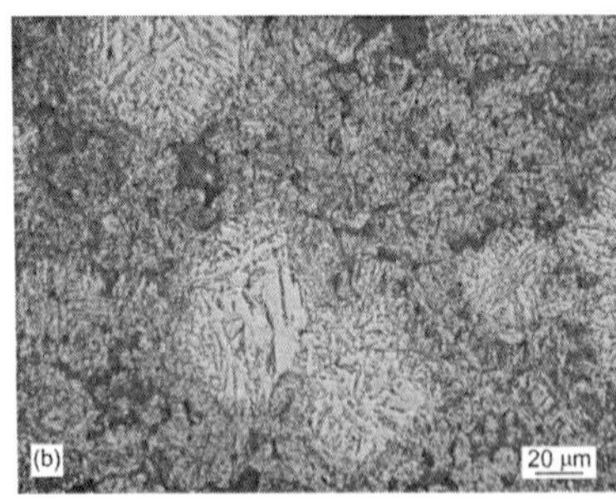

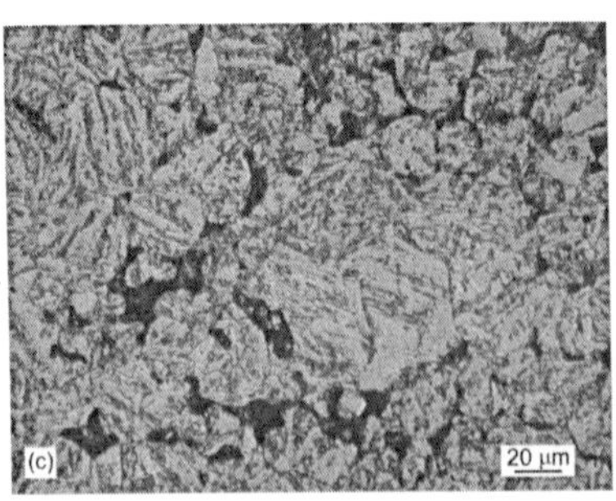

Fig. 8.66. Microstructures of hybrid Fe–0.85Mo–(1,3,0)Mn–0.5C steels sintered at 1120°C for 60 min. (a) – Fe–0.85Mo–1Mn–0.5C, (b) – Fe–0.85Mo–3Mn–0.5C, (c) – Fe–0.85Mo–0Mn–0.5C. Nital etched.

Fe–1.5Mo–0Mn–C steel, Fig. 8.65(c), is formed with smaller areas of bainite and martensite (Rm = 643 MPa).

The microstructures of the Fe–0.85Mo–Mn–0.5C steels are shown in Fig. 8.66.

The microstructures of this steel group (0.85Mo) are more heterogeneous compared to the previous steel group (1.5Mo). This was caused by different diffusivity of manganese in materials alloyed with 0.85% Mo and with 1.5% Mo. All three shown microstructures are formed predominantly by bainite of different fineness, especially in surface layers of powder particles in the sintered specimens.

No ferrite grains were observed in the microstructures of both groups steels based on the Mo-prealloyed powders. On the other side, in the mixed steels of the same composition the presence of ferrite grains was observed in all steels.

This is the proof of higher diffusivity of manganese in the Mo-prealloyed powders compared to the plain iron powder.

Fracture. Figure 8.67 shows the fractures of Fe–0.85Mo–xMn–0.5C steel samples after the impact energy test. The fracture of the steel without Mn is characterised by dimple morphology, partly with transcrystalline facets, Fig. 8.67(c). Singular brittle facets were observed on the fracture surface of steels with 1% Mn, even if this steel exhibited the highest impact energy values compared to the steels without Mn and with higher Mn contents. A higher proportion of brittle transcrystalline facets was observed in the steel with 2% Mn, Fig. 8.67(b).

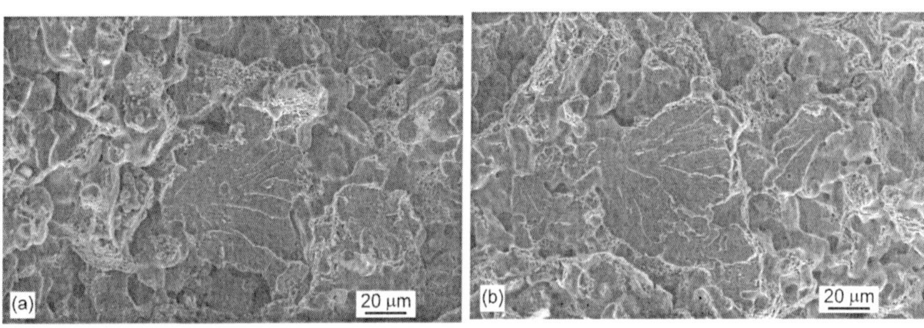

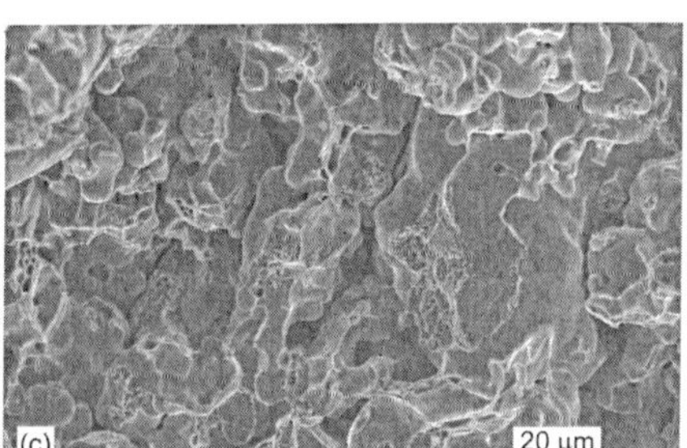

Fig. 8.67. Fractures of Fe–0.85Mo–(1,2)Mn–0.5C steel impact test samples. (a) – Fe–0.85Mo–0.5C, (b) – Fe–0.85Mo–2Mn–0.5C, (c) – Fe–0.85Mo–2Mn–0.5C. SEM.

Part II

This part of the book presents information on the sintered manganese-containing steels processed under various conditions of alloying, powder compaction and sintering of compacts based on various kinds of metal powders obtained by other authors starting from the first studies dealing with manganese in powder metallurgy since 1948. The processing of manganese steels in this case is studied taking into account the results obtained regardless of whether the effect of the high manganese vapour pressure on sintering in the selected sintering atmosphere was taken into account.

9

Basic characteristics of manganese steels from the year 1948

9.1. Thermodynamic and physical characteristics of chromium and manganese as alloying elements in powder metallurgy

The detailed thermodynamic and physical characteristics of chromium and manganese and sintering conditions of compacts containing manganese in different atmospheres are presented in advance to evaluate their ability to reduce oxides of both elements. This allows to compare them with previous results and subsequent studies dealing with the use of these elements in the development of sintereted high-strength structural parts and for other applications with respect to the type and purity of the atmosphere [15] (see also Chapter 2).

In order to produce high-strength sintered steels, the steels are alloyed mainly with Cu, Ni and Mo. These alloying elements can be utilized in the sintering process without oxidation problems. The objective of investigations in this work has been to establish conditions under which easily oxidizable elements as Cr and Mn can be utilized as alloying elements in sintered steels.

Alloying with chromium. The specimens for testing were prepared on the basis of sponge and atomised iron powders and ferrochromium carburé (FeCr, 64.6% Cr, 6.2% C). The mixture containing Fe–2% Cr–0.7% C was prepared on the basis of the atomised iron powder ASC100.29, FeCr and graphite. The powder was compacted at 600 MPa and then sintered at 1120°C for 30 min in an $5N_2/95H_2$ atmosphere in a pusher furnace with the possibility of controlling the dew points in the range from –30°C to + 20°C. However, they are difficult to produce in a production furnace. The carbon content from 0.5% decreased to 0.3% and the oxygen content increased from 0.06% to 0.09% with increasing dew point.

For testing the influence of the sintering temperature in the range 1100–1150°C, compacts were prepared as in the previous case except that the sponge iron powder SC100.26 was used. The sintering atmosphere was $5N_2/95H_2$ + 1% CH_4 with a dew point of –15°C. The resulting carbon content for all series was between 0.47 and 0.51%. It was stated that chromium oxide can be reduced when graphite is present in the compact when sintering at a temperature of 1130°C and higher. Within a compact there are local 'microclimates' created by locally high concentrations of carbon that can reduce the oxides present in the material. The oxygen content bonded to chromium decreased from 0.13% at 1100°C to 0.03% at 1150°C. The presented oxygen contents show how temperature affects the chromium oxidation–reduction process of Cr oxides.

Alloying with manganese. It has already been established that manganese cannot be prevented from oxidation and that the manganese oxide cannot be reduced in hydrogen, endogas or any nitrogen-based atmospheres at the dew point that are realistic in common sintering furnaces. Furthermore, it was concluded that manganese cannot oxidise to a greater extent than that determined by the amount of oxygen present. An atmosphere with a low content of oxygen, like $95N_2/5H_2$, will only oxidise it partly and may reduce the manganese oxide locally where the carbon concentration is high enough.

Manganese, contrary to chromium, does not form stable nitrides during sintering at 1120°C in a nitrogen-based atmosphere. When no nitrides form it might be easier to use Mn rather than Cr under these sintering conditions event though it has a higher affinity for oxygen than chromium.

When sintering compacts with additions of alloying particles containing Cr and Mn at 1120°C in commercial atmospheres with low dew points, the oxidation of these elements cannot be completely prevented. When the sintering temperature of 1250°C was used, this temperature was considered moderately high, still economically interesting and should be high enough for reduction of oxides of the alloying elements also when nitrogen-based atmospheres with the composition $5N_2/95H_2$ are used.

Cr and Mn as alloying elements increase hardenability, which makes the properties of such material dependent on the cooling rate after sintering. If no liquid phase or liquid phase which does not wet iron is formed, the particle size of the alloying additives is also of greatest importance for the homogenisation process. Under the mentioned conditions, Cr and Mn will always to some extent take part in gas–metal reactions. The degree to which these elements will be excluded from being dissolved in the iron matrix and thereby not contribute to the formation of strength-increasing phases in the sintered steel will determine whether they can be utilized in a certain sintering process or not.

During the sintering of manganese steel, the conditions for the reduction with solid carbon do not form because at temperatures over 900°C, solid carbon (graphite) cannot be present in the sintered steel due to its dissolution in austenite.

In terms of the mentioned thermodynamic calculations, the presence of MnO in the admixed sintered Fe–Mn compacts is *apriori* assumed. There are only two possibilities for the presence of some manganese oxides in sintered systems which were not analysed in the literature. The surface of the starting manganese carrier particles (electrolytic manganese, ferromanganese grades, special Mn-containing master alloys) could be covered with some manganese oxide. The MnO, which is green in colour and by this easy discernible also by examination, can be excluded due to its reducibility only at the temperature exceeding the common sintering temperatures, as follows from the previous data. In addition, there are no literature data regarding the Mn oxides on the particles of the manganese carriers. Manganese particles can only oxidise in the compacts during the preheating period under laboratory as well as industrial sintering conditions up to the temperature at which the equilibrium occurs at the dew point of the used atmosphere. For instance, when using an H/N sintering atmosphere with the dew point of –54°C, the equilibrium conditions the for Mn–O system correspond to the temperature of 1100°C.

9.2. Properties of sintered manganese steels studied in the initial stage

Manganese has been the object of systematic interest in PM since 1948–1950. An incomplete list of published works up to 1999 (228) concerning manganese in powder metallurgy [200] confirms a large extent of research activities in this field. A survey of the results by the use of manganese in powder metallurgy with 118 references was prepared [69].

The following results describe the mechanical properties of sintered manganese steels attained on the basis of various materials and under different processing conditions, which make it possible to consider the extent to which the thermodynamic requirements to avoid the oxidation of manganese in sintering were fulfilled in subsequent studies.

9.2.1. Starting knowledge about sintered manganese steels with high manganese and carbon contents

The preparation of sintered 1.5% Mn steel on the basis of carbonyl iron powder with the aim to prepare the porous semiproduct for forging or rolling to full density was tested as early as in 1936 without some basic results [201]. Later, manganese became the object of systematic interest in PM since 1948. Owing to the shortage of nickel in ~1970, there was considerable interest by PM manufacturers in substituting manganese for nickel. The following results represent the mechanical and other properties of sintered manganese steels attained under most various materials and processing conditions, which make it possible to characterise them completely.

The book [202], published in 1948, describes the processing conditions used at that time for the preparation of sintered alloy steels in terms of physical–metallurgical and technical aspects. The manganese carried used for the preparation of manganese steels was ferromanganese or pyrolusit with carbon as a reduction agent. Pyrolusite reduction also took place in sintering in hydrogen only in the presence of another metal, in this case iron powder. The use of pyrolusite or ferromanganese had shortcomings, *i.e.* at 1.5% carbon content the amount of free carbon in the sintered compacts increased sharply already after adding 0.5% manganese. In doing so, the amount of non-metallic inclusions increased with increasing addition of manganese. It was evident that manganese was not fully bound to iron or carbon. The properties of sintered manganese steels were not listed in [202].

The first more comprehensive data on the properties of sintered manganese steels depending on the addition of manganese, carbon, and compacting pressure after sintering at two temperatures are listed in [203]. Samples were prepared using Hametag iron powder and low-carbon ferromanganese (90% Mn, 0.5% C). Manganese steels with a manganese addition of 2–16% and 0–2% graphite were studied. The powder mixtures were pressed to samples with a diameter of 25 mm, 10 mm height at a pressure of 460 to 800 MPa, and tensile test specimens were compacted at a pressure of 400, 600 and 800 MPa. The samples were sintered in graphite boxes under alumina powder. The samples containing carbon up to 1% were sintered at 1200°C and those with 2% added carbon were sintered at 1140°C, both systems for 4 h in technical purity nitrogen. High-manganese steels require a somewhat lower sintering temperature.

Table 9.1 gives the values of the green and sintered density of the tested steel compacts in relation to the manganese and graphite addition.

At low manganese contents there is a usual increase in density after sintering. The density of the compacts with high manganese content decreased due to diffusion porosity.

Figure 9.1 shows the tensile strength and elongation of the sintered manganese steel in dependence on Mn (0–16)% and two carbon levels. The tested range of the compositions includes the manganese contents for structural and partly austenitic steels. The highest tensile strength (670 MPa) was recorded for the martensitic steel at 4% Mn and 0.8% C with zero ductility. The tensile strength of the steels with 8% Mn and 0.8% C was 280 MPa and ductility 14.9%.

Table 9.1. Green (ρ_g) and sintered density (ρ_s) of manganese steels in dependence on manganese and carbon content [203]

Mn [%]	ρ_g [g cm^{-3}] at C content [%]			ρ_s [g cm^{-3}] at C content ~[%]		
	0.1	0.5	1.0	0.3	0.4	0.8
4	6.83	6.75	6.73	6.95	6.83	6.79
8	6.81	6.64	6.65	6.77	6.73	6.56
12	6.76	6.60	6.60	6.63	6.46	6.46
16	6.70	6.58	6.58	6.54	6.36	6.33

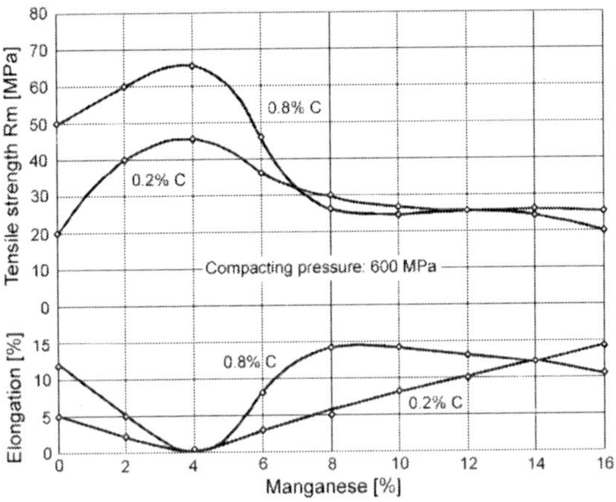

Fig. 9.1. Tensile strength and elongation of sintered steels in dependence on manganese and carbon addition. Sintering 4 h at 1200°C in hydrogen under alumina in closed graphite boxes; Hametag iron powder, 600 MPa [203].

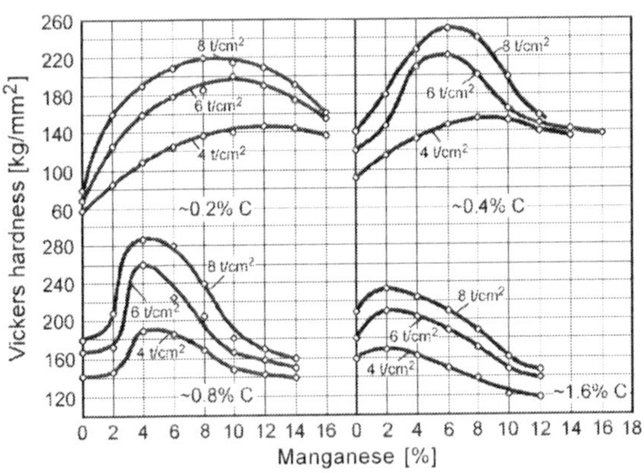

Fig. 9.2. Hardness HV of sintered manganese steels [203]. Other as in Fig.9.1.

Figure 9.2 shows the hardness values of the investigated manganese steels in relation to the manganese and graphite addition and compacting pressure.

As can be seen, the sintered manganese steels with a manganese content of 40 to 10% achieved, due to the martensitic structure, a significant maximum hardness. With increasing carbon content the highest hardness value shifts to lower values, of the width of martensite region decreased.

Figure 9.3 shows the hardness of copper-saturated sintered manganese steels. Saturation of copper was carried out by placing

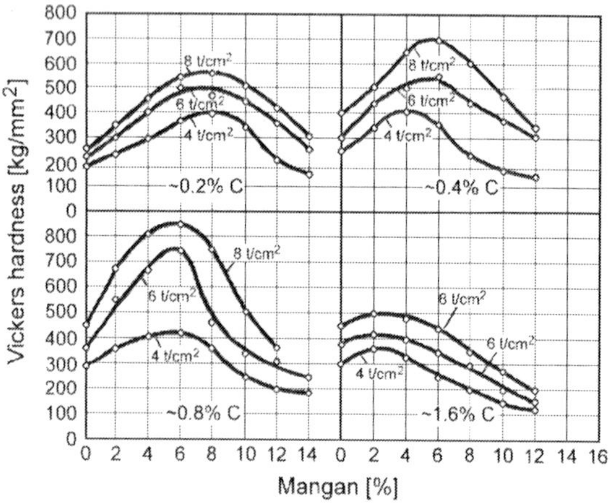

Fig. 9.3. Hardness HV of copper-saturated manganese steels in dependence on manganese and carbon content and on compacting pressure [203].

an appropriate amount of Cu on the porous sample. In this condition, the samples were heated in a pusher furnace under hydrogen to the appropriate temperature for 10 to 15 min and then slowly cooled.

The hardness curve of the copper-saturated samples are similar to the curves of the unsaturated samples, Fig. 9.2. The hardness of the saturated samples with about 0.8% C and 6% Mn reached 700 to 800 HV 40. In spite of this hardness, the saturated materials were quite tough, because the hard but brittle martensitic area was surrounded by a tough network of the copper alloy.

The non-saturated and copper-saturated steels were quenched from 900°C in water. The hardness values of these samples are shown in Table 9.2.

The hardness values of the copper-saturated samples were higher compared to the values of the non-saturated samples. Increase of the hardness was also dependent on the composition of the matrix.

9.3. Processing conditions and properties of sintered austenitic manganese steels

Austenitic manganese steels are characterised by high abrasive resistance. They have a large area for use in special machine parts exposed to high abrasive wear and in other industrial branches.

The authors of [242] used for the preparation of the compacts containing corresponding high manganese and carbon contents the electrolytic and reduced iron powder MH100.24, as a manganese

Table 9.2. Hardness HV 40 of the copper-saturated and unsaturated quenched manganese steels in dependence on manganese and carbon addition [203]

Mn [%]	Unsaturated				Saturated			
	Carbon content [%]				Carbon content [%]			
	0.2	0.4	0.8	1.6	0.2	0.4	0.8	1.6
2	184	220	308	280	410	465	565	245
4	224	232	294	224	465	605	367	274
6	264	274	221	184	513	367	175	265
8	250	221	184	156	465	297	166	206
10	224	168	166	130	432	175	150	175
12	203	156	145	141	336	175	150	n. d.
14	184	141	151	n. d.	245	166	n. d.	n. d.
16	164	113	n. d.	n. d.	n. d.	n. d.	n. d.	n. d.

carrier electrolytic manganese and ferromanganese and natural graphite. The powder mixtures containing 0, 5, 10, 15 and 20% of manganese and 0, 0.4, 0.8, 1.2, 1.6, and 2% of graphite were prepared and then compacted at 600 MPa into test compacts for tensile and transverse rupture strength. These specimens were sintered for 1 h at 1000–1300°C in dry hydrogen.

Figure 9.4 shows the dimensional changes of the compacts in the upper part of the diagram in dependence on: a) reduced iron powder (MH100.24) and electrolytic manganese, b) reduced iron powder and ferromanganese, and c) electrolytic iron powder and electrolytic manganese after sintering temperature of 1100°C. In the lower part of the diagram are shown the dimensional changes of the compacts after sintering at 1000°C, 1100°C, 1200°C and 1300°C, all in dependence on the manganese addition

As follows from the diagram, Fig. 9.4, sintering temperature exerted a considerable influence on the dimensional change, and sintered compacts had the tendency to shrink with increase in temperature and showed significant dimensional changes above 1200°C. This indicated that sintering between iron and manganese above 1200°C was accelerated.

Figure 9.5 shows the effects of manganese contents, sintering temperature and raw materials on the sintered density. The sintered density decreased with manganese content. This is due to fact that the ultimate dimensions were larger when mixed with manganese powders, and that the theoretical density of manganese is less than that of iron, and that the density of green compacts decreased with manganese contents. The density of compacts from the reduced iron powder was higher than that from the electrolytic iron powder.

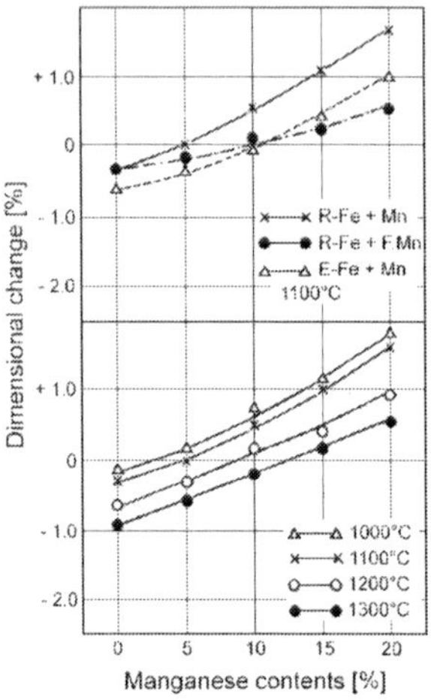

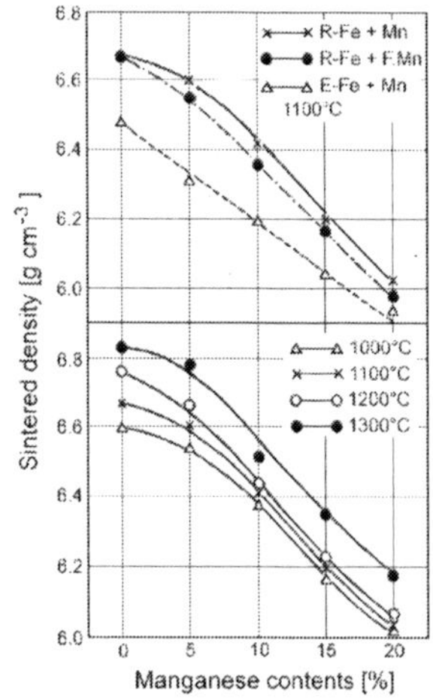

Fig. 9.4 (left). Effects of manganese contents, sintering temperature and raw materials on the dimensional changes of manganese steels [204].
Fig. 9.5 (right). Effects of manganese contents, sintering temperature and raw materials on sintered density [204].

The effects of manganese content and sintering temperature on tensile strength are shown in Fig. 9.6, and the effects of manganese contents, sintering temperature and raw materials on transverse rupture strength are shown in Fig. 9.7.

The tensile strength shows a very rapid increase up to 10% manganese followed by a rapid decline. A maximum tensile strength of about 440 MPa can be obtained for the sintered iron–manganese alloys, sintered at 1300°C, with the value for 10% manganese. Above 10% manganese, there would be an increase of the porosity owing to the growth of the compacts and the microstructure of sintered compacts would be austenitic. The increase of manganese addition of 10% caused a reduction in tensile strength.

For the same case as can be seen in Fig. 9.6, transverse rupture strength shows a very rapid increase up to 10% manganese followed by a rapid decline. Fig. 9.7. But when mixed with ferromanganese in place of electrolytic manganese powder, maximum transverse rupture strength can be obtained for 5% manganese content, and this maximum value of about 900 MPa is higher than that of the compacts from the electrolytic manganese powder. This indicates that here would be an influence by differences of concentrations of manganese oxide films on the surface which affects alloying between iron and manganese.

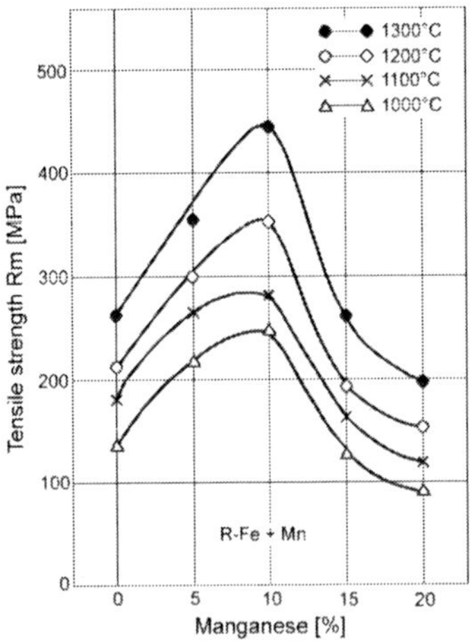

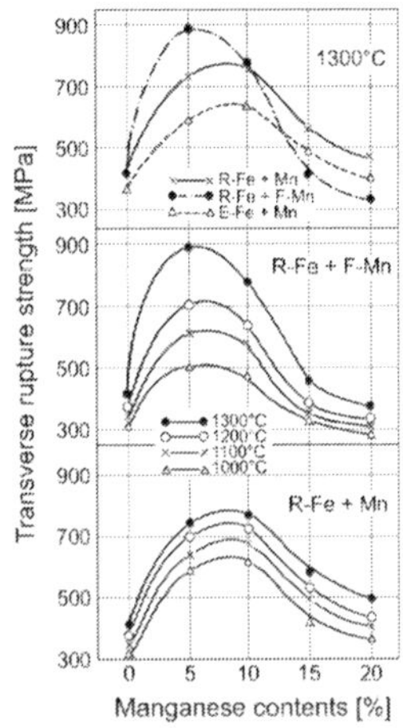

Fig. 9.6 (left). Effects of manganese contents and sintering temperature on the tensile strength (R–Fe – reduced iron powder MH100.24), 600 MPa [204].

Fig. 9.7 (right). Effects of manganese contents and sintering temperature on the transverse rupture strength [204].

It follows from the previous data that the sintered manganese steel with the austenitic structure had good mechanical properties, particularly high impact energy, though the porosity of sintered compact was higher than 15%. In order to obtain a homogeneous austenitic microstructure, it is necessary to add high manganese and high carbon content, for example 2% C for 10% Mn or 1.2% C for 20% Mn with sintering at 1200°C.

The characteristics of pearlite in a Fe–6% Mn–0.52% C steel were studied in [63]. The test specimens were prepared from a water-atomised iron powder and milled high-carbon ferromanganese with particle sizes ranging from 24 to 129 μm. The high level of high-carbon ferromanganese (5.5%) was chosen to form a liquid phase during during sintering for different times at 1160°C under pure hydrogen.

A large variation of microhardness values was observed for the different sintering times studied, which is listed in Table 9.3, suggesting the presence of ferrite, pearlite, bainite and martensite.

The large variation in microhardness values (90–780 HV 0.02) was attributed to the effect of manganese and carbon on the resulting microstructure analysed and the prersence of the mentioned constituents. A reduction of the pearlite volume fraction with increasing sintering time was observed.

Table 9.3. Results of the microhardness (HV 0.02) measurements in the analysed regions of the Fe–6% Mn–0.52% alloy for 0, 30, 120 and 300 min of sintering

Microhardness HV 0.02	Constituents
90 – 193	ferrite
217 – 297	pearlite
309 – 441	pearlite, bainite
528 – 780	bainite, martensite

9.3.1. High-alloyed sintered manganese steels

The findings in properties of austenitic manganese steels and in properties of sintered structural low-alloyed manganese steels became an impetus for study of the preparation of high-alloyed manganese sintered steels. The microstructure, kinetics of martensite transformations and mechanical properties of powder Fe–(4–40)Mn steels were investigated [72, 73]. The aim was to study the feasibility to prepare the Mn steels of those compositions by the PM route. The dependence of the yield and tensile strength of hot extruded high-alloyed powder Fe–Mn steels on manganese content in shown in Fig. 9.8. (The nitrogen-atomised powders could not be used for cold compacting due to high hardness). The high strength properties at –196°C and elongation of ~65% for Fe–30Mn were regarded as unexpected.

The main advantage of the production of the parts from high-manganese powder steel is the manufacturing of the parts which cannot be produced from melting steel of the same composition due to zero machinability.

The authors of [207] studied the sintered austenitic Fe–14Mn–8.5Ni–3Cr and Fe–14Mn–3Cr–2.5Ni–0.2C steels with the aim to increase the stability of austenite with the addition of TiC or Al_2O_3 (dispersion phase) regarding the thermally induced martensite transformation. Further were investigated the properties of sintered Fe–12Mn–1.2C (Hadfield) steel and Fe–18Mn–1Cr–0.3C steel (ASC 100.29 iron powder, electrolytic manganese, 600 MPa, 30 min at 1280°C, hydrogen + getter). The Hadfield steel reached the tensile strength of 341 MPa, yield strength of 248 MPa, elongation of 11%, hardness of 248 HV 010 and impact energy of 64 J. [208] presented that the sintered Fe–14Mn–1.4C steel with the porosity of 12–13% exhibited as porous by higher abrasive wear resistance (sintering for 2 h at 1200°C) compared with conventional high-manganese steel.

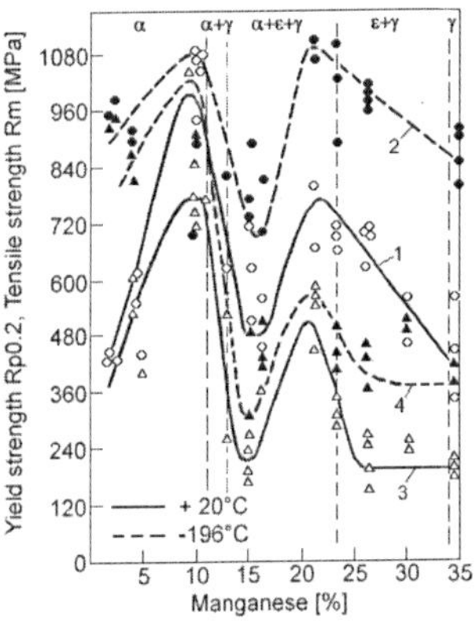

Fig. 9.8. Yield and tensile strength of sinter extruded high-alloyed manganese steels in dependence on manganese content. Nitrogen-atomised high- Mn alloyed powders, apparent density 4.6–5.6 g cm^{-3}, direct hot extrusion, water quenching from 1100°C, tempering. Testing of strength properties at +20°C and −196°C, 1,2 – Rm, 3,4 – Rp0.2 [205,206].

The tools for mining wear were manufactured from this steel. The bushes for locomotives were manufactured by powder forging from high-manganese steels [205,206].

Figure 9.9 shows the tensile strength and elongation of sintered high-manganese steel in dependence on density and manganese carrier grade. The tensile strength as well as elongation increased with the density increase. The highest strength of 380 MPa was recorded for the steel to which manganese was added as high-carbon ferromanganese. The elongation attained the highest value of 10% with high-carbon ferromanganese. The lowest values of both properties were achieved with the addition of manganese as electrolytic manganese.

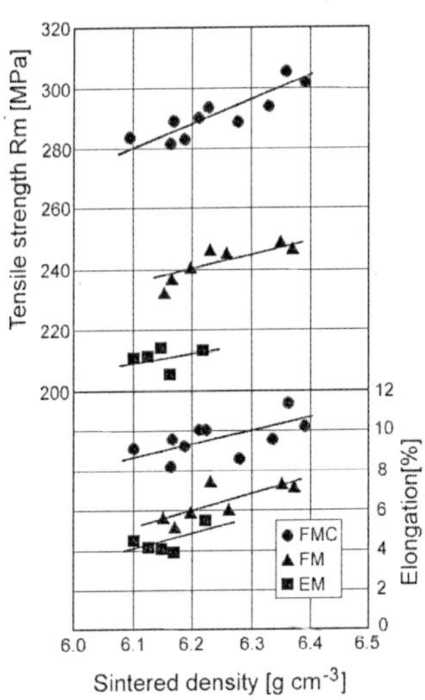

Fig. 9.9. Tensile strength and elongation of sintered high manganese steels (13% Mn, 1.2% C) in dependence on sintered density and manganese carrier grade (sintering at 1150°C for 60 min and water-quenched from 1050°C). FMC – high-carbon ferromanganese, FM – medium-carbon ferromanganese, EM – electrolytic manganese [209,210].

9.4. Alloying of manganese steels with master alloys

9.4.1. Alloying of manganese steels by master alloys in carbide form composed from elements with high oxygen affinity

The object of the investigation was to improve the mechanical properties of sintered steels by introduction in PM alloying the elements commonly used in ingot metallurgy, such as manganese, chromium, vanadium and silicon. The high oxygen affinity of these elements presents recently the known problem for their use in PM. Since the desired alloying elements Cr and Mn are known as carbide-formers, and since carbon, as shown in the previous results, was also studied in alloying of manganese steels, alloys were selected which contained the alloying elements in the carbide form. Regarding the high affinity of chromium and manganese for oxygen, these master alloys were produced by melting metallurgical high-carbon ferroalloys in vacuum induction furnaces using a susceptor followed by cleaning, jaw crushing, fine grinding (up to 0.15 mm) and attritor milling or milling in a planetary mill under cyclohexane. Special measures were taken during fine and ultrafine grinding (average particle size 5 μm) [48,211,212].

The following advantages were attributed to these master alloys: carbon in combined form, at least two of the elements Mn, Cr, Mo and/or V, which are stable during the heating up to isothermal sintering, dissociate under sintering condition, should minimize production costs, contain additional molybdenum. Further, the amount of added free carbon was minimized and they possess sufficient oxygen stability.

The following master alloys, denoted as MCM, MVM and MM, were developed and their chemical and phase compositions is listed in Table 9.4.

The mechanical properties of steels alloyed with MCM master alloy are shown in Fig. 9.10 and in Fig. 9.11 with MVM master alloy. The steel with 7% MCM reached the maximum UTS of 650 MPa, Rp0.2 of 490 MPa at a hardness of 86 HRB and 2.5% elongation. The mechanical properties of the steels alloyed with addition of 5 % up to 10% of MVM master alloy were on the same level. By adding free carbon to obtain a total content of 0.6% C, the content of the master alloy added was reduced. In spite of their optimum tensile

Table 9.4. Chemical and phase compositions of carbide master alloys [1,212–214]

MCM [mass %]					MVM [mass %]					MM [mass %]			
Mn	Cr	Mo	Fe	C	Mn	V	Mo	Fe	C	Mn	Mo	Fe	C
25	23	22	22	7	25.5	23	25.5	20	5	40	20	32	7
80% $(Cr,Mn,Fe,Mo)_7C_3$ 10% $(Fe,Mo,Cr,Mn)_6C$ 5% $(Fe,Mo,Cr,Mn)_3C$ 5% α–(Fe,Mn,Cr,Mo) O – ~0.18%					25% (V,Mo)C 25% $(V,Mo_2)C$ 20–25% α–(Fe,Mn,Mo) 20–25% γ–(Fe,Mn,Mo) MnO (Percentage depending on oxygen content, O –~0.2%					Three carbides phases with M_7C_3, little metallic α phase O – ~0.3%			

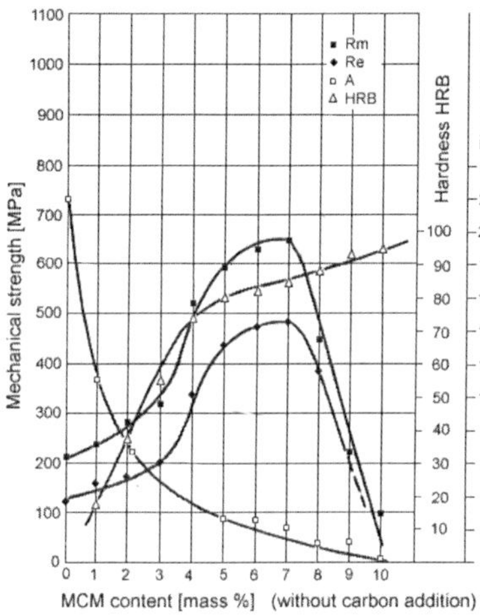

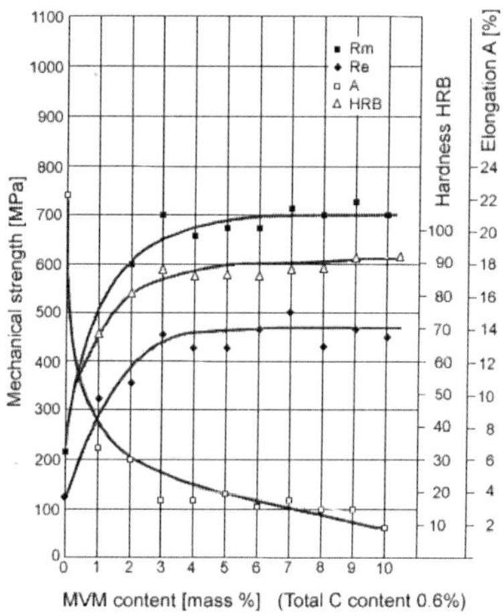

Fig. 9.10 (left). Mechanical properties of sintered steels alloyed with MCM master alloy without carbon addition. Atomet 28 iron powder, 600 MPa, 1h at 1280°C, cracked ammonia, getter – Fe+8% Al, walking beam furnace [61,65,212].
Fig. 9.11 (right). Mechanical properties of sintered steels alloyed with MVM–0.6C (total) master alloy. Sintering as in Fig.9.10 [61,65,212].

strength, the 4% MCM-containing steels, under normal sintering conditions, had a martensitic microstructure.

These alloys were used in practice in a limited number of sintered parts. The hardenability of MCM and MVM alloys in dependence on the distance of the quenched end of the tested specimens is shown in Figs. 9.12 and 9.13.

Figure 9.14 shows the mechanical properties of sinter-forged heat-treated 1% MCM steels and Fig. 9.15 the mechanical properties of heat-treated 2% MCM sintered steel, both in dependence on annealing temperature.

Figure 9.16 shows the mechanical properties of heat-treated 2% MVM sinter-forged steels and Fig. 9.17 those of heat-treated conventional Ck 45 wrought steel, both in dependence on annealing

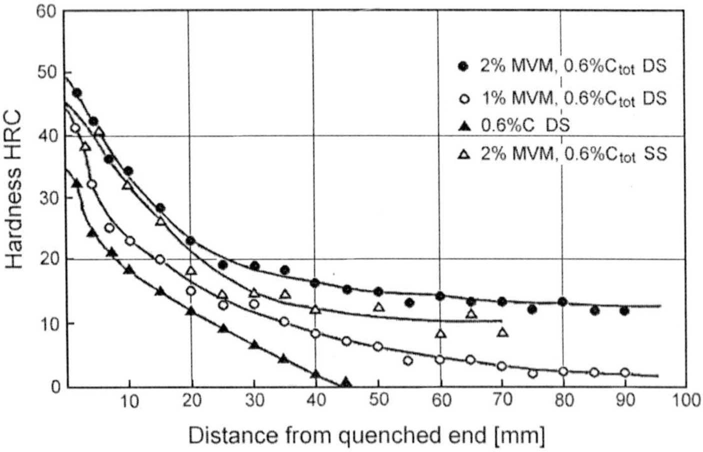

Fig. 9.12. Hardenability of MCM alloys [215,216].

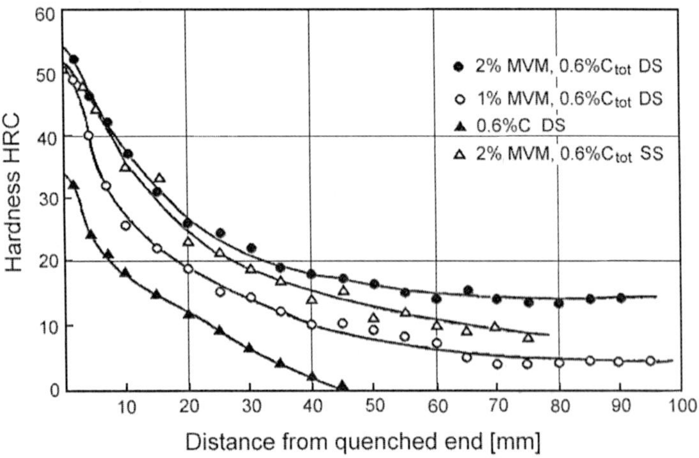

Fig. 9.13. Hardenability of MVM alloys [215,216].

temperature. The group of special master alloys (MCM, MVM, MM) completed a ternary Fe–Si–Mn alloy (called SM) developed for the same use, which consisted of the ternary silicides $(Fe,Mn)_3Si$ and $(Fe,Mn_5)Si_3$, which was able to form a transient liquid phase at sintering temperature. Figure 9.18 shows the mechanical properties of Fe–Si–Mn–C steels attained by the optimum addition of the SM master alloy. Good dimensional accuracy, high strength and wear resistance were expected from these steels. The highest tensile strength of 920 MPa was attained with 0.4% C.

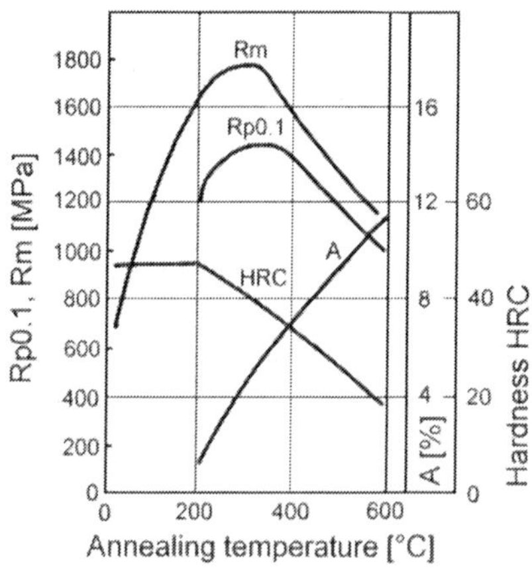

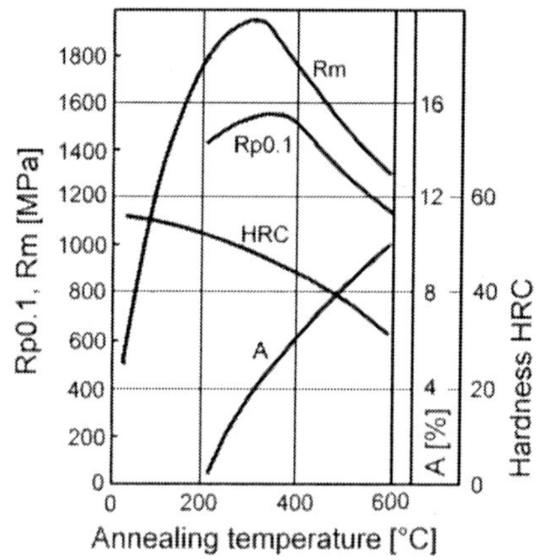

Fig. 9.14 (left). Mechanical properties of heat-treated sintered steels, forging technique in dependence on annealing temperature: 1% MCM, 0.6% C_{tot}, SF [216].
Fig. 9.15 (right). Mechanical properties of heat-treated sintered steels in dependence on annealing temperature; 2% MCM, 0.6% C_{tot}, SF [216].

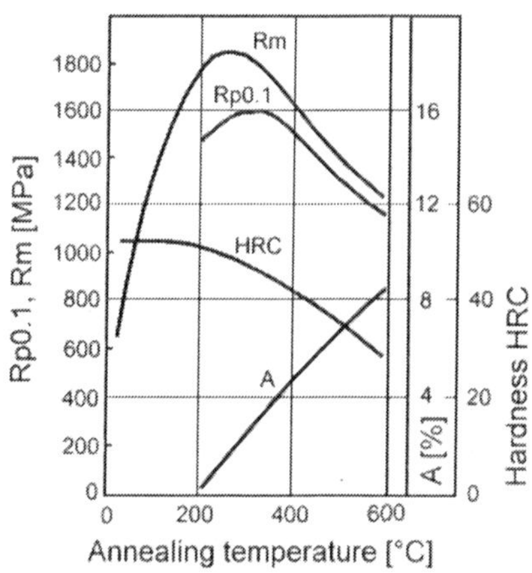

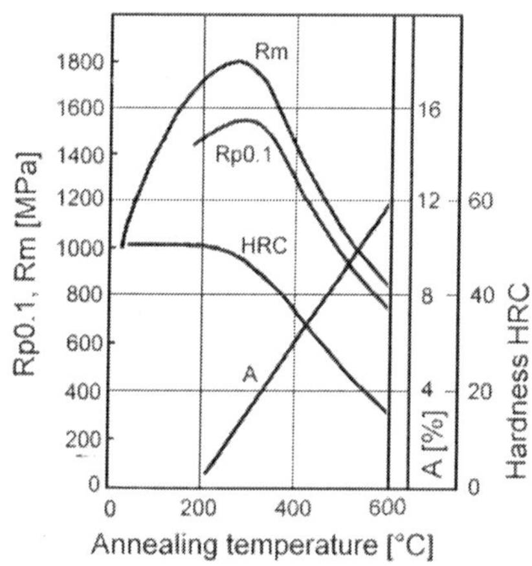

Fig. 9.16 (left). Mechanical properties of heat-treated sintered steels in dependence on annealing temperature: 2% MVM, 0.6% C_{tot}, SF [216].
Fig. 9.17 (right). Mechanical properties of heat-treated Ck 45 conventional wrought steels in dependence on annealing temperature, SF [216].

Figure 9.19 shows the dependence of the mechanical properties of Fe–Si–Mn steels on manganese content for 2% Si. At Mn contents of more than 2% strength and hardness increased strongly, due to the formation of Mn martensite. Elongation after fracture dropped to values of approx. 2–3%. Tensile strength reached 600 MPa, regarded as a good value for the carbon-free sintered steel.

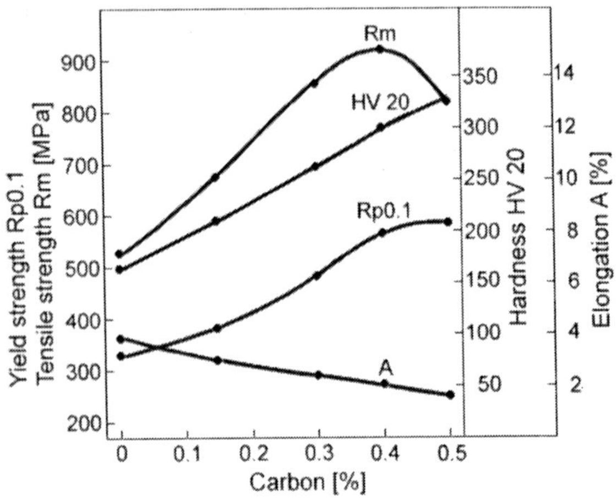

Fig. 9.18. Mechanical properties of Fe–Si–Mn–C sintered steels containing 1.4% Si + 3.2% Mn (ternary SM–master alloy, SM) in dependence on carbon content. Atomet 28 iron powder, 600 MPa, 1 h at 1250°C, hydrogen, d.p.–30°C, cooling rate ~40°C /min [26,215–217].

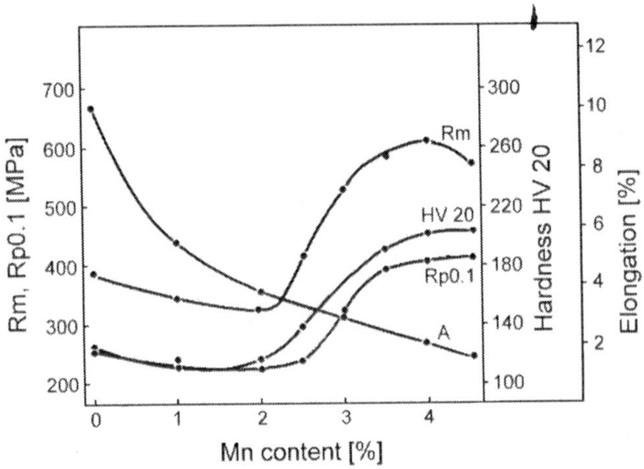

Fig. 9.19. Influence of Mn content on properties of Si–Mn sintered steels (2% Si) [213,214].

Addition of carbon (0.25%) can further increase strength, as shown in Fig. 9.20. Tensile strength increased to values higher than 700 MPa. Elongation after fracture was approx. 2%. Maximum strength was reached with 1–1.5% Si and 2–3% Mn (the density being 7.0 g cm^{-3}).

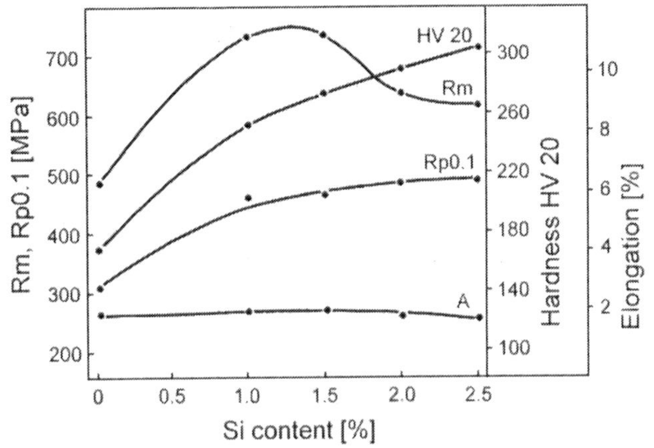

Fig. 9.20. Influence of Si content on the mechanical properties of Si–Mn–C sintered steels (3% Mn, 0.25% C) [213,214].

9.4.2. Alloying of manganese steels by Fe–Mn–Si–C master alloy

The aim of the following study was to use low alloying element contents to improve the mechanical properties of Fe–Mo–Mn–Si–C sintered steels with ternary Fe–Mn–Si master alloy addition [218].

The prealloyed Astaloy 85Mo powder (Fe–0.85% Mo) was used as a base powder. Fe–Mn–Si master alloy powders with different compositions were produced by mechanical milling in the laboratory. In the cast milling process, iron, manganese and silicon were molten down in an induction furnace in an argon atmosphere, and the ingots were cast. The ingots were crushed and ball milled in n-Heptane (C_7H_{16}) for 90 min. Four Fe–MnSi master alloy powders with the compositions Fe–35% Mn–14% Si, Fe–35% Mn–20% Si, Fe–45% Mn–20% Si and Fe–60% Mn–14% Si were produced, and denoted M1, M2, M3 and M4. C-F10 graphite was used in this study.

Two groups of tests were made. In the first group, different amounts of the M2 master alloy powders and graphite were added to Astaloy 85Mo powder. The content of M2 master alloy was 1%, 2%, 3% and 4%, and that of carbon was 0.3,5, 0.5% and 0.7%, respectively. The specimens were sintered at 1250°C for 30 min in hydrogen–30% nitrogen with a dew point –55°C.

In the second group, 4% of the four Fe–Mn–Si master alloy powders and 0.5% of graphite were mixed with Astaloy 85Mo powder, corresponding to 1.90%, 2.20%, 2.60% and 2.96% of total contents of manganese and silicon in the M1, M2, M3 and M4

Table 9.5. Compositions of the tested sintered steels

Master alloy	Steel composition
Fe–35% Mn–20% Si (M2)	Fe–0.85% Mo–0.35% Mn–0.2% Si–(0.3, 0.5, 07)% C*
Fe–35% Mn–20% Si (M2)	Fe–0.85% Mo–0.70% MN–0.4% Si–(0.3, 0.5, 07)% C*
Fe–35% Mn–20% Si (M2)	Fe–0.85% Mo–0.1.05% Mn–0.6% Si–(0.3, 0.5, 07)% C*
Fe–35% Mn–20% Si (M2)	Fe–0.85% Mo–1.40% Mn–0.8% Si–(0.3, 0.5, 07)% C*
Fe–35% Mn–14% Si (M1)	Fe–0.85% Mo–1.40% Mn–0.56% Si–0.5% C**
Fe–35% Mn–20% Si (M2)	Fe–0.85% Mo–1.40% Mn–0.8% Si–0.5% C**
Fe–45% Mn–20% Si (M3)	Fe–0.85% Mo–1.80% Mn–0.8% Si–0.5% C**
Fe–60% Mn–14% Si (M4)	Fe–0.85% Mo–2.40% Mn–0.56% Si–0.5% C**

*The steels sintered at 1250°C for 30 min in H_2–30% N_2 with a dew point of –55°C.
** The steels sintered at 1120°C for 30 min in H_2 with the dew points of –5°C and –45°C, and in N_2–10% H_2 with the dew points –25°C and –43°C

master alloy steels, respectively. The specimens were sintered at 1120°C for 30 min in two types of pure hydrogen and two types of nitrogen–10% hydrogen. The dew points were –45°C and –5°C for the pure hydrogen atmospheres, –43°C and –25°C for the nitrogen–10% hydrogen gases. The compositions of the steels are shown in Table 9.5. Standard specimens for testing of mechanical properties were prepared by compaction at 640 MPa. Green density was ~7.1 g cm^{-3}.

Mechanical properties of steels sintered at 1200°C. The ultimate tensile strength, yield strength, elongation and hardness are shown in Fig. 9.21. The three properties (except elongation) increase with increasing the M2 master alloy content as well as carbon content. These properties were regarded for the synergetic effect of Mn+Si and of as sintered carbon content. The elongation of Fe–Mn–Si-sintered steels decreases with increasing amount of Mn+Si and carbon. The tensile strength of 985 MPa, yield strength of 679 MPa, hardness of 332 HV 10 and elongation of 1.32% were obtained.

Mechanical properties of steels sintered at 1120°C. The tensile strength, yield strength, elongation and hardness of the four master alloy sintered at 1120°C are shown in Fig. 9.22. The tensile and yield strengths do not change significantly with the dew point of the sintering atmospheres. The steels sintered in hydrogen show higher tensile, yield strength and hardness than those in nitrogen–10% hydrogen. The highest tensile strength and yield strength were obtained for Mn+Si = 2.2% in both sintering atmospheres.

The elongation decreases with total manganese and silicon contents. The higher elongation values were obtained in the compacts sintered in the gas with a lower dew point.

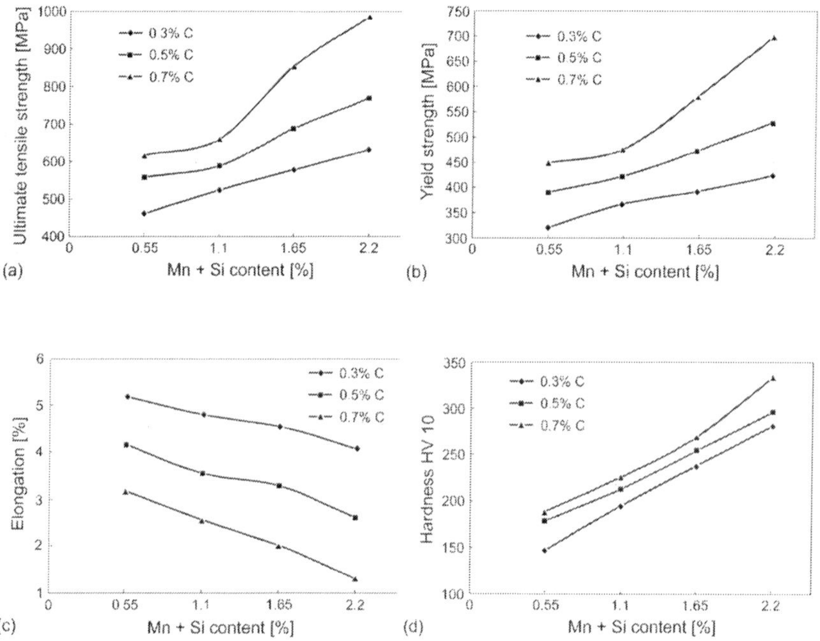

Fig. 9.21. Effect of manganese and silicon contents on ultimate tensile strength (a), yield strength (b), elongation (c) and hardness (d) of Astaloy 85Mo–M2 steels with different carbon contents sintered at 1250°C for 30 min in H_2–30% N_2 [218].

The tensile strength of 825 MPa, yield strength of 705 MPa, hardness of 321 HV 10 have been obtained. The best combinations of properties appeared for Mn+Si = 2.2% in both sintering atmospheres.

The mechanical properties of the M2 master alloy steels sintered at 1250°C were higher than for those sintered at 1120°C in both cases for 30 min.

9.4.3. Alloying of manganese steels by master alloys containing Cr and Mn

This section presents the base characteristics of a newly developed master alloy called ML for the use of manganese and chromium as alloying elements for the production of Mn–Cr alloyed sintered steels [219–221]. The master alloy containing 10% Cr and 10% Mn in the iron base was kerosene-atomised to prevent pick-up of the alloy constituents.

Figure 9.23 shows the characteristic microstructure. The master alloy particles were rather round and had no adverse effect on the compacted properties, unlike in other MCM powders.

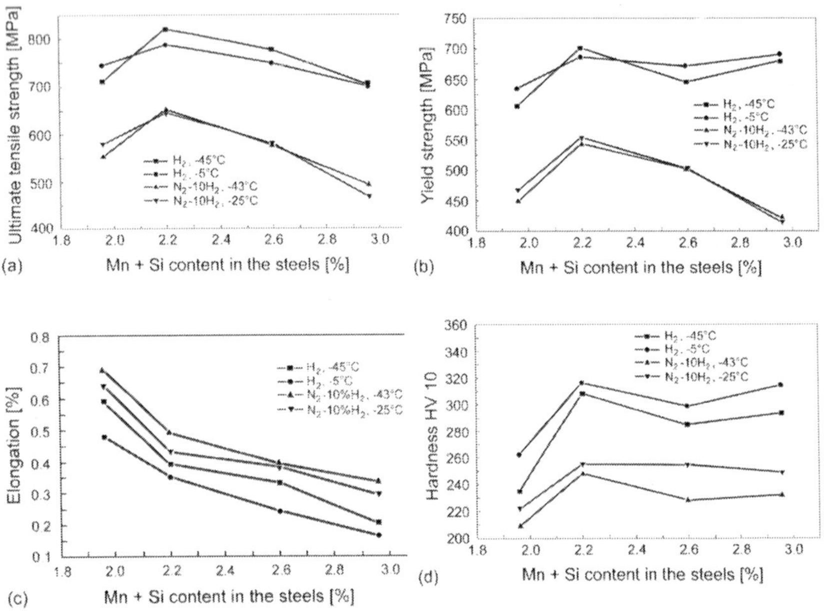

Fig. 9.22. Effect of manganese and silicon contents on ultimate tensile strength (a), yield strength (b), elongation (c) and hardness (d) of Astaloy 85Mo–M2 steels with different carbon contents sintered at 1120°C for 30 min in H_2 and N_2–10% H_2 with different dew points [218].

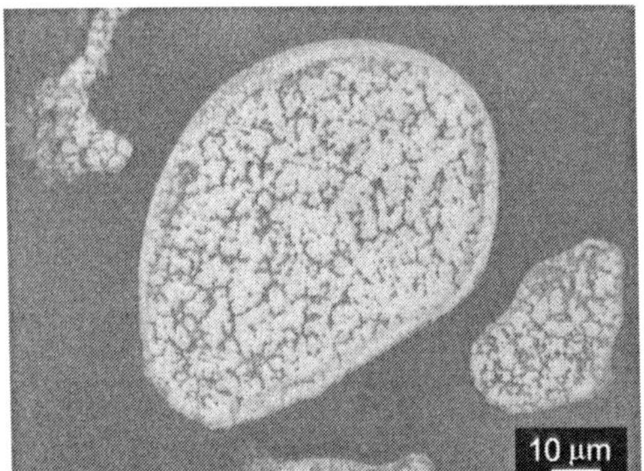

Fig. 9.23 Coarse master alloy particles after glyceregia etching [220].

The average carbon content in the alloy was about 0.9%, the microhardness was about 370 HV 0.05. Since alloying of the base material should take place via solid state diffusion, the particle size was <50 μm. The final mixtures were prepared on the basis of the water-atomised iron powder Atomet 1011HP and Atomet 4401. The nominal chemical of the mixtures is given in Table 9.6.

Table 9.6. Nominal chemical composition of two mixtures in [mass %]

Alloying elements	C	Mn	Cr	Mo
Atomet 1001HP + 12.5% master alloy + 0.4% graphite + 0.8% microwax	0.51	1.32	1.25	–
Atomet 4401 + 12.5% master alloy + 0.4% graphite + 0.8% microwax	0.51	1.38	1.25	0.74

Table 9.7. Mechanical properties and density of tensile strength test bars compacted at 700 MPa

Property	Density [g cm^{-3}]	Hardness HB (2.5/6.25)	Yield strength Rp0.2 [MPa]	Tensile strength [MPa]
Atomet 1001HP + 12.5% master alloy + 0.4% graphite + 0.8% microwax	7.09	139	339	539
Atomet 4401 + 12.5% master alloy + 0.4% graphite + 0.8% microwax	7.06	204	523	724

Table 9.7 gives the mechanical properties and density of tensile strength test bars compacted at 700 MPa with the chemical composition presented in Table 9.6.

The tensile strength test bars were compacted at 700 MPa. Sintering took place in an industrial walking beam furnace at 1280°C for about 30 min in an atmosphere of 95% N_2 and 5% H_2.

For investigation of the potential of these steels as quenched and tempered steels, both materials were tempered at temperatures between 300 and 600°C. Figures 9.24 and 9.25 compare the yield strength values with the sintered state.

In particular, the yield strength Rp0.2 benefited significantly from the tempering treatment. Since all values were averages of five tests, it was quite clear that 500°C was a good tempering compromise between strength and hardness. The gain in yield strength from tempering amounts to ~10 to 20% compared with the values after sintering. After high-temperature sintering at 1280°C and tempering, the resultant tensile properties surpassed the yield and tensile strengths by the better compressibility of the lower alloyed base powders, even if the properties were normalized to identical densities, the performance of the lower Mo base powder is superior [221].

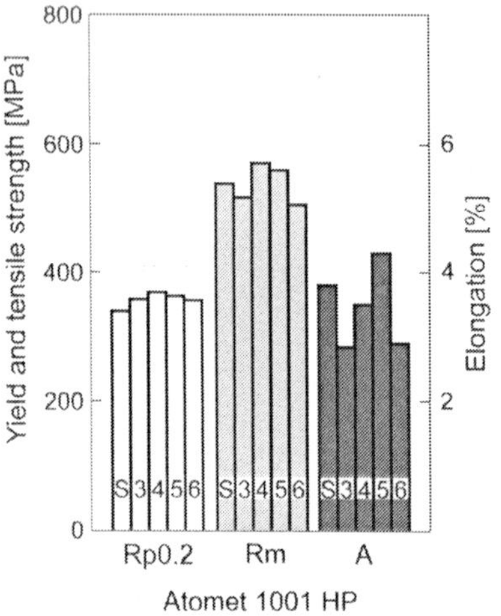

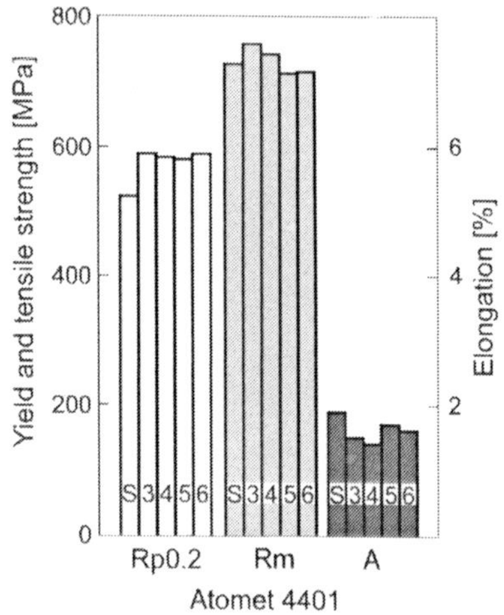

Fig. 9.24 (left). Yield and tensile strength and elongation of sintered steels based on Atomet 1001HP powder: S – as sintered, 3 – 300°C, 4 – 400°C, 5 – 500°C, 6 – 600°C tempered for 1 h [220].

Fig. 9.25 (right). ield and tensile strength and elongation of sintered steels based on Atomet 4401 HP powder: S – as sintered, 3 – 300°C, 4 – 400°C, 5 – 500°C, 6 – 600°C tempered for 1 h [220].

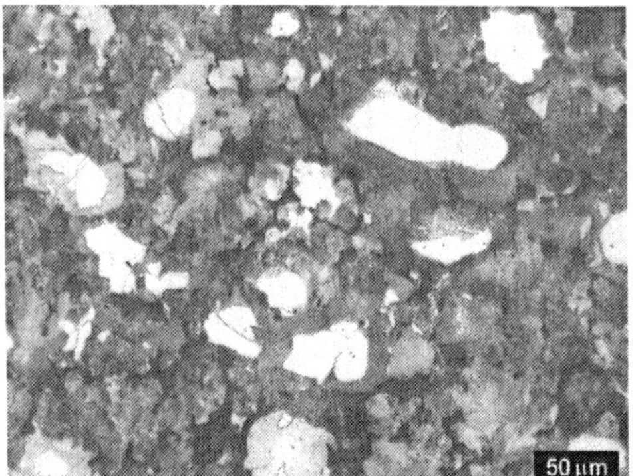

Fig. 9.26. Microstructure with Atomet 1001HP as a base material [219,220].

The microstructure of the base powder was determined. With the pure iron, Fig. 9.26, a heterogeneous material consisting of ferrite, pearlite, bainite, martensite and retained martensite was obtained. The amount of martensite and retained austenite was reduced by tempering, and martensite decomposed. In bainite and martensite some 1.3% Cr and 1.6% Mn were found, whereas in the unetched regions with higher retained austenite content the Cr content amounted to 3.3% and that of Mn to 2.8%. The time–temperature

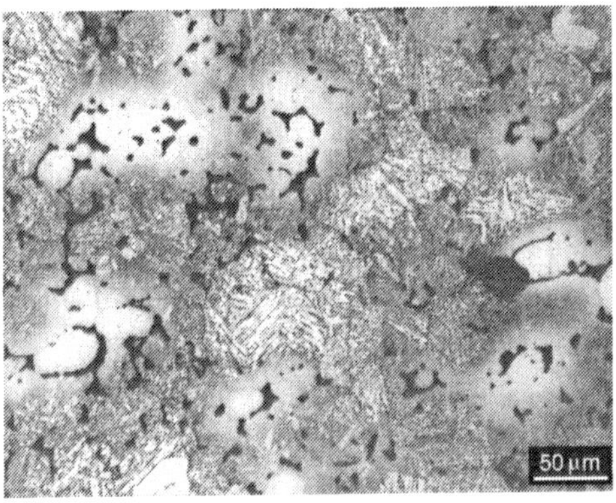

Fig. 9.27. Microstructure with Atomet 4401 as a base material [219,220].

combination during sintering was obviously not sufficient to fully homogenize the alloy content of the master alloy particles.

The microstructure of the Mo steel base material is shown in Fig. 9.27. Ferrite was missing here because of the 0.5% Mo content. The centre of larger particles consisted of upper bainite, which seemed to contain with around of 0.7% Cr and 1.2% Mn a higher alloy content than the ferritic phase in Mn-free material. The unetched areas in this figure had similar contents around 3% Cr and Mn as in the first material.

The strength level obtained with 0.5% total carbon in pure iron to which 12.5% master alloy was added was comparable with Fe–2% Cu–0.6% C.

Addition of 12.5% of a rather round and soft master alloy containing 10% Cr and 10% Mn did not significantly change the powder characteristics nor the processing behaviour of the mixtures with 0.5% carbon. Because of the high chromium and manganese content, high-temperature sintering was inevitable to reduce the oxides of the master alloy powder. Tempering at around 500°C further improved significantly the yield strength but also added to the total cost.

The properties of sintered steels alloyed with (1–4)% MCM master alloy with additions of 0.3 and 0.6% phosphorus are presented in [222,223]. According to the results, phosphorus increased the yield and tensile strength. In [224] was recommended for production of high strength steels between others also 40Ni–20Mn–18Mo balance Fe master alloy.

9.4.3.1. Alloying of manganese steels with atomised Fe–Mn–C master alloy

Mechanical properties. The composition of the ternary Fe–Mn–C master alloy was designed using thermodynamic modelling techniques to produce a transient liquid phase upon sintering at conventional temperatures, Fig. 9.28, and to be suitable for production by atomisation to give a fine spherical powder with a low oxygen content in order to maximize its compressibility (atomizing conditions and composition not described). Manganese added in this way has reduced the oxidation potential and vapour pressure compared to elemental manganese and to manganese in the form of ferromanganese. This allowed steels alloyed in this way to be sintered at 1120°C.

For the preparation of the test specimens, the master alloy was added to the base iron powder (Atomet 1001E) in order to give overall manganese contents of 1.0, 1.5 and 1.8%. Graphite was added to give total carbon contents of 0.4, 0.6 and 0.8% and 0.6% EBS wax to all mixtures. Tensile strength specimens were compacted from the mixtures at 700 MPa. The specimens were sintered at 1120°C in a $95H_2/5N_2$ atmosphere for 20 min in an industrial belt furnace [225].

Table 9.7 gives the values of the mechanical properties of sintered samples, each value representing the mean of five samples.

As follows from the Table 9.7, relatively high strength values were obtained compared with the results published for ferromanganese-alloyed steels sintered at 1120°C.

The characteristic microstructures of two steels are shown in Figs. 9.29 and 9.30.

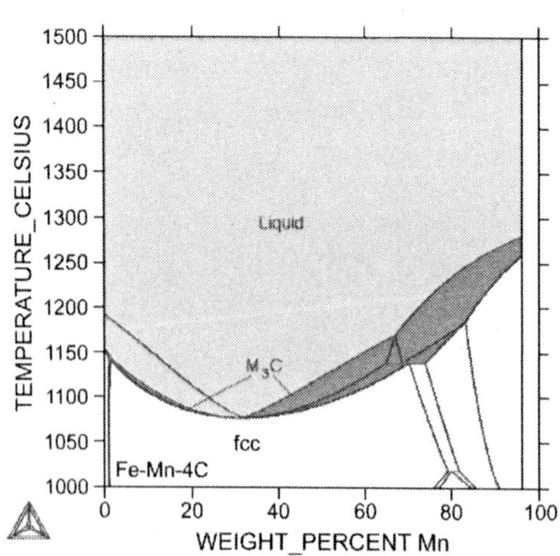

Fig. 9.28. Ternary isopleth for the Fe–Mn–C system [225].

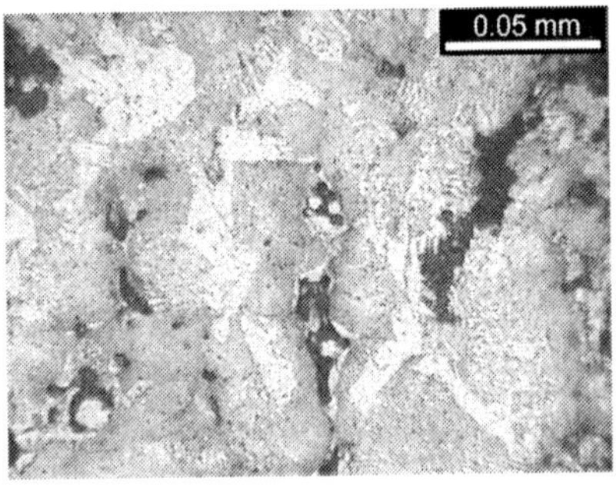

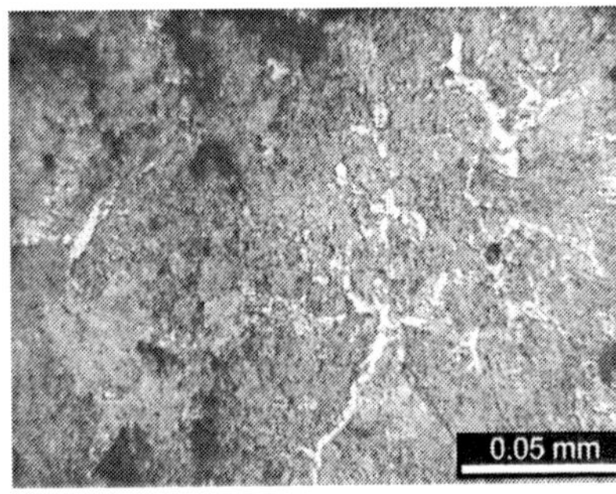

Fig. 9.29 (left). Microstructure of Fe–1.5Mn–0.4C steel [225].
Fig. 9.30 (right). Microstructure of Fe–1.8Mn–0.6C steel [225].

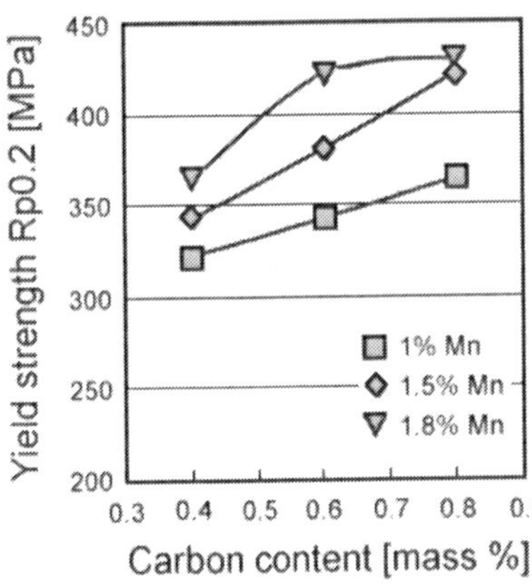

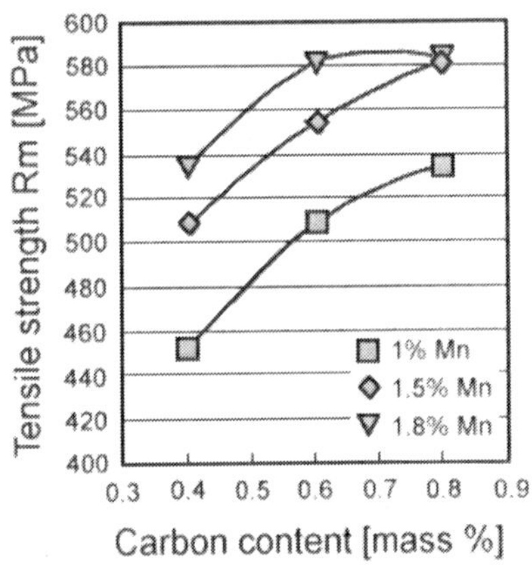

Fig. 9.31 (left). Yield strength vs. carbon content at the three levels of manganese added as Fe–Mn–C master alloy [225].
Fig. 9.32 (right). Tensile strength vs. carbon content at the three levels of manganese added as Fe–Mn–C master alloy [225].

The shown microstructures of the specimens were ferritic/pearlitic with the proportion of pearlite increasing with increasing carbon content and the interlamellar spacing of the pearlite.

Figures 9.31 and 9.32 show the yield and tensile strength values in dependence on carbon content for 1Mn, 1.5Mn and 1.8Mn steels. The highest strength properties were obtained in the steels with 1.8% Mn and 0.8% C.

Metallographic analysis confirmed the presence of islands of undissolved master alloy particles in the microstructure, Fig. 9.33.

Bearing in mind the presented mechanical properties which were obtained using the Mn-alloy steel component, in that time manufactured from FD-0205 powder. The component was an oil

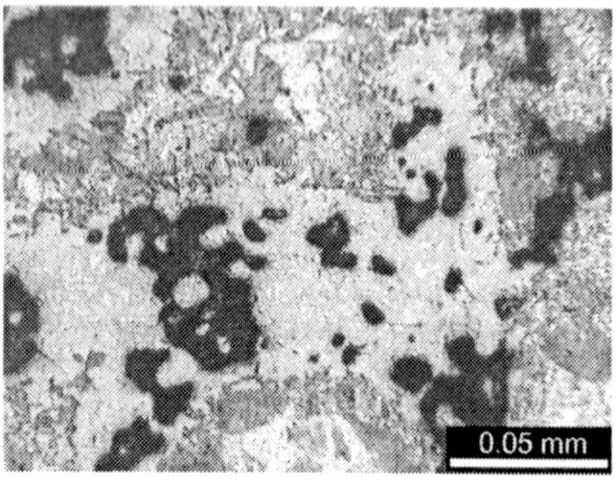

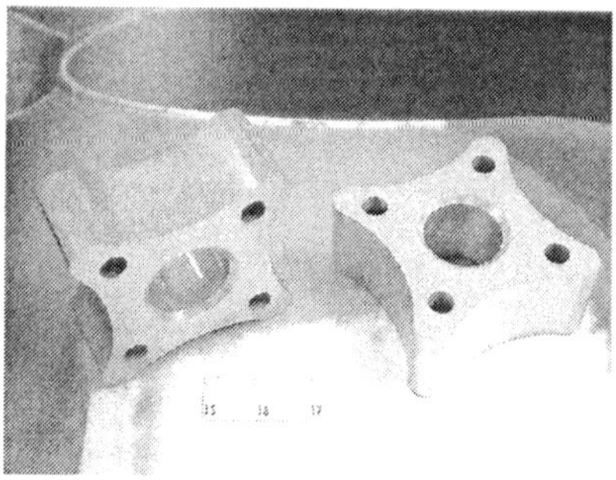

Fig. 9.33 (right). Microstructure of Fe–1.8Mn–0.6C steel with undissolved master alloy 'islands' [225].
Fig. 9.34 (right). Oil pump rotor part produced from Mn alloyed steel [225].

pump rotor, Fig. 9.34, a part requiring hardness and wear resistance. The parts were compacted from a binder-mixed Fe–1.5Mn–0.8C composition and were sintered together with the FD-0205 parts at 1120°C under the conditions as the tensile strength bars. After sizing the series parts were characterised for surface microhardness (298 HV 0.1 against 306 HV 0.1 for the parts from FD-0205 powder), and an indication of the wear resistance by testing them in a microtest ball-in disc test ring, using a 1100Crb steel ball 3 mm in diameter. No indication was seen that there was any increased wear in the Mn-alloyed parts compared with the Ni–Mo–Cu alloyed series parts.

Tensile and fatigue properties of manganese-containing steels alloyed with Fe–Mn–C master alloy. The materials were prepared from powder blends containing Fe-based powders (Ancorsteel 85HP or Astaloy CrM) to which Mn was added as a Fe–Mn–C master alloy sieved under 20 μm. Ni-additions were made using fine, Inco 123 (<8 μm) and ultrafine particles <0.5 μm. Tensile testing was carried out using standard specimens. Axial fatigue testing was carried out on hourglass specimens, ejected under pressure from the die pressing cavity. Ten specimens per material were tested. The nominal compositions for the powder mixtures are summarized in Table 9.8 [226].

Based on the two Fe-based powders used a set of tensile bars prepared with additions of 0.8% Ni, either in fine or ultrafine form, and a constant amount of the Fe–Mn–C master alloy to give a total concentration in the material of 1.1% Mn.

The results of the tensile tests after sintering at 1120°C followed by accelerated cooling and tempering are summarized in Figs. 9.35 and 9.36. Figure 9.35 corresponds to those combinations based on

Table 9.8. Mechanical properties of Mn steel specimens

Powder mix	Rp0.2 [MPa]	UTS [MPa]	A [%]	Hardness HRB
Fe–1Mn–0.4C	322	451	5.5	73
Fe–1Mn–0.6C	343	208	4.7	79
Fe–1Mn–0.8C	364	534	2.5	84
Fe–1.5Mn–0.4C	343	508	4.7	78
Fe–1.5Mn–0.6C	381	564	3.3	82
Fe–1.5Mn–0.8C	421	581	2.0	86
Fe–1.8Mn–0.4C	364	534	2.5	84
Fe–1.8Mn–0.6C	421	581	2.0	86
Fe–1.8Mn–0.8C	428	583	1.7	89

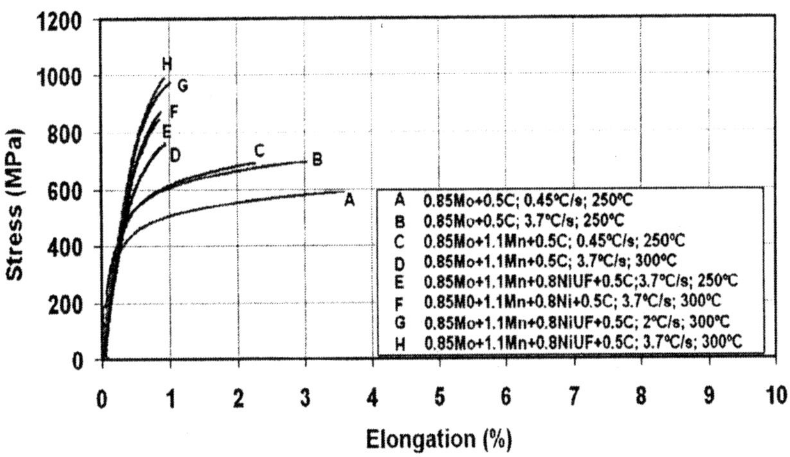

Fig. 9.35. Relationship between the stress and elongation of sintered steel based on Fe–0.85Mo base powder [226].

Ancorloy 85HP, while Fig. 9.36 includes the data for materials based on Astaloy CrM. In both cases the influence of the combined action of Mn and Ni on tensile strength and yield stress is evidenced. As noticed from the tensile curves in Fig. 9.35 and Fig. 9.36 the highest strengths (Fig. 9.36, curve D) were obtained for a chemical composition involving Cr, maximum cooling rate below 2°C/s and tempering at 200°C. In the same figure (curves F and E) increasing

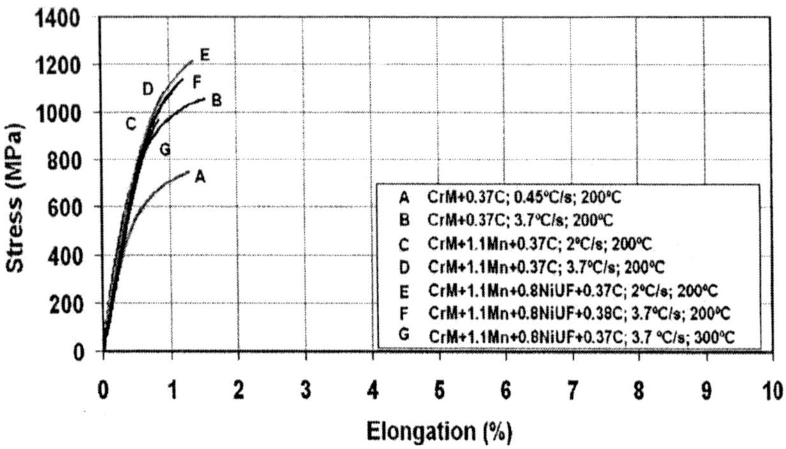

Fig. 9.36. Fe–3Cr–0.5Mo [226].

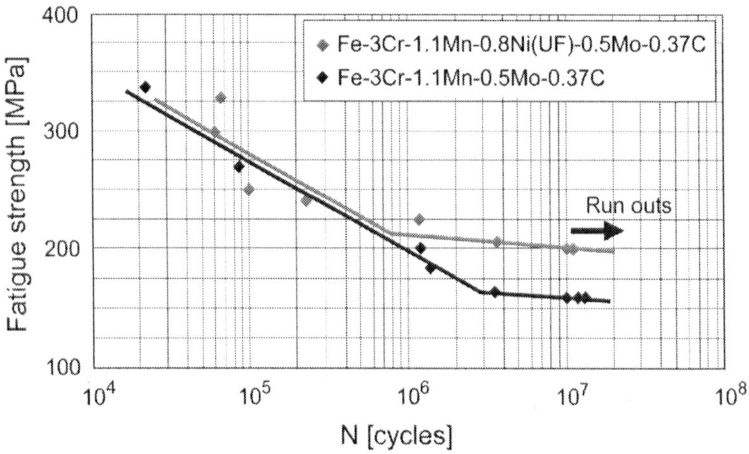

Fig. 9.37. S–N curves for the materials under study after sintering at 1120°C, cooled at 2°C/s and tempered at 200°C for 1 h [226].

the cooling rate produced a marked decrease in strength even after tempering at 200°C or 300°C.

The material with the highest tensile strength (curve D in Fig. 9.36) was subjected to axial fatigue testing under constant amplitude loading. For comparison, the material giving curve C in Fig. 9.36 was also tested since after processing under identical conditions informations. In Fig. 9.37, the endurance limit for CRMNUF is higher than that exhibited by the material without nickel. Nevertheless it was important to mention that for CRMNUF three run outs were obtained at a stress of 196 MPa, for material CrM were obtained in

Table 9.9. Nominal chemical composition (mass %) of the experimental steels used and a type of nickel added (balance to given compositions was Fe)

Material designation	Mo	Mn	Ni	Cr	Graphite	Ni grade
85M	0.85	1.1	–	–	0.5	–
85MN	0.85	1.1	0.8	–	0.5	Inco 123
85MNUF	0.85	1.1	0.8	–	0.5	Ultrafine
CRM	0.5	1.1	–	3.0	0.37	–
CRMN	0.5	1.1	0.8	3.0	0.37	Inco 123
CRNMUF	0.5	1.1	0.8	3.0	0.37	Ultrafine

Table 9.10. Mechanical properties of Mn steels alloyed with 'point sources' in the sintered condition and after forging and heat treatment. Steel A – (0.8–1.1)Mn–(0.35–0.45)C–0.5Si, steel B – (0.98–1.1)Mn–(0.60–0.70)C–0.15Si; P – porosity, ρ – density [227]

Steel	As sintered					As powder forged and heat-treated					
	P [%]	Rm [MPa]	TRS [MPa]	A [%]	K [J]	ρ [%]	Rm [MPa]	TRS [MPa]	A [%]	K [J]	HRC
A	13	284	–	4.8	2.5	7.75	1100	2200	7.0	5.3	41
B	13	463	756	2.4	1.4	7.74	1245	2245	3.0	5.2	43

tests at 150 MPa. The criterion for a run was a survival of $>10^7$ cycles.

9.4.3.2. Manganese carrier – diffusion manganese-prealloyed powders

The preparation of diffusion-alloyed manganese powders, which can be called master alloys (in the original called as 'point sources'), was carried out by the following method. Manganese or ferromanganese powder and 1–2% of ammonia chloride (NH_4Cl) were admixed to the iron powder. The mixture was in a closed pot heated for 2–4 h at the temperature of 1000–1100°C. (The process was activated by the addition of NH_42CO_3 in the amount of 10%). The iron powder saturated by manganese during this process in the form of a cake was ground (good grindability) to powder and used as a manganese carrier for alloying of sintered manganese steels. The content of manganese in the powder was adjustable, usually at 15–30% Mn. The chemical composition of the experimental steels is listed in Table 9.9 and Table

9.10 gives the mechanical properties of Mn steels alloyed with 'point sources'.

9.5. Liquid phase sintering of Fe–Mn steels

9.5.1. Liquid phase sintering of Fe–Mn steels through low-melting alloys

Many low-temperature melting master alloys containing manganese were developed and/or determined on the basis of the equilibrium diagrams for an increase the density and the homogeneity of the microstructure by liquid phase sintering. This was expected to result in the trouble-free sintering of manganese steels due to lower chemical activity of the master alloys. The largest complex of Mn-containing master alloys offers Table 9.11.

The master alloys were produced by melting under argon followed by gas atomisation with nitrogen into a chamber filled also with nitrogen. The master alloys were screened through a 0.071 mm sieve.

A water-atomised forging quality steel powder (Ancorsteel 1000 B) was mixed with the desired amount of master alloy, graphite, and pressing lubricant. The powder mixture was pressed into performs 76 mm in diameter and about 6.8 g cm^{-3} in density, and sintered at 1232°C in hydrogen. Sintered performs were shot peened to seal pores at the surface and then induction heated to about 982°C under nitrogen, before being transferred to a press for forging into cylinders 102 mm in diameter by 30 mm thick. Near full density was reached. The packs were sectioned into blanks for test bars and normalized at 1000°C. Jominy and tensile strength bars were machined after a subsequent quench and temper heat treatment to obtain hardness of 26 HRC.

Table 9.12 gives the chemical composition of sintered steels from admixtures of 2.5% master alloy to pure iron powder for master alloy 342 and 400. The mechanical properties of steels with the composition listed in Table 9.12 are listed in Table 9.13.

As follows from Table 9.13, the tensile strength was 820 MPa for both alloys with ductility approaching values for conventional steels, as reflected by elongation of 24–26% and reduction of area of 48–53%.

Master alloys can be admixed also with prealloyed powders.

Table 9.11. Chemical composition of master alloys with maximal concentrations of manganese for alloying powder steels with manganese in combination with other elements through liquid phase sintering. T_l – liquidus temperature, T_s – solidus temperature [1,228,229]

Master alloy mix	Chemical composition [mass %]							T_l [°C]	T_s [°C]
	Mn	Ni	Cr	Mo	Fe	Cu	Si		
342	40	30	15	5	10	–	–	1171	999
400	44	25	–	11	19	–	–	1204	1166
524	55	18	3	8	14	–	2	1129	643
533	56	24	3	6	11	–	–	1157	1032
533S	*	*	*	*	*	–	2.5	1166	993
533M	*	*	*	*	*	–	2.5	1104	1010
534	52	22	8	6	12	–	–	1154	1132
534S	*	*	*	*	*	–	2.5	1132	1071
534M	*	*	*	*	*	–	2.5	1149	1016
535	47	20	13	6	14	12.5	2.5	1210	1166
535S	*	*	*	*	*	–	2.5	1149	1071
535M	*	*	*	*	*	–		1174	1054
528	75	25	–	–	–	–	–	1060	927
527	74	12.5	–	–	–	–	1	1207	1096
344	36	30	18	6	10	–	–	1216	1077
345	41	25	18	6	10	–	–	1229	1093
346	38	23	18	6	15	–	–	1232	1088
506	64	16	0	10	10	–	–	1227	1102
508	56	14	0	15	15	–	–	1260	1093
509	56	15	15	5	10	–	–	1188	1132
510	56	14	10	10	10	–	–	1227	1102
511	59	11	15	5	10	–	–	1249	1116
512	53	17	15	5	10	–	–	1216	1093
513	56	14	22	8	–	–	–	1216	1143
514	50	20	15	5	10	–	–	1335	1049
515	46	24	15	5	10	–	–	1204	1143
531	72	14	–	–	–	14	2	1204	1088
532	72	14	–	–	–	14	–	1104	977
532	71	14	–	–	–	–	–	1043	966

*Same percentages as immediately above, except reduced proportionally for the presence of silicon

Table 9.12. Chemical composition of sintered steels produced from premixes of 2.5% master alloy with pure iron powder

Master alloy	Chemical analysis [mass %]						
	C	Si	Mn	Cr	Mo	Ni	Cu
342	varied	0.01	1.02	0.42	0.14	0.42	–
400	varied	0.01	1.00	0.04	0.26	0./13	0.03

Table 9.13. Tensile strength, hardness, and Charpy impact properties of sintered, forged, and quenched and tempered steels, produced from premixes of 2.5% master alloy with pure iron powder. ψ – reduction of area

Master alloy	C [%]	HRC	UTS [MPa]	Rp0.2 [MPa]	A [%]	ψ [%]	Impact energy [J]		
							23°C	–18°C	–51°C
342	0.31	26	820	696	24	48	61	53	46
400	0.31	25	820	717	26	53	68	65	31

9.5.2. *Alloying of manganese steels with atomised Mn–Cu and Mn–Ni master alloys*

Tests of alloying of sintered steels with Mn were carried out on a low-melting Mn–Cu alloy with 35% Cu (melting point 870°C) and Mn–Ni low-melting alloy with 40% Mn (melting point 1018°C). The powders of these alloys with high compressibility were produced by dispersion.

The properties of these sintered steels alloyed with Mn–Cu and Mn–Ni master alloys are listed in Table 9.14.

Because of their high compressibility in the powder form, these low-melting alloys were recommended for large-series production of high-strength sintered parts.

In [231] it is recommended to use for sintering manganese steels the master alloys 89Mn–11Si (melting point 1000°C), 64.2Mn–35.8Si (melting point 1020°C) and 61Al–39Mn (melting point 880°C). The 40Ni–20Mn–18Mo, balance Fe, extremely fine master alloy enhancing diffusion of alloying elements into the iron matrix for sintering at temperatures in excess of 1180°C was recommended for use in [232]. The data about the production and use of some of the mentioned low-melting master alloys for alloying sintered manganese steels are lacking.

Table 9.14. Properties of sintered with addition of Mn, Cu and Ni. RZ iron powder, 800 MPa, sintering for 1 h at 1120°C, hydrogen, dew point –32°C [230]

Steel No.	Chemical composition [mass %]			Density [g cm^{-3}]	Rm [MPa]	Hardness HB
	Cu	Mn	Ni			
1	2	1.6	0.4	7.21	430	145
2	2	1.2	0.8	7.22	420	148
3	2	1.0	1.0	7.39	413	148
4	2	0.6	2.4	7.22	421	150
5	2	1.8	1.8	7.21	399	145
6	2	2.0	2.0	7.17	415	153
7	2	0.8	3.2	7.24	423	148

10

Effect of variable processing conditions and materials on properties of sintered Mn–C steels

10.1. Alloying with low-melting elements and sintering in high-purity atmospheres

10.1.1. Alloying with low-melting elements

The goal of the following research and development work in terms of thermodynamics was to reduce the high demands on the purity of the sintering atmosphere at sintering of manganese alloy steels, because when the samples are covered with a getter manganese oxides can be reduced what could occur only at sintering temperatures over 1200°C. Therefore, tests were carried out to determine whether the addition of some low-melting elements would allow sintering and homogenisation of the structures of sintered iron–manganese samples also at lower temperatures.

10.1.1.1. Effect of tin

Samples were prepared from SC100.26 iron powder (0.02% C, 0.5% O), electrolytic manganese <63 μm, and a fine tin powder, particle size ~4 μm. To determine the mechanical properties, standard test bars were pressed from the powder mixture at a pressure of 600 MPa, which were then sintered at different temperatures for 1 h. The heating and cooling rate was 10°C/min. All sintering processes were carried out in hydrogen with a dew point of –70°C.

Mechanical properties. Figure 10.1 shows the tensile strength and ductility of Fe–2Mn and 2Mn–1Sn as a function of sintering temperature. The effect of accelerated formation of the alloying agent, in this case the liquid Mn–Sn–rich phase, can be seen in the results. While high strength of the Fe–2Mn compacts was achieved at 1280°C, the alloy Fe–2Mn–1Sn obtained comparable strength after sintering at 1050°C. The same is true for elongation. Fe–2Mn samples after sintering at temperatures between 1100 and 1200°C due to incomplete formation of solid solution showed approximately 20% higher elongation values than those for plain sintered iron of comparable density. Alloy Fe–2Mn–1Sn showed at all the sintering temperatures ductility values that correspond to hardening with a homogeneous solid solution.

In the sintered materials, the role of tin as an alloying element forming solid solution is not significant. Tin affects much more the course of the reaction of iron and manganese compacts by the formation of a manganese-rich liquid phase. This allows production of structurally relatively homogeneous sintered materials based on manganese-alloyed iron with the addition of tin at sintering temperatures around 1050°C.

In [235], the effects of more processing factors on the properties of iron–tin–manganese alloys were studied as compared to [233,234]. Iron powder SC100.26 and air-atomised tin powder (<53 μm) were

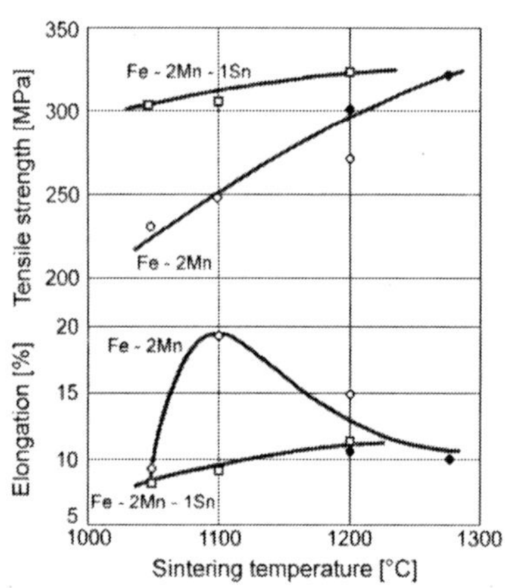

Fig. 10.1. Tensile strength and elongation of Fe–2Mn and Fe–2Mn–1Sn steel compacts in dependence on sintering temperature between 1050 and 1280°C sintered for 1 h; SC100.26 iron powder, EMn, 600 MPa, hydrogen, dew point –70°C [233,234].

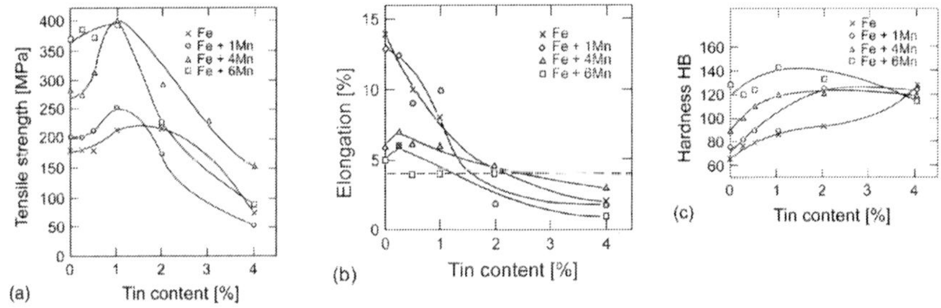

Fig. 10.2. Effect of tin content on tensile strength (a), elongation (b) and hardness (c) of Fe–(1,4,6)Mn steels; SC100.26 iron powder, EMn, 600 MPa, 1 h at 1150°C, dry hydrogen [235].

used for the preparation of the specimens for investigation of the mechanical properties of this sintered powder system. Zinc stearate in the amount of 0.5% was added as a pressing lubricant. The mixtures were compacted at 620 MPa in the form of standard tensile strength bars.

The compacts were sintered in a tube furnace in an oxygen-free dry hydrogen atmosphere. All compacts were first heated in the cooler part of the furnace at about 400°C for 5 min to vaporize the lubricant and then slowly moved into the hot zone where they were sintered at the stated temperature for the required time. After the end of the sintering operation the specimens were moved to the water-cooled end of the tube for a short period.

The effect of increasing tin content to 4% on the tensile strength of iron compacts containing 0, 1, 4 and 6% manganese and sintered for 1 h at 1150°C is shown in Fig. 10.2. The peak strength was obtained for the alloy containing 1% tin. Maximum tensile strengths of 400 MPa are shown by both the 4% and the 6% manganese alloys. The elongation of the 4% and 6% manganese alloys fell steadily as the tin content increased. This effect was accentuated in the case of iron–1% manganese compacts with a dramatic drop in ductility for compositions up to 1% tin, Fig. 10.2(b). Hardness increased with tin content up to 2% tin, Fig. 10.3(b), except in iron–6% manganese compacts where the peak hardness was achieved with 1% tin addition followed by progressive softening.

Effect of manganese content. Figure 10.3 shows the effect of manganese addition up to 20% in iron compacts containing 1% tin on tensile strength and hardness. The tensile strength reached a maximum at 4–6% manganese and 1% tin, while the alloys with an even lower tin content alloys showed comparable values at 6%

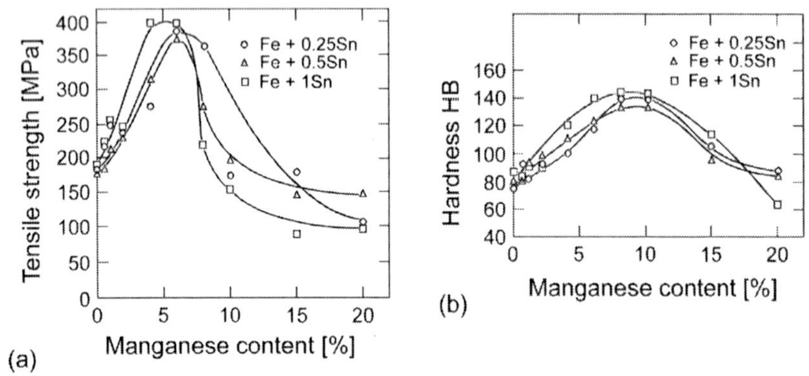

Fig. 10.3. Effect of manganese content (up to 20%) on tensile strength (a) and hardness (b) of three iron–tin compositions [235].

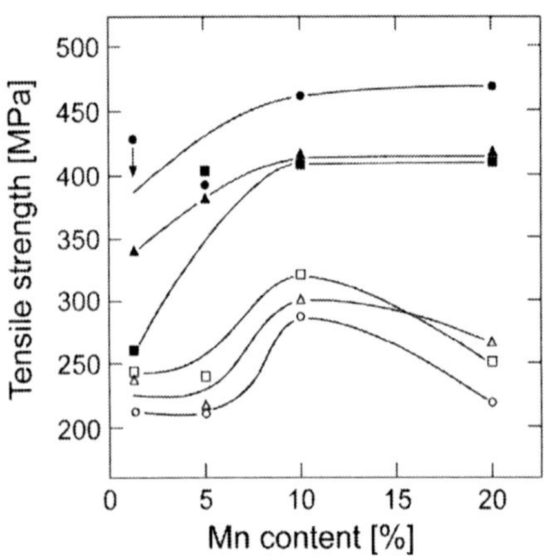

Fig. 10.4. Effect of Mn content on tensile strength of three compositions selected for optimum tensile strength. For designation of the composition of corresponding alloys see Fig. 10.5 [235].

manganese. The hardness values obtained were mostly governed by the level of manganese addition and maximum hardness was obtained with 8–10% manganese. For the iron–1% tin series of compacts, maximum hardness values of 140 HB were reached over the wider range of 6–10% manganese addition. The effect of Mn content on tensile strength in shown in Fig. 10.4.

Effect of sintering time. The samples were sintered for up to 4 h at 1150°C in hydrogen. The optimum tensile strength was obtained after about 2 h after which the high tensile strength values were maintained. All alloys selected as having maximum tensile properties behaved similarly in their response to increased sintering times.

Effect of sintering temperature. To investigate the effect of sintering temperatures, the samples were sintered for 1 h at 950,

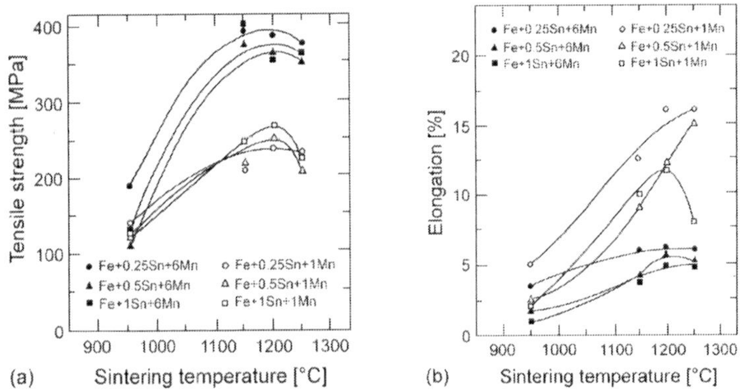

Fig. 10.5. Effect of sintering temperature on three compositions containing 6% Mn and 1% Mn selected for optimum tensile strength (a) and for elongation (b) [235].

1150, 1200 and 1250°C. In most cases sintering at 1150 and 1200°C was found to be necessary to obtain high strength, Fig. 10.5. The ductility, Fig. 10.5 (b), of the iron+0.25% tin+1% manganese alloy was best after sintering at 1250°C although the improvement brought about by this slightly increased temperature was less than that produced by the increased sintering time.

Effect of sintering atmosphere. In this case the samples were sinteretd in N_2, $75N_2/25H_2$, $50N_2/50\ H_2$, $25N_2/75\ H_2$ and in H_2. The alloys shown in Fig. 10.5(a) and (b) were sintered in the mentioned atmospheres. Regarding the tensile strength values, the most marked effect of sintering atmosphere was with the Fe–1% Sn–4% Mn and Fe–1% Sn–6% Mn alloys. In these cases the best properties were obtained after sintering in pure hydrogen. In other cases, marginally higher tensile strength values were obtained when sintering in pure nitrogen. The effect of sintering atmosphere composition on the alloys which exhibited maximum tensile strength was negligible, varying only from slightly about 400 MPa to just below this value for the complete range of sintering atmospheres. The hardness values were in all cases highest for materials sintered in pure nitrogen.

Effect of heat treatment. For the investigation of the heat-treated compacts they were quenched after sintering for 1 h in hydrogen at 1150°C and then either tempered for 1 h at 250°C or given a post–sintering heat treatment for 2 h in nitrogen at 1150°C followed by water quenching. In most cases the highest strength values were obtained after post-sintering heat treatment in nitrogen for 2 h at 1150°C. The compositions of the alloys giving best tensile strength and hardness were iron+0.25% tin+6% manganese or iron+0.5%

tin+6% manganese. The composition of the alloy giving highest ductility was iron+0.25% tin+1% manganese. The results showed that the most useful range of tin addition was between 0.25% and 1% whereas additions of manganese of up to 6% gave best results.

10.1.1.2. Effect of zinc

The aim for the use of zinc as the alloying element in sintered manganese steels was the same as in the case of tin – sintering under the effect of a low-melting Zn–Cu master alloy (45Zn–35Mn–20Cu, melting temperature of ~850°C, coded as A45Zn). The investigation of the mechanical properties of Fe–2.3Mn–1Cu–1.1Zn–0.35C (6.7% MA45Zn) steel due to the high vapour pressure of zinc was real only as sinter-forged at a low temperature of 900°C. After quenching and tempering at 500°C the tensile strength of 1050 MPa and hardness of 700 HV were attained [236].

10.1.1.3. Effect of boron

For this study the samples 15 mm in diameter and 3 mm long with a green density of 6.50 g cm^{-3} were compacted. The samples without boron addition (A), with boron addition (A1) and with the addition of $Na_2B_4O_7$ (A2) on the basis of iron powder NC100.24 were prepared. Manganese was added to the powder mixtures in the form of medium-carbon ferromanganese with a particle size of <16.2 μm, and natural graphite was also added. Amorphous boron and burnt $Na_2B_4O_7$ with the average particle size of <10 μm were used as additives. Sintering was carried out in a vacuum furnace with heating and cooling rates of 5°C/min. The samples were sintered at temperatures of 850, 950, 1050 and 1150°C for 5, 10, 15, 60 and 90 min.

The mechanism of structure formation was the same for all investigated samples and it was not influenced by the presence or absence of the activator addition. Both alloying elements C and Mn penetrated into the iron matrix by their diffusion on the surface of the particles in the compact. The process of structure formation at all examined sintering regimes started with pearlite formation. The introduction of microadditions of B and $Na_2B_4O_7$ influenced the rate of tested processes and quantity of formed phases. The formation of the structure depended on whether sintering was carried out at the temperature less or higher than the temperature of phase transformation.

Sintering in the α-phase resulted in the formation mostly of pearlite. The microhardness of the pearlite changed insignificantly from 146 HV 0.05 after 5 min sintering increased with increasing sintering time by about 50 HV 0.05 after 90 min soaking.

At sintering in the γ-phase the beginning of iron matrix saturation with alloying elements was ahead of carbon diffusion and pearlite phase formation. After an incubation period, the formation of a phase different from pearlite was observed. This gave evidence for the occurrence of intensive manganese diffusion.

Large quantities of phases characterised by a higher degree of alloying – basically martensite and austenite – were formed in the presence of boron. The measurements of TEF and microhardness showed higher homogeneity of the samples with the $Na_2B_4O_7$ addition.

The boron addition accelerated the processes of structure formation. Its influence was more significant at temperatures higher than the temperature of the alpha–gamma polymorphous transformation. The boron influence on the kinetics of the structure formation is due to a complicate set of reasons.

The results show that the boron addition accelerated the process of structure formation in the Fe–13Mn–1C PM system. The homogeneity of the structures of the investigated samples was higher when boron was introduced in the form of a low-melting compound [237].

10.2. Sintering of Fe–Mn–C steels in high-purity atmospheres

The Fe–3% Mn–0.8% C specimens in these studies were made on the basis of sponge iron powder NC100.24 and atomised iron powder grade ABC100.30. Manganese was added to the iron powders in the form of medium- and high-carbon ferromanganese powders.

The specimens from the powder mixtures were compacted at 600 MPa and sintered in a dried hydrogen atmosphere in a laboratory tube furnace. The sintering furnace had a convective water-jacketed cooling zone. The dew point of the hydrogen atmosphere was controlled from –40 to –60°C. Sintering was performed at 1120, 1150, 1180 and 1200°C for 60 min. To prevent oxidation, the specimens were set in a stainless steel semi-closed boat and were removed only after the boat had cooled to room temperature. Properties of the investigated manganese steels are listed in Table 10.1. Elongation was in the range of 0.2 to 0.7%.

Table 10.1. Properties of sintered Fe–3Mn–0.8C steel in dependence on iron powder grade, manganese carrier grade, sintering temperature and dew point of hydrogen. HP – high-carbon ferromanganese, El – medium-carbon ferromanganese, graphite added to 0.8% C [238]

Iron powder, Mn carrier	Sint. temp. [°C]	Dew point [°C]	Sint. dens. [g cm^{-3}]	Rm [MPa]	TRS [MPa]	KC [J]	HV 10
NC100.24, El	1120	–40		452	976	8.3	147
NC100.24, EL		–50	6.52	492	1128	8.2	153
NC100.24, EL		–60		433	1243	7.9	172
NC100.24, EL		–40		385	1236	6.1	193
NC100.24, EL	1150	–50	6.58	490	1136	8.8	147
NC100.24, EL		–60		477	1254	7.4	168
NC100.24, EL		–40		451	832	3.7	112
NC100.24, EL	1180	–50	6.70	423	1007	4.4	165
NC100.24, EL		–60		583	1253	7.2	168
NC100.24, EL		–40		469	1240	3.5	268
NC100.24, EL	1200	–50	6.60	527	1091	5.6	239
NC100.24, EL		–60		445	1016	5.5	193
ABC100.30, EL	1180	–60	6.82	429	–	–	–
ABC100.30, EL		–60	6.88	468	–	–	–
ABC100.30, HP		–60	6.6	334	–	–	–
ABC100.30, HP	1180	–60	6.9	340	–	–	–

Table 10.2. Mechanical properties of sintered Fe–3Mn–0.8C steel specimens differently sintered and cooled; NC100.24 iron powder, 600 MPa, hydrogen, dew point –60°C, El – medium-carbon ferromanganese, HP – high-carbon ferromanganese, graphite added to 0.8% C [238]

Steel	Sint. temp. [°C]	Cool. rate [°C/min]	Density [g cm^{-3}]	Rm [MPa]	HV 10
Fe–3Mn–0.8C, HP	1120	40	6.8	516	217
Fe–3Mn–0.8C, El	1120	40	6.7	559	276
Fe–3Mn–0.8C, El	1120	10	6.7	480	183
Fe–3Mn–0.8C, El	1250	10	6.8	590	189
Fe–3Mn–0.8C, El	1250	40	6.8	614	239
Fe–3Mn–0.6C, El	1280	7	6.7	360	169
Fe–3Mn–0.8C, El	1250	40	6.8	614	239

Table 10.3. Mechanical properties of Fe–(3,4)Mn–0.8C steels in dependence on sintering temperature and cooling rate; NC100.24 iron powder, sintered density ~6.9 g cm^{-3}, sintering for 60 min, hydrogen, dew point –60°C, n.d. – not determined [238]

Steel	Sintering temperature [°C]	Cooling rate [°C/min]	Rp0.2 [MPa]	Rm [MPa]	TRS [MPa]	HV 30 core	KC [J]
Fe–3Mn–0.8C (C_c – 0.6%)	1120	4.8	330	478	1270	224	17.2
		5.5	brittle	535	1000	292	5.2
Fe–3Mn–0.8C (C_c – 0.5%)	1250	4.7	440	595	1330	254	11.5
		5.7	brittle	491	1090	335	4.9
Fe–4Mn–0.8C (C_c – 0.6%)	1120	4.8	422	669	1021	n.d.	n.d.
		55	brittle	580	1097	n.d.	n.d.
Fe–4Mn–0.8C (C_c –0.5%)	1250	4.7	434	636	1467	n.d.	n.d.
		5.7	brittle	428	847	n.d.	n.d.

Tables 10.2 and 10.3 give the mechanical properties of manganese steels sintered with the special aim in high-purity hydrogen (dew point –60°C). Table 10.2 presents the properties of manganese steel sintered at three temperatures in high-purity hydrogen and cooled at different rates. The effect of cooling rate in the range of 1.4 to 65°C/min on the mechanical properties of Fe–(3,4)Mn–0.8C steels sintered in high-purity hydrogen at two temperatures was investigated [98–100]. Table 10.3 presents the mechanical properties attained under two cooling rates. The elongation of the steels was in the range of 0.2 to 0.5%. The results can be used to deduce the effect of the applied variable factors.

Higher cooling rates caused a decrease of tensile and transverse rupture strength and impact energy values compared to those cooled at lower cooling rates. On the other hand, the hardness increased after cooling with high cooling rates.

10.3. Effect of processing conditions on properties of manganese steels

10.3.1. Effect of iron powder and ferromanganese grades

The effect of processing conditions on the properties of the sintered Fe–Mn–C steel in dependence on the manganese carrier grade, iron powder grade and sintering time was investigated by [239].

Samples were prepared from sponge iron powders SC100.26 (SC) and NC100.24 (NC) and water-atomised powders ASC100.29 (ASC). The manganese content of 3% was introduced in the system in the form of high-carbon ferromanganese (FMC), medium-carbon ferromanganese (FM) and electrolytic manganese. When high-carbon ferromanganese was used, the corresponding carbon content in the systems Fe–3Mn–C was 0.26%. In other systems graphite in the amount of 0.26, 0.6 and 0.9% and 0.8% HW wax as a lubricant were added to the mixtures.

Three different types of specimens were prepared. The tensile test specimens were prepared by compacting at 400, 600 and 800 MPa to densities of ~6.0, ~6.5 and ~7.0 g cm^{-3}, respectively; rectangular three-point bend strength (TRS) and impact energy (KC) test specimens with the density ~7.0 g cm^{-3} were also prepared. Sintering was carried out under following conditions:

1. 1120°C for 90 min; cracked ammonia with the dew point of –33°C, a semi-closed container with protection by Al_2O_3 + FMC + graphite powder mixtures, cooling rate ~40°C/min;
2. 1120°C for 180 min; the same conditions as in 1;
3. 1130°C for 60 min; cracked ammonia with the dew point of –23°C, a semi-closed container with ceramic protection, cooling rate ~10°C/min.

The effect of both the green density and the Mn content on the tensile strength of Fe–3Mn–0.26C steels in relation to the iron

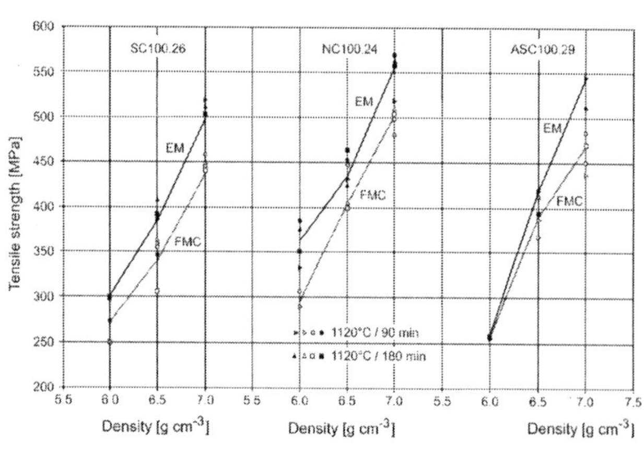

Fig. 10.6. Effect of green density, iron powder grade and Mn alloying component (EM and FMC) on tensile strength of Fe–3Mn–0.26C specimens sintered at 1120°C for 90 min and 180 min and cooled at 40°C/min [239].

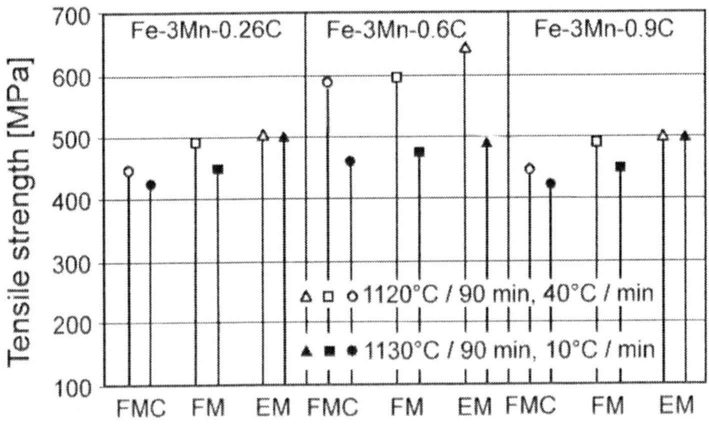

Fig. 10.7. Effect of graphite addition, Mn carrier grade and cooling rate on tensile strength of Fe–3Mn–C specimens based on SC100.26 iron powder, sintered at 1220°C for 90 min and cooled at 40°C/min and 10°C/min [239].

powder grade, manganese carrier grade (EM) and (FMC) and the sintering time (conditions 1 and 2) is presented in Fig. 10.6.

According to Fig. 10.6, the electrolytic manganese is more effective in diffusion and it does not act as a source of such impurities such as Si, Al etc., further residual pores left behind the Mn particles were smaller than those left behind the ferromanganese particles.

The tensile strength increased with increasing density. Depending on the Mn content, the tensile strength of 205–413 MPa at a density of ~6.0 g cm^{-3} was achieved. At the density of ~6.5 g cm^{-3} the tensile strength ranged 300 to 460 MPa, and for ~7.0 g cm^{-3} the tensile strength was between 440 and 580 MPa.

The effect of graphite addition, Mn content, manganese carrier grade and cooling rates of 40°C/min and 10°C/min (the sintering conditions 1 and 3) on the tensile strength of Fe–3Mn–C steel specimens based on SC100.26 iron powder is listed in Fig. 10.7.

The effect of sintering conditions, iron powder grade, manganese carrier grade and graphite addition on tensile and three-point bending strength, impact energy and hardness measured in the core of the tested specimens at a constant cooling rate are listed in Table 10.4.

The attained tensile strength for all the presented results was 422–706 MPa and impact energy values were in the range from 4 to 22 J. The hardness values measured in the core differed from those measured on the surface of the specimens. Higher tensile strength

Table 10.4. Basic steel compositions, sintered density, sintering parameters. 600 MPa, $75H_2/25N_2$, dew point –30°C and mechanical properties of investigated Fe–Mn–C steels; A = FMC+graphite, B = EL+graphite; cooling rate for all steels = 40°C/min; A – high carbon ferromanganese, E – electrolytic manganese

Steel Fe–3Mn+C (%)	Iron powder + A, B, C	Sint. dens. [g cm^{-3}]	Sintering temp./time [°C/min]	UTS [MPa]	TRS [MPa]	KC [J]	HV 30 core
0.26	SC+A	6.86	1120/90	446	820	15	188
0.60	SC+A	6.82	1120/90	591	1074	12	205
0.90	SC+A	6.81	1120/90	614	1175	9	208
0.26	SC+A	6.84	1130/60	425	894	5	161
0.60	SC+A	6.73	1130/60	462	739	7	226
0.90	SC+A	6.73	1130/60	358	766	10	209
0.26	SC+B	6.88	1120/90	504	1012	13	166
0.60	SC+B	6.93	1120/90	652	1184	12	199
0.90	SC+B	6.94	1120/90	652	1182	15	197
0.26	SC+B	6.93	1130/60	502	826	12	182
0.60	SC+B	6.93	1130/60	494	864	7	208
0.90	SC+B	6.91	1130/60	511	983	6	213
0.26	NC+A	6.74	1120/90	502	871	14	192
0.60	NC+A	6.79	1120/90	619	1124	8	210
0.90	NC+A	6.72	1120/90	625	1018	10	205
0.26	NC+A	6.84	1130/60	460	816	17	175
0.60	NC+A	6.79	1130/60	564	867	12	208
0.90	NC+A	6.70	1130/60	531	916	10	234
0.26	NC+B	6.86	1120/90	561	1035	17	193
0.60	NC+B	6.92	1120/90	706	1231	12	222
0.90	NC+B	6.77	1120/90	705	1225	14	216
0.26	NC+B	6.87	1130/60	570	952	15	195
0.60	NC+B	6.89	1130/60	569	917	14	214
0.90	NC+B	6.75	1130/60	537	880	12	201
0.26	ASC+A	6.88	1120/90	475	890	25	209
0.60	ASC+A	6.81	1120/90	488	898	6	209
0.90	ASC+A	6.77	1120/90	453	857	5	218
0.26	ASC+A	6.99	1130/60	386	839	7	204
0.60	ASC+A	6.85	1130/60	443	750	5	227
0.90	ASC+A	6.81	1130/60	439	762	4	242
0.26	ASC+B	6.87	1120/90	542	1011	9	214
0.60	ASC+B	6.91	1120/90	424	806	16	226
0.90	ASC+B	6.71	1120/90	406	764	5	234
0.26	ASC+B	6.85	1130/60	511	816	10	237
0.60	ASC+B	6.77	1130/60	422	612	8	207
0.90	ASC+B	6.72	1130/60	434	597	8	206

values were achieved after sintering at 1120°C for 90 min than after sintering at 1130°C for 60 min. The highest tensile strength values of 706 and 705 MPa were recorded for the Fe–3Mn steel with the addition of 0.6 and 0.9% graphite based on NC100.24 iron powder alloyed with electrolytic manganese (B). The Fe–3Mn–C sintered steels showed the bend strength values higher than 1000

Table 10.5. Mechanical properties of the sintered Fe–3Mn–0.7C steel specimens [239]

Alloy	Rp0.2 [MPa]	UTS [MPa]	TRS [MPa]	Hardness surf./core HV 10	Impact energy [J]
1	327	478	860	153/168	12.5
2	490	642	1256	180/207	11.0
3	433	522	948	182/231	7.0

MPa. The NC100.24 and SC100.26 sponge iron powders when compared to water-atomised ASC100.29 iron powder seem to be more advantageous for alloying with manganese [239].

10.3.2. Effect of sintering conditions

The effect of the sintering conditions on the microstructure and mechanical properties was studied on specimens based on SC100.26 iron powder, medium-carbon ferromanganese (particle size <10 μm) and natural graphite. Tensile strength bars were prepared by compaction at 600 MPa from the mixture of these materials with the addition of a lubricant to produce the Fe–3% Mn–0.7% C steel composition. Sintering was carried out at 1120°C for 60 min, in a $75H_2/25N_2$ atmosphere, dew point –60°C (alloy 1), b) 1200°C for 60 min, $75H_2/25N_2$ atmosphere, dew point –60°C (alloy 2), and c) 1180°C for 40 min, $75N_2/2H_2$ atmosphere, dew point –30°C (alloy 3). The mechanical properties of the investigated steels are listed in Table 10.5.

Higher strength properties were obtained after sintering at 1200°C when compared to the values after sintering at 1120°C. The increase was approximately 163 MPa for both the tensile, yield and fracture strength. The hardness measured in the core also increased by 23%. The relative increase in the three-point bend strength was about 40%. The impact energy values differed only slightly – from 12.5 to 11.1 J. The positive effect of industrial sintering at 1180°C compared to 1120°C is reflected in higher yield and tensile and transverse rupture strengths.

The characteristic microstructure constituents of Fe–3Mn–C alloys sintered under the mentioned three sintering conditions are listed in Table 10.6 and their microhardness values are listed in Table 10.7.

Table 10.6. Microstructure constituents of Fe–3Mn–0.7C sintered alloy in dependence on sintering conditions [240]

1120°C/60 min, dew point –60°C		1200°C/60 min, dew point –60°C		1180°C/60 min, dew point –30°C	
Constituent	Area [%]	Constituent	Area [%]	Constituent	Area [%]
ferrite	~3.5	ferrite	~1.5	ferrite	~1.5
pearlite	~77.5	pearlite	~48.5	pearlite	~43.5
austenite/ martensite	~9	austenite/ martensite	~40	austenite/ martensite	~45

Note: Pores represent 10% of the fraction of the ferrite area

Table 10.7. Microhardness values and Mn content in the microstructure constituents of Fe–3Mn–0.7C steels (EDX analysis) [240]

Constituent	Microhardness HV 0.02	Mn [%]
Ferrite	100 – 170	<0.5
Coarse pearlite	215 – 270	1 – 2
Fine pearlite (+bainite)	240 – 280	~3
Retained austenite	not detemined	~4.9
Martensite	580 – 860	3.5 – 4

Specimens sintered at 1120°C exhibited predominantly a pearlitic microstucture with ~3.5% ferrite area and with ~9% of a mixed austenite/martensite microstructure. The microstructure of the specimens sintered at 1200°C was more heterogeneous. The microstructure composition of the specimens sintered at 1180°C was similar to that sintered at 1200°C, only the austenite–martensite area increased to 45%.

10.3.3. Effect of base powder grades

The effect of base iron powders on the properties of sintered manganese steels was studied in [241,242]. For this investigation, sponge NC100.24, and water-atomised ABC100.30, and ASC100.29 iron powders were used as the starting materials. 0.8% of carbon was introduced as fine CU–F graphite and 3% of manganese as medium-carbon ferromanganese with a mean particle size of 15 μm. Using only die lubrication, the ISO 2740 dogbone specimens were compacted

Table 10.8. Densities before and after sintering of Fe–3Mn–0.8C

Sintering	Density [g cm^{-3}]					
	NC100.24		ABC100.30		ASC100.29	
	green	sintered	green	sintered	green	sintered
1120°C, H_2	6.86	6.89	7.11	7.11	6.95	6.93
1120°C, N_2	6.88	6.91	7.10	7.09	6.95	6.94
1250°C, H_2	6.87	6.91	7.12	7.09	6.95	6.93
1250°C, N_2	6.87	6.91	7.12	7.09	6.95	6.92

Note: 10% of the fraction of the ferrite area

at 660 MPa, followed by sintering in dry hydrogen or nitrogen in a laboratory tube furnace included a water-jacketed rapid convective cooling zone. The dew point of the sintering atmospheres was –60°C (15 ppm moisture). To produce a Mn-rich microclimate (self-gettering effect), the specimens were sintered in a semi-closed stainless steel container with a labyrinth seal. Compacts were heated to the sintering temperature at a rate of 75°C/min and held at 1120 or 1250°C for 60 min. The convective cooling rate, determined in the temperature range of 1100–500°C, was approximately 64°C/min. After sintering, for stress relief, all specimens were tempered at 200°C for 1 h in argon. Oxygen content in all starting iron powders was ~0.2%. The carbon in the starting powder mixes was 0.8% and after sintering typically: 1120°C/H_2 – 0.65%, 1120°C/ N_2 – 0.74%, 1250°C/ H_2 – 0.54% and 1250°C/N_2 – 0.62%.

Densification. The densities of Fe–Mn–C compacts remained unchanged in dependence on sintering, as shown in Table 10.8. For ABC100.30 based steel it was thus ~7.1 g cm^{-3}, more than 0.2 g cm^{-3} larger than for the sponge iron powder-based alloy.

Metallographic and microstructural observations. Until recently, presence of manganese oxides was characteristic of PM manganese steels, a relevant feature of the microstructures was the absence of oxide networks, whether nitrogen or hydrogen was the sintering atmosphere. All microstructures comprised a mixture of martensite, retained austenite, pearlite and bainite, in different proportions depending principally on the sintering temperature (1120°C or 1250°C), with insignificant differences between the two steels based on sponge and atomised powders shown in Fig. 10.8. In general, the structures of steels sintered in nitrogen were less inhomogeneous than those sintered in hydrogen, which also showed evidence of surface

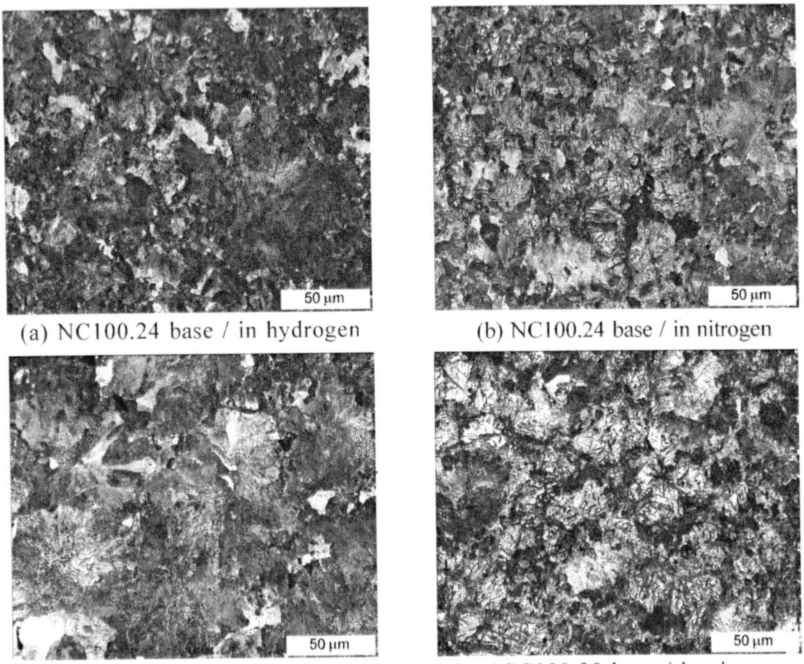

(a) NC100.24 base / in hydrogen (b) NC100.24 base / in nitrogen

(c) ASC100.29 base / in hydrogen (c) ASC100.29 base / in nitrogen

Fig. 10.8. Microstructures of specimens sintered at 1120°C. The considerable similarity of all non-uniform microstructures containing predominantly lower bainite, pearlite and some austenite (light areas) should be noted [241,242].

Table 10.9. 0.2% offset yield, $R_p0.2$, and fracture, Rm strength of Fe–3Mn–0.8C sintered steel

Sintering	Strength [MPa]					
	NC100.24		ABC100.30		ASC100.29	
	yield	fracture	yield	fracture	yield	fracture
1120°C, H_2	366±53	542±62	427±63	524±74	413±53	466±52
1120°C, N_2	441±28	677±50	493±57	603±74	493±67	565±38
1250°C, H_2	440±37	713±68	522±79	650±76	497±60	578±64
1250°C, N_2	467±59	732±53	518±66	651±67	485±88	617±55

± standard deviation measured on 5 samples

decarburisation. In specimens sintered at 1120°C the microstructures were inhomogeneous.

The structures were coarser after sintering at 1250°C, consisting mainly of bainitic and martensitic regions. The specimens sintered in nitrogen, upper and lower bainite contained lath-shaped fine aggregates of ferrite and cementite.

Table 10.10. Transverse rupture strengths, TRS, of Fe–3Mn–0.8C steel

Sintering	Stress [MPa]		
	NC100.24	ABC100.30	ASC100.29
1120°C, H_2	1120±138	1073±111	959±141
1120°C, N_2	1357±113	1128±89	1057 ±152
1250°C, H_2	1337±150	1250±196	996±93
1250°C, N_2	1354±121	1244±184	1109±134

± standard deviation measured on 15–26 samples

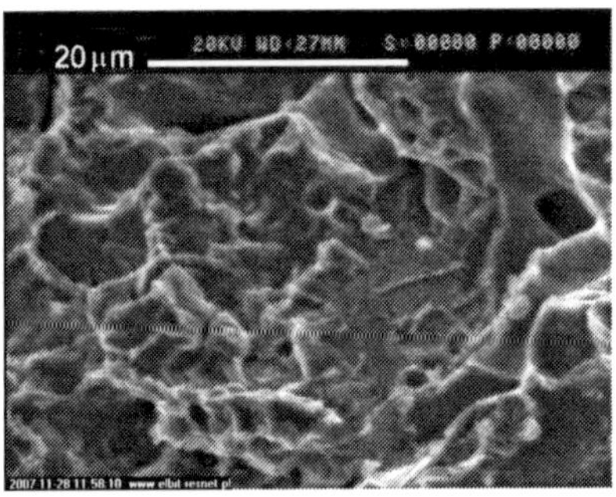

(a) NC100.24 sintered in nitrogen at 1120°C

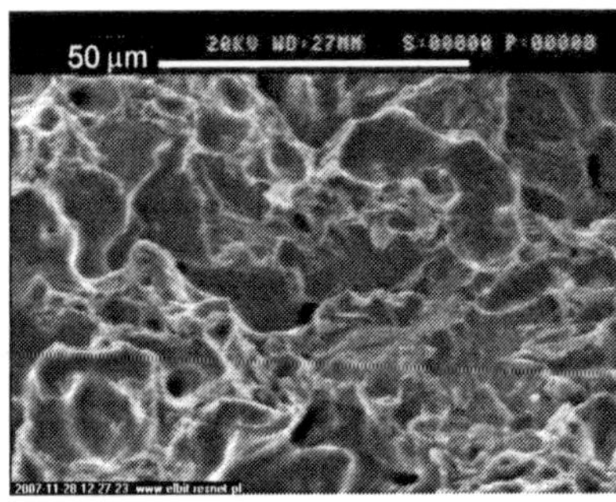

(b) NC100.24 sintered in nitrogen at 1250°C

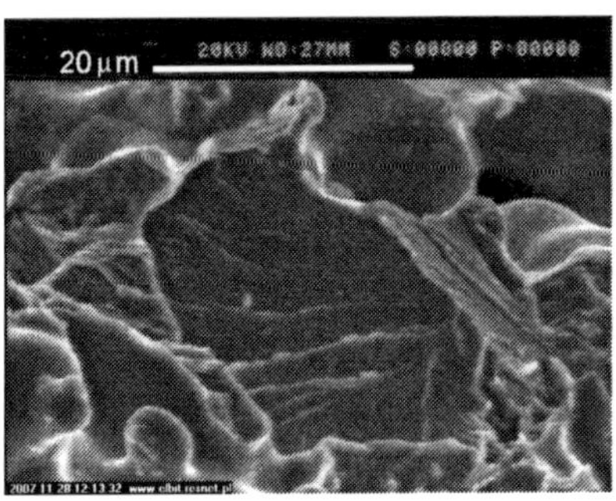

(c) ASC100.29 sintered in nitrogen at 1120°C

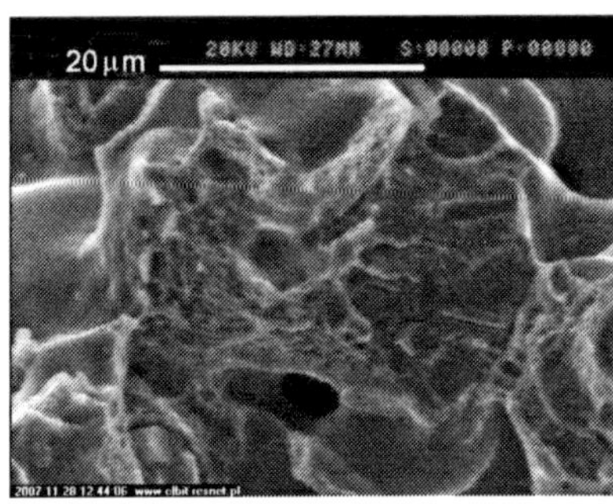

(d) ASC100.29 sintered in nitrogen at 1250°C

Fig. 10.9. Fractographs of specimens emanating from NC100.24 and ASC100.29 powders sintered at 1120°C and 1250C in nitrogen. To be noted are the regions of intergranular and cleavage fracture in all the figures, but only in sponge-based material, (a) and (b), fairly extensive regions of dimple rupture, which appear also in (d), material emanating from atomised powder sintered at 1250°C. SEM [241,242].

Mechanical properties. The values of 0.2% offset yield, tensile and transverse rupture strengths and tensile failure strain were determined, usually for 15 specimens of each batch. The results were averaged and are presented in Tables 10.9 and 10.10.

Fractographic observations. All fractographic observations, examples of which for the two material types are given in Fig.

10.9, revealed regions of intergranular and cleavage failure. It should be noticed, however, that only in the steel emanating from sponge powder are there extensive regions of dimple rupture. After sintering at 1250°C material emanating from the atomised powder, tensile failure also included operation of some local dimple failure mechanism. It should be added that sinter-neck areas in fractographs of sponge-based material also appear larger, thus with their reduced Mn contents, favour ductile rather than brittle fracture on this microscale. The fractures of steels based on ASC100.29 powder sintered at 1120°C and 1120°C exhibited transgranular cleavage facetes.

It was stated: a) the suitability of sintering Fe powder mixed with ferromanganese and graphite to Fe–3Mn–0.8C steel in dry nitrogen or hydrogen in semi-closed containers, b) the properties of steels emanating from ABC100.30 were superior to those from ASC100.29, c) tensile strength values and transverse rupture strengths were always highest for the sponge iron-based steels, to those based on atomised iron powders, reaching 730 and 1350 MPa, respectively.

10.3.4. Effect of some processing conditions and of different materials

To study the effect of processing variables on the properties of the manganese steels, the data were processed in three groups. The first group pertained to the influence of iron and ferromanganese powder grades on the mechanical properties of steels and had to do with the observed variations in the mechanical properties caused by changes in sintering temperature and the dew point of the sintering atmosphere. The second one dealt with the effect of cooling rate on the structure and mechanical properties of PM Mn steels. The third group of data was related to the influence of the chemical composition of the sintering atmosphere and additional post-sintering heat treatment on the structure and mechanical properties of PM Mn steels [234].

Two iron powders grades have been used in the investigation: sponge iron powder grade NC100.24 and atomised iron powder grade ABC100.30. Manganese was added in the form of medium-carbon ferromanganese powders (FMn) and high-carbon ferromanganese HP III grade. Additions of ultrafine graphite were made to all materials.

Eight different compositions of powders were prepared. Two types of compacts were pressed: 5 × 10 × 55 mm transverse rupture test bars and tensile strength test bars. The green densities of specimens were in the range 6.85–7.17 g cm^{-3} and 6.70–7.06 g cm^{-3} for rectangular and ISO 2740 specimens, respectively. Sintering was carried out in a laboratory furnace. The ranges of as sintered densities were 6.70–7.07 g cm^{-3} and 6.60–7.04 g cm^{-3} for rectangular and ISO specimens, respectively.

Tensile tests were carried out for the as sintered or sintered and tempered steels on dogbone tensile specimens. The resulting stress–strain curves were analysed to identify the 0.2% offset yield strength ($R_{0.2}$ offset), tensile strength (UTS) and tensile elongation at fracture (A_t). Transverse rupture strength (TRS), impact test and surface hardness were determined on rectangular specimens, respectively.

The results of mechanical testing are presented in Tables 10.11–10.15. Tables 10.11 and 10.12 show the results of mechanical investigations of Fe–3Mn–0.8C and Fe–4Mn–0.8C steels. The mechanical properties of these steels in dependence on iron powder grades and cooling rate are given in Table 10.11. In general, the UTS and TRS of the Fe–3Mn–0.8C PM steel slightly increase with decreasing dew point of the sintering atmosphere. Their dependence on the dew point is listed in Table 10.12. The data indicate that the effects of Mn, the sintering atmosphere dew point and cooling rate

Table 10.11. Mechanical properties of sintered manganese steels based on different iron and ferromanganese powder grades in dependence on cooling rate; 100% H_2, –60°C dew point [234].

Sintering temperature [°C]	Cooling rate [°C/min]	Speci-men	UTS [MPa]	TRS [MPa]	HV10	KC [J]	Rp0.2 [MPa]
1180	1.5	3A	467	1176	148	10.06	313
	60		320	799	242	2.99	nd
	1.5	4A	479	1136	151	7.39	343
	60		256	646	341	1.90	nd
	1.5	3B	456	880	153	6.97	318
	60		340	795	195	4.07	n. d.
	1.5	4B	491	922	157	2.80	335
	60		286	459	227	2.66	nd
1200	60	3C	421	919	184	5.56	290
		4C	312	597	215	3.75	280
		4D	289	788	242	2.12	nd

3A – ABC100.30+3Mn+0.8C (FMn), 4A – ABC100.30+4Mn (FMn)+0.8C, 3B – ABC100.30+3Mn (HPIII)+0.8C, 4B – ABC100.30+4Mn (HPIII)+0.8C, 3C – NC100.24+3Mn (HPIII)+0.8C, 4C – ABC100.30+ 4Mn (HPIII) + 0.8C, 4D – NC100.24+4Mn (FeMn) + 0.8C, nd – not defined.

Table 10.12. Mechanical properties of Fe–3Mn–0.8C PM steels based on NC100.24 and medium-carbon ferromanganese as Mn carrier powder grades, sintered at different temperature in H_2 atmosphere with a different dew point [243]

Sintering temperature [°C]	Dew point [°C]	UTS [MPa]	TRS [MPa]	HV10	KC [J]	Rp0.2 [MPa]
1120	–40	452	1036	150	8.17	400
	–50	492	1189	150	8.23	380
	–60	433	1291	179	7.36	330
1150	–40	385	1253	179	5.82	n.d.
	–50	490	1161	156	8.27	395
	–60	427	1312	181	8.15	410
1180	–40	401	845	157	2.72	n.d.
	–50	423	1046	184	4.46	470
	–60	583	1354	166	6.78	490
1200	–40	469	1280	211	3.01	n.d.
	–50	527	1186	217	4.87	n.d.
	–60	445	1194	187	3.43	n.d.

n.d. – not defined

on the mechanical properties of the investigated alloys are relatively predictable.

The cooling rate had a strong effect on the UTS, TRS, KC and apparent surface hardness, which is shown in Table 10.13. Decreasing the cooling rate to 5°C/min in some cases increased the mechanical properties by a factor of up to ~2. Slow cooling rates favoured plasticity, which ceased to be significant for cooling rates above ~40°C/min. The increase in both Mn concentration and sintering temperature raised the plasticity in PM Mn steels. In both cases – sintering at 1120°C and 1250°C – the specimens with the highest tensile properties were found for the alloys cooled with cooling rates from 9 to 30°C/min. The highest UTS and TRS values are not found in those specimens exhibiting the highest hardness values but in the moderately cooled specimens.

The cooling rate had a strong effect on the UTS, TRS, KC and apparent surface hardness, which is shown in Table 10.13. Decreasing the cooling rate to 5°C/min in some cases increased the mechanical properties by a factor of up to ~2. Slow cooling rates favoured plasticity, which ceased to be significant for cooling rates above ~40°C/min. The highest UTS and TRS values were found in the moderately cooled specimens.

Table 10.13. Mechanical properties of Fe–3Mn–0.8C PM Mn steels based on NC100.24 iron and medium-carbon ferromanganese as Mn carrier powder grade, sintered in H_2 and cooled to room temperature with different cooling rates [243]

Sintering temperature [°C]	Cooling rate [°C/min]	UTS [MPa]	TRS [MPa]	HV30	KC [J]	Rp0.2 [MPa]
1120	5	478	1272	174	17.22	380
	9	506	1230	183	10.86	414
	16	568	1195	185	11.06	461
	30	557	1124	197	8.22	516
	47	559	1100	202	5.84	478
	55	535	954	213	5.24	518
	67	475	1066	206	4.68	528
1250	5	595	1327	189	11.50	461
	9	614	1311	189	9.92	477
	20	583	1210	192	7.66	511
	32	586	1193	204	9.38	545
	41	521	1080	218	4.59	442
	57	491	1087	229	4.92	404
	64	458	1060	210	5.04	477

Table 10.14. Mechanical properties of Fe–3Mn–0.8C PM Mn steels based on NC100.24 and medium-carbon ferromanganese as Mn carrier powder grade, sintered in H_2 and tempered at different temperatures

Sintering temperature [°C]	Tempering temperature [°C]	UTS [MPa]	TRS [MPa]	HV 30	KC [J]	Rp0.2 [MPa]
1120	200	740	1745	250	12.08	412
	300	740	1507	238	10.40	430
	400	744	1462	233	11.51	460
1250	200	830	1479	247	10.10	474
	300	798	1448	252	4.87	472
	400	854	1755	257	15.69	499

Table 10.15. Mechanical properties of Fe–3Mn–0.8C PM Mn steels based on NC100.24 and medium-carbon ferromanganese powder grade, sintered in H/N atmosphere with different H_2 content; tempering 200°C [243]

Sintering temperature [°C]	H_2 content [%]	UTS [MPa]	TRS [MPa]	HV 30	KC [J]	Rp0.2 [MPa]
1120	100	642	1279	196	4.68	386
	75	640	1291	195	5.09	400
	25	650	1299	201	7.44	420
	5	657	1234	197	7.27	421
	0	655	1259	212	14.31	410
1250	100	717	1296	212	5.04	426
	75	749	1343	212	13.30	443
	25	778	1411	222	10.83	459
	5	725	1420	221	12.45	451
	0	776	1458	228	13.23	464

Table 10.16. Mechanical properties of Fe–3Mn–0.8C PM Mn steels based on NC100.24 and Elkem powder grade, sintered in H_2 and tempered at different temperatures [243]

Sintering temperature [°C]	Tempering temperature [°C]	UTS [MPa]	TRS [MPa]	HV30	KC [J]	Rp0.2 [MPa]
1120	200	740	1745	250	12.08	412
	300	740	1507	238	10.40	430
	400	744	1462	233	11.51	460
1250	200	830	1479	247	10.10	474
	300	798	1448	252	4.87	472
	400	854	1755	257	15.69	499

Specimens sintered at 1120°C and 1250°C and cooled at 67°C/min and 64°C/min, respectively, were also tempered at 200°C, 300°C and 400°C, Table 10.14. Significant improvement in properties was attained as a result of the 200°C stress relief. The results were similar for the 400°C temper and there was an indication of temper embrittlement at 300°C, more severe for the specimens sintered at 1250°C.

The results of tensile and bend tests carried out on the Fe–3Mn–0.8C steels, sintered in the atmosphere with a different H_2 content completed with N_2 show that it is possible to achieve high properties of PM Mn steels during sintering in a nitrogen atmosphere, Table 10.15. Sintering in this atmosphere resulted in the highest values of UTS and TRS. It means it is possible to produce PM Mn steels in something other than a hydrogen-rich atmosphere without decreasing their mechanical properties.

The impact properties of investigated steels are shown in Tables 10.15 and 10.16. It can be seen that the toughness is high particularly in view of the strength level involved. As might be expected the impact toughness in slowly (1.5°C/min) cooled conditions is superior

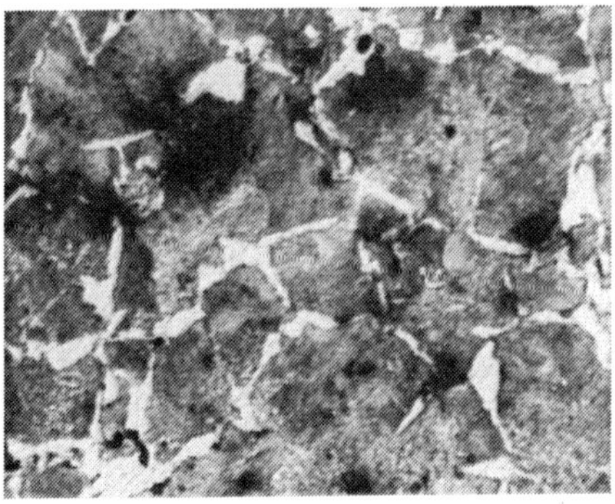

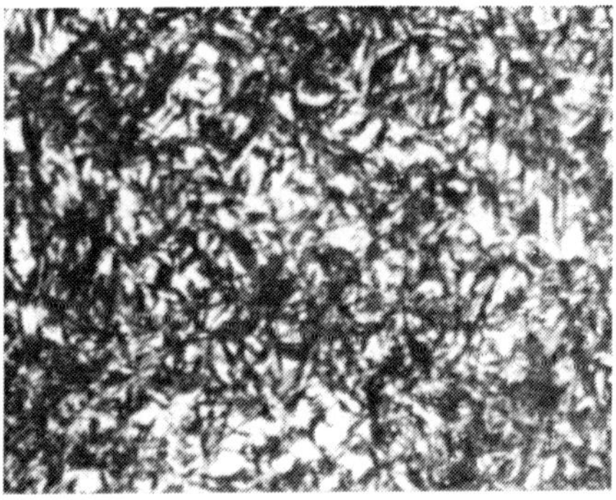

Fig. 10.10 (left). Micrograph of Fe–3Mn–0.8C steel sintered at 1250°C; H_2, slow cooled (5°C/min), original magnification ×430; pearlite and ferrite [243].

Fig. 10.11 (right). Micrograph of Fe–3Mn–0.8C steel sintered at 1120°C; H_2, rapid cooled (>55°C/min), original magnification ×430; bainite, martensite and retained austenite [243].

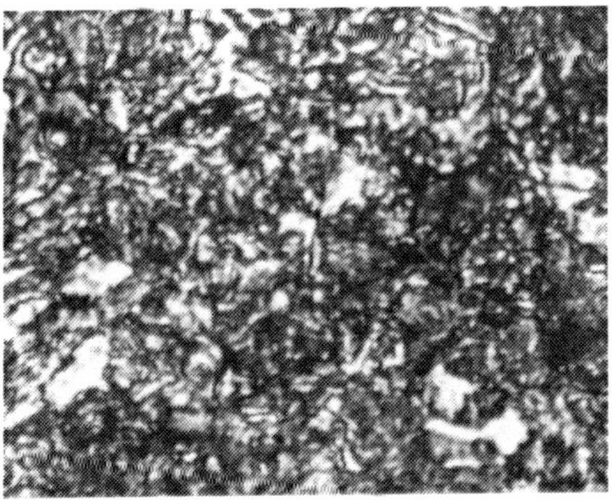

Fig. 10.12. Micrograph of Fe–3Mn–0.8C steel sintered at 1250°C and tempered at 200°C; H_2, original magnification ×830; martensite and retained austenite [243].

to that of the convective cooled condition. The impact strength of the specimens sintered in the atmosphere with a dew point of –40°C is somewhat lower, probably because of oxides introduced during alloy preparation.

Specimens sintered at 1120°C and 1250°C and cooled at 67°C/min and 64°C/min, respectively, were also tempered at 200°C, 300°C and 400°C, Table 10.16. Significant improvement in properties was attained as a result of the 200°C stress relief. Results were similar for the 400°C temper and there was an indication of temper embrittlement at 300°C, more severe for the specimens sintered at 1250°C.

Structural constituents observed in Fe–Mn–C steels are presented in Figs.10.10–10.12.

The structure differences are connected with the different manganese content. In the convective cooled condition (60°C/min) the microstructure of the sinter hardened steel consisted of a bainite/martensite matrix containing some retained austenite. The convectively cooled specimens showed a very heterogeneous microstructure consisting of martensite, upper and lower (coarse and fine) bainite, retained austenite and in some cases ferritic cores of the Fe powder particles. This is due to the poorly distributed Mn that decreased in concentration towards the centres of the powder particles. Carbon was retained in the Mn-rich areas, and interiors of the particles were thus mainly ferritic. Faster Mn diffusion and thus better microstructural homogeneity was achieved by sintering at a higher temperature.

It was recognised that a not significant amount of Mn, an unexpensive alloying element which is required, plus tight process control to enable the phenomenon to be used in a cost-effective manner, make it possible to introduce PM manganese steels into practical production. The optimum combination of the mechanical properties appeared to be obtained with a total 3% Mn addition at 0.8% C level [244].

10.3.5. Effect of cooling rate

The effect of cooling rate on Fe–(3,4)Mn–0.8C steels prepared on the basis of NC100.24 iron powder and on mediun-carbon ferromanganese (<20 μm) as manganese carrier, and graphite powder grade C–UF was investigated. The tensile strength bars and rectangular specimens, 55 × 10 × 5 mm, were compacted at 660 and 820 MPa. Sintering was carried out for 1 h in pure hydrogen with –60°C dew point in a laboratory tube furnace at 1120°C and 1250°C. The specimens were moved to the water-cooled end of the tube for 'convective' cooling. The ranges of cooling rates (CR), determined between 800°C and 300°C, were investigated [244].

Table 10.17. The effect of cooling rate on the mechanical properties of Fe–(3,4)% Mn–0.8% C steels sintered at 1120°C

	Fe–3Mn–0.8C				Fe–4Mn–0.8C			
CR [°C/min]	UTS [MPa]	Rp0.2 [MPa]	A [%]	TRS [MPa]	UTS [MPa]	Rp0.2 [MPa]	A [%]	TRS [MPa]
5	478	380	1.1	1272	478	422	1.87	1168
9	506	414	1.0	1230	506	417	1.63	1144
16	568	460	1.1	1195	508	374	2.22	1189
30	557	516	0.6	1124	557	414	1.62	1232
47	559	478	0.7	1100	559	379	0.83	1313
55	535	518	0.6	954	535	403	0.86	1175
64	486	ND	0.6	1060	486	392	0.80	1107

CR – cooling rate, ND – not defined

Table 10.18. The effect of cooling rate on mechanical properties of Fe–(3,4)% Mn–0.8% C steels sintered at 1250°C

	Fe–3Mn–0.8C			Fe–4Mn–0.8C		
CR [°C/min]	UTS [MPa]	Rp0.2 [MPa]	TRS [MPa]	UTS [MPa]	Rp0.2 [MPa]	TRS [MPa]
5	595	460	1327	636	434	1490
9	614	477	1310	600	460	1475
20	583	511	1210	547	459	1389
32	586	545	1193	497	480	1419
41	521	442	1080	469	ND	1206
57	491	404	1097	428	ND	923
62	507	ND	993	485	ND	1002

CR – cooling rate, ND – not defined

The mechanical properties of Fe–(3,4)Mn–0.8C steels sintered at 1120°C in relation to the cooling rate are listed in Table 10.17 and of those sintered at 1250°C in Table 10.18.

It was pointed out that no retained austenite was found in the 4% Mn specimens cooled rapidly. The changes in the properties between characteristic features in 4% Mn specimens cooled slowly (5 and 9°C/min) and with the cooling rate ~40°C/min were recorded. The higher sintering temperature resulted also in more rounded pores and a slight improvement in the mechanical properties. For higher manganese content this trend was not clear. The tensile strength and elongation were higher for lower sintering temperature, bend strength was also lower, as opposed to the 4% Mn specimens sintered at 1250°C.

According to the presented results the sintered Fe–(3–4)% Mn–0.8% C steels with high hardenability may be used for sinter hardening treatment. At both low and high sintering temperatures it is possible to produce a high-strength manganese steel. After cooling at rates below 10°C/min mainly fine pearlite and feathery bainite were observed. When cooling at commercial rates above 10°C/min, the investigated sintered manganese steels were partially hardened (bainite/martensite structure) and tempering is required to improve the plasticity of these steels.

The effect of cooling rate and of base iron powders was also studied in [245]. Sponge NC100.24 and water-atomised ABC100.30 iron powder were used for this investigation. Carbon was introduced as graphite and manganese as milled medium-carbon ferromanganese (El) and high-carbon ferromanganese (HP3). Alloys with manganese in the range 2–4% Mn and carbon contents between 0.4 and 0.8% were investigated.

The specimens from the mixtures were compacted at 600 MPa followed by sintering in hydrogen and in a hydrogen-rich atmosphere in the temperature range 1120–1300°C. Temperatures for laboratory sintering were generally in excess of 1180°C for a dew point in the range –40 to –60°C.

Metallographic observation. The most striking feature of the microstructures of the steels sintered in the atmosphere with a dew point of –60°C was the absence of oxide networks. It was seen that the specimens made from ABC100.30 iron powder showed larger, more rounded ferrite area cores than those made using NC100.24 iron powder.

Mechanical properties. After laboratory sintering of leaner, Fe–2Mn–0.5C steel, the tensile strength specimens based on sponge iron powder, sintered at 1140°C, with the slow cooling rate, produced densities of only 6.4–6.5 g cm^{-3}, with the tensile strength of 230–380 MPa. The apparent bend strength was in the range 570–770 MPa.

A number of alloys were screened for a promising combination of strength and ductility: 3 and 4% Mn, up to 0.8% C, and Si contents up to 0.55%, was investigated. The recorded apparent bend strength values were in the range 600–1180 MPa and tensile strength values

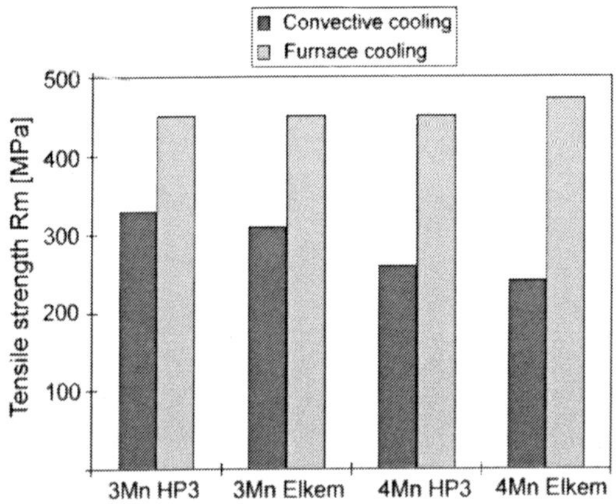

Fig. 10.13. Histogram of mean tensile strength of 3Mn–0.8C and 4Mn–0.8C steels processed using high-carbon ferromanganese (HP3) and medium-carbon ferromanganese (El) powders and cooled from sintering temperature of 1180°C in hydrogen with dew point better than –40°C [245].

in the range 260–620 MPa. There was significant increase in apparent hardness as a result of relatively convective cooling, compared to furnace cooling. At the surface, hardness increased from 120–170 HV 10 to 160–390 HV 10. The corresponding increase in apparent cross-section hardness was from 130–160 to 160–290 HV 10. The hardness of 4% Mn alloy specimens was higher than that the 3% Mn ones, sometimes by more than 100 HV 10. The effects of increased cooling rate and increasing manganese content had the tendency to lower the tensile strength, Fig. 10.13.

The effect of using different types of ferromanganese, in the laboratory processing conditions, was not significant. All apparent bend strength values of TRS specimens, regardless of composition and processing parameters, were in the range 800–1200 MPa for sponge iron alloys and 640–1120 MPa for the water-atomised alloys.

10.3.6. Effect of sintering conditions

For the study the effect of sintering behaviour in a large range of sintering temperatures, 0.5% natural graphite and electrolytic manganese as a manganese carrier with the content of 1 and 4%, respectively, were admixed to the atomised iron powder ASC100.29. The impact energy bars were compacted at 600 MPa with die wall lubrication. The samples were sintered for 60 min between 700°C and

Table 10.19. Physical and mechanical properties of atomised iron powder (ASC100.29) with admixed 0.5% C and different Mn addition (1%, 4% Mn) at different sintering temperatures; green density for 1Mn steel was 7.11 g cm^{-3} and for 4Mn steel was 7.03 g cm^{-3}

Mn addition	Sintering temperature [°C]	Sintered density [g cm^{-3}]	Dimensional change [%]	Impact energy [J]	Hardness HV 30
1% Mn	700	7.09	0.13	0.53	63
	750	7.09	0.25	0.79	63
	800	7.07	0.40	0.87	84
	850	7.06	0.37	1.1	94
	900	7.05	0.37	1.42	96
	950	7.06	0./34	2.21	104
	1000	7.07	0.32	3.45	100
	1050	7.09	0.21	4.56	115
	1100	7.10	0.11	7.35	117
	1150	7.11	0.06	12.01	125
	1200	7.14	–0.07	17.74	135
	1250	7.15	–0.10	16.65	129
	1300	7.15	–0.15	19.06	148
4% Mn	700	6.88	0.63	0	59
	750	6.81	0.95	0.21	59
	800	6.79	1.09	0.32	71
	850	6.75	1.13	0.45	80
	900	6.76	1.15	0.56	84
	950	6.75	1.22	0.75	101
	1000	6.75	1.15	0.87	113
	1050	6.75	1.08	1.37	131
	1100	6.81	0.95	1.37	169
	1150	6.79	0.92	1.56	171
	1200	6.82	0.76	2.80	211
	1250	6.84	0.70	2.52	213
	1300	6.82	0.68	2.76	227

1300°C with 50°C interval in two push type tube furnaces in a high-purity atmosphere (5.0 quality = 99.999%) [246].

The physical and mechanical properties of the investigated steels are shown in Table 10.19.

Dimensional changes. Comparing the dimensional changes in 4% and 1% Mn content specimens, swelling was observed at 4% Mn at all temperatures, and a maximum was reached at 950°C. At a low Mn content shrinkage was observed at temperature from 1200°C to 1300°C. Thus, the dimensional change is clearly related to the density – the higher the expansion during sintering, the lower is the sintered density, which is the regular behaviour for sintered materials.

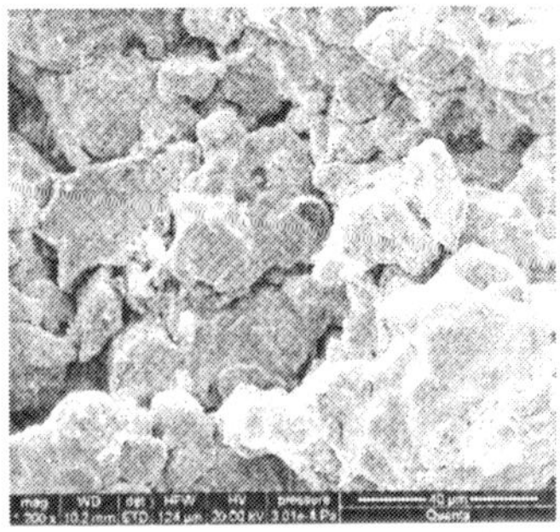

(a) ASC+0.5% C+1% Mn, 700°C, ×1200 (b) ASC+0.5% C+4% Mn, 700°C,×1200

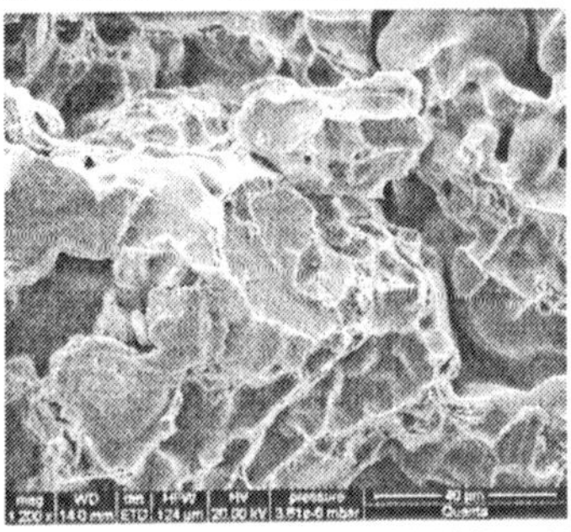

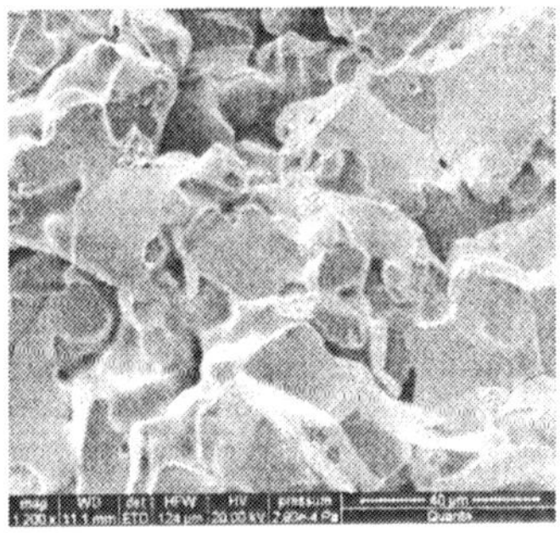

(c) ASC+0.5% C+1% Mn, 1300°C, ×1200 (d) ASC+0.5% C+4% Mn, 1300°C, ×1200

Fig. 10.14. Characteristic fractures of manganese steels with 1% and 4% Mn sintered at 700°C and 1300°C. SEM [246].

Impact energy. As follows from Table 10.20, the impact energy increases with increasing temperature. However, in alloys with the low manganese content the impact energy was about 9 times higher than that for the high Mn content alloy.

Hardness. Hardness shows the inverse behaviour for the mentioned samples. The hardness decreased at sintering temperatures from 700°C to 950°C with increasing manganese content, while at higher temperatures the higher manganese content became more effective and hardness increased for the 4% Mn manganese material much stronger from 1000°C to 1300°C compared to the 1% manganese material.

Fracture. Figure 10.14 shows the characteristic fractures of the samples sintered at 700°C and 1300°C.

Fractographic investigations carried out showed that sintering at 700°C did not result in visible interparticle necks being formed. The materials sintered at 1300°C showed the largest difference while in 1% Mn a mix of ductile rupture and transgranular cleavage fracture was found, and the material with 4% Mn showed pronounced brittle

Table 10.20. Nominal composition (%) of the powders

Powder	% Mn	% Cr	% Ni	% Mo
E	0 .8	1	–	–
F	1 .2	0 .2	–	–
G	0 .7	1 .2	0 .4	0 .4

Table10.21 Cooling rate as a function of dimension and nitrogen pressure

Dimension [mm]	N_2 pressure [bar]	Cooling rate [K/s]
25	2	2.2
	5	4 .6
	8	6.1
10	2	4.6
	5	8.2
	8	10.6
5	2	16.9
	5	26
	8	31.5

transgranular fracture, thus explaining the very poor impact energy of these steels.

10.4. Hardenability study

10.4.1. Effect of manganese on hardenability of prealloyed powders

In this work the hardenability of three new prealloyed powders containing manganese was investigated under the typical conditions of vacuum furnaces with gas cooling, and of sinter hardening in atmosphere furnaces. The aim of this work was to study the minimum cooling rate which has to be guaranteed in the parts to produce a hardened microstructure, in dependence on the carbon content [247].

The nominal composition of the prealloyed powders used in the this study is reported in Table 10.20.

To the powders a corresponding extra amount of graphite was then added to obtain the sintered carbon contents of 0.30%, 0.50% and 0.80%. Two types of specimens were produced for hardenability tests: disks 5 mm in diameter and of 10 mm height and Charpy impact bars 10 × 10 × 55 mm in size. Disks 25 mm in diameter and 25 mm in height were also produced, to determine the correlation between the cross-section and the actual cooling rate over a wide range of diameters. Specimens were compacted to 6.8 g cm^{-3} and sintered in a $90N_2/10H_2$ atmosphere at 1250°C, 30 min. Heat treatment was carried out in a vacuum furnace.

The austenitisation temperature was set up at $A_3(A_{cm})$+20°C and the isothermal holding time was 30 min. The cooling rate was varied by changing the pressure of cooling nitrogen. The combination of three nitrogen pressures, 2, 5 and 8 bars, with the two cross-sections

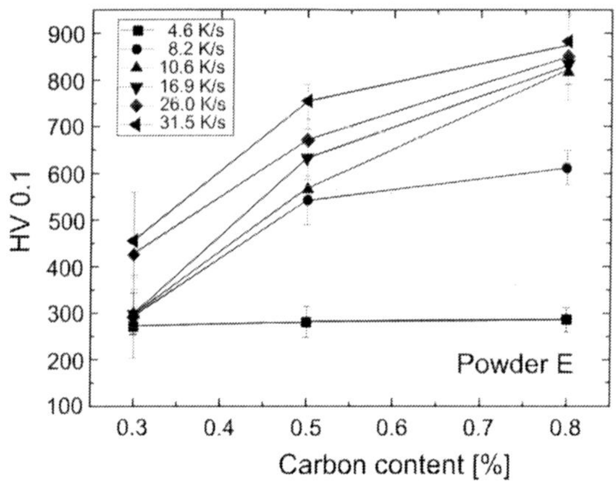

Fig. 10.15. Correlation between microhardness, carbon content and cooling rate for the heat-treated material E [247].

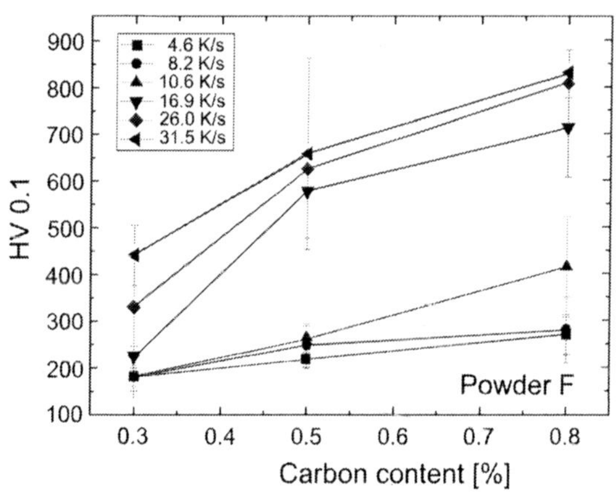

Fig. 10.16. Correlation between microhardness, carbon content and cooling rate for the heat-treated material F [247].

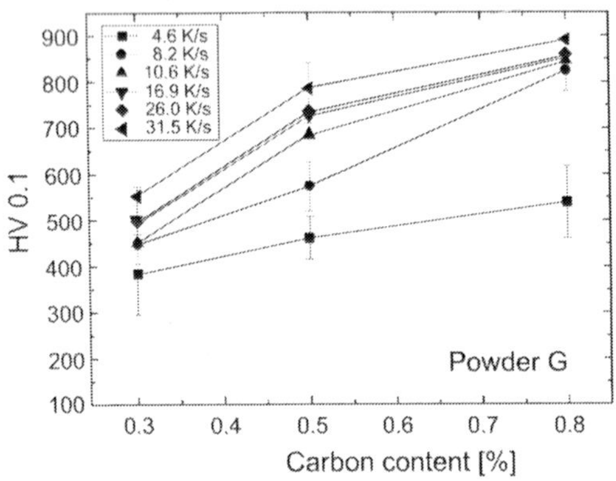

Fig. 10.17. Correlation between microhardness, carbon content and cooling rate for the heat-treated material G [247].

of the specimens resulted in the cooling rates reported in Table 10.21; they range between 4.6 K/s (impact bars quenched with the lowest pressure) and 31.5 K/s (5 mm diameter cylinders quenched with the highest pressure).

Since the minimum cooling rate exceeds that of industrial sinter hardening processes in atmosphere furnaces, some materials were heat-treated with a cooling nitrogen pressure of 1.5 bar using four impact bars bonded together. This resulted in an actual cooling rate in the centre of the 20 × 20 mm cross-section ranging between 1.8 and 2.9 K/s.

The microstructural analysis was carried out on etched (50% Nital – 50% Picral) metallographic sections. The HV 0.1 microhardness was measured in the centre of the median cross-section. The data reported in the following are the mean values of the experimental results.

Figures 10.15–10.17 report the microhardness (HV 0.1) *vs.* the sintered carbon content at the different cooling rates for materials based on powder E, F and G, respectively. As expected, the increase in the carbon content and in the cooling rate increases microhardness.

The figures can be used to determine the minimum cooling rate to obtain a hardened microstructure (under the assumption of 50% of martensite) for each carbon content. In particular:

- materials based on powder E: with 0.5% C and 0.8% C, they can be hardened with a cooling rate over about 8 K/s,
- materials based on powder F: with 0.5% C and 0.8% C, they can be hardened with a cooling rate over about 15 K/s,

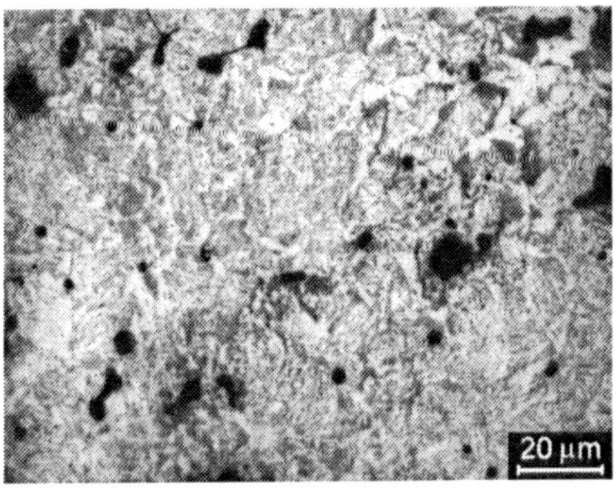

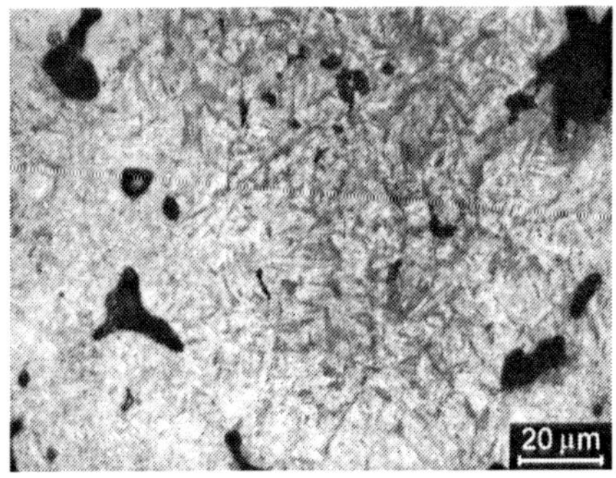

Fig. 10.18 (left). Powder E 0.3%C, 4.6 K/s – mixture of ferrite and pearlite [247].
Fig. 10.19 (right). Powder E 0.8%C, 31.5 K/s – fully martensite [247].

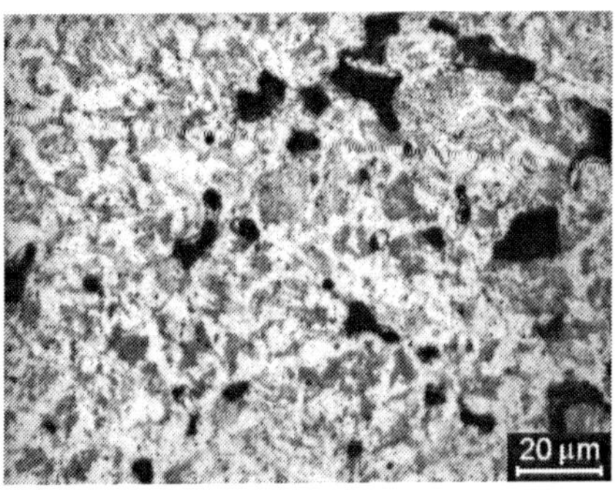

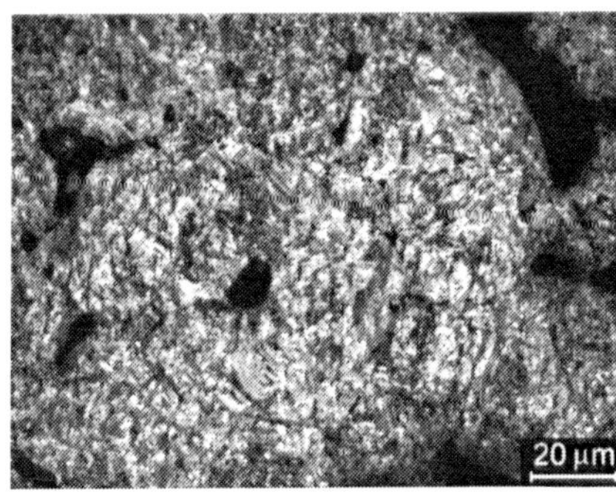

Fig. 10.20 (left). Powder F 0.3%C, 4.6 K/s – mixture of ferrite and pearlite [247].
Fig. 10.21 (right). Powder F 0.8%C, 31.5 K/s – lower bainite and martensite [247].

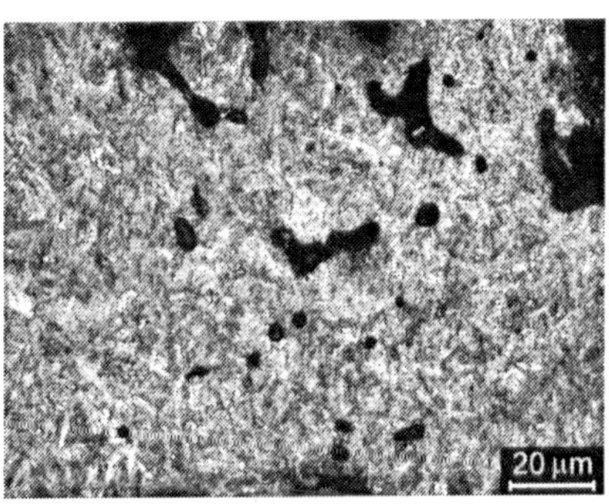

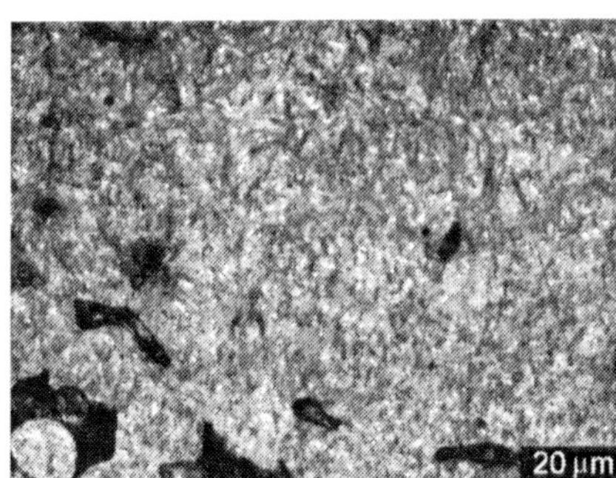

Fig. 10.22 (left). Powder G 0.3%C, 4.6 K/s – upper bainite and martensite [247].
Fig. 10.23 (right). Powder G 0.8%C, 31.5 K/s – fully martensite [247].

- materials based on powder G: they can be hardened with a cooling rate over 4.6 K/s for all the carbon contents investigated.

Microstructural analysis was carried out on all the specimens. On increasing carbon content and cooling rate, microstructures evolve:

- from a mixture of ferrite and pearlite to a full martensite in materials based on powder E, Figs. 10.18 and 10.19,
- from a mixture of ferrite and pearlite to a mixture of lower bainite and martensite in materials based on powder F, Figs. 10.20 and 10.21,
- from a mixture of upper bainite and martensite to a full martensite in materials based on powder G, Figs. 10.22 and 10.23.

On the basis of the microhardness data and of the microstructural analysis, hardenability of the three materials at the same carbon content results G>E>F.

The results of the present investigation may be used to design material for any specific application.

10.4.2. Effect of tempering temperature

Samples for the investigation of the effect of tempering temperature on the structure and mechanical properties of sinter hardened manganese steels were prepared from powder mixtures for Fe–3% Mn–0.8% C composition on the basis on NC100.24 iron powder and manganese as medium-carbon ferromanganese (FMn), <40 μm, and natural graphite.

Table 10.22. Post–tempering properties of Fe–3% Mn–0.6% C sinter hardened steels; sintering temperature 1120°C, hydrogen, cooling rate 66°C/min (sintered densities after tempering at all temperatures were 6.93–6.97 g cm^{-3}) (ND not defined)

Mechanical properties	Tempering temperature [°C]			
	None	200	300	400
Ultimate tensile strength–UTS [MPa]	475	740	740	774
$R_{0.2}$ offset strength [MPa]	ND	411	430	460
Elongation [%]	0.10	2.32	1.89	1.82
Transverse rupture strength–TRS [MPa]	1070	1478	1507	1462
UTS/TRS	0.44	0.50	0.49	0.53
Impact energy [J]	4.7	12.08	10.40	11.51
Surface hardness HV 30	ND	250	238	233
Cross-sectional hardness HV 30	301	227	219	216

Table 10.23. Post–tempering properties of Fe–3% Mn–0.6% C sinter hardened steels; sintering temperature 1250°C, hydrogen, cooling rate 66°C/min (sintered densities after tempering at all temperatures were 6.93–7.07 g cm^{-3}) (ND – not defined)

Mechanical properties	Tempering temperature [°C]			
	None	200	300	400
Ultimate tensile strength–UTS [MPa]	507	830	798	854
$R_{0.2}$ offset strength [MPa]	ND	474	472	499
Elongation [%]	0.10	2.0	1.4	2.0
Transverse rupture strength–TRS [MPa]	990	1479	1448	1755
UTS/TRS	0.51	0.56	0.56	0.49
Impact energy [J]	4.9	10.1	4.9	15.7
Surface hardness HV 30	ND	247	252	257
Cross-sectional hardness HV 30	276	278	274	245

The blended powders were compacted in lubricated dies at pressures in order to achieve the green density of 6.9–7.0 g cm^{-3} for transverse rupture strength/impact and for tensile strength test bars. Sintering was carried out in hydrogen with a dew point of –60°C at 1120 and 1200°C for 60 min. At the end of the sintering operation the specimens were moved to the water-cooled end of the furnace (sinter hardening effect). To prevent oxidation, the specimens were placed in a semi-closed container. After conventional cooling, the specimens were subsequently tempered in hydrogen at 200, 300 and 400°C for 1 h. The carbon contents of specimens sintered at 1120 and 1250°C were ~0.6 and 0.5% C, respectively [248].

The mechanical properties of the examined Fe–3Mn–0.8C steel sintered at 1120 and 1250°C are shown in Tables 10.22 and 10.23.

The results indicate that manganese, as an austenite stabilizer, has contributed appreciably to hardenability. The Fe–3% Mn–0.6% C alloy has sufficient hardenability such that the mechanical properties have a low sensitivity to the cooling rate. However, there was a limit to the amount of manganese which could be introduced because of the bainite reaction itself became so retarded that steel became martensitic after air cooling in the smaller section zone. The rapid cooled structure consisting of a bainite/martensite matrix, which contains some retained austenite, gives much higher hardness.

Tables 10.22 and 10.23, which compare tensile and yield strength of the alloys in both sinter hardened and tempered conditions, demonstrate this point clearly: for all the specimens tempering leads to a marked increase in tensile and bend strength. The hardness of the Fe–3% Mn–0.6% C alloy, which is very sensitive to tempering temperature in terms of tensile properties, is greatly adversely affected by tempering temperature.

The tempering temperature of manganese steels may result in temper embrittlement – a reduction of toughness as measured by the impact test.

This shows the importance of combining high strength and ductility by understanding the sinter hardening process.

10.5. Sintering in semi-closed containers

The sintering in double boxes containing a getter – alumina + graphite – is used in laboratory conditions to prevent decarburisation of the carbon-containing steels. The aim of sintering of manganese steels in semi-closed containers is:

- processing to avoid/minimize formation of oxide veins in a dry hydrogen-rich atmosphere and sintering in the alpha phase,
- reduction of surface oxides, because the flowing atmosphere can no longer easily interact with the components surfaces in the presence of Mn vapour,
  - to form the source of Mn vapour in a getter in the semi-closed containers (local dew point), which reacts with the atmosphere and thereby dries it – this enables processing of Mn alloys at temperatures as low as 1140°C at inlet dew points no better than –30°C,
  - to reduce the extent of Mn sublimation and oxidation [249].

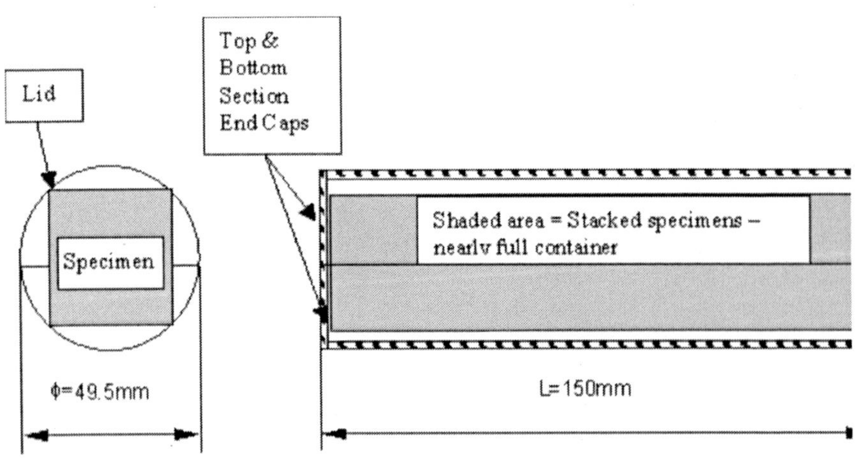

Fig. 10.24. Schematic of semi-closed sintering container [249].

Sponge NC100.24 iron powder was used as the base iron powder for the preparation of the test specimens for sintering of manganese steels in semi-closed containers. Carbon was introduced as fine graphite and 3 and 4% of manganese as medium-carbon ferromanganese (FMn) with a particle size of <20 μm was added. Starting compositions chosen were Fe–(3,4)Mn–0.8C. The compacts for testing the tensile and transverse rupture strength were compacted at 660 MPa (green density ~6.9 g cm^{-3}) followed by sintering in nearly full semi-closed stainless steel containers with a labyrinth seal shown in Fig. 10.24. The container did not have any ferromanganese added to it, but was loaded to capacity, ~20 specimens, with any remaining space in the container filled with ferrous ballast. Therefore the green compacts were the carriers of Mn and C in the container. This arrangement provided sufficient manganese vapour and free carbon and improve the local dew point (self-gettering). It means that the source of Mn vapour can be a getter, which reacts with the atmosphere and thereby dries it. This enables processing of Mn alloys at temperatures as low as 1140°C at an inlet dew point no better than –30°C. The container was pushed into the hot zone of the stainless steel tube of the furnace. The furnace atmospheres were pure hydrogen or pure nitrogen with a gas flow of ~0.2 cm^3/min per cm^2 of the furnace tube cross-sectional area and an inlet dew point no worse than –60°C. Sintering times were 1 h and sintering temperatures 1120 and 1250°C. After sintering, the specimens were withdrawn from the hot zone to the cool zone of the furnace at 1.5–65°C/min. Some specimens cooled at a rate of 65°C/min were tempered at 200–450°C for 1 h.

Table 10.24. Results of Leco analyses of Fe–3Mn–0.8C compacts after sintering at 1120°C and 1250°C; average densities ~7.0 g cm^{-3}

Sintering temperature 1120°C			Sintering temperature 1250°C		
Cooling rate [°C/min]	Carbon content [%]	Oxygen content [%]	Cooling rate [°C/min]	Carbon content [%]	Oxygen content [%]
1.3	0.62	0.25	1.4	0.49	0.13
55	0.55	0.19	57	0.53	0.16
65	0.66	0.20	65	0.53	0.15

Table 10.25. Effect of cooling rate on the mechanical properties of Fe–3Mn–0.8C (final 0.6% C) specimens sintered at 1120°C. CR – cooling rate

CR [°C/min]	YS, 0.2% [MPa]	UTS [MPa]	TRS [MPa]	Hardness HV 30	KC [J]
1.3	315	447	1050	169	16.5
4.8	330	478	1270	224	17.2
8.7	455	506	1230	217	10.8
15.5	450	568	1200	148	11.0
30.1	485	557	1120	172	8.2
47.0	510	559	1100	176	5.8
55.0	brittle	535	1000	292	5.2
60–65	brittle	486	1050	303	6.5

Table 10.26. Effect of cooling rate on the mechanical properties of Fe–3Mn–0.8C (final ~0.5% C) specimens sintered at 1250°C; CR – cooling rate

CR [°C/min]	YS, 0.2% [MPa]	UTS [MPa]	TRS [MPa]	Hardness [HV30]	KC [J]
1.4	385	587	1240	166	21.3
4.7	440	595	1330	254	11.5
9.4	410	614	1310	239	9.9
20.4	325	583	1210	265	7.7
31.8	570	586	1190	282	9.4
41.1	brittle	521	1080	328	4.6
57.0	brittle	491	1090	335	4.9
60–65	brittle	507	990	276	4.9

The following properties were measured: green and sintered density, dimensional change, transverse rupture strength (TRS), impact energy (KC), yield strength (YS), tensile strength (UTS), elongation (A) and hardness (HV 30).

Chemical analyses of Fe–(3,4)Mn–0.8C alloys, Table 10.24, show, however, that decarburising took place at actual carbon contents to ~0.6 and ~0.5% C, respectively. The analyses further indicate a minor manganese loss.

Table 10.27. The effect of cooling rate on the mechanical properties of Fe–4Mn–0.8C (final 0.6% C) specimens sintered at 1120°C; CR – cooling rate

CR [°C/min]	YS,0.2% [MPa]	UTS [MPa]	TRS [MPa]
4.8	422	669	1021
8.7	417	680	1116
15.5	374	651	1227
30.1	414	659	1121
47.0	brittle	482	1151
55.0	brittle	580	1097
65	brittle	561	1079

Table 10.28. Effect of cooling rate on the mechanical properties of Fe–4Mn–0.8C (final 0.5% C) specimens sintered at 1250°C. CR – cooling rate

CR [°C/min]	YS, 0.2% [MPa]	UTS [MPa]	TRS [MPa]
4.7	434	636	1467
9.4	460	600	1451
20.4	459	547	1362
31.8	brittle	496	1364
41.1	brittle	468	1124
57.0	brittle	428	847
65	brittle	485	891

Table 10.25 shows the mechanical properties of Fe–3Mn–0.8C sintered at 1120°C, and in Table 10.26 the properties of the same steel sintered at 1250°C. The effects of increasing Mn content and sintering temperature were to lower the tensile elongation, with the reverse effect on the yield and tensile strengths of slowly cooled specimens. Slow cooling rates favoured plasticity, which ceased to be significant at cooling rates above 30–40°C/min. After sintering Fe–3Mn–0.8C steel at 1120°C with cooling below 30°C/min UTS values were up to 568 MPa and TRS up to 1230 MPa. The best combinations of mechanical properties were attained at the lowest cooling rates. The tensile strength of the specimens cooled at 65°C/min was only ~486 MPa, and ~1050 MPa in TRS.

The same for strength properties is approximately valid also for Fe–3Mn–0.8C steel sintered at 1250°C compared to those sintered at 1120°C. The highest UTS of 614 MPa and TRS of 1310 MPa were

Table 10.29. Average properties of 60 TRS and 54 UTS sintered at 1120°C and 1250°C and sinter hardened Fe–3Mn–0.8C specimens tempered at 300°C for 1 h

Sintering temperature [°C]	Hardness HV30	UTS [MPa]	TRS [MPa]	KC [J]
1120	227	751	1481	11.3
1250	265	827	1630	10.2

achieved after cooling rates lower than 20°C/min. Hardness increased up to 335 HV 30.

Tables 10.27 and 10.28 present the properties of Fe–4Mn–0.8C steel sintered at 1120°C and 1250°C in dependence on cooling rate.

Table 10.29 lists the average properties of UTS, TRS, hardness and impact energy values of sinter hardened Fe–3Mn–0.8C steel.

The specimens sintered at 1250°C being lower in final carbon, their exhibits higher hardenability. The lower C content of the 1250°C material reduces the amount of retained austenite, but increases the martensite fraction. The 1250°C sintered material has greater austenite grain size compared with the 1120°C material.

Some of the sinter hardened alloys (cooled at 65°C/min) were subsequently tempered in hydrogen at temperatures ranging from 200 to 450°C for 1 h. Significant improvement in properties was attained as a result of the 200°C stress relief, *e.g.* 830–850 MPa tensile strength with >2% ductility for the 1250°C sintering temperature.

The mechanical results versus tempering temperature suggest possibly two temperatures where embrittlement was occurring. The first, at 250°C, is probably due to carbide film formation at grain boundaries and, in the case of the 1120°C material, at 350°C due to tramp element and/or oxygen segregation. On tempering at 400°C there was a recovery of toughness due to martensite and retained austenite transforming to bainite, which gives the highest impact value for the 1250°C material.

[250] investigated the strength properties of Fe–3Mn–0.8C steel sintered in semi-closed containers with a labyrinth seal in hydrogen and nitrogen with a dew point of –60°C to produce 'Mn rich–microclimate' in dependence on three iron powder grades and two sintering temperatures (1120 and 1250°C).

10.5.1. Effect of manganese on microstructure formation

Test specimens were prepared from sponge iron powder SC100.26,

Table 10.30. The thickness of Mn enriched areas

Material	T = 900°C, thickness [μm]	T = 1050°C, thickness [μm]	T = 1200°C, thickness [μm]
Fe–3Mn	4.76	7.22	17.85
Fe–3Mn–0.3C	5.46	7.81	18.05*

Table 10.31. Diffusion coefficients and diffusion distances for Mn in Fe matrix

T [°C]	T [K]	t [s^{-1}]	D [cm^2/s]	$X_{calculated}$ [μm]	$X_{measured}$ [μm]
900	1173	180	$5.72 \cdot 10^{-11}$	0.29	4.76
1050	1323	180	$7.46 \cdot 10^{-12}$	0.80	7.22
1200	1473	180	$8.41 \cdot 10^{-12}$	1.24	17.85

medium-carbon ferromanganese (FMn) with particle size <40 μm and natural graphite and HW wax as a lubricant. The powder mixtures of Fe–3Mn and Fe–3Mn–0.3C composition were prepared. Cylindrical samples 13 mm in diameter and 7 mm long were pressed to the density of 7.0 g cm^{-3}. Sintering was carried out in a laboratory tube furnace in a $25N_2/75H_2$ atmosphere with the dew point of –30°C in semi-closed containers with a getter of Al_2O_3+10% FMn+5% graphite. Sintering was carried out at 750, 900, 1050, 1150 and 1200°C for 3 min in a $75N_2/25H_2$ atmosphere with the dew point –30°C. Heating rate was 10°C/min and cooling rate was 75°C/min.

The aim of the study was the effect of manganese on microstructure formation and the depth of manganese penetration into the iron powder particles during sintering [250].

The measured values of alloyed regions at 900, 1050 and 1200°C are presented in Table 10.30.

Microstructure formation. It was stated that the diffusion-induced grain boundary migration (DIGM) was a recognized phenomenon that led to unexpected motion of grain boundaries. The microstructures of the investigated materials were complex and heterogeneous, consisted of austenite, martensite, bainite and fine pearlite depending on the local Mn and C contents. Intensive sublimation of manganese with the temperature increasing promoted Mn redistribution in the vapour

Table 10.32. The results of Mn penetration depth

T [°C]	a_{Fe-3Mn} [μm]	$a_{Fe-3Mn-0.3C}$ [μm]
900	1.01	76.7
1050	43.9	37.5
1200	71.7	70.2

phase within the powder compact and increased Mn content on the prior surfaces of powder particles.

Calculations of diffusion coefficients and diffusion distances for Mn in Fe matrix after sintering at different temperature during 3 min values are summarized in Table 10.31.

It is clearly evident that the data of apparent diffusion coefficients in bulk manganese diffusion in the Fe matrix are lower than the data for sintered manganese steels.

Manganese penetration depth. Mn penetration depths for sintering temperatures of 900, 1050 and 1200°C in Fe–3Mn are listed in Table 10.32.

The DIGM mechanism, together with intensive sublimation of manganese (occurring at high temperatures and promoting Mn redistribution in the vapour phase) in the presence of a liquid phase

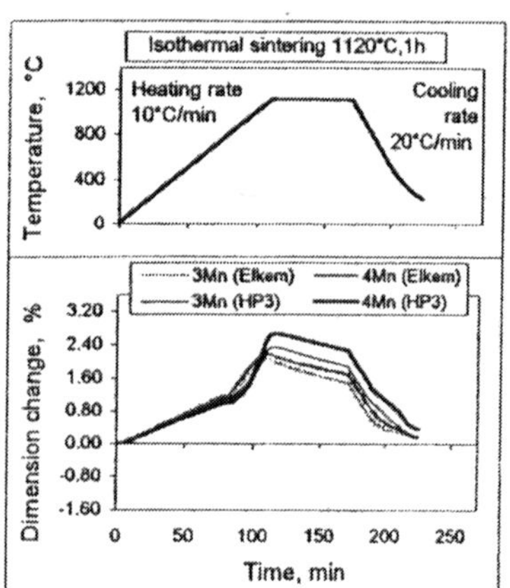

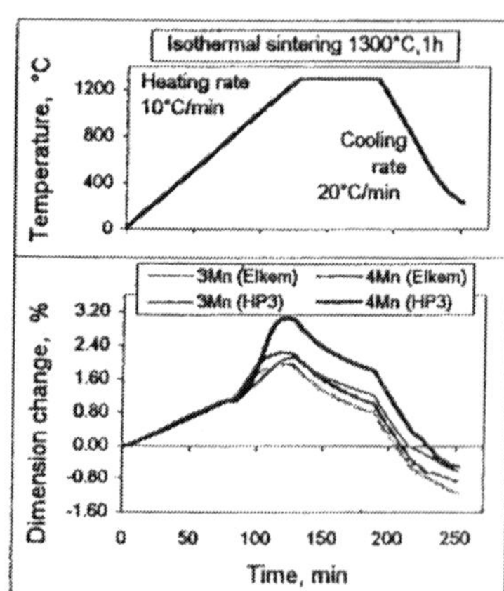

Fig. 10.25 (left). Dilatometric curves and temperature profile. Isothermal sintering at 1120°C [251].
Fig. 10.26 (right). Dilatometric curves and temperature profile. Isothermal sintering at 1300°C [251].

formed at temperatures of 1200°C, improved manganese distribution around the original Fe particle boundaries.

10.5.2. Dimensional changes of Fe–Mn–C steels

For the preparation of the test samples the following starting materials in powder form were used:

- sponge iron powder NC100.24,
- ferromanganese grades with particle size <40 μm:
  medium-carbon ferromanganese (EL–ELKEM, 0.23% O),
  high-carbon ferromanganese (HP3, 0.08% O),
- natural graphite C–UF.

Two mixtures for each ferromanganese powder were prepared:

- Fe–3% Mn–0.8% C and Fe–4% Mn–0.8% C.

The mixtures were compacted at 600 MPa into specimens of size 4 × 4 × 15 mm what resulted in green densities of ~6.7 g cm^{-3}. Sintering was carried out in a horizontal push rod dilatometer. Pure, dry hydrogen with a dew point below –60°C was used as a sintering atmosphere. Isothermal sintering was carried out for 1 h at temperatures: 1120, 1240 and 1300°C [251].

The sintering behaviour during sintering was displayed in the dilatometric curves shown in Figs. 10.25 and 10.26. Specimens heated from room temperature expanded in accordance with the linear expansion coefficient of α-iron, but at temperatures of about 800–850°C a small transient shrinkage was observed due to the α→γ transformation. There is a decrease from the temperature for pure iron of 910°C caused by alloying elements, especially Mn, going into solution. After the α→γ transformation was completed, the specimens expanded thermally due to continuing alloying processes.

Up to the α→γ transformation temperature compacts containing HP3 ferromanganese showed similar thermal expansion in comparison with the samples alloyed with medium-carbon ferromanganese (EL). In the higher temperature region Fe+HP3 samples showed larger total expansion compared to the behaviour of the Fe+EL compacts. For samples sintered at 1120°C and 1240°C, the maxima of the dimensional changes according to the dilatometric curves lie at the beginning of isothermal sintering. Processes causing swelling (thermal expansion, dissolution of ferromanganese, decarburizing, chemical reactions between components and or sintering temperature) are dominant up to these temperatures. Shrinkage of almost all

Table 10.33. Processing characteristics of sintered specimens

Composition	Density [g cm^{-3}]		$\Delta l/l_0$ [%]	Δm [%]	C [%]	R_{FR} [MPa]
	green	sintered				
A+0.5Mn–0.5C	7.03	6.98	–0.31	–0.19	0.38	364
B+0.5C	7.07	7.04	–0.32	–0.05	0.47	401

investigated compacts heated to the sintering temperature of 1300°C began at about 1240°C.

These results will contribute to the commercial objective of obtaining consistent dimensional changes after sintering of mixtures based on NC100.24 iron powder with different ferromanganese + graphite additions. The optimal composition studied was a mixture containing 3% Mn added as EL or HP3 ferromanganese.

10.6. Effect of different processing variables on properties of Fe–Mn–C steels

The sintering behaviour of Fe–Mn–C powder system. For this study the following raw materials were employed: prealloyed iron powders **A** and **B** with a manganese content of 0.3% and 0.8%, respectively: medium-carbon ferromanganese powder with the particle size under 45 μm, graphite UF4 and wax (AW) powder as a lubricant. For experiment were prepared two mixtures (mass %): **A**+0.5Mn+0.5C and **B**+0.5C to obtain admixed and prealloyed systems with the same Mn content of 0.8%. (Composition of **A** and **B** powders was not declared). Cylindrical specimens (10 mm diameter, 12 mm long) were

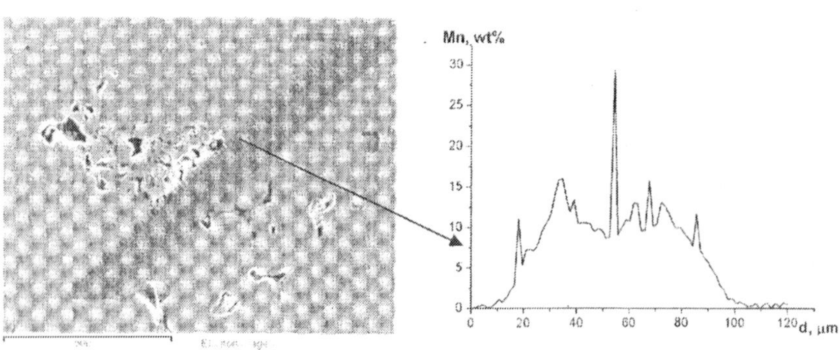

Fig. 10.27. Mn profile of FeMn residue in **A**+0.5Mn+0.5C sintered specimen [252].

pressed at 600 MPa to a green density of 7.03–7.07 g cm^{-3}. Sintering was carried out in a laboratory tube furnace at 1120°C or 1200°C for 30 min in an atmosphere of 93%N_2/7%H_2 with an inlet dew point of –60°C and then cooled (~50 K/min) to room temperature. The heating and cooling rates were ~10°C/min. The specimens were assessed in a stainless container. The heating and cooling rates were ~10°C/min [252].

The processing characteristics of the sintered specimens are listed in Table 10.33.

The R_{FR} values were obtained utilizing a 'button' tensile test on the same cylindrical samples machined to the 'button' shape by the electrospark method. The mass changes, Δm, are in accordance with the carbon loss values. Higher R_{FR} was attained for the steel based on powder **B**.

The microstructure of the **A**+0.5Mn+0.5C consisted mostly of ferrite and pearlite. The complicated sequence of microstructure constituents around FeMn residuals – austenite, which in direction of the particle centre changed to martensite and bainite – is a consequence of the Mn concentration gradient due to diffusion processes. Ferrite, which forms the centre of the prior Fe particles, was surrounded by small areas (10–20 μm) of pearlite. The results of SEM+WDX analysis show that Mn travels through the iron matrix up to ~60 μm, Fig. 10.27.

The negative effect of the oxidised residuals of FeMn particles in the sintered microstructure of the admixed system was reflected in lower rupture strength values than those achieved for the prealloyed system. The high microstructure inhomogeneity near the ferromanganese residue, which is a characteristic feature of admixed system due to the manganese distribution effect during sintering cycle, has a significant affect on the mechanical properties as well.

Table 10.34. Studied materials with their designation codes and nominal compositions

Material code	Mixture composition [%]	Mn carrier	Alloy composition [%]
CEM	**C**+0.5Mn+0.5C	electrolytic Mn	Fe–0.8Mn–0.5C
CFM	**C**+0.5Mn+0.5C	FeMn	Fe–0.8Mn–0.5C
D	**D**+0.5C	prealloyed	Fe–0.8Mn–0.5C

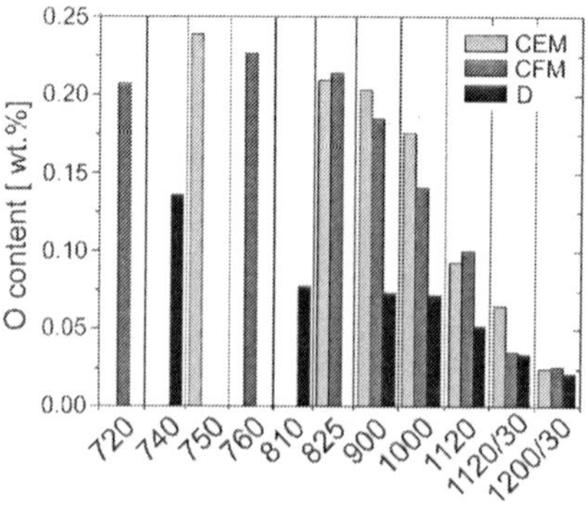

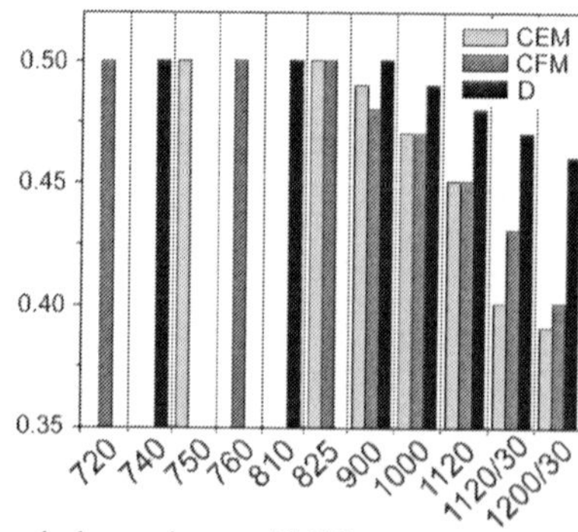

Fig. 10.28 (left). Oxygen content for sampled specimens [253].
Fig. 10.29 (right). Carbon content for sampled specimens [253].

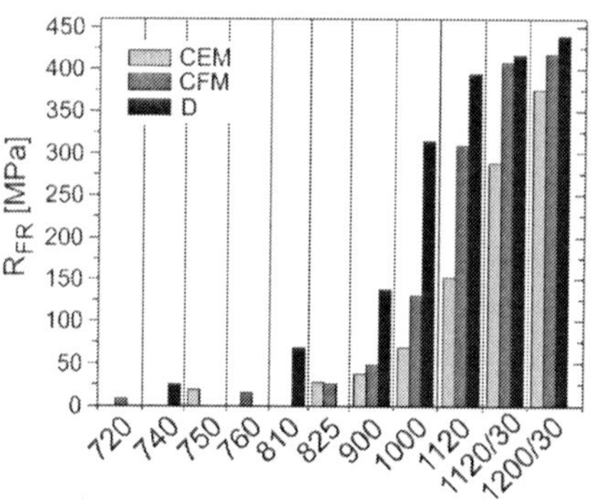

Fig. 10.30. Rupture strength of specimens [253].

Alloying mode. For the study, three different powder mixtures with the same manganese content of 0.8% were prepared on the basis of water-atomised steel powders **C** (Fe–0.3Mn) and **D** (Fe–0.8Mn). Powders were admixed with 0.5% of graphite and 0.8% of lubricant (amide wax). Additionally, 0.5% Mn was admixed to powder **C** in order to obtain 0.8% Mn in total. As the manganese carrier in the alloy **CEM,** the high-purity electrolytic manganese powder (99.89% Mn, 0.076% Si–Fe–Se, 0.0013% P, 0.018% S) was used. Medium-carbon ferromanganese powder (FeMn) (81.39% Mn, 1.48% C, 0.29% Si, 0.17% P, 0.01% S) was used as the manganese carrier in the alloy **CFM.** Both kinds of manganese carrier were sieved, and the size fraction below 45 μm was used. The materials compositions and designations are presented in Table 10.34. The first numbers for the specimens designations indicate the sampling temperature, second number – sintering time.

Cylindrical specimens (10 mm diameter, 12 mm long) were pressed at 600 MPa to a green density of ~7.0 g cm^{-3}. Sintering was carried out in laboratory tube furnace at 1120°C for 30 min in an atmosphere $10H_2/90\%N_2$ with inlet dew point in the range of –55°C to –60°C. The specimens were assessed in a stainless steel container. The heating and cooling rates were ~10°C/min. Sintering was carried out at 1120 or 1200°C for 30 min followed by cooling (~50 K/min) down to room temperature.

Oxidation of the samples during heating stage in the case of both admixed specimens is clearly evident on bulk oxygen content, Fig. 10.28 – 2–3 times higher values of oxygen content than for prealloyed material. Admixed with electrolytic manganese material also shows higher oxygen values during heating and after sintering at lower temperature. After high temperature sintering all three materials show the same level of oxygen. Good correlation between oxygen and carbon content can be seen, much lower carbon loss is evident for prealloyed material during all stages, Fig. 10.29.

Prealloyed material shows much higher strength in terms of rupture strength values at all temperatures, Fig. 10.30. The lowest strength was obtained in the case of material admixed with high-purity electrolytic manganese (**CEM** alloy), even admixed with ferromanganese material has a much higher strength. The rupture

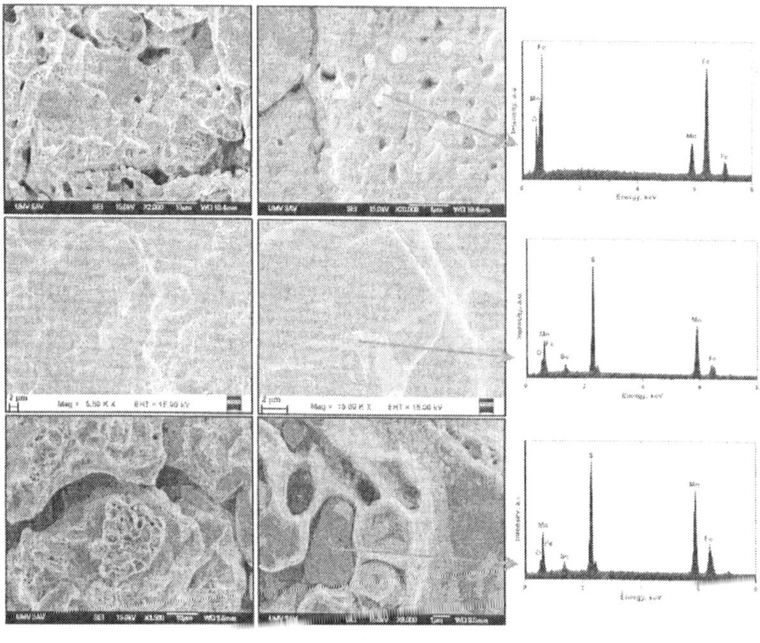

Fig. 10.31. SEM+EDX analysis of combinations observed on the fracture surface of interrupted sintered specimens **CFM**1120/30 (top). **CEM**1200/30 (middle) and **D**1120/30 (bottom) [253].

strength value for **CEM** alloy is still much lower, even lower obtained for prealloyed material after sintering at lower temperature.

Fractography. High resolution SEM imaging combined with X–ray microanalysis indicates that the particulate features observed around manganese carrier residues at low temperatures are manganese oxides in the case of both **CEM** and **CFM** materials after processing at all temperatures, Fig. 10.31. Manganese carrier residues are formed mostly by manganese oxide. However, after sintering at high temperature, together with the essential decrease of the size of manganese residues the oxygen content decreases as well, while a higher sulphur content is registered. Concerning the tendency of sulphur enrichments in particulate features after high-temperature sintering, observed for prealloyed material **D**, sulphur segregation from the base powder is assumed as a possible mechanism.

The analysis of manganese-containing sintered steels based on the fully prealloyed powder and the admixed system using high-purity electrolytic manganese and medium-carbon ferromanganese indicates high brittleness of admixed materials, which is the main obstacle for their industrial utilisation. The lower strength of the admixed materials is described based on the oxidation of manganese vapour and further condensation of the surrounding iron particles that suppress interparticle necks development. The worst results were obtained for the material admixed with electrolytic manganese. This is explained by more intensive manganese evaporation and accordingly higher specimen oxidation at low temperatures.

Brittleness of structural PM steels. The starting materials employed for this study were water-atomised iron powder

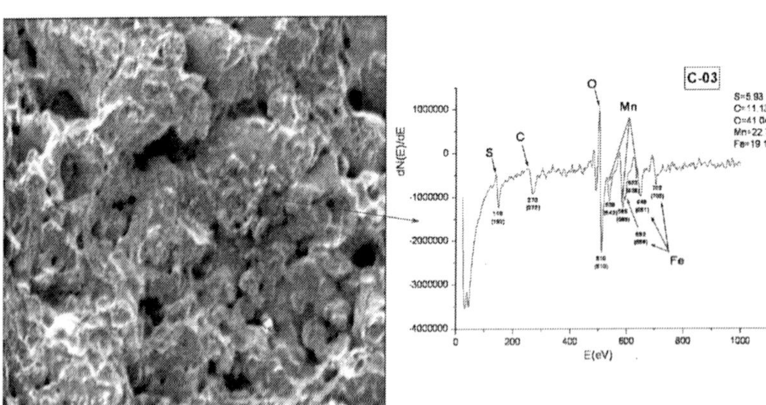

Fig. 10.32. Auger spectra of the ferromanganese residue on the fracture surface for the sintered specimen A [254].

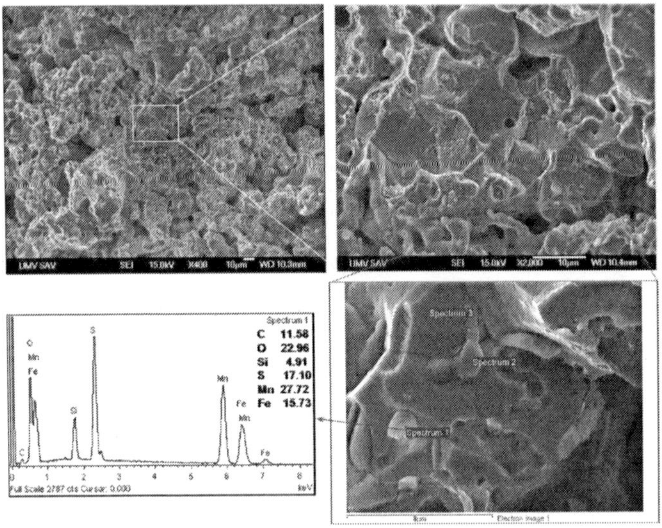

Fig. 10.33. SEM+EDX analysis of fracture surface of specimen A in the areas around ferromanganese residue [254].

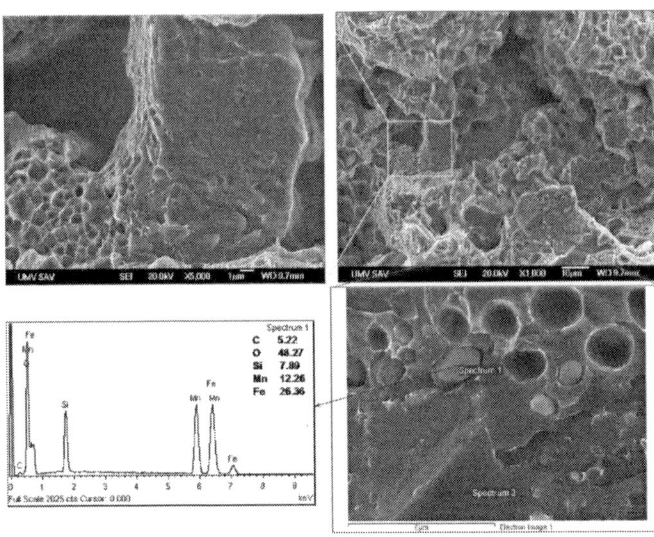

Fig. 10.34. SEM+EDX analysis of fracture surface of specimen B in the areas around ferromanganese residue [254].

ASC100.29, medium-carbon ferromanganese powder with the particle size under 45 μm, graphite powder UF4 and amide wax powder as a lubricant. The powder mixture of the composition (%): ASC100.29+0.8Mn+0.5C+0.8AW was prepared and cylindrical specimens (10 mm diameter, 12 mm long) were pressed at

600 MPa to a green density of about 7 g cm^{-3}. Sintering was carried out in a laboratory tube furnace for 30 min in an atmosphere of $90\%N_2/10\%H_2$. Specimen A was sintered at 1120°C in an atmosphere with a dew point of –40°C and a cooling rate of 10°C/min. Specimen B was sintered at 1200°C in an atmosphere with a dew point ~ –60°C and a cooling rate of 50°C/min. The sintered specimens were dry machined to final dimensions (4 × 4 × 10 mm) and fractured (bending) directly in ultrahigh vacuum.

Examples of characteristic features of the fracture surfaces for specimens A and B are presented in Figs.10.32–10.34. The fracture surfaces in both cases are rather rough, showing ductile interparticle fracture with good pronounced lines and dimples of different size, initiated by bridge porosity and non-reduced refractory oxides on the prior particles surfaces. A small portion of transgranular cleavage facets was registered as well, being more frequent for the high-temperature sintered specimen B. The FeMn residues were surrounded by intergranular decohesion facets with a huge amount of point inclusions on them, especially in the case of low-temperature sintered specimen A, Figs. 10.33–10.34. Auger analysis of specimen A, Fig. 10.32, indicates high sulphur, oxygen and carbon contents, together with a very high manganese content in all analysed places of ferromanganese residue localisation. The SEM+EDS analyses of such locations in the case of both specimens, Figs.10.33–10.34, also indicate a high manganese content. The agglomeration of such point inclusions forms so-called point oxides, as indicated by surface chemical analysis by X-ray photoelectron spectroscopy, and Auger electron spectroscopy. The area analysed in XPS analysis was about 0.8 mm in diameter. High-resolution SEM combined with X-ray microanalysis analysis was performed on the fracture surfaces of the specimens A and B after surface chemical analysis.

As a result, much better interparticle connections were developed in the case of specimen B. The lower atmosphere purity and lower sintering temperature in the case of specimen A lead to continuous manganese oxidation and formation of a thick oxide layer around the ferromanganese particles.

The slower cooling rate at low atmosphere purity (long exposure to strongly oxidising conditions) also results in higher oxidation. All these effects result in more than 10 times higher oxygen content in the case of specimen A in comparison with B (0.307 and 0.025 wt.% respectively).

The analysis of defective areas in the admixed Fe–0.8Mn–0.5C sintered steel by SEM+EDX, Auger and XPS indicates that the brittleness of studied material is caused by the weakness of boundaries of the base matrix particles around the manganese carrier residues.

The presence of complex oxides and manganese sulphide was detected on the intergranular decohesion facets. The composition and morphology of these mentioned products are dictated by the type of the manganese carrier used and sintering conditions (temperature profile, sintering atmosphere purity, etc.). Sintering at higher temperatures with good atmosphere purity (dew point ~60°C) in a N_2/H_2 atmosphere leads to a significantly decreased content of oxides but a higher portion of MnS, supposed to be less harmful for the mechanical properties

Table 10.35. Characteristic data of sintered gear wheels; manganese added in form of medium-carbon ferromanganese

Powders and processing	Composition [%]		Density	Surface hardness A* B**	Core hardness	Tooth bend strength	Microstructure
	Mn	C	[g cm^{-3}]	HV 10	HV 10	daN	
NC100.24	2.19	0.47	6.78	A–115	155	416	Fine pearlite, islands of ferrite
NC100.24	2.18	0.38	6.77	A – 114	157	417	Coarse pearlite, islands of ferrite
NC100.24	2.68	0.55	6.77	A – 151	164	506	Fine pearlite, very small amount of ferrite, small amount of retained austenite in Mn–rich areas
ASC100.29	2.79	0.55	6.73	A – 127	150	452	Martensite/retained austenite, bainite
NC100.24 carbonitrided and heat-treated (SH)	2.82	1.46	6,89	A–534 B – 340	280	457	Martensite/retained austenite, troostite
ASC100.29 carbonitrided and heat-treated (AH)	2.70	1.70	6.85	A – 443 B – 333	277	402	Martensite/retained austenite, bainite
NC100.24 sized, carbonitrided and heat-treated (SS)	2.76	1.36	6.91	A – 498 B – 324	270	488	Martensite/retained austenite, troostite
ASC100.29 sized, carbonitrided and heat-treated (AS)	2.56	1.49	6.90	A – 505 B – 336	279	427	

A* – surface of the heat-treated gear, B**– surface of the gear ground off in approx. 0.2 mm, SH... etc., gear wheel batch deisgnation

10.7. Processing conditions and properties of sintered manganese steel for structural parts

10.7.1. Preparation of gear steel

For the study the basic characteristics of iron powders in relation to the production method (sponge, atomised) demonstrated by the microstructure and by this in properties of manganese steels also in dependence on processing conditions used. For this purpose the sintered manganese steels on two iron powder grades were prepared. [255]. The aim of the study was to prepare the gear steel fulfilling the static and dynamic and wear resistance on gear tooth. Eight batches of manganese steel gear were studied. Gear tooth bend tests were made on three gears from each batch. The test were performed on the 18 tooth of the gear randomly chosen from each batch.

The representative results from the test are given in Table 10.35.

From the manganese steel production viewpoint, the atomised powder, which has a coarse-grained structure and low specific surface area, was less suitable for alloying with manganese. In the material based on ASC100.29 powder, the microstructure was more heterogeneous and in as sintered gears consisted of pearlite with islands of ferrite and even retained austenite with martensite needles. In the material based on NC100.24 powder the microstructure consisted of pearlite with small islands of ferrite. Because of rapid sublimation of manganese from ferromanganese particles and manganese vapour condensation on the surface of iron particles in the sintered compact, the characteristics of the iron particles surface played an important role.

The density and hardness of the manganese steel gears were found to be similar to Distaloy DC–1+0.5% C gears. The bending strength of the gear wheel teeth with manganese steel was generally superior to DC–1+0.5% C steel. The material based on NC100.24 iron powder

Fig. 10.35. 14-tooth development gear; static tooth bending test (3 teeth fixed) [256].

had higher strength as compared with that based on ASC100.29. The most important aspects of the gear wheels quality are seen as ensuring a uniform structure free of segregation and a consistent response to heat treatment. Experience shows that the defects most commonly encountered in sintered gears are decarburizing, the manganese loss due to sublimation and evaporation, and poor carbon and manganese distribution.

10.7.2. Innovative processing of manganese steel gears

In fine-pitch gearing application, gear trains are sometimes subject to high static loads. An example would be given when a mechanism is driven into a mechanical stop. It is extremely important that the gears be capable of withstanding the maximum static torque [256].

The following section contains method by which static strength of gears can be statistically evaluated, in addition to the Weilbull representation of the strength of different types of gears. Strength variations are verified quantitatively by using a standard production gear for baseline comparisons.

The object of investigation was a fourteen-tooth Ritzel spur gear shown in Fig. 10.35. The overall length (blank width was 17 mm and the outside diameter was 34 mm).

Several batches of laboratory and industrially sintered Fe–3% Mn–0.8% C steel gears were studied. The base iron powder NC100.24, atomised ASC100.29 and atomised Fe(PFeR) iron powders were used. Manganese was introduced as ground medium-carbon ferromanganese with the particle size below 20 μm. Carbon was added as F10 graphite. Ferromanganese and copper powders were mixed with a Kenolube lubricant with iron powders. The mixtures were compacted at 600 MPa in the form of a Ritzel gear to a density 6.75–6.85 g cm^{-3}. The green compacts were sintered in a laboratory tube furnace at 1120°C and in an industrial furnace at 1150°C, both for 60 min under: a) hydrogen/nitrogen mixtures (0–100% H_2) in the laboratory furnace with the dew point of the mixtures of –60°C (10 ppm moisture), b) in dissociated ammonia with the dew point of –30°C (375 ppm moisture) in the industrial furnace. At the end of the laboratory sintering period the specimens were moved to the water-cooled end of the furnace tube. The cooling rate was ~10°C/min.

The gears were sintered in a special micro-atmosphere to maintain the tight oxygen, carbon and manganese control required for strength and hardness. In the industrial furnace, the green compacts were

Table 10.36. Characteristic data of Ritzel Fe–3% Mn–0.8% C and Fe–1.5% Cu (reference material) sintered gears. 1 – industrial sintering, 2 – laboratory sintering in H_2, 3 – laboratory sintering in N_2, industrial sintering in dissociated ammonia. Manganese added as medium-carbon ferromanganese, n.d. – not defined

Powders and processing	Chemical composition [%]		Sint. density [g cm^{-3}]	Cross-section hardness	Max. static tooth stress [MPa]
	Mn	C			
1. Fe(PFeR)–1.5% Cu	n. d.	nd	6.8	95 HB	528
1. NC100.24	2.8	0.63	6.8		1002
1. ASC100.29	2.8	0.61	6.9		774
2. NC100.24	2.8	0.62	6.8	164–180 HV 10	997
2. ASC100.29	2.8	0.65	6.9		761
2. NC100.24	2.8	0.65	6.8		1015
3. ASC100.29	2.8	0.65	6.9		780
Fe–3Mn–0.8C 1. Fe(PFeR)	n.d.	n.d.	6.8	168 HB	783

placed in a semi-closed container with a getter powder (alumina with 12% ferromanganese powder).

The static teeth bending test, because of its severity, was particularly useful in discriminating the difference between material and heat treatments. The consistency and level of static strength in PM gears are listed in Tables 10.36 and 10.37.

The standardised materials tests gave parameter values, which were strictly valid for practical application, only the sintered component in question was stressed in use in the same way as the test specimens.

Sinter hardened Mn steels with higher strength lost their sizeability so that the dimensional precision can only be achieved with an additional mechanical finishing step.

Processing Mn steels in flowing hydrogen–rich atmospheres required a dew point –55°C at 1120°C or attainment of a reducing microclimate. This was achieved through the use of semi-closed containers, which ensured the presence a cloud of manganese vapour. The furnace atmosphere of nitrogen with a dew point of –55°C proved successful as sinter hardening in an atmosphere of equally dry hydrogen.

Table 10.37. Ritzel gear. Results of the Weibull statistical data reduction for the tooth bending test results

Powders and processing	Sintering	Weibull modulus m	Characteristic static tooth root stress [MPa]
1. Fe(PFeR)– 1.5% Cu (reference material)	1150°C, dissociated ammonia, –30°C dew point	11.4	551
Fe–3% Mn–0.8% C (nominal composition)			
1. NC100.24	1150°C, dissociated ammonia, –30°C dew point, protective powder and semi–closed container	23.1	1024
1. ASC100.29		21.5	792
2. NC100.24	1120°C, hydrogen, –60°C dew point, sinter hardened and tempered	25.9	1018
2 ASC100.29		16.4	788
2. NC100.24	1120°C, nitrogen, –60°C dew point, sinter hardened and tempered	28.8	1033
3. ASC100.29		19.2	804
Fe–3Mn–0.8C 1. Fe(PFeR)–	1150°C, dissociated ammonia, –30°C dew point	20.4	804

It was increasingly believed that the properties of PM steels processed from the atomised powders were superior to those from the identically processed sponge iron-based powders. The mechanical properties of sponge-based gears were somewhat superior.

In the present study, a new technique for sintering of manganese steel was developed, and an extended characterisation of Fe–3% Mn–(0.6–0.8)% C alloy was presented. Fe–3% Mn–0.8% C steel gears had strength far superior to Fe(PFeR)–1.5% Cu gears, and had contributed greatly to the expansion of the use.

10.7.3. Mechanical properties of sintered manganese steel gears

[257] studied the production of Celma and Ritzel gears under the laboratory as well under industrial conditions. Several batches of manganese steel gears were studied together with a composition based

Fig. 10.36 (left). Celma hand-held powered drilling machine spur gear.
Fig. 10.37 (right). Illustration of the tooth bending strength (crushing load) test [257].

on the industrial powder DC–1 (Fe–2Ni–1.5Mo) and water-atomised iron powder Fe(PFeR) admixed to 1.5% Cu.

Prototypes manufactured were of a spur pinion gear for a concrete mixer, referred to as the Ritzel gear, and of a spur gear for a drilling machine, referred to as the Celma gear. The Ritzel gear caries 14 teeth with a 28.0 mm standard pitch diameter. The Celma gear carries 39 teeth with a 39.0 mm standard pitch diameter. The Celma gear blank width was 9 mm and the outside diameter was 41.3 mm; for the Ritzel gear blank width was 17 mm and the outside diameter was 34.0 mm. Internal configurations were formed simultaneously with the gear profile, eliminating subsequent machining operations.

Manganese was introduced in the system as medium-carbon ferromanganese with the particle size below 20 μm. Carbon was introduced as Trimrex F10 graphite.

Sintering was carried out under industrial conditions at 1150°C for 55 to 65 min in a cracked ammonia atmosphere at a cooling rate of 10°C/min in the water-cooled zone, both with and without protective powder or laboratory sintering in H_2/N_2 at 1120°C for 60 min in a semi-closed container. The container did not have any ferromanganese added to it, but was loaded to capacity ~20 gears, with the green compacts being the carriers of Mn and C. This arrangement provided sufficient manganese vapour to both improve the local dew point (self–gettering) and minimise the loss of manganese from the compacts. The furnace atmosphere was of a technical purity dissociated ammonia atmosphere with an inlet dew point no worse than –30°C. The green compacts under industrial conditions were sintered in semi-closed, double-box containers with a labyrinth seal.

After sintering, the gears had a final composition Fe–3Mn–0.6C. Average sintered densities of the gears varied from 6.72 to

Table 10.38. Characteristic data of Celma Fe–3Mn–0.8C and DC–1+0.5% C (reference material) sintered gears. Manganese added as medium-carbon ferromanganese

Batch number, powders and processing	Chemical composition [%]			Sintered density	Cross-section hardness	Max. static tooth root stress
	Mn	C	O	[g cm^{-3}]	[HV10]	[MPa]
1. DC–1 +0.5% C, sintered and heat-treated (reference material)	–	0.42	0.296	6.81	266	1017
2. NC100.24, ind. sintering	2.7	0.620	0.315	6.80	171	726
3. NC100.24, ind. Sintering under protective powder	2.7	0.619	0.337	6.76	175	875
4. NC100.24 ind. Sintering semi-closed container	2.8	0.630	0.216	6.78	178	913
5. NC100.24, ind. Sintering, protective powder and semi-closed container	2.8	0.631	0.227	6. 77	164	988
6. ASC100.29, ind. sintering, protective powder and semi-closed container	2.8	0.625	0.260	6.72	150	854
7. NC100.24, lab. sintering in H_2 in semi-closed container	2.8	0.640	0.316	6.89	280	986
8. ASC100.29, lab. sintering in H_2 semi–closed container	2.8	0.635	0.301	6.86	277	860
9. NC100.24, lab. sintered in N_2 in semi–closed container	2.8	0.712	0.495	6.91	270	918
10. SC100.29, lab. sintering in N_2 in semi-closed container	2.8	0.704	0.465	6.90	279	706

6.91 g cm^{-3}. The sintered (and optionally sized) gears were hardened (rapidly cooled from 850°C) in a furnace with a protective atmosphere and tempered at 200°C for 1 h. Figure 10.36 shows both gears.

Both Celma and Ritzel gears are shown in Fig. 10.36 and the method of determination of the tooth bending strength is shown in Fig. 10.37. The data for the Celma Fe–3Mn–0.8C and DC–1+0.5% sintered gears are given in Table 10.38.

The strength of a gear wheel is indeed the strength of its weakest tooth; accordingly statistical aspects of gear tooth fracture were analysed using Weibull two-parameter methodology, as there were no indications that more than one failure mechanism was operating. Densities of ~6.7 and ~6.8 g cm^{-3} were recorded for steels emanating from sponge and water-atomised iron powders, respectively.

There was a significant increase in apparent hardness as a result of a relatively rapid cooling at 60°C/min (sinter hardening), compared to furnace cooling. At the surface, HV 10 increased from 120–170 to 160–390 and the corresponding increase in (apparent) cross–section hardness was from 130–160 to 160–290.

The gear teeth were tested to failure; Table 10.38 summarises the data. The scatter of maximum static tooth root stress of nitrogen-sintered gears is at least as good as that of comparable hydrogen- and dissociated ammonia-sintered gears.

The density and hardness of the manganese steel gears were similar to Distaloy DC–1+0.5% C sintered gears. The bending strength of the manganese steel teeth was generally superior to those made of DC–1+0.5% C. The material based on NC100.24 iron powder had a higher strength compared to that based on ASC100.29 powder.

The characteristic static tooth root stresses for the Celma and Ritzel gears made of ASC100.29+FMn powders, identically laboratory sintered, were 879 and 788 MPa, respectively, giving a ratio 1.11 to be compared with the prediction of 1.08 for m of 18. This agreement appears satisfactory, but it should be noted that very small differences in strength were considered and the applicability of this approach needs testing on much larger strength differences.

10.8. Effect of Mn addition on strain micromechanism in as sintered 316L steel

Taking the advantage of the self-cleaning Mn effect, one objective of this study was to analyse the viability of sintering stainless steels in

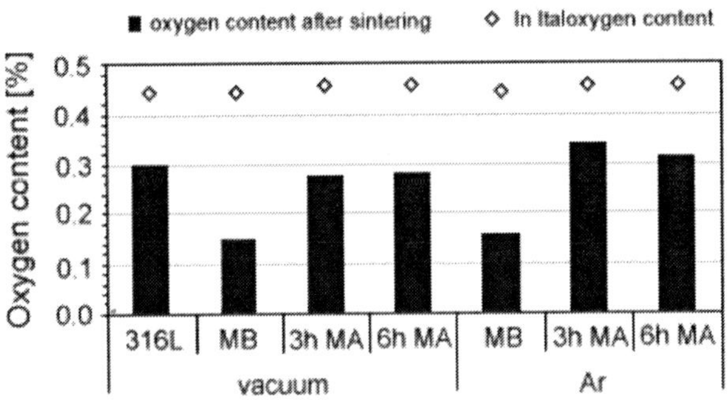

Fig. 10.38. Oxygen content before and after sintering [258].

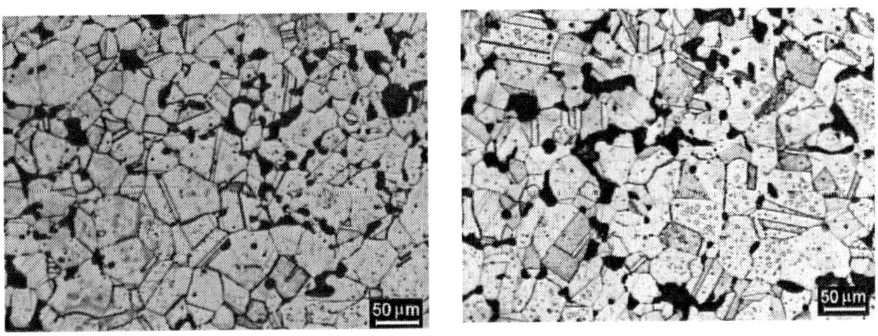

Fig. 10.39. Microstructures of 316+1Mn mixed in as master alloy mechanically alloyed [258].

argon atmospheres of technical purity with addition of Mn powder to the base steel powder, what it would allow carrying out the sintering process in continuous furnaces [258].

The modification of the pore system due to the presence of Mn, introduced a new goal in this work, to determine the effect of pores size and shape on the SCF and how the additions of Mn can modify this mechanical response, by reducing the 'stress raising' effect of the pores.

The starting powders for the study were 316L water-atomised austenitic stainless steel (150 µm) and electrolytic Mn (5 µm) to achieve a final composition of 316+1Mn. Zinc stearate was used as a lubricant (0.1%). With Mn additions by mechanical blending, it was expected to slightly round the pores because of Mn sublimation and condensation on the particle surface. An alternative mixing method was proposed; the idea was to create a master alloy (50% base material – 50% Mn) by mechanical alloying, and to dilute it into the base material 316L by mixing for 20 min. Two milling

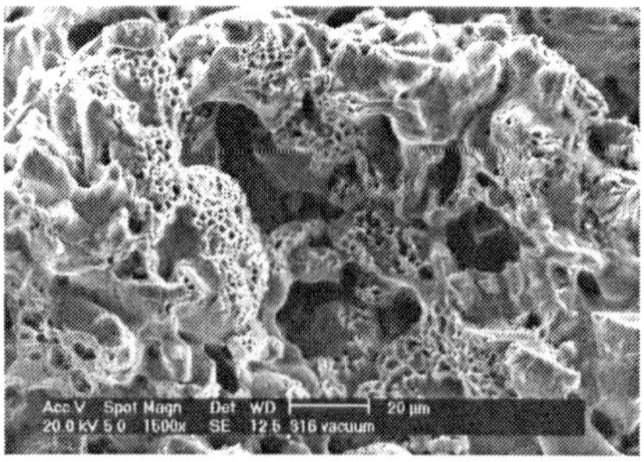

316 sintered in vacuum, ×1500

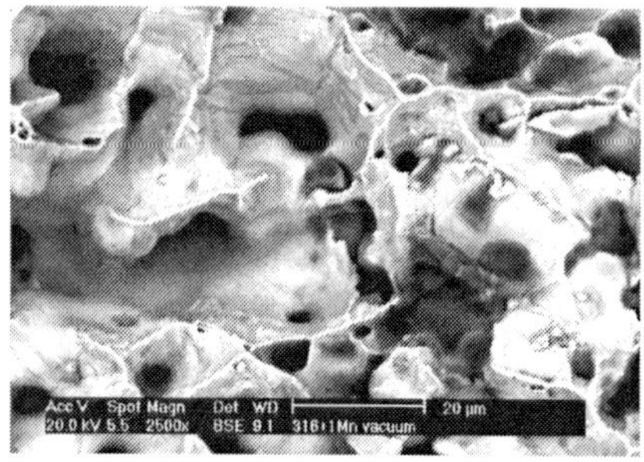

M.B. sintered in vacuum, ×2500

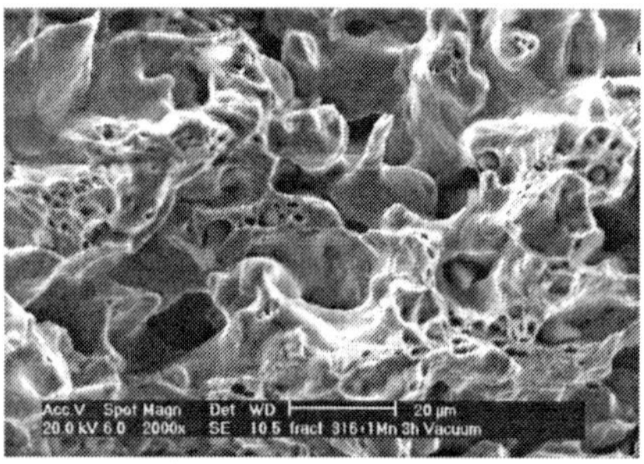

M.B. 3 hours, vacuum, ×2000

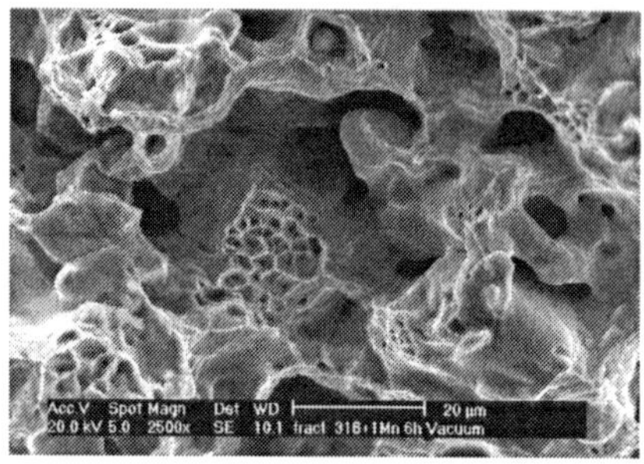

M.A. 6 hours, vacuum, ×2500

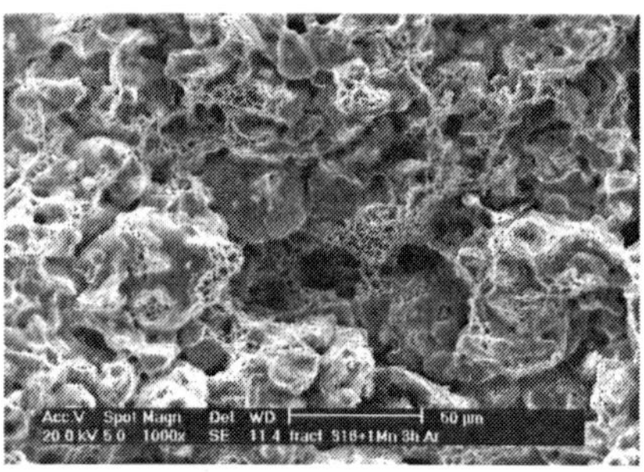

M.A. 3 hours, argon, ×1000

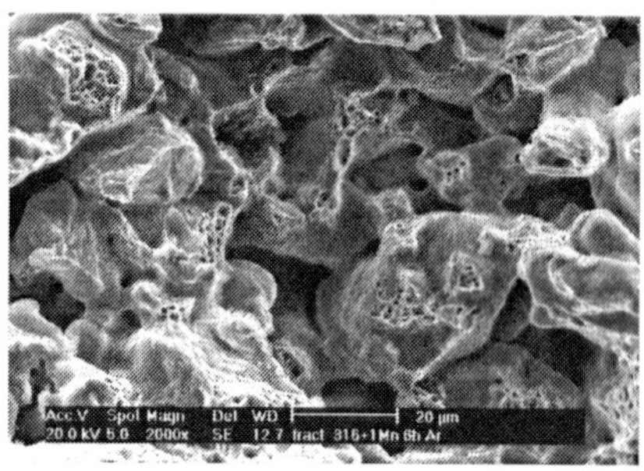

M.A. 6 hours, argon, ×2000

Fig. 10.40. Fracture surfaces by SEM [258].

times were chosen with the aim of having different alloying grades: 3 hours, when Mn is already homogeneously distributed but there is still free Mn, and 6 hours, when Mn is in solid solution and it is not possible to distinguish from base material. Tensile and bending tests specimens were compacted at 700 MPa, and those sintered in vacuum and argon atmospheres at 1250°C for 40 min.

The mechanical properties were analysed by tensile tests. Metallographic study as well as oxygen content variation were considered to better understand the plastic deformation behaviour of the material. This study was completed with the examination of fracture surfaces. All samples were referred to 316L stainless steel.

All materials showed an acceptable oxygen content level after sintering, but the difference between adding Mn by blending (MB) or by milling (MA) was clearly visible, Fig. 10.38. The self–cleaning effect of Mn was higher in the mechanically blended samples.

The analysed microstructure of the sintered parts showed austenite without secondary phases, Fig. 10.39. The grain size of the specimens sintered in argon slightly smaller than for those sintered in vacuum.

Manganese addition modified the pore morphology and hence the function of pores as 'stress raisers'.

Figure 10.40 shows the SEM micrographs of the fracture surface of the different materials. In the reference 316L sintered in vacuum, fracture placed at necks showed certain ductility with microvoids and some deformation of contacts.

The most ductile micromechanism was found for the mechanically blended parts, which showed large deformation, with a high strain degree in the neck. When Mn was added as elemental powder, it was taking better the advantage of Mn sublimation. In the sintered parts, there was found condensed manganese on particle surfaces. The consequences are better contacts between particles and lower oxygen content, affecting both factors controlling the deformability.

When Mn was added as a master alloy, there was no Mn on the particle surfaces, so the inhibition of the sublimation phenomena when Mn is not free is confirmed. Also, plastic deformation is not as considerable as it was in the mechanically blended parts. Anyway, for shorter milling times, when Mn is not alloyed in the 316L particles, the fracture micromechanism appears to be more ductile and strain in necks higher.

Mn additions protect stainless steels from oxidation in argon atmospheres during sintering. The microstructure does not show differences between 316L or 316L containing Mn. With master alloys generated by mechanical alloying, the sintering process is improved because of the high lattice defects concentration and the sintering activation, but the sublimation phenomenon is inhibited and the reduction effect of Mn vapour is lower.

11

Effect of processing conditions and materials on properties of sintered Fe, Cr, Mo, C steels containing manganese

11.1. Effect of various processing condition on properties of Fe–Mn–C steels with addition of molybdenum

11.1.1. Effect of different addition mode of molybdenum on properties of manganese steel

The study [259] investigated possibility to determine the optimum addition of manganese and molybdenum for preparing the combination of compressibility and mechanical properties of sintered steels.

The following raw materials were used: water-atomised WPL–200 iron powder, prealloyed Astaloy 85Mo powder, molybdenum oxide reduced to powder, manganese in the form of medium-carbon ferromanganese <63 μm and water-atomised prealloyed Ultrapac (Ni) powder. Five different compositions of the mixtures based on previously used raw materials were prepared and are shown in Table 11.1. Some samples were heat-treated from the austenitisation temperature of 830°C (hold 20 min) and then quenched in oil.

Table 11.1. Nominal composition of admixed manganese with prealloyed molybdenum powder for Mn–Mo–C steel alloy

Powder mixture code					
Mn0	Mn1	Mn2	Mn3	Mn4	Ult–Ni
Iron powder +0.85Mo, 0.5C	Iron powder +0.85Mo, 0.5C, 1Mn	Iron powder +0.85Mo, 0.5C, 2Mn	Iron powder +0.85Mo, 0.5C, 3Mn	Iron powder +0.85Mo, 0.5C, 3Mn	Ultrapac powder, 1.75Ni, 1.5Cu, 0.5Mo, 0.5C

The effect of manganese addition on the as sintered density and sintered and heat-treated hardness of sintered steels is shown in Fig. 11.1.

The as sintered density decreased with addition of Mn from 0 to 3%. The increase of the hardness of sintered Mn steels up to 140 HV 5 for 2Mn steel was followed by a small decrease for 3Mn steel specimens. The highest hardness of the heat-treated steel specimens was obtained in the 1Mn steel specimens (475 HV 5) and the relatively lowest value of ~380 HV 5 in the 3 Mn steel specimen.

The diffusion interaction between molybdenum in the prealloyed powder and admixed manganese was studied in [260]. The test specimens were prepared from NC100.24 iron powder, Astaloy Mo powder (Fe–1.5% Mo), high-carbon ferromanganese and natural graphite to achieve the 0.8% C content. The compacts were compacted at pressure of 600 MPa and sintered at 1120°C for 60 min in purified hydrogen with dew point –35°C. The studies were carried out with Fe–3% Mn–0.6% C–0.5% Mo, Fe–3% Mn–0.8% C–0.5% Mo, Fe–4% Mn–0.6%–0.5% Mo and Fe–4% Mn–0.8% C–0.5% Mo samples.

The properties determined were density, apparent surface hardness, apparent cross-section hardness, microhardness, UTS, TRS and impact toughness. The results are presented in Table 11.2.

The data listed in Table 11.2 show improvement in the hardness of Mo-alloyed manganese steels and also a dramatic increase in microhardness. The results show that molybdenum increases both the apparent surface and cross section hardness of sintered steels. Manganese and molybdenum in combination have been found to be substantially more effective than manganese individually.

Maximum tensile strength (516 MPa), transverse rupture strength (1137 MPa) and impact toughness (11.5 J) were obtained for Fe–3% Mn–0.8% C steel. Although the Mo-alloyed manganese steels

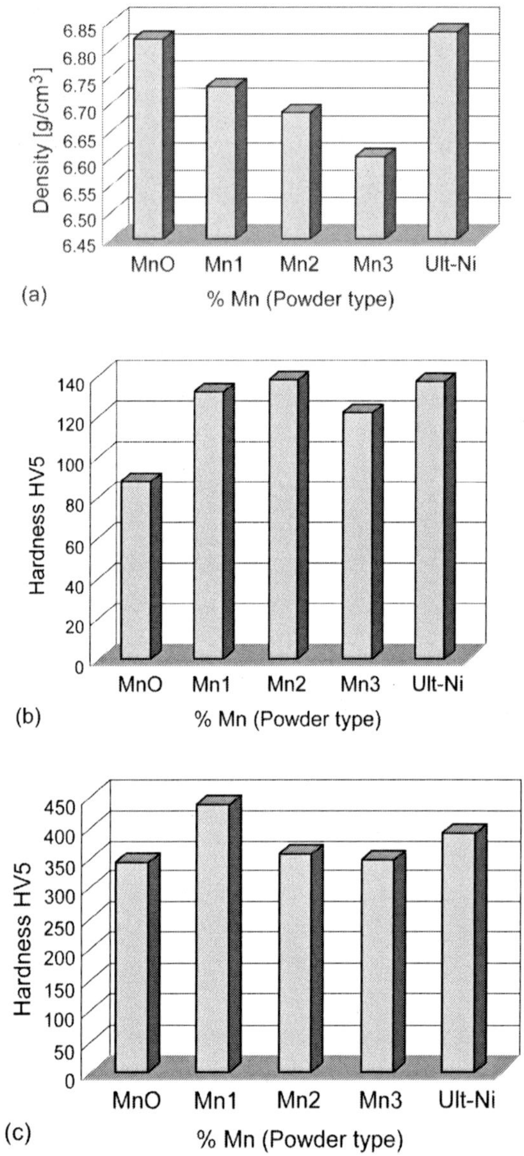

Fig. 11.1. Effect of Mn addition on: (a) density of sintered steels after compaction at 600 MPa, (b) hardness of sintered steel samples, (c) hardness of heat-treated steel samples [259].

showed little change or worse mechanical properties: UTS, TRS and impact strength. The study has paved the way for the development of high-strength Fe–Mn–Mo–C sintered steels, using admixed ferromanganese powder and molybdenum prealloyed powder.

Table 11.2. Density and mechanical properties of sintered Mo–alloyed manganese steels; sintering at 1120°C for 60 min in hydrogen with dew point –35°C

Chemical composition	Density [g cm^{-3}]	Surface hardness HV 10	Cross-section hardness HV 10	Micro-hardness HV 0.02	Tensile strength [MPa]	Transverse rupture strength [MPa]	Impact energy [J]
Fe–3% Mn–0.6% C–0.5% Mo	6.83	217	281	732	462	749	4.3
Fe–3% Mn–0.8% C–0.5% Mo	6.84	273	273	894	374	586	6.2
Fe–4% Mn–0.6% C–0.5% Mo	6.76	262	281	1109	386	534	3.3
Fe–4% Mn–0.8% C–0.5% Mo	6.82	255	177	869	283	544	4.8
Fe–3% Mn–0.8% C	6.99	170	216	987	516	1137	11.5
Fe–4% Mn–0.8% C	6.75	131	230	697	395	674	4.5

[261] investigated the effect of addition of molybdenum by two methods on the properties of sintered Fe–Mn–Mo–C steels. In the first route, Astaloy Mo prealloyed powder was a Mo carrier (A), and in the second route (I) reduced Mo powder was used. Sintering was carried out at 1120°C and 1250°C in pure hydrogen with –60°C dew point (A) and at 1280°C in the $75H_2/25N_2$ atmosphere with –65°C dew point (I).

The starting materials were: sponge iron powder NC100.24, water-atomised powder ABC100.30, Astaloy Mo (Fe–1.5% Mo), Astaloy 0.85Mo (Fe–0.85% Mo), high-carbon ferromanganese HPII <40 μm, medium-carbon ferromanganese (Ell) (77% Mn, 1.3% C, 0.4% Si, 1.5% O) <20 μm, medium-carbon ferromanganese (El2) (87.6% Mn, 1.0% C, 0.46% O) <40 μm and fine graphite. The compositions of the various batches investigated are given in Table 11.3.

The mixtures were compacted at 500 MPa into standard dogbone tensile specimens and into rectangular specimens (55 × 10 × 5 mm). The green densities were in the range 6.5–6.8 g cm^{-3}. Sintering was carried out in a laboratory tube furnace. The specimens were placed

Table 11.3. Composition of the powder mixtures and the sintering temperature

Powder mixture	Nominal composition	Formulation	Sintering temperature [°C]
A 1	Fe–3Mn–0.8C	NC100.24+HPII+C	1120
A 2	Fe–3Mn–0.8C	ABC100.30+HPII+C	1120
A 3	Fe–3Mn–0.8C	NC100.24+Elkem II+C	1120
A 4	Fe–3Mn–0.8C	NC100.24+Elkem II+C	1250
A 5	Fe–3Mn–0.6C–0.5Mo	NC100./24+HPII+Astaloy Mo+C	1120, 1250
A 6	Fe–3Mn–0.8C–0.5Mo	NC100./24+HPII+Astaloy Mo+C	1250
A 7	Fe–4Mn–0.6C–0.5Mo	NC100./24+HPII+Astaloy Mo+C	1120, 1250
A 8	Fe–4Mn–0.8C–0.5Mo	NC100./24+HPII+Astaloy Mo+C	1120, 1250
I N–51	Fe–3Mn–0.6C	NC100.24+El S+C	1280
I N1–51	Fe–3Mn–0.6C–0.85Mo	NC100.24+El S+Mo+C	1220
I N2–51	Fe–3Mn–0.6C–1.5Mo	ASC100.29+El S+Mo+C	1280
I A–51	Fe–3Mn–0.6C	ASC100.29+El S+C	1280
I A1–51	Fe–3Mn–0.6C–0.85Mo	ASC100.29+El S+Mo+C	1280
I A2–51	Fe–3Mn–0.6C–1.5Mo	ASC100.29+Elkem S+Mo+C	1280

Table 11.4. Chemical composition of Fe–3Mn–0.8C (A4) specimens after sintering

Sintering temperature [°C]	Cooling rate [°C/min]	Mn [%]	C [%]	O [%]	Fe
1120	1.5 (furnace cooling)	3.85	0.66	0.20	Bal.
1120	1.5 (furnace cooling)	3.75	0.50	0.13	Bal
1250	60	3.71	0.53	0.16	Bal.

in semi-closed containers. After sintering the specimens were furnace cooled or moved to the water-cooled end of the tube for convective cooling. Some of the convectively (rapidly) cooled (sinter hardening) alloys were subsequently tempered in hydrogen at 200°C for 60 min. The chemical analysis of the sintered Mn steels showed some decarburisation of the steels, Table 11.4.

The mechanical properties of the specimens (A) cooled at 40°C/min are summarised in Table 11 5.

Table 11.6 lists the basic characteristics of sintered steels cooled at 7°C/min.

Table 11.5. Properties data for Mn steel specimens as sintered and cooled at 40°C/min (A)

Powder mixture	Sintering temperature [°C]	Sintered density [g cm^{-3}]	Apparent hardness HV 30	Tensile strength [MPa]	Elongation [%]
A 1: Fe–3Mn–0.8C	1120	6.8	217	516	0.2
A 2: Fe–3Mn–0.8C	1120	6.8	214	491	0.2
A 3: Fe–3Mn–0.8C	1120	6.7	276	539	0.3
A4: Fe–3Mn–0.8C	1250	6.8	239	614	0.5
A 5: Fe–3Mn–0.6C–0.5Mo	1120	6.7	273	460	0.3
A 5: Fe–3Mn–0.6C–0.5Mo	1120	6.8	315	596	0.5
A 6: Fe–3Mn–0.8C–0.5Mo	1120	6.7	281	451	0.2
A 6: Fe–3Mn–0.8C–0.5Mo	1250	6.8	324	531	0.4
A 7:Fe–4Mn–0.6C–0.5Mo	1120	6.7	262	383	0.3
A 7: Fe–4Mn–0.6C–0.5Mo	1120	6.8	316	416	0.6
A 8: Fe–4Mn–0.8C–0.5Mo	1120	6.7	255	283	0.2
A 8: Fe–4Mn–0.8C–0.5Mo	1250	6.8	314	319	0.4

Table 11.6. Properties data for Mn steel specimens as sintered and cooled at 7°C/min (I); all specimens sintered at 1280°C

Powder mixture	Sintered density [g cm^{-3}]	Apparent hardness HV 30	Tensile strength [MPa]	Elongation [%]
I N–51:Fe–3Mn–0.6C	6.8	169	360	0.5
I N1–51: Fe–3Mn–0.6C–0.85Mo	6.5	175	460	0.7
I N2–51: Fe–3Mn–0.6C–1.5Mo	6.6	189	520	0.8
I A–51: Fe–3Mn–0.6C	6.7	171	340	0.5
I A1–51: Fe–3mn–0.6C–0.85Mo	6.7	184	430	0.5
I A2–51: Fe–3Mn–0.6C–1.5Mo	6.6	205	590	0.5

The transverse rupture (TRS) strength and impact energy values for the specimens tempered at 200, 300 and 400°C are given in Table 11.7.

It follows from the data that green densities of ~6.6 g cm^{-3} after compaction at 500 MPa are inadequate if high strengths are to be obtained. Increasing the sintering temperature to 1250 or 1280°C (not necessarily desirable) has a little significant effect at this low

Table 11.7. Effect of sintering and tempering temperature on mechanical properties of Fe–3Mn–0.5Mo–0.6C steel cooled at 65°C/min and tempered for 60 min

Powder mixture	Sintering temper–ature [°C]	Hardness HV 30	Tempering temper– ature [°C]	Tempered hardness HV 30	TRS [MPa]	Impact energy [J]
A 6: Fe–3Mn–0.8C–0.5Mo	1120	291	200	256	930	3.8
A 6: Fe–3Mn–0.8C–0.5Mo	1120	304	300	236	900	–
A 6: Fe–3Mn–0.8C–0.5Mo	1120	285	400	252	800	5.4
A 6: Fe–3Mn–0.8C–0.5Mo	1250	315	200	255	1180	5.1
A 6: Fe–3Mn–0.8C–0.5Mo	1250	319	300	253	1230	–
A 6: Fe–3Mn–0.8C–0.5Mo	1250	318	400	247	960	5.5
A 8: Fe–4Mn–0.8C–0.5Mo	1120	370	200	268	–	–
A 8: Fe–4Mn–0.8C–0.5Mo	1120	312	300	242	–	–
A 8: Fe–4Mn–0.8C–0.5Mo	1120	324	400	254	790	4.5
A 8: Fe–4Mn–0.8C–0.5Mo	1250	352	200	297	–	–
A 8: Fe–4Mn–0.8C–0.5Mo	1250	368	300	273	–	–
A 8: Fe–4Mn–0.8C–0.5Mo	1250	387	400	305	890	3.1

density. The dew point requirements become less demanding with increasing temperature.

To utilise the potential of alloying elements, higher sintering temperatures, around 1250°C, are required. Furnace cooling produced significant ductility at the expense of hardness and convective cooling resulted in much higher hardness. The mechanical properties (hardness, transverse rupture strength) show the positive effect of Mo addition to Fe–3Mn–C steel.

The effect of 0.5% Mo addition on the processing, and strength of PM Fe–3.5Mn–0.7C steel was investigated in [262]. The following starting materials were used for the preparation of the test compacts:

- water-atomised iron powder ASC100.29,
- sponge iron powder NC100.24,
- Astaloy Mo (Fe–1.5% Mo) powder,
- high-carbon ferromanganese (HP) <40 μm,
- medium-carbon ferromanganese (EI) <40 μm,
- high-purity graphite.

Powder mixtures were based on the reference material Fe–3.5% Mn–0.7% C and the standard addition of 0.5% Mo,

Table 11.8. Compositional and sintering parameters of alloys formulated by addition of ferromanganese and graphite to plain iron and Astaloy Mo powder compacted at 600 MPa

Alloy	Nominal composition	Formulation	Sintering atmosphere/ dew point	Sintering temperature [°C]	Sintered density [g cm^{-3}]
A	Fe–3.5Mn–0.7C	ASC100.29+E+C+liquid paraffin	$75H_2/25N_2$ –60°C	1140	7.0
B	Fe–3Mn–0.5Mo–0.6C	NC100.24+Astaloy Mo+HP+C	H_2 –35°C	1120	6.9
C	Fe–3Mn–0.5Mo–0.8C	NC100.24+Astaloy Mo+HP+C	H_2 –35°C	1120	6.8
D	Fe–4Mn–0.5Mo–0.6C	NC100.24+Astaloy Mo+HP+C	H_2 –35°C	1120	6.7
E	Fe–4Mn–0.5Mo–0.8C	NC100.24+Astaloy Mo+HP+C	H_2 –35°C	1120	6.7
F	Fe–3Mn–0.5Mo–0.8C	NC100.24 + Astaloy Mo + HP+C	$75H_2/25N_2$ –60°C	1120	6.8
G	Fe–3Mn–0.5Mo–0.8C	NC100.24+Astaloy Mo+HP+C	$75H_2/25N_2$ –60°C	1150	6.6
H	Fe–3Mn–0.5Mo–0.8C	NC100.24+Astaloy Mo+HP+C	$75H_2/25N_2$ –60°C	1180	7.1
I	Fe–3Mn–0.5Mo–0.7C	ASC100.29+EL+C+liquid paraffin	$75H_2/25N_2$ –60°C	1180	7.1

Note: to the alloys B – H zinc stearate was added as lubricant, liquid paraffin was used as binder

accomplished by mixing Astaloy Mo with iron powder. Mn was varied between 3 and 4% and C between 0.6 and 0.8%.

Standard tensile bars, squared samples of size 15 × 15 × 4 mm and rectangular test specimens 55 × 11 × 5 mm, were compacted at pressures of 300, 400, 500, 550, and, generally of 600 MPa. Some samples were sintered in semi-closed containers containing a mixed getter powder. Sintering was carried out in a laboratory tube furnace at 1120, 1140, 1150 and 1180°C, generally for 1 h in hydrogen with the dew point of –62°C. Cooling rates of ~4, 20, and 35 K/min were employed. The compositions of the various batches investigated are given in Table 11.8.

The strength data for the various compositions and processing routes are summarised in Table 11.9.

Table 11.9. Transverse rupture and tensile strength of alloys characterised in Table 11.8

Alloy	Nominal composition	Transverse rupture strength [MPa]	Tensile strength [MPa]
A	Fe–3.5Mn–0.7C	860	420
B	Fe–3Mn–0.5Mo–0.6C	750	460
C	Fe–3Mn–0.5Mo–0.8C	590	320
D	Fe–4Mn–0.5Mo–0.6C	530	310
E	Fe–4Mn–0.5Mo–0.8C	540	280
F	Fe–3Mn–0.5Mo–0.8C	670	–
G	Fe–3Mn–0.5Mo–0.8C	650	–
H	Fe–3Mn–0.5Mo–0.8C	690	–
I	Fe–3Mn–0.5Mo–0.7C	1130	530

Note: Failure strain, %, for alloys A and I was 1.5–2.5 and for other alloys >1

The highest tensile and transverse rupture strength were obtained for the steel I with a paraffin binder sintered at 1180°C.

As follows from the presented data, when processing at 1140°C in an atmosphere with a dew point –60°C, it would be possible to obtain acceptable microstructures and good tensile strength of 500 MPa with 2–3% ductility of the sintered 3Mn–0.5Mo–C steels. The effect of Mo was beneficial.

11.1.2. Effect of cooling rates on properties of Fe–Mn–Mo–C steels

The following studies deal with the effects of cooling rates on the microstructure and mechanical properties of Fe–3Mn–0.8C steel (green compact composition) and concurrently starts to explore the effect of tempering and of incorporated 0.5% Mo. Even at low cooling rates such Mo-containing steels frequently contain low temperature transformation products associated with residual stresses. As Mn promotes temper embrittlement at ~300°C, tempering and stress relieving were also studied [263,264].

The following materials were used for the preparation of test specimens: NC100.24 iron powder, Astaloy Mo (Fe–1.5% Mo), medium-carbon ferromanganese separated to <20 μm, ultrafine graphite powder and zinc stearate as a lubricant. The mixtures were compacted into tensile specimens and rectangular specimens (55 × 10 × 5 mm) at pressures between 600 and 700 MPa to give green

Table 11.10. Effect of cooling rate on the mechanical properties of Fe–3Mn–0.5Mo–0.5C steel specimens sintered at 1120°C

Cooling rate [°C/min]	0.2% offset strength [MPa]	Tensile strength [MPa]	Elongation [%]	TRS [MPa]	Cross-section hardness HV 30	Impact energy [J]
1.3	315	447	1.8	1050	169	16.5
4.8	330	478	1.0	1270	224	17.2
8.7	455	506	1.0	1230	217	10.8
15.5	450	568	0.6	1200	248	11.0
30.1	485	557	0.6	1120	272	8.2
47.0	510	559	0.5	1100	276	5.8
55.0	–	535	0.2	1000	292	5.2
63.1	–	498	0.2	1080	289	8.9
64.0	–	486	0.1	1000	318	5.9
66.6	–	475	0.1	1070	301	4.7

Table 11.11. Effect of cooling rate on the mechanical properties of Fe–3Mn–0.5C steel specimens sintered at 1250°C

Cooling rate [°C/min]	0.2% offset strength [MPa]	UTS [MPa]	Elongation [%]	TRS [MPa]	Cross-section hardness HV 30	Impact energy [J]
1.4	385	587	3.1	1240	166	21.3
4.7	440	595	1.7	1330	254	11.5
9.4	410	614	1.25	1310	239	9.9
20.4	325	583	1.2	1210	265	7.7
31.8	570	586	0.3	1190	282	9.4
41.1	–	521	0.2	1080	328	4.6
57.0	–	491	0.1	1090	335	4.9
60.0	–	507	0.1	990	276	4.9

densities of ~7.0 g cm^{-3}. Sintering was carried out in hydrogen with dew point –60°C for 60 min in a laboratory tube furnace at 1120°C or 1250°C. In order to reduce manganese sublimation and oxidation, the specimens were placed in semi-closed containers. The specimens were furnace cooled or moved to the water-cooled end of the tube for convective cooling. A range of the cooling rates (determined between 800 and 300°C), approximately 5, 10, 40 and 65°C/min, was investigated. The actual carbon contents of specimens sintered

Table 11.12. Properties of sintered and sinter hardened Fe–3Mn–(0.5,0.6)C steel specimens in dependence on tempering temperature

Sintering temperature [°C]	Tempering temperature [°C]	0.2% offset strength [MPa]	UTS [MPa]	Elongation [%]	TRS [MPa]	Impact energy [J]
1120	None	–	470	0.1	1070	4.7
	200	410	740	1.9	1480	12.0
	250	440	700	1.0	1400	8.0
	300	430	740	1.9	1510	10.4
	350	450	700	0.9	1420	8.9
	400	460	740	1.9	1460	11.5
	450	440	680	1.7	1550	10.9
1250	None	–	460	0.1	1060	5.1
	200	470	830	2.3	1480	10.1
	250	500	790	0.9	1370	4.1
	300	470	800	1.4	1450	4.9
	350	520	840	1.0	1530	8.8
	400	500	850	2.0	1750	15.7
	450	500	700	1.2	1740	15.3

Table 11.13. Effect of the sintering temperature of 1120 and 1250°C on mechanical properties of sinter hardened Fe–3Mn–0.5Mo–(0.5,0.6C) steel in dependence on tempering temperatures. n.d. – not determined

Property	Sintering temperature 1120°C	Plus stress relief 1 h at 200°C	Plus stress relief 1 h at 300°C	Plus stress relief 1 h at 400°C	Sintering temperature 1250°C	Plus stress relief 1 h at 200°C	Plus stress relief 1 h at 300°C	Plus stress relief 1 h at 400°C
TRS [MPa]	750	930	960	800	1110	1180	1230	960
Impact energy [J]	3.3	3.6	n.d.	5.4	4.3	5.1	n.d.	5.5
Tensile strength [MPa]	320	450	n.d.	n.d.	460	570	n.d.	n.d.
Elongation [%]	0.5	2.0	n.d.	n.d.	0.7	2.0	n.d.	n.d.

at 1120°C and 1250°C were ~0.6 and 0.5%, respectively. Table 11.10 lists the mechanical properties of Fe–3Mn–0.5Mo–(0.5,0.6) C specimens sintered at 1120°C, Table 11.11 – sintered at 1250°C,

Table 11.12 – sinter and sinter hardened, and Table 11.13 – sinter hardened in dependence on tempering temperature.

According to the results, the higher sintering temperature also resulted in more rounded porosity and in a slight improvement in the mechanical properties, Tables 11.10 and 11.11. Slow cooling rates favoured plasticity, which ceased to be significant for cooling rates of ~40°C/min. Specimens cooled at ~60°C/min were also tempered at the temperatures of 200 to 450°C, Table 11.12. Significant improvement in the properties was attained as a result of the 200°C stress relief, *i.e.* 830 MPa tensile strength with >2% ductility for the 1250°C sintering temperature. The results were similar for the 400°C temper and there was an indication of temper embrittlement at 300°C, more severe for the specimens sintered at 1250°C. Preliminary studies on the effect of alloy modification by 0.5% Mo, Table 11.13, did not indicate any improvement in the mechanical properties resulting from only this addition. The best combination of the mechanical properties, attained at the lowest cooling rates, Table 11.10, can be exceeded with the fast cooling and tempering, Table 11.12.

11.1.3. Dimensional changes of the compacts based on plain iron powders with addition of prealloyed molybdenum powders and of manganese tested by dilatometry

The task of the present work was to consider both plain iron powders, in different ratios, in order to control and reduce dimensional changes occurring during sintering of manganese steels also with molybdenum addition in dilatometry tests [265].

The following raw materials were used for the preparation of the test samples:

- sponge iron powder NC100.24,
- atomised iron powder ASC100.29,
- prealloyed Aastaloy 85Mo (0.85% Mo) powder,
- prealloyed Astaloy Mo (1.5% Mo),
- medium-carbon ferromanganese, fraction <40 μm,
- fine graphite powder C-UF and Kenolube powder as lubricant.

Three series by 5 powders mixtures were prepared by powder mixing:

I. Fe + 3% Mn + 0.6% C + 0.6% Kenolube, where Fe indicates:

  A. 100% NC100.24,
  B. 75% NC100.24 + 25% ASC100.29,
  C. 50% NC100.24 + 50% ASC100.29,

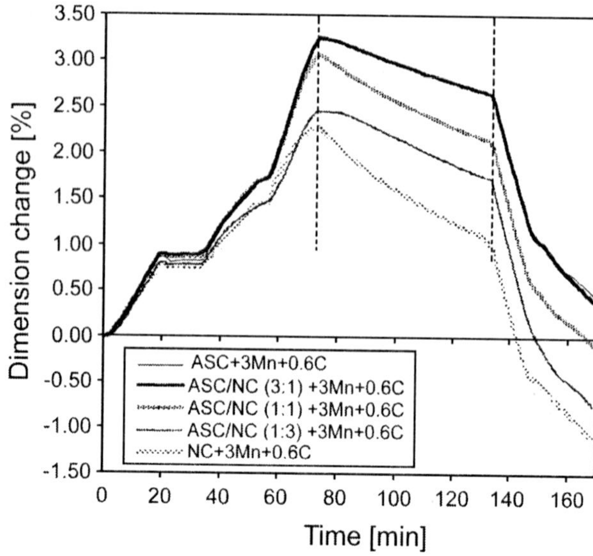

Fig. 11.2. Dilatomertric curves for Fe+3Mn+0.6C materials isothermally sintered at 1150°C [265].

D. 25% NC100.24 + 75% ASC100.29,

E. 100% ASC100.29,

II. Fe + 3% Mn + 0.6% C + 0.85% Mo + 0.6% Kenolube, and

III. Fe + 3% Mn + 0.6% C + 1.5% Mo + 0.6% Kenolube.

The mixtures were pressed into dogbone specimens for mechanical tests, and rectangular compacts of size 4 × 4 × 15 mm for compressibility and dilatometry investigation at a pressing pressure of 500 and 600 MPa. Specimens for mechanical tests were sintered in a laboratory tube furnace in semi-closed containers. The sintering behaviour of the investigated powder mixtures was controlled using a push rod dilatometer. Pure dry hydrogen and nitrogen with the dew points below –70°C were used to prepare gas mixtures containing 75% hydrogen +25% nitrogen as the sintering atmosphere. Isothermal sintering temperature was 1150°C or 1280°C for 1 h. The changes in the slope of dilatometric traces were examined especially for (NC100.24 + ASC100.29 + 3% Mn + 0.6% C powder mixtures) (I. series).

The whole dilatometric traces associated with sintering of investigated materials are shown in Figs. 11.2–11.5.

The more NC100.24 powder was present, the higher shrinkage was observed. The close to zero dimensional changes were obtained

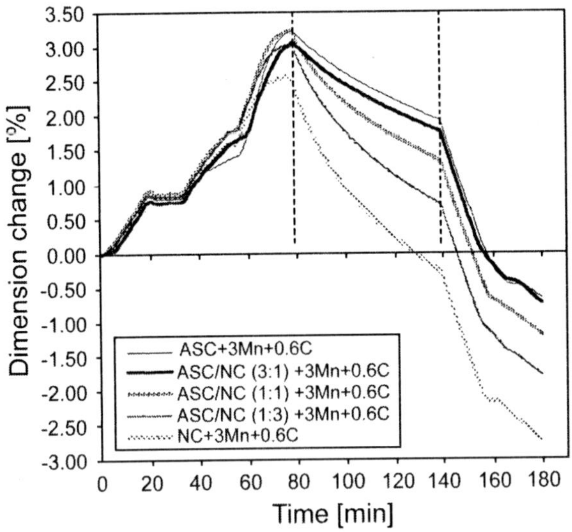

Fig.11.3. Dilatomertric curves for Fe+3Mn+0.6C materials isothermally sintered at 1280°C [265].

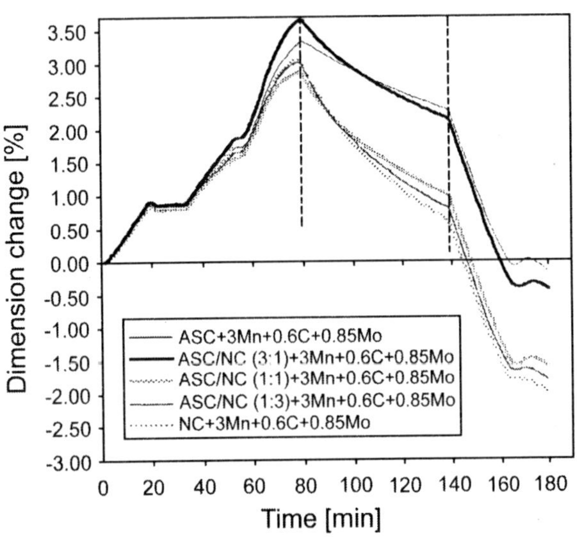

Fig. 11.4. Dilatomertric curves for Fe+3Mn+0.85Mo+ 0.6C materials isothermally sintered at 1280°C [265].

for materials based on ASC100.29 or ASC100.29/NC100.24 = 3 : 1 plain iron powder mixture and sintered at 1150°C.

The molybdenum addition to Fe+3Mn+0.6C powder mixtures modified the sintering behaviour. The addition of 0.85% Mo

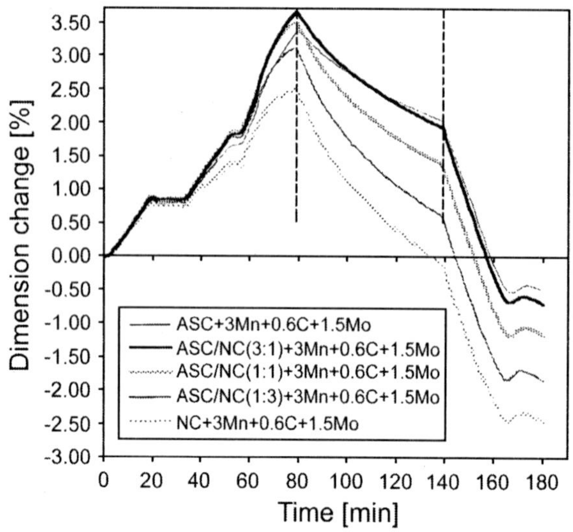

Fig. 11.5. Dilatomertric curves for Fe+3Mn+1.5Mo+0.6C materials isothermally sintered at 1280°C [265].

Table 11.14. Density and mechanical properties of investigated materials

No.	Material	Density [g cm^{-3}]	Tensile strength [MPa]	Hardness HV 10
1	ASC100.29+3% Mn+0.6% C	6.69	404	138
2	ASC100.29+3% Mn+0.6% C+0.85% Mo	6.75	545–632	258
3	ASC100.29+3% Mn+0.6% C+1.5% Mo	6.74	555–750	260
4	NC100.24+3% Mn+0.6% C	6.68	450–497	145
5	NC100.24+3% Mn+0.6% C+0.85% Mo	6.54	725–778	253
6	NC100.24+3% Mn+0.6% C+1.5% Mo	6.69	737–749	265

suspended shrinkage, while the effect of 1.5% Mo addition on dimensional changes may be neglected.

The sintered density and mechanical properties values are given in Table 11.14. The specimens made from NC100.24 powder show higher tensile strength and hardness compared with the compacts based on ASC100.29 powder.

These results will contribute to the commercial objective of obtaining consistent dimensional changes after sintering of mixtures based on NC100.24 and ASC100.29 plain iron powders with the addition of ferromanganese and molybdenum.

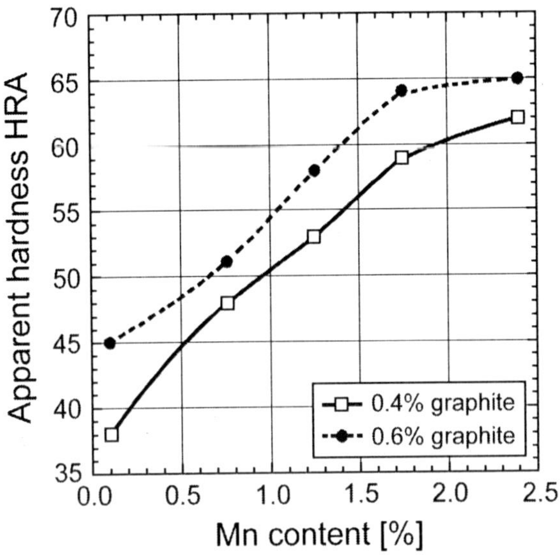

Fig. 11.6. Effect of Mn and graphite content on apparent hardness [266].

11.1.4. Properties of diffusion-alloyed steels affected by manganese

The objective of the current study was to determine if Mn-containing PM steels could be produced and sintered using conventional practices [266].

The developed manganese alloys were produced in the industrial pilot plant for this study. It was found that the presence of Mo in Mn-containing PM steels greatly improves the alloy properties and therefore, the Mo content was targeted at 0.5%. The Mn content of the alloys was varied from 0 to 2.5%. Graphite levels of 0.4 to 0.9% were studied and 0.75% lubricant was added to all mixes. Diffusion alloys FD-0205 and FD-0405 were also tested with 0.5% graphite as a comparison with the new developed alloys.

Standard transverse strength and tensile specimens were pressed at 415, 550 and 690 MPa and sintered in a 90% nitrogen/10% hydrogen atmosphere in a continuous belt furnace. The primary temperature and time was 1120°C for 15 min, the experiments were also conducted at lower time (5 min) at temperatures of 1065°C and 1095°C. The cooling rate in the samples measured between 650°C and 315°C was either 0.7°C/s or 1.6°C/s. All samples were tempered at 205°C for 60 min.

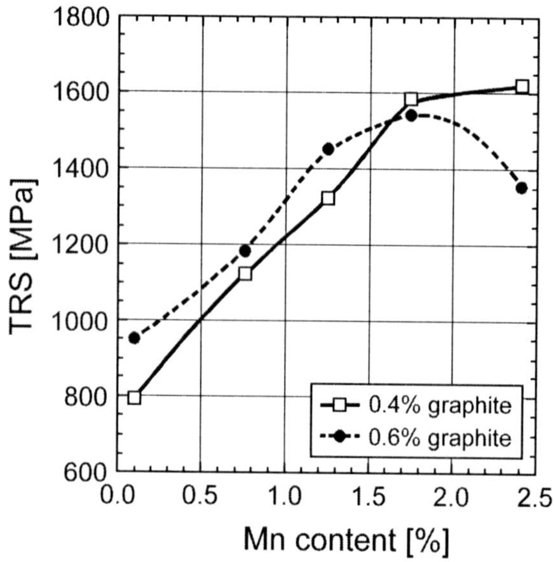

Fig. 11.7. Effect of Mn and graphite content on TRS [266].

Table 11.15. Mechanical properties of three alloys made with 0.6% graphite, compacted at 690 MPa and sintered at 1120°C for 15 min

Alloy	Sintered density [g cm^{-3}]	Dimensional change [%]	Transverse rupture strength [MPa]	Hardness HRA	Yield strength [MPa]	Tensile strength [MPa]	Elongation [%]
MnMo	7.02	0.32	1407	58	544	757	2.1
FD-0205	7.07	0.31	1262	52	437	660	3.2
FD-0405	7.12	0.13	1462	56	478	781	3.3

The increase in the Mn addition led to a positive increase in the mechanical properties. In Fig. 11.6 it can be seen that the hardness increases with Mn content up to 1.8% Mn and begins again to level off at higher levels with 0.6% graphite addition at a cooling rate of 0.7°C/s. In Fig. 11.7 a similar trend is apparent for transverse rupture strength. The TRS increases to 1.8% Mn and then declines with higher Mn contents.

Table 11.15 contains the mechanical property data for the developed Mn–Mo alloy and the diffusion alloys compacted at 690 MPa. The hardness and yield strength of the developmental alloy is superior to both diffusion alloys. The lower ductility with the Mn–Mo alloy results in a similar UTS to FD-0405. The trends

Table 11.16. Mechanical properties of three alloys made with 0.6% graphite, compacted at 690 MPa and sintered at 1120°C for 5 min

Alloy	Sintered density [g cm^{-3}]	Dimensional change [%]	TRS [MPa]	Hardness HRA	Yield strength [MPa]	Tensile strength [MPa]	Elongation [%]
Mn–Mo	7.02	+0.39	1324	58	537	723	1.7
FD-0205	7.06	+0.40	1214	51	412	620	2.8
FD-0405	7.10	+0.22	1331	53	425	675	2.8

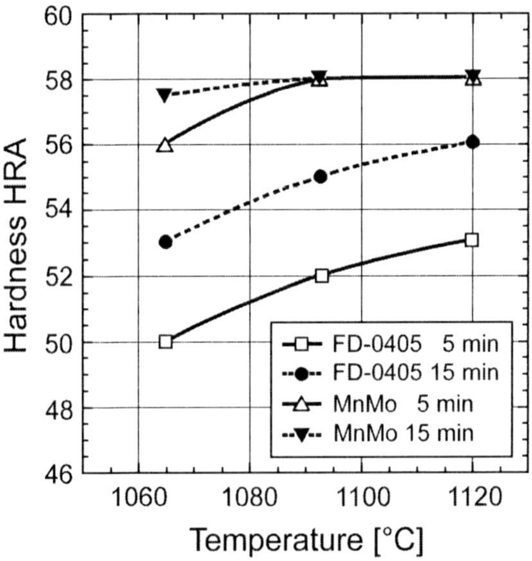

Fig. 11.8. The effect of sintering time and temperature on apparent hardness [266].

in the mechanical properties continue at a short sintering time of 5 min, indicating good sinterability, as seen in Table 11.16.

Figures 11.8 and 11.9 show the effect of sintering temperature and time on the hardness and TRS of the Mn–Mo and FD-0405 alloys, respectively. The hardness of FD-405 alloy is quite low at the lower temperatures. The Mn–Mo alloy has a much more robust hardness response with the sintering conditions. Both alloys show similar TRS behaviour with time and temperature.

The development of Mn-containing steels appears feasible in conjunction with Mo-containing steels. The hardness values are comparable with diffusion alloys FD-0205 and FD-0405. The developmental Mn–Mo alloy containing 1.35% Mn had good

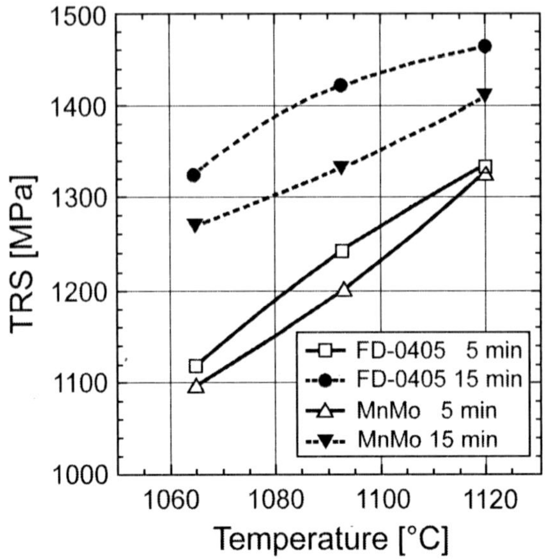

Fig. 11.9. The effect of sintering time and temperature on TRS [266].

Table 11.17. Nominal composition of the base alloys studied

Base alloy	Fe [%]	Mo [%]	Ni [%]	Cu [%]	Mn [%]
Alloy 1	bal.	0.5*	–	–	1.3
Alloy 2	bal.	0.8*	–	–	1.3
FD-0205	bal.	0.5**	1.8*	1.5**	0.1*
FD-0405	bal.	0.5**	4.0**	1.5**	0.1*
FLD-49DH'	bal.	1.5*	–	2.0**	0.1*

*prealloyed, **diffusion-alloyed, 'not official MPIF designation

sinterability under conventional sintering conditions. Extended sintering times and high temperatures are not required. The strength and hardness of Mn–Mo steels is similar to that of FD-0405 and superior to FD-0405. This lean alloy system is appropriate for sinter hardening of small and medium sized parts.

The study of [267] was built on some earlier research that showed that Mn steels can be produced and sintered using conventional practices. A proprietary additive is used as the Mn source and when added to Mo-based steels, shows promising results. The combination of Mo and Mn provides good nominal hardenability. This study discussed the properties of Mn steels and

Table 11.18. Mechanical properties data for a 7.0 g cm^{-3} sintered density (interpolated) of studied alloys. (DC – dimensional change)

Base alloy	Cooling rate [°C/s]	DC [%]	Hardness HRA	0.2% YS [MPa]	UTS [MPa]	Elongation [%]	Impact strength [J]
Alloy 1	0.7	+0.36	58	485	640	1	12
Alloy 2	0.7	+0.40	64	570	675	1	11
FD-0205	0.7	+0.29	53	380	600	2	21
FD-0405	0.7	+0.10	57	400	670	2	24
FLD-49DH$^{\times}$	0.7	+0.36	57	530	670	1	13
Alloy 1	1.6	+0.38	64	650	785	1	13
Alloy 2	1.6	+0.42	68	780	845	1	12
FD-0205	1.6	+0.30	55	440	660	2	22
FD-0405	1.6	+0.07	57	460	730	2	23
FLD-49DH$^{\times}$	1.6	+0.44	66	750	785	<1	13

compared them with traditional alloys, including several diffusion-alloyed materials.

The nominal composition of the Mn alloys is given in Table 11.17. The manganese alloys were compared with diffusion-alloyed FD-0205, FD-0405 and FLD-49DH in premixes containing 0.6% graphite and 0.75% EBS wax. TRS, dogbone tensile and impact specimens were compacted at pressures from 415 to 690 MPa.

The specimens were sintered in a belt furnace for 15 min at temperature of 1120°C in an atmosphere of $90N_2–10H_2$. Two cooling rates, measured between 660 and 315°C, were 0.7°C/s and 1.6°C/s.

Jominy end quench hardenability specimens prepared were 100 mm by 25 mm diameter bars at a nominal density of 7.0 g cm^{-3} and were end quenched from 900°C. An additional test was run at 950°C for alloy FLD-49DH. The hardenability of alloys FD-0205 and FD-0405 was derived from MPIF standard 35 data and estimated for a density of 7.0 g cm^{-3}.

Mechanical properties. Table 11.18 contains the mechanical property data for the five alloys. As seen in the table, the apparent hardness of the manganese-containing alloys rivals that the high alloy sinter-hardening FLD-49DH at both conventional and accelerated cooling hardening at both conventional and accelerated rates. The

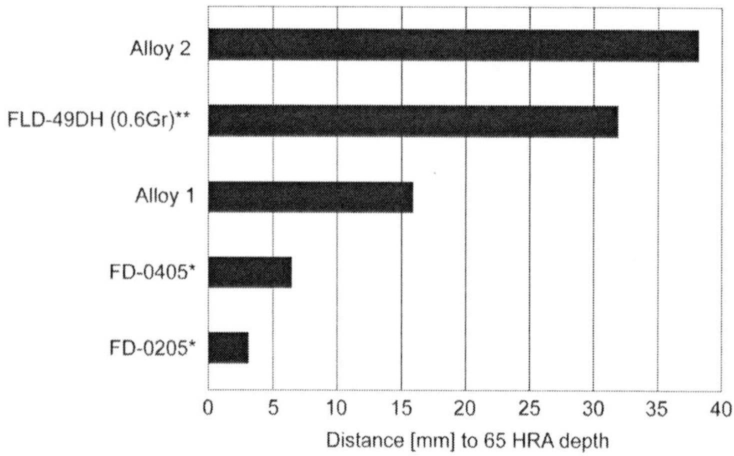

Fig. 11.10. Jominy end-quenched hardenability of the 5 alloys. *Values for FD-0205 and FD-0405 are taken from MPIF standard 35. **FLD-49DH austenitized at 950°C [267].

values in the table are interpolated from the plots of the sintered density of 7.0 g cm^{-3}.

Considerably lower hardness have the alloys with faster cooling rates. The yield strength of both Mn-containing steels is far superior to FD-0205 and FD-0405. The ultimate tensile strength (UTS) of the manganese alloys is equivalent to that of FD-0405 and FLD-49DH at conventional cooling rates. The higher Mo alloy exhibits the highest UTS with accelerated cooling. The ductility and impact properties of the Mn alloys are similar to that of Ni-free diffusion alloy FLD-49DH, whereas the Ni-containing FD-0405 and FD-0405 have much higher values.

Jominy end quench tests were performed to quantify the hardenability of the Mn–Mo alloys. Figure 11.10 shows the results, where a greater distance from the quenched end (J depth) indicates greater hardenability. The iron-based diffusion alloys have poor hardenability, as evidenced by the short distances and the low apparent hardness and yield strength. Alloy 2 hardenability may be under-reported with an austenitizing temperature at 900°C, as it was found that the Jominy depth of another hybrid PM alloy based on 0.8% Mo increased when the austenitizing temperature was increased above 900°C.

The Mn-containing steels based on Mo prealloys described above can be utilised as cost-effective alternatives to Ni and/or Cu-containing diffusion-alloyed materials for many applications.

The behaviour of these Ni-free alloys is quite similar to that of the higher molybdenum FLD-49DH alloy with respect to mechanical properties. Excellent hardness and yield strength can be achieved with accelerated cooling, as these lean alloys sinter harden with 0.6% graphite addition. No special sintering practices are required outside of N_2–H_2 based atmospheres. The Mn alloys are good candidates to replace more highly alloyed diffusion-based products in many applications where higher hardness and yield strength are required.

11.1.5. Properties of industrially sintered Fe–Mn–Mo–C steels

The objectives of the following study were to investigate and correlate the effect of composition and processing conditions (including semi-closed containers), heat treatment and sinter hardening with the microstructure and mechanical properties of Fe–(2–4)Mn–(0.3,0.7)C and Fe–2Mn–0.85Mo–(0.5,0.7)C steels sintered in an industrial pusher furnace [268].

The raw materials employed were sponge iron powder SC100.26, water-atomised prealloyed powder Fe–0.85% Mo, medium-carbon ferromanganese (FeMn, 56% of particles under 10 μm), graphite and HW wax as a lubricant.

Four different specimen types, from the powder mixtures compacted at 600 MPa to a green density of ~7.0 g cm^{-3}, the tensile, rectangular three point bend (55 × 10 × 5 mm), unnotched impact energy (55 × 10 × 10 mm) and fatigue (ISO 3928) compacts were prepared.

Sintering was carried out at 1180°C for 40 min in an atmosphere of 25% H_2–75% N_2 with the inlet dew point of –55°C, which corresponds to an 'effective' dew point –42°C. Specimens were furnace cooled at ~10 K/min. The specimens to be sinter hardened were reheated to 900°C in a laboratory furnace and air cooled at 55 K/min (in the range 900 – 400°C) and then tempered at 200°C for 60 min. Specimens were sintered in semi-closed containers with a getter Al_2O_3+10% FeMn powder+5% graphite. Specimens for heat treatment were austenitised at 850°C for 30 min in an atmosphere of 75% N_2–25% H_2 in semi-closed containers with a getter Al_2O_3+15% FeMn powder+10% graphite, quenched in oil and then tempered at 200°C for 60 min.

Mechanical properties. The mechanical properties of specimens which were slow cooled, sinter hardened, heat-treated and sintered in semi-closed containers, are recorded in Table 11.19.

Table 11.19. Mechanical properties of Fe–(2,3,4)Mn–(0.3,0.7)C and Fe–2Mn–0.85Mo–(0.5,0.7)C alloys. 1 – Fe–2Mn–0.3C, 2 – Fe–2Mn–0.7C, 3 – Fe–3Mn–0.3C, 4 – Fe–3Mn–0.7C, 5 – Fe–4Mn–0.3C, 6 – Fe–4Mn–0.7C, 7 – Fe–2Mn–0.85Mo–0.7C, 8 – Fe–2Mn–0.85Mo–0.7C – slow cooled, b – sinter hardened, c – sintering in semi–closed container, d – heat-treated

Alloy	$R_{p0.2}$ [MPa]	Max. tensile strain in bend [%]	TRS [MPa]	UTS [MPa]	Max. tensile stress in bend [MPa]	Hardness HV 10		Impact energy [J]
						surface	core	
1a	388	2.90	958	473	761	170	172	20
2a	428	2.31	1136	580	925	186	181	16
3a	449	1.30	994	524	812	190	195	9
4a	435	2.30	1123	569	917	189	208	9
5a	B	B	966	501	966	212	210	8
6a	412	0.96	867	481	690	229	253	7
7a	489	1.57	1280	637	769	200	199	18
8a	430	B	747	413	747	210	214	11
1b	557	3.40	1091	498	870	168	173	20
2b	570	2.93	1148	572	900	177	203	15
3b	471	2.70	1108	557	965	202	194	17
4b	518	1.90	1151	602	999	222	234	13
5b	328	1.80	1263	664	1027	230	287	14
6b	403	1.62	1066	540	873	236	273	15
8b	451	1.36	1196	603	1040	260	287	19
1c	B	2.43	983	451	567	176	162	21
2c	504	2.21	1125	530	691	180	181	17
3c	498	0.73	882	489	647	202	193	8
4c	B	1.39	1011	549	813	206	206	13
5c	B	B	733	423	733	225	236	4
6c	404	0.91	1058	528	838	208	220	7
7c	498	1.20	1216	542	749	209	197	10
2d	B	B	1071	640	1071	346	336	8
4d	B	B	903	554	903	402	360	10
6d	604	0.90	805	422	643	326	297	11
8d	B	B	1015	602	1015	377	356	13

*B means that the specimens were brittle

In terms of the effect of Mn content: 2, 3 and 4% comparing slow cooled specimens with 0.3% C, increasing Mn from 2 to 4%

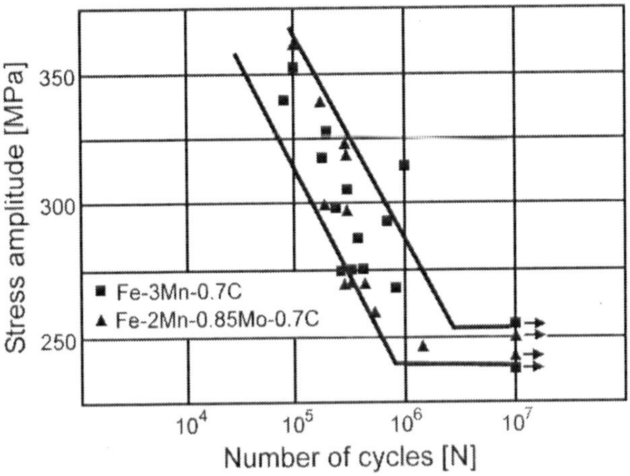

Fig. 11.11. Wöhler curves in plain bending of sintered hardened specimens of Fe–3Mn–0.7C and Fe–2Mn–0.85Mo–0.7C alloys. R = 1, frequency = 1450/min, $N_c = 10^7$ [268].

(batches 1a, 3a, 5a) resulted in a slight strength increase only for 3% Mn and a decrease for 4% Mn alloy, which exhibited a completely brittle behaviour. For 0.7% C, the highest strength, ductility and impact energy were achieved for 2% Mn. Comparing sinter hardened specimens with 0.3% C, increasing Mn from 2 to 4% (batches 1b, 3b, 5b) resulted in a strength increase of about 20%, whereas ductility and impact energy decreased.

Increasing C content from 0.3 to 0.7% C (batches 1a and 2a) for the specimens slowly cooled with 2% Mn, resulted in a strength increase of about 30% and a slight decrease of ductility and impact energy. For 3% Mn (batches 3a and 4a) had no such pattern, the effect of C was negligible. For 4% Mn (batches 5a and 6a) the C effect was reversed, whereas the low C specimens were brittle, the 4Mn–0.3C steel recorded some plasticity at the same strength level.

The fatigue strength tests results in the form of Wöhler curves for the plain fatigue tests on unnotched specimens of selected slow cooled and sinter hardened alloys are presented in Fig. 11.11. The fatigue strength at $N_c = 10^7$ was 140, 160 and 175 MPa for the Fe–2Mn–0.3C, Fe–3Mn–0.3C and Fe–3Mn–0.7C slow cooled alloys, respectively, and 230 and 255 MPa for the Fe–3Mn–0.7C and Fe–2Mn–0.85Mo–0.7C sinter hardened alloys.

The highest mechanical properties in the entire range of composition investigated in the slow cooled state (yield, tensile and bend strengths of 499, 637 and 1200 MPa), respectively, and an impact energy of 18 J were achieved for the Fe–2Mn–0.85Mo–0.5C alloy.

Sinter hardening led to some improvement in the mechanical properties; the highest yield, tensile and bend strengths of 570, 664, and 1263 MPa, respectively, at acceptable impact energy of 14 J for the Mo free Fe–4Mn–0.3C alloy.

11.2. Processing and properties of sintered Mn–Cr–Mo–C steels

11.2.1. Sinterability and hardenability of Mn–Cr–Mo steels

11.2.1.1. Sinterability of Mn–Cr–Mo steels

The objective of this study was the sinterability of potential high-strength ferrous alloys determined on compacts of Fe–(1,3)Mn–(0–0.85)Mo–(0–4Cr)–(0–1)Si–(0.4–0.8)C [269].

Water-atomised, ABC100.30, sponge, NC100.24, and super compressible, SC100.26, iron powders were used in this investigation. Elemental additions were molybdenum, silicon, chromium (electrolytic), manganese and carbon as fine graphite. Prealloyed powders used in place of or in combination with iron powders were Astaloy CrM and Astaloy 85Mo. Milled manganese and chromium ferroalloys were employed, as were molybdenum and chromium carbides. The medium-carbon ferromanganese was milled in nitrogen predominantly to the 10–20 μm particle size. Alloys with Mn in the range of 1–3% Mn, Mo of 0–0.85%, Cr of 0–4%, Si of 0–1% and C between 0.4 and 0.8% were investigated.

Mixtures with the admixed lubricant were compacted at 600 MPa to give green densities of the compacts in the range of 6.5–6.7 g cm^{-3}, followed by sintering in a hydrogen or hydrogen-rich (75% minimum) atmosphere with the dew points in the range of −40 to −70°C in the temperature range of 1140–1280°C. Laboratory sintering at temperatures in excess of 1180°C was perfected.

The dilatometric traces for Fe–2.5Mo–1Cr–0.7C–(0 and 1)Si indicate that the alpha–gamma transformation, a peak in the plots, was appreciably depressed in the silicon-free alloy. For the alloy containing 1% Si a significantly smaller thermal expansion above 1150°C was observed, Fig. 11.12, caused by the rapid formation of a liquid phase, identified metallographically.

Specimens were tested in tension or simple bending. The results are summarised in Table 11.20.

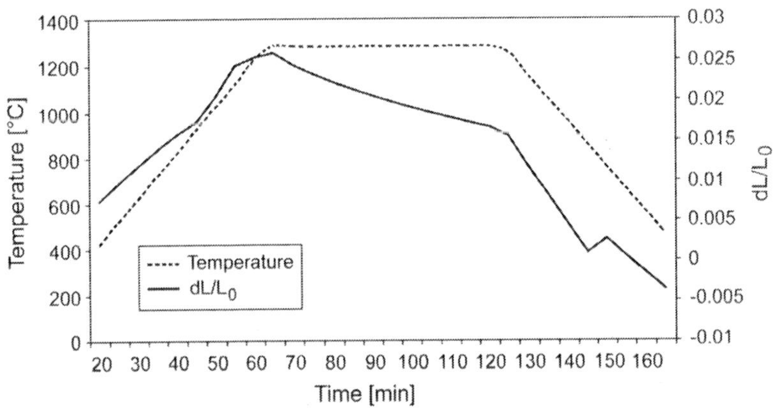

Fig. 11.12. Dilatometric trace for Fe–2.5Mn–1Si–0.7C alloy [269].

Table 11.20. Preliminary mechanical properties data as sintered specimens and following heat treatment for alloys B and C

Alloy	Composition	Heat treatment	3-point bend strength [MPa]	Tensile strength [MPa]
A	Fe–3Mn–0.8C		1080	450
B	Fe–1.5Mn–1Cr–0.45Mo–0.5C		750	420
		900°C – 10 min 180°C – 1 h	900	450
C	Fe–2.5Mn–1Cr–0.7C		1170	560
		850°C – 30 min 175°C – 1 h	1310	800
D	Fe–2.5Mn–1Cr–0.45Mo–0.7C		1120	
E	Fe–2.5Mn–1Cr–1Si–0.7C		950	
I	Fe–3Cr–0.5Mo–0.4C			720
J	Fe–3Cr–1Mn–0.5Mo–0.4C			840

11.2.1.2. Hardenability of Mn–Cr–Mo steels

The objective of the following study was the effect of residual oxygen on the mechanical behaviour and hardenability of Mn–Cr–Mo alloyed steels [270].

For the preparation of test specimens two conventional alloy powders and two newly developed Mn–Cr powders were used. The former contained more than 0.2% oxygen, while the latter contained

Table 11.21. Composition of atomised alloy powders

Powder	Chemical composition [%]					
	Mn	Cr	Mo	C	Fe	O
1	0.89	1.02	0.25	0.11	bal.	0.60
2	0.85	1.11	0.32	0.02	bal.	0.30
3	0.65	1.05	0.25	0.07	bal.	0.14
4	0.85	1.11	0.32	0.01	bal.	0.09

No. 1 and 2 are conventional powders, No. 3 and 4 are newly developed powders

less than 0.2%. Table 11.21 shows the composition of these powders [270].

The specimens were compacted to a density of ~7.1 g cm^{-3}. The carbon content of these specimens before sintering was 0.5%. Several types of sintering atmospheres, such as vacuum (300 – 600 ppm), H_2 (600 – 800 ppm), and cracked ammonia (900 – 1200 ppm) were used for sintering at 1250°C for 30 min. Heat treatment consisted from austenisation at 840°C for 1 h, quenching in oil and tempering at 250°C for 1.5 h.

Figure 11.13 shows the effect of the oxygen content after sintering on the tensile and impact strength. The impact strength is remarkably affected by oxygen after sintering, and is improved as the oxygen content decreases and Fig. 11.14 shows the effect of the oxygen content on the hardenability of sintered steels demonstrated as maximum hardness values.

Figure 11.15 shows the relationship between the carbon content after sintering and tensile strength and impact energy values of the investigated sintered steels. The impact strength improves as the carbon content decreases.

As follows from the results, the mechanical properties and hardenability of the Fe–Mn–Cr–Mo sintered steels strongly depend on the oxygen content, and they improve as the oxygen content decreases. In order to obtain high strength and ductility of the Fe–Mn–Cr–Mo sintered steels, it is necessary to maintain a low oxygen content in both powders and in the sintered body.

11.2.1.3. Sintering of Mn–Cr–Mo steels based on prealloyed powder in nitrogen

In [271] both sponge iron powder NC100.24 and Astaloy CrM powders were used. Carbon was introduced as fine graphite and 1.5% manganese as medium-carbon ferromanganese with particle size

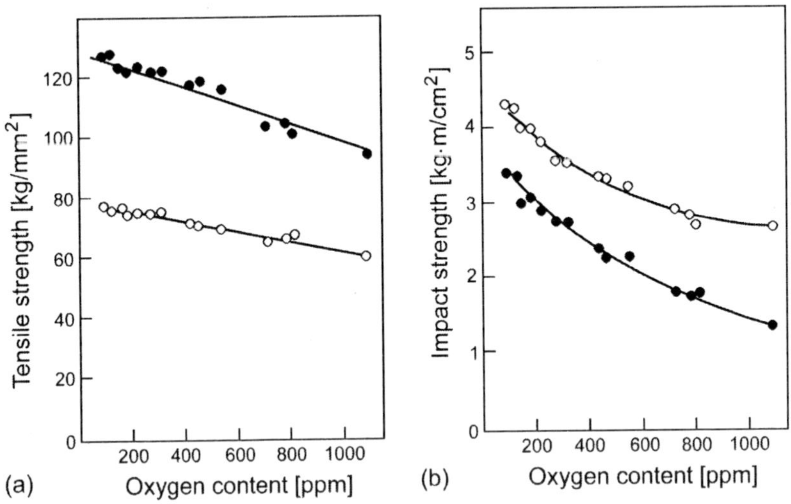

Fig. 11.13. Effect of oxygen content after sintering on: (a) tensile strength and (b) impact strength of Fe–Mn–Cr sintered steel (using powder 3), as sintered (white points), as heat-treated (black points) [270].

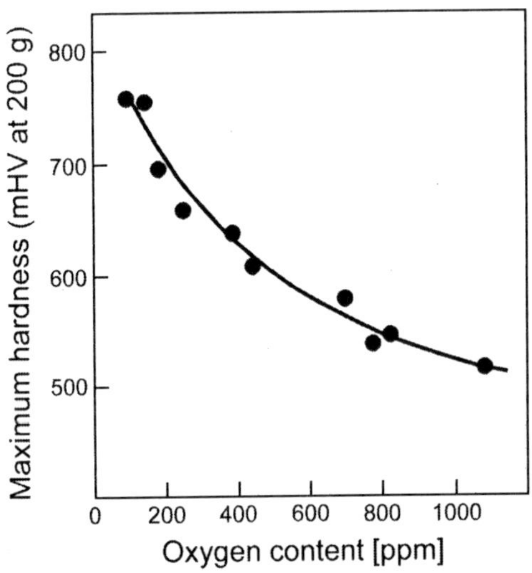

Fig. 11.14. Effect of oxygen content after sintering on hardenability of Fe–Mn–Cr–Mo sintered steel (using powder 3) [270].

less than 20 μm, with a median of 8 μm. The tensile and transverse rupture strength (55 × 11 × 5 mm) specimens were compacted at 600 MPa followed by sintering in a semi-closed container with a labyrinth seal. The furnace atmospheres were pure hydrogen or pure nitrogen with an inlet dew point of –60°C. The container did not

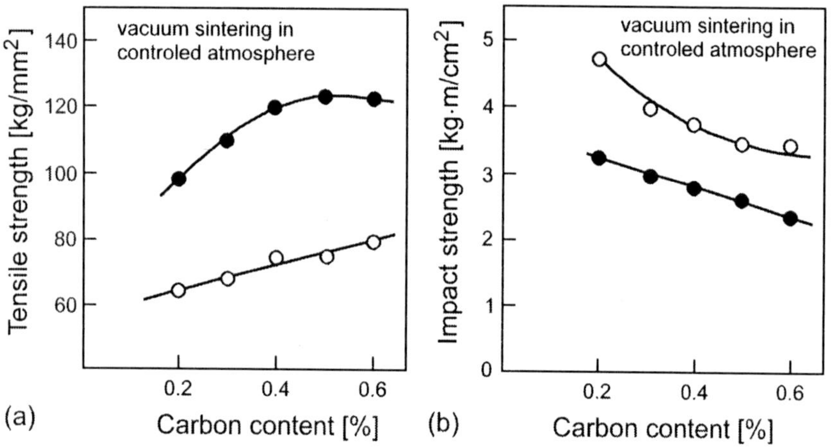

Fig. 11.15. Effect of carbon content on (a) tensile strength and (b) impact strength of Mn–Cr–Mo sintered steel. (White points – as sintered, black points – as heat-treated) [270].

have any ferromanganese added to it. Sintering was carried out in a tube furnace at 1120°C for 60 min. After sintering the specimens were withdrawn from the hot zone to the cool zone of the furnace at 65°C/min. The sintered specimens were tempered at 200°C for 60 min.

The results of the chemical analyses are presented in Table 11.22. Irrespective of the sintering atmosphere, both oxygen and carbon contents were near 0.4% and, though nitrogen was an order of magnitude larger for N_2 sintering, at 0.04% it was still insignificant. Table 11.23 records the volume decrease and the tensile and bend strength of the sinter hardened specimens sintered in technical nitrogen and in technical hydrogen, respectively. The mean tensile strength was 586 MPa for hydrogen and 633 MPa for nitrogen sintering.

Sintered specimens had heterogeneous microstructures; the most significant feature of all microstructures being the absence of chromium and manganese oxide networks. There was evidence of poor Cr and Mo diffusion into sponge iron and, conversely, poor Mn diffusion into Astaloy CrM particles. A tempering treatment then carried out at 200°C for 60 min in nitrogen caused the martensite microhardness to decrease. The pearlite structure had an average hardness of 200 HV 0.05, while the martensite 390 HV 0.05.

Figures 11.16 and 11.17 show Weibull plots of the data for UTS and TRS, respectively. Weibull modulus m values of ~15 specimens indicate a relatively good reproducibility for sintered specimens.

Table 11.22. Chemical analyses of Fe–1.5% Mn–1.5% Cr–0.25% Mo–0.4% C steel

Sintering conditions: 1120°C, 60 min	Chemical analysis [%]		
	carbon	oxygen	nitrogen
H_2	0.445	0.401	0.004
N_2	0.412	0.380	0.044

Table 11.23. Volume change and mechanical properties of Fe–1.5% Mn–1.5% Cr–0.25% Mo–0.4% C steel

Sintering atmosphere	$\Delta V/V_0$ [%]	HV 30	TRS [MPa]	UTS [MPa]	A [%]	K [J]
H_2	–0.66	197	1178	586	0.6	11.9
N_2	–0.62	215	1293	633	0.9	13.5

This experimental work has indicated the potential of utilising the local microclimate attained in the semi-closed container, to sinter Cr-containing ferrous powders in nitrogen. It is essential, before industrial sintering, to determine the necessary conditions regarding the presence of Mn and Cr carriers, which are adequate for Cr alloy containing Mn.

11.2.1.4. Properties of Fe–Mn and developed Fe–Mn–Cr–Mo steels

In [272], the aim of research work was to study the implementation of good drying facilities and/or use of a gettering system which creates a local dry micro-climate around the parts being sintered. Therefore, the relevant atmosphere quality should be improved dramatically.

The aim of the project was to to concentrate, if at all possible, on maintaining sintering temperatures below 1150°C, which is the maximum allowable temperature in industrial metal link belt furnaces.

A laboratory tube furnace was modified to allow the introduction of gases to be carried out under reducing conditions.

Study of 2–4Mn alloys. As given in Table 11.24 recorded hardness values were in the range of 183–276 HV 30 and tensile strength values of 190–610 MPa. The highest tensile strength of 610 MPa was obtained for the steel 4 and the lowest one for the steel 2.

Figure 11.18 presents the fracture of the steel with 0.8% C and Fig. 11.19 the microstructure of the steel with 0.5% C, even of the

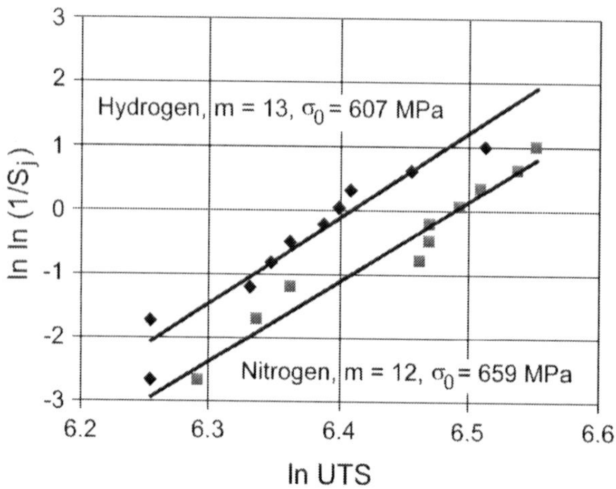

Fig. 11.16. Weibull plot of UTS, in MPa, of Fe–1.5% Mn–1.5% Cr–0.25% Mo–0.4% C steel specimens. For specimens sintered in hydrogen the Weibull modulus was $m = 13 \pm 3$ and characteristic UTS 607 MPa. For specimens sintered in nitrogen was $m = 12 \pm 3$ and characteristic UTS 659 MPa [271].

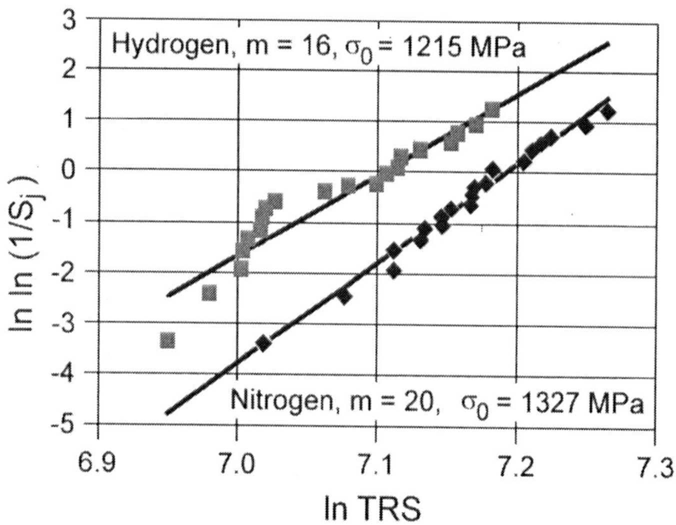

Fig. 11.17. Weibull plot of TRS, in MPa, of Fe–1.5% Mn–1.5% Cr–0.25% Mo–0.4% C steel specimens. For specimens sintered in hydrogen Weibull modulus was $m = 16 \pm 3$ and characteristic TRS 1215 MPa. For specimens sintered in nitrogen was $m = 20 \pm 3$ and characteristic TRS 1327 MPa [271].

predominantly pearlitic furnace cooled specimens were complex and inhomogeneous and additionally comprised ferrite and retained austenite at the pores.

Table 11.24. Properties data for Mn steel specimens: as sintered and cooled at 40 and 10°C/min, latter results in bold; C added as graphite

Powder mixture	Formulation	Sintering temperature [°C]	Hardness HV 30	Tensile strength [MPa]
1 Fe–3Mn–0.8C	NC100.24+HP+C	1120	217	520
2 Fe–3Mn–0.8C	ABC100.30+HP+C	1120	214	190
3 Fe–3Mn–0.6C	NC100.24+EL+C	1120	276	560
3 Fe–3Mn–0.6C	NC100.24+EL+C	**1120**	**183**	**480**
3 Fe–3Mn–0.6C	NC100.24+EL+C	**1250**	**189**	**590**
4 Fe–3Mn–0.8C	NC100.24+EL+C	1250	239	610

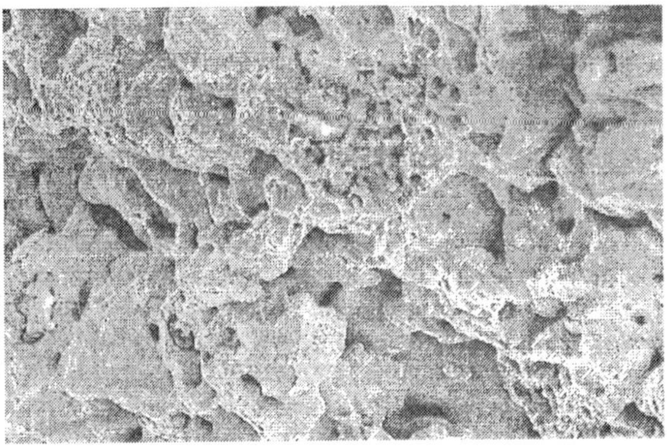

Fig. 11.18. Fracture of Fe–3% Mn–0.8% C steel sintered at 1120°C, slow cooled sample [272].

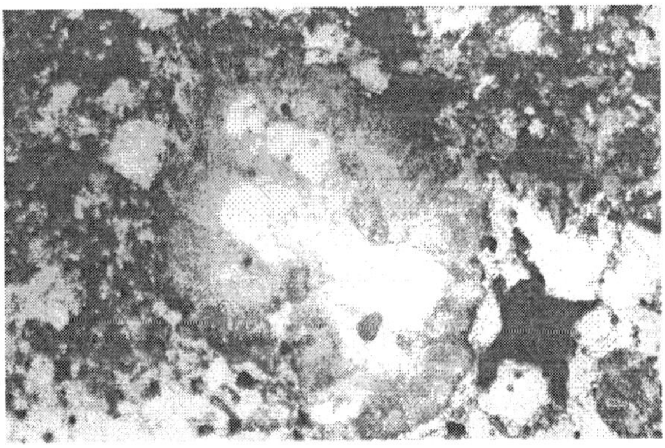

Fig. 11.19. Optical micrograph of the microstructure of Fe–3% Mn–0.5% C steel showing that carbon was retained in Mo-rich areas, and interiors of the larger particles were ferritic [272].

Table 11.25. Composition, sintering temperature and mechanical properties data for alloy specimens A – G and following heat treatment for alloy B and C using ferromanganese (FeMn), chromium carbide and graphite addition to iron and Astaloy powders, pressed at 625–700 MPa. C added as graphite

Alloy	Formulation	Sintering temperature [°C]	Heat treatment	3-point bend strength [MPa]	Tensile strength [MPa]
A – Fe–3Mn–0.8C	ABC100.30+FeMn+ Cr_3C_2+ Mo_2C+C	1180		1080	450
B – Fe–1.5Mn–1Cr–0.45Mo–0.5C	NC100.24+FeMn+ Cr_3C_2+Mo_2C+C	1280		750	420
		1200	900°C – 10 min 180°C – 1 h	910	450
C – Fe–2Mn–1Cr–0.7C	NC100.24+FeMn+ Cr_3C_2+C	1250		1170	560
			800°C – 30 min 175°C – 1 h	1310	800
D – Fe–2Mn–1Cr–0.7C	Astaloy0.85Mo+white paraffin+NC100.24+ Cr_3C_2+FeMn+C	1200		1120	
E – Fe–2.5Mn–1Cr–1Si–0.7C	NC100.24+FeMn+ SiC+Cr_3C_2	1280		950	
F – Fe–3Cr–0.5Mn–0.8C	Astaloy CrM+0.4C	1200			720
G – Fe–3Cr–1Mn–0.5Mo–0.6C	Astaloy CrM+1Mn+04C	1280			840

Development of PM Mn–Cr–Mo steels. It is known that whilst Mo tends to reduce the rate of C diffusion in gamma Fe, Mn raises the rate, so that the alpha–gamma transformation occurs earlier in the heating cycle. There is also added benefit that the dew point is locally improved due to manganese volatility, which allows chromium oxide to be reduced at a lower temperature. The mechanical properties are summarized in Table 11.25. TRS values were in the range 750–1180 MPa (as sintered) and after heat treatment 910 and 1310 MPa. The tensile strength reached 420–840 MPa without heat treatment and after heat treatment it was 450–800 MPa. The beneficial effect of 1% Si addition was also recorded.

11.3. Processing and properties of hybrid sintered Mn–Cr–Mo–C steels

11.3.1. Properties of sintered 3Cr–0.5Mo steel promoted by manganese in form of liquid phase

In [273] prealloyed Astaloy CrM grade (3Cr–0.5Mo) with 0.35 and 0.5% C was taken into account as a reference material; F-119 master alloy to promote the liquid phase and a plain manganese powder grade with a particle size below 4 μm (nomenclature is detailed in Table 11.26) were used. In case of the F-119 master alloy, a high-energy grind process was performed before the normal mixture process with the Astaloy CrM powder and graphite. When manganese was incorporated, the milling time did not exceed 60 min. All materials were compacted at 700 MPa. Sintering was performed in an industrial furnace under dissociated ammonia at 1120°C for 45 min. Table 11.26 lists, beside the mixture nomenclature, also the composition of the master alloy.

Compared with manganese, the effect of F119 promoting a transient liquid phase can be seen in Fig. 11.20 with the dimensional changes in dependence on base alloys and carbon content. Higher combined carbon seems counteract better the contraction effect, thus, it leads to better dimensional stability.

In Table 11.27 the sintered densities and mechanical properties of sintered steels based on Astaloy CrM powder are listed. The mechanical properties are influenced by the carbon content and sintered density. The Astaloy CrM+0.5% C+2% Mn (B5) is closer to reference Astaloy CrM with 0.35% C.

Figures 11.21 and 11.22 show the relation of hardness vs. tensile strength: (a) with 0.35% C added, (b) with 0.5% C added.

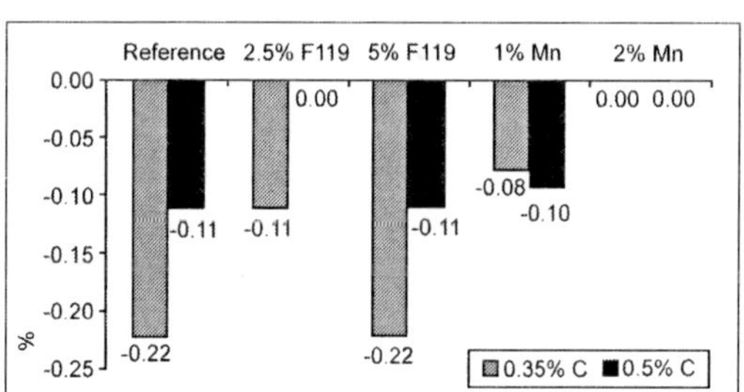

Fig. 11.20. Dimensional changes as difference between compacted and sintered materials [273].

Table 11.26. Mixture nomenclature. Master alloy composition: Fe–1.74Mn–0.68P–0.93–named F–119. Astaloy CrM as base powder

C [%]	Master alloy [%]	Mn [%]	Name
0.35	–	–	1
	2.5	–	A1
	5	–	A3
		1	B1
		2	B3
0.5		–	2
		–	A2
		–	A4
		1	B2
		2	B4

Table 11.27. Mechanical properties of Astaloy CrM+x% C with different additions

Alloy	ρ [g cm^{-3}]	HRA	Rm [MPa]	$R_{p0.1}$ [MPa]	A [%]	E [GPa]
2r	7.10	58	483	392	0.895	96
3r	7.04	61	587	482	1.01	78
A2	6.94	58	493	418	0.89	90
A3	7.14	60	582	428	6.86	96
A4	6.87	53	483	332	0.80	97
A5	6.84	58	532	392	0.88	111
B2	6.97	65	448	388	0.41	112
B3	7.09	70	447	414	0.51	98
B4	7.07	71	531	456	0.64	114
B5	7.10	71	552	488	0.64	111

Promoting the liquid phase by adding the master alloy Fe–1.74Mn–6.98P–0.93 (F–119) did not offer expected improvements, although it has provided some valuable contribution. Thanks to manganese additions, carbon diffusion improved: reduction provided by manganese better removed the oxide particle layer on Astaloy CrM powder.

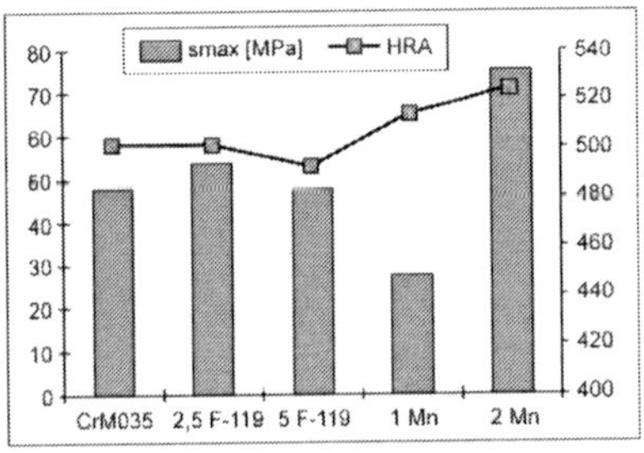

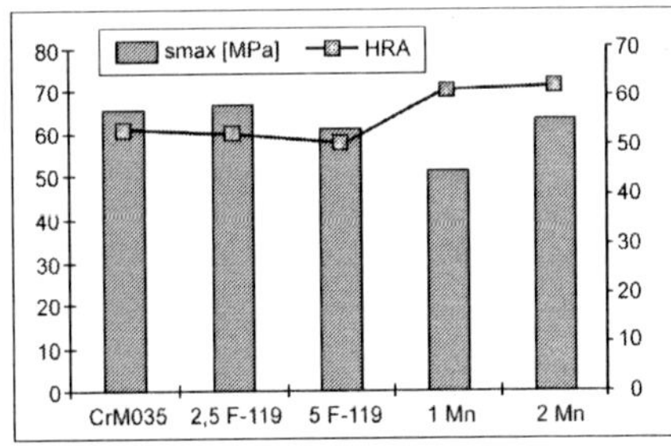

Fig. 11.21 (left). Hardness vs. tensile strength, 0.35% C [273].
Fig. 11.22 (right). Hardness vs. tensile strength, 0.5% C [273].

11.4.2. Properties of hybrid Mn–Cr–Mo steels coated with hydrocarbons

The aim of this work was to study the effect of a carbon–hydrogen substance, CHS, as a carbon donor instead of graphite [274].

For coating of metal powders two techniques of coating with CHS were employed:

coating from solution – method SC, coating without utilising solvents – method WM. Compacts for tensile strength bars, unnotched Charpy specimens were pressed at 600 or 800 MPa from the powder mixtures with different carbon contents; some mixtures, method SC + G, containing also added graphite as shown in Table 11.28. The mechanical properties are shown in Table 11.29.

High-carbon sintering creates liquid cementite (Fe_3C) networks at the prior particles boundaries, which leads to good densification. Mechanical tests show good tensile strength coupled with excellent impact resistance.

The use of a hydrocarbon donor to replace the majority of graphite minimises the problem of carbon segregation which can lead to poorly distributed densification.

The addition of 1% Mn improves the sinterability of CrM based materials as it lowers the $\alpha \rightarrow \gamma$ transition temperature by ~100°C.

The microstructures were homogenous, particularly for specimens with no added graphite and/or sintered at higher temperatures, Fig. 11.23(a),(b) and Fig. 11.24(a),(b), in contrast to conventionally cooled Cr–Mn steels processed also in semi-closed containers, but at 1120°C.

Sintered microstructures comprised fine pearlite, bainite and a continuous network consisting of both primary eutectic and proeutectoid precipitates.

Table 11.28. Composition (carbon contents after sintering), compacting pressure, green density (GD), sintered density (SD) and sintering temperature (Sinter T)

Composition	Designation	Pressure [MPa]	GD [g cm^{-3}]	Sinter T [°C]	SD [g cm^{-3}]	Sinter T [°C]	SD [g cm^{-3}]
Fe–3Cr–0.85Mo–1Mn–1.28C	A–SC	600	6.9	1285	7.2		
		800	6.9	1285	7.4		
Fe–3Cr–0.5Mo–1.48C	G–WM	800	6.3	1180	6.9		
Fe–3Cr–0.5Mo–1.48C	H–WM	800	6.3	1180	6.9		
Fe–3Cr–0.5Mo–1Mn–0.98C	EB–SC	600	6.7			1295	6.9
		800	6.9			1295	7.0
Fe–3Cr–0.85Mo–1Mn–1.48C	EA:SC+G (0.9% G)	600	6.8	1285	7.6	1295	7.7
		800	7.0	1285	7.5	1295	7.5

Table 11.29. Mechanical properties of A, EB and EA alloys

Alloy	Compacting pressure [MPa]	Surface hardness HV 30		0.2% offset strength [MPa]	Tensile failure strength [MPa]	Impact energy [J]
		after sintering	after heat treatment			
EB(SC)	600	222–240	200–215	554	653	18
EB(SC)	800			615	761	26
EA(SC+G)	600	420–436	298	730	753	24
EA(SC+G	800		–320	758	778	36
A/SC	600	253–264	194–205	722	761	29
	800			777	788	36

Note: Max. tensile strain was in the range of 2.08 (alloy EB(SC), 600 MPa to 4.84% (alloy EA(SC+G), 800 MPa

11.4.3. Dimensional changes of Fe–Cr–Mo–Mn steels tested by dilatometry

11.4.3.1. Dimensional changes of Fe–Cr–Mo–Mn steels with manganese incorporated in the form of medium-carbon ferromanganese

The following starting materials in powder form were used in [275] for testing the dimensional changes of Fe–Cr–Mo–Mn steels:

- prealloyed Fe–3Cr–0.5Mo (Astaloy CrM) grade powder,

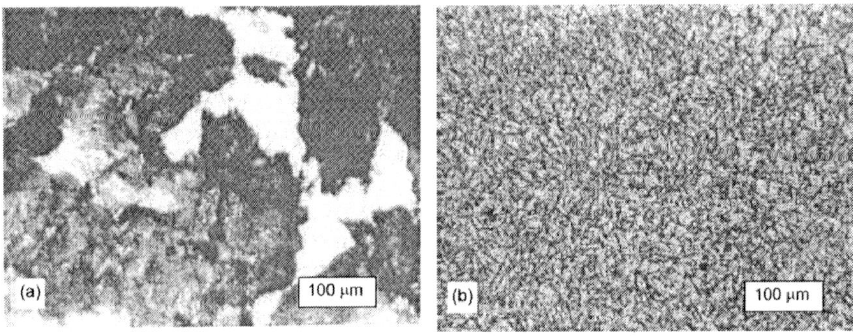

Fig. 11.23. Relatively homogenous microstructures of (a) Astaloy CrM+1% Mn+0.83% C sintered at 1280°C and (b) Astaloy CrM+1Mn+1.45C sintered at 1295°C, austenitized at 950°C and quenched to give martensite + retained austenite [274].

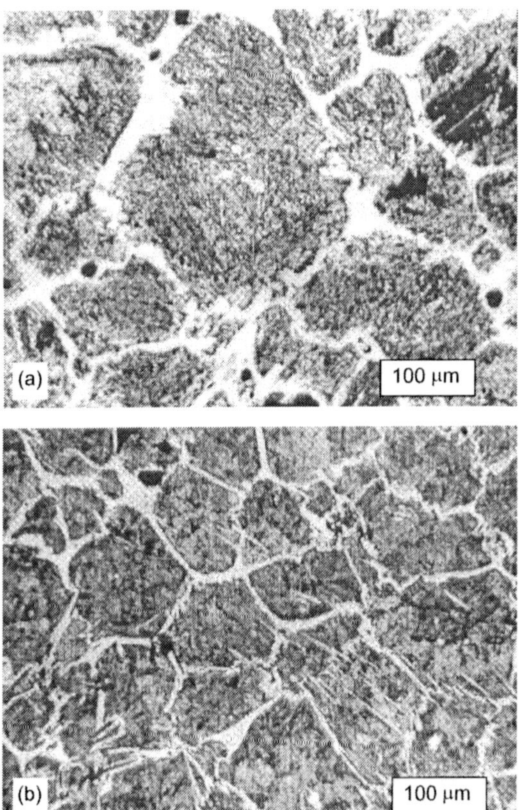

Fig. 11.24. Microstructures of steels based on (a) Astaloy 0.85Mo and (b) Astaloy CrM, sintered at 1180°C. Fine pearlitic matrix with network consisting of primary (eutectic) and proeutectoid cementite [274].

- medium-carbon ferromanganese, smaller fraction than 40 μm,
- C-UF fine graphite powder and Kenolube powder as lubricant.

Following mixtures were prepared:

- Astaloy CrM+0.6% C,
- Astaloy CrM+0.6% C + 1.0 % Mn,
- Astaloy CrM+0.1% C + 3.0 % Mn,
- Astaloy CrM+0.3% C + 3.0 % Mn,
- Astaloy CrM+0.5% C + 3.0 % Mn,
- Astaloy CrM+0.6% C + 3.0 % Mn.

The mixtures were compacted into rectangular specimens of size 4 × 4 × 15 mm at apressure of 600 MPa, resulting in green densities of about 6.8 g cm^{-3}. Sintering was carried out in a horizontal push rod dilatometer. Pure dry hydrogen and nitrogen with a dew point of –70°C were used to prepare two gas mixtures:

- 75 % hydrogen + 25% nitrogen,
- 10% hydrogen + 90% nitrogen.

Samples were firstly heated at 20°C/min up to 350°C, at which temperature they were left for 15 min to remove the lubricant, and then heated at 10°C/min to the sintering temperature. Isothermal sintering was carried out for 1 h. For comparison, the as received Astaloy CrM powder was also used.

Figures 11.25–11.27 are examples of complete dilatometric plots obtained during investigation. The expansion rate then decreases slightly because of the initiation of the $\alpha \rightarrow \gamma$ transformation. Mn and Cr lower the temperature at which this transformation starts, Fig. 11.28. The lowest temperature for the initiation of the $\alpha \rightarrow \gamma$ transformation was thus ~774°C for the Astaloy CrM+0.6% C+ 3.0% Mn with finish occurring at ~834°C. There was only a minor effect due to the composition of the sintering atmosphere, including the undetectable influence of N_2/H_2 ratio, on the $\alpha \rightarrow \gamma$ transformation. In the case of the unmodified CrM powder the shrinkage due to the $\alpha \rightarrow \gamma$ transformation began at ~874°C and finished at ~918°C.

After completion of the $\alpha \rightarrow \gamma$ transformation the specimens expand until the sintering plateau. All specimens sintered at 1150°C expanded continuously during heating, Fig. 11.25, whereas those heated to 1280°C reached their maximal length at about 1200°C, Figs. 11.26 and 11.27. Only the unmodified Astaloy CrM samples tended to expand continuously, which indicates that thermal expansion dominated the sintering shrinkage.

The addition of carbon and of ferromanganese to the Astaloy CrM powder, which enhance sinter hardening effects, can be related to the

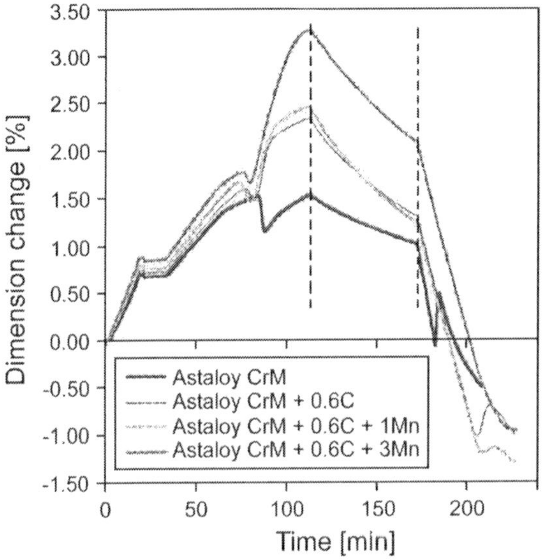

Fig. 11.25. Dilatometric curves for 1150°C and $25N_2/75H_2$ furnace atmosphere [275].

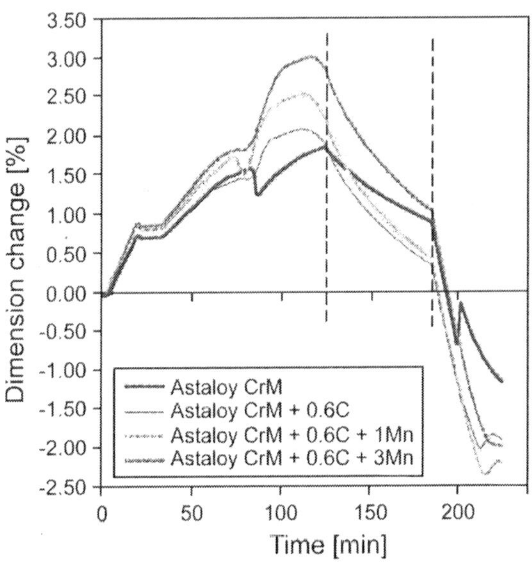

Fig. 11.26. Dilatometric curves for 1280°C and $25N_2/75H_2$ furnace atmosphere [275].

heating and cooling segments of the dilatometric curves, Fig. 11.28 and Fig. 11.29. The approximate temperature of the start of bainitic transformation appears to be ~460°C for Astaloy CrM+0.6% C and 430°C for Astaloy CrM+0.6% C+3% Mn, respectively.

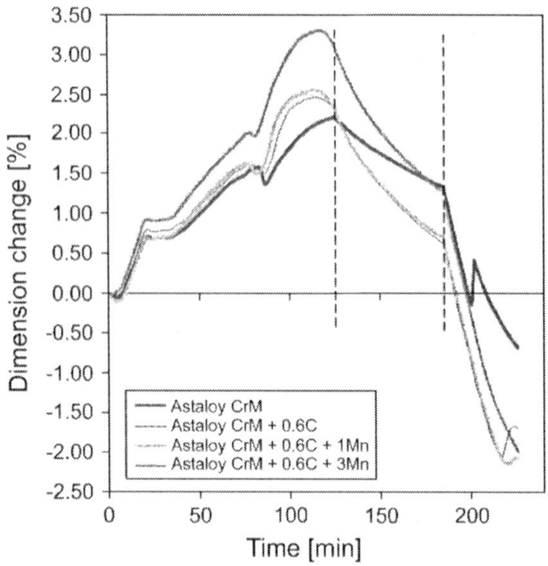

Fig. 11.27. Dilatometric curves for 1280°C and $90N_2/10H_2$ furnace atmosphere [275].

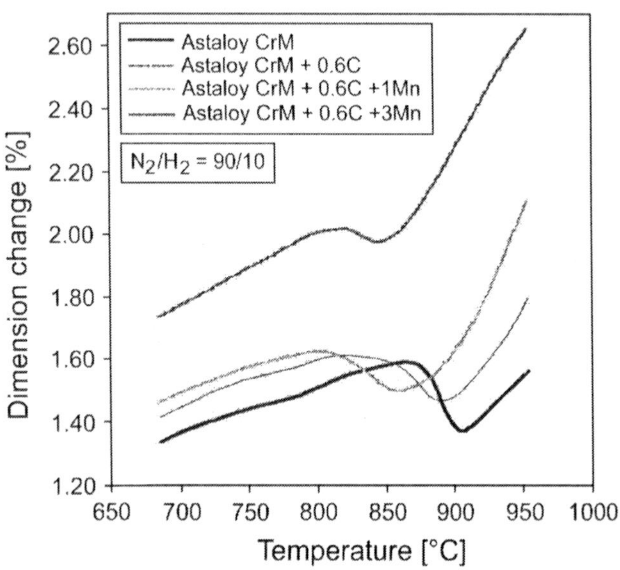

Fig. 11.28. Detailed dilatometric heating segments [275].

Thus, Astaloy CrM with carbon and ferromanganese additions are powder mixtures suitable for sinter hardening (possibly without a temper or stress relief even at relatively low cooling rates of 0.3–0.5°C/s). There are also indications that high sintering temperature

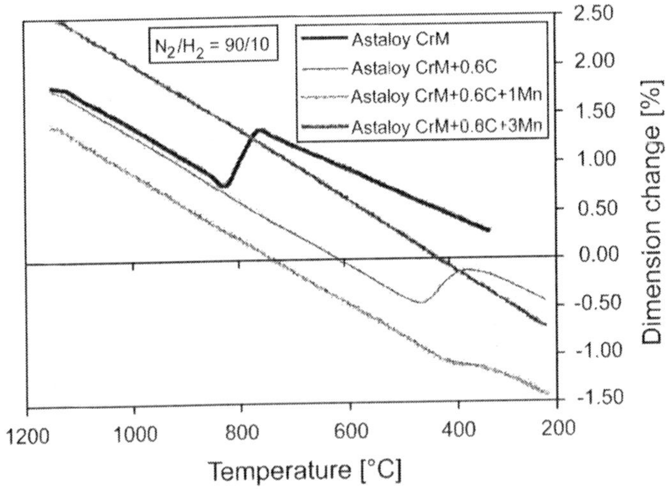

Fig. 11.29. Dilatometric cooling segments [275].

is not necessary and that chemical homogenisation is sufficiently progressed even at a sintering temperature of 1150°C, to ensure sinter hardening effects during post sinter cooling.

11.4.3.2. Dimensional changes of Fe–Cr–Mo–Mn steels with manganese incorporated in the form of medium-carbon ferromanganese and electrolytic manganese

In [276] the following starting materials in powder form were used for testing the dimensional changes Fe–Cr–Mo–Mn steels:

- prealloyed Fe–3Cr–0.5Mo (Astaloy CrM) grade powder,
- medium-carbon ferromanganese smaller fraction then 40 μm,
- elemental electrolytic manganese powder,
- C-UF fine graphite powder and Kenolube powder (0.6%) as lubricant.

Following mixtures were prepared:

- Astaloy CrM+1.45% C,
- Astaloy CrM+1.45% C + 1.0 % Mn (ferromanganese),
- Astaloy CrM+4.45% C + 3.0 % Mn (electrolytic manganese),
- Astaloy CrM+0.6% C + 3.0 % Mn.

For comparison, as received Astaloy CrM was also used.

The mixtures were compacted into rectangular specimens of size 4 × 4 × 15 mm at pressure of 600 MPa, resulting in green densities of about 6.6 g cm^{-3}. Sintering was carried out in a horizontal push rod dilatometer. Pure dry hydrogen and nitrogen with a dew point of –70°C were used to prepare two gas mixtures:

- 75 % hydrogen + 25% nitrogen, and
- 10% hydrogen + 90% nitrogen.

Samples were firstly heated at 10°C/min up to 600°C, at which temperature they were left for 30 min to remove the lubricant, then heated at 10°C/min to a temperature of 1100°C at which they were left for 1 h. Cooling rate was 30°C/min.

In an attempt to increase the ductility of the high-carbon alloyed steels, some sintered samples were reheated at 10°C/min to 180°C and to 710°C, at which temperature they were left for 2 h. Then they were cooled at a rate of 30°C/min to 180°C and left at this temperature for 1 h.

Dilatometric traces in Fig. 11.30–Fig. 11.32 suggest liquid phase formation from ~1240°C.

During further heating to an isothermal sintering temperature of 1280°C, only unmodified Astaloy CrM samples tended to expand continuously. In contrast, in the specimens containing carbon and manganese addition, sintering shrinkage appears to dominate thermal expansion, starting at about 1200°C at which temperature these samples reached their minimal length.

The sintering behaviour was not significantly influenced by the type of manganese carrier, elemental or ferromanganese powder, as shown in Fig. 11.32.

The addition of carbon or carbon and ferromanganese to the Astaloy CrM powder, which influences sinter hardening, can be related to the cooling segments of the dilatometric curves, Figs. 11.33 and 11.34. The approximate temperature of the start of bainitic transformation is ~410°C for Astaloy CrM+1.45% C. The M_s appears to be ~200°C for Astaloy CrM+1.42% C, and ~190°C for Astaloy CrM+1.45% C = 1% Mn, respectively.

The presence of the liquid phase during sintering is supported by observation of rounded pores and carbide networks at grain boundaries in Astaloy CrM+1.45% C and Astaloy CrM+1.45% C +1% Mn specimens, as shown in Fig. 11.35.

In agreement with segments cooling curves (Figs. 11.33 and 11.34) the microstructure is predominantly baitinic with a small amount of martensite for Astaloy CrM+1.45% C, whilst with the addition of 1% Mn the hardenability is increased and the microstructure contains more martensite with some retained austenite, Fig. 11.36.

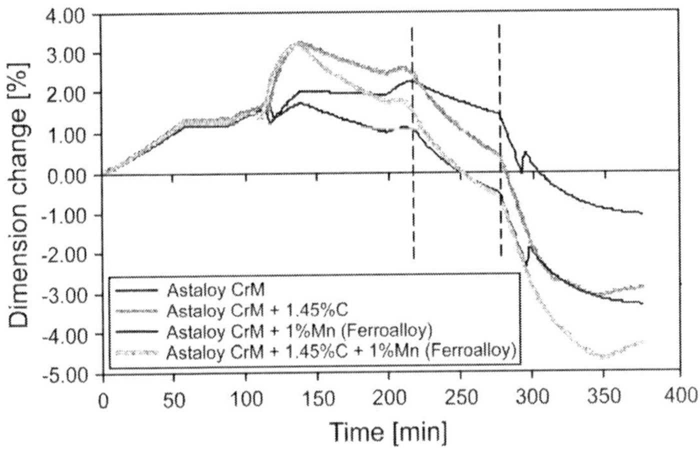

Fig. 11.30. Dilatometric curves for 75 hydrogen/25 nitrogen sintering atmosphere. Isothermal sintering 1280°C, 1 h, $75H_2/25N_2$ [276].

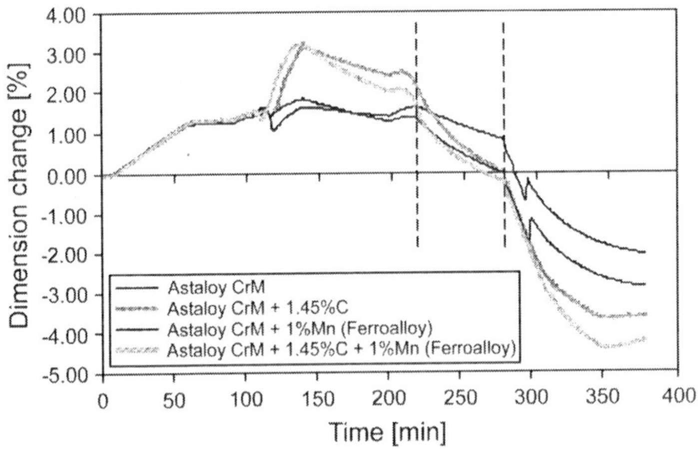

Fig. 11.31. Dilatometric curves for 75 hydrogen/25 nitrogen sintering atmosphere. Isothermal sintering 1280°C, 1 h, $10H_2/90N_2$ [276].

Manganese added, especially in ferroalloy form, minimise alpha phase sintering and the attendant retained oxide problem. A sintering atmosphere rich in nitrogen (10% H_2/90% N_2), if sufficiently dry, is more suitable for sintering Astaloy CrM+graphite (+manganese) mixtures than dissociated ammonia. Graphite and graphite + manganese additions strongly enhance sinter hardening of the samples based on the Astaloy CrM powder.

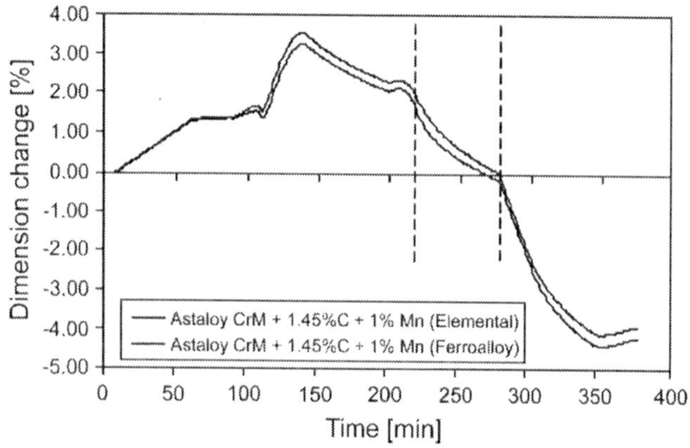

Fig. 11.32. Dilatometric curves for specimens containing different manganese carriers and for $10H_2/90N_2$ sintering atmosphere. Isothermal sintering 1280°C, 1 h [276].

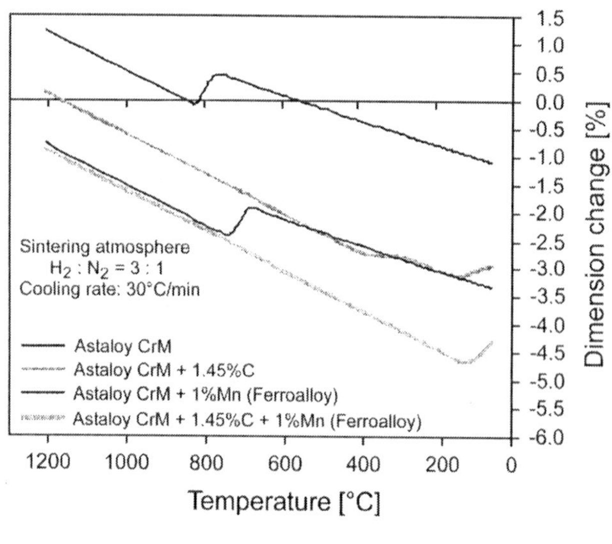

Fig. 11.33. Dilatometric cooling segments for 30°C/min cooling rate and for $75H_2/25N_2$ sintering atmosphere [276].

11.5. Properties of sintered steels based on prealloyed Cr–Mo–Mn powders

11.5.1. Properties of high temperature sintered steels based on prealloyed Cr–Mo–Mn powders

The PM industry has resorted to relatively high alloy contents to produce high strength properties, sintered at temperatures up to 1150°C. In spite of its suitability to high volume production and good

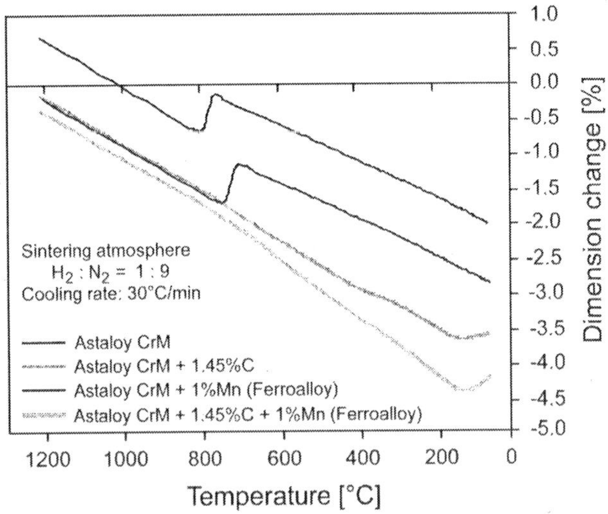

Fig. 11.34. Dilatometric cooling segments for 30°C/min cooling rate and for $10H_2/90N_2$ sintering atmosphere [276].

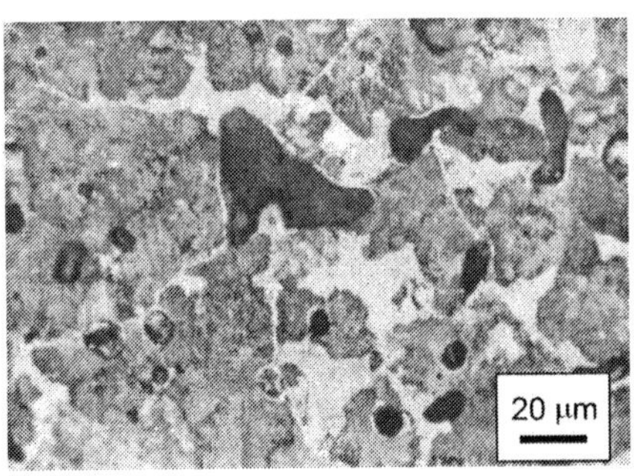

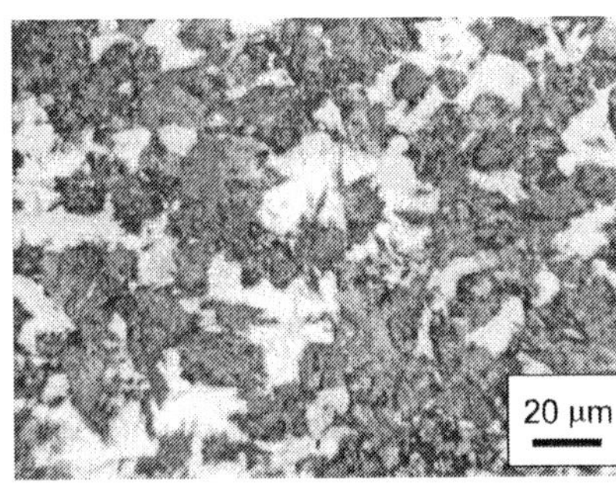

Fig. 11.35 (left). Micrograph of Astaloy CrM+1.45% C steel sintered in 10 hydrogen/90 nitrogen atmosphere and cooled at 6°C/min. Rounded pores and continuous grain boundary carbides [276].

Fig. 11.36 (right). Micrograph of Astaloy CrM+1.45% C steel sintered in 10 hydrogen/90 nitrogen atmosphere and cooled at 30°C/min. Martensite, bainite and retained austenite [276].

strength levels, property deficiencies are obvious when comparison is made with wrought steels. This is particularly the case with regard to fracture toughness, a property governed primarily by density but also to a large extent, by alloying. As a consequence, the adoption of high-temperature sintering (HTS) becomes the fundamental requirement [277].

Process disadvantage and advantage. The major disadvantage of HTS is that it requires expensive furnaces. An operating practice to be recommended is the dedication of furnaces to a strict temperature

Table 11.30. Alloy types

Alloy type	Application
FeMnC	as sintered, high endurance
FeMnMoC	high hardenability, low distortion
FeMn(Cr)MoC	high ductility, formable
FeMnCrMoC	very high strength

and speed regime which is not governed by the part being sintered. Conventional and new developed alloys are processed at 1280°C. The primary motivation is the superior properties. HTS gives rise to fewer nickel-rich zones in the microstructure.

As sintered FeMnC grades. The main requirement of a large proportion of parts produced a fall into this category. Clutch backing and apply plates for automatic transmission for example, require good high stress–low cycle endurance properties, which in turn means good fracture toughness. Table 11.30 lists the alloy types according to requirements for individual application areas. It follows from the literature that copper and nickel would not be obvious choices when designing an alloy for maximum dynamic endurance. Ferrite solution strengthening is more effectively achieved by addition of manganese. Chromium and molybdenum would be obvious choices for developing hardenability.

For examples, Fig. 11.37 shows a clutch backing plate durability which was attained in the year 1980 up to 1995. Figures 11.38–11.40 show the mechanical properties of FC-0208 compared to Fe–1Mn–0.7C steel which are clearly higher.

The process has promted the development of tailored alloys such as sintered FeMnC grades, soft gas pressure quenched FeMnMoC and carburising grades in FeCrMnMoC, with a significantly superior balance of the properties compared to conventional PM alloys.

The main advantage of HTS is the scope which offers in alloy development, in particular when coupled with new densification methods.

11.5.1.1. Effect of nickel addition on properties of manganese steels based on prealloyed powders

The present work analyses the influence of nickel and manganese on alloying and microstructural development during sintering of PM steels. This work also compares the influence of Mn, either externally

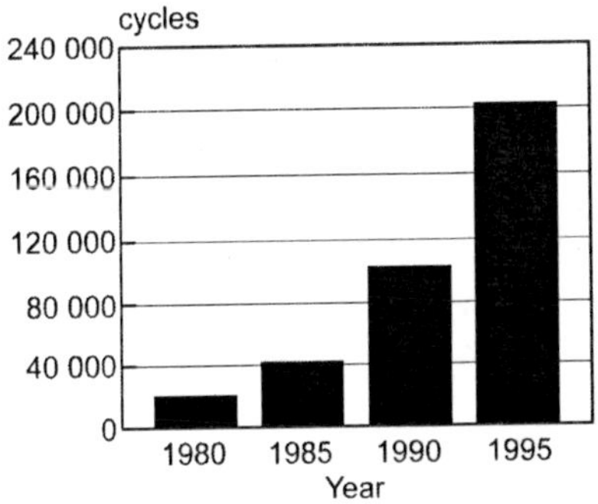

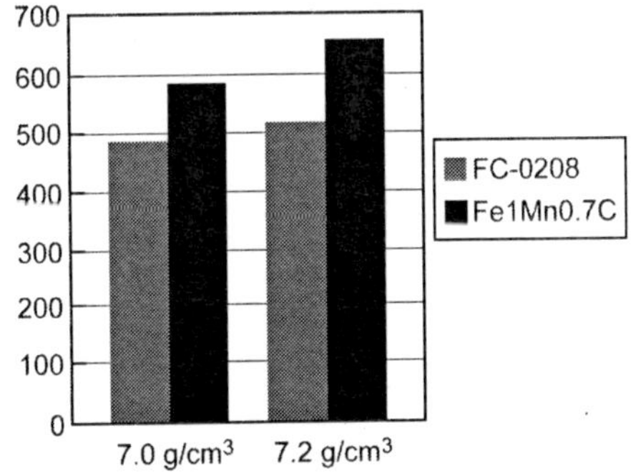

Fig. 11.37 (left). Clutch backing plate durability [277].
Fig. 11.38 (right). Tensile strength [MPa] [277].

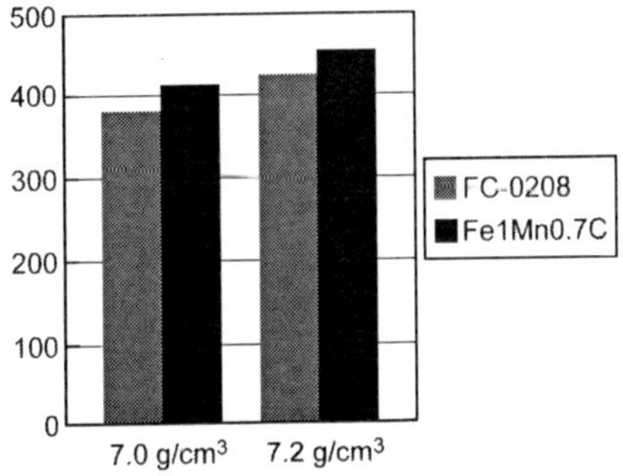

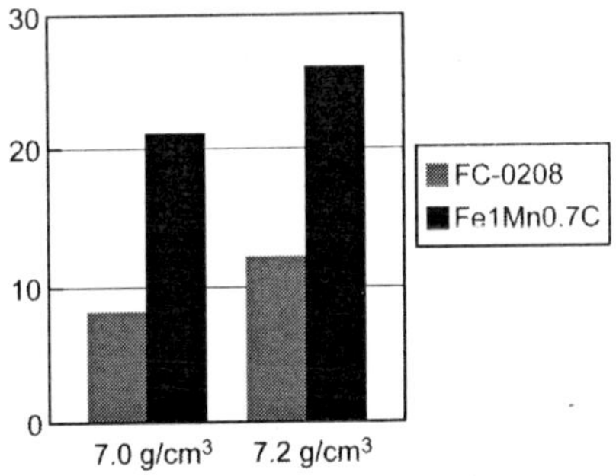

Fig. 11.39 (left). Yield strength [MPa] [277].
Fig. 11.40 (right). Impact energy [J] [277].

added as a master alloy or as part of a prealloyed Fe-base powder, on the microstructure of selected sintered steels [278].

To this purpose, a prealloyed powder containing Fe–1.4C–0.8Mo–0.2Ni–0.6Mn was water-atomised under laboratory conditions. The powder mixture with a graphite addition was compacted at 750 MPa to reach a density of 7.2 g cm^{-3} and sintered at 1240°C for 30 min under a $90N_2$–$9H_2$–$1CH_4$ gaseous atmosphere followed by controlled cooling and tempering at 200°C for 1 h. Other Fe-based powders used were Atomet 1001, Ancorsteel 30HP and Astaloy CrL. In the admixed conditions Mn was added as an atomised master alloy. Sintering cycles were performed in a tube furnace at temperatures between 1120 and 1240°C with holding times between 15 and 60 min. The study was concentrated on the microstructure characterisation.

Figures 11.41 and 11.42 illustrate that using prealloyed Fe-based powders of quite modest hardenability the addition of Mn results in the formation of martensite even after conventional cooling rates

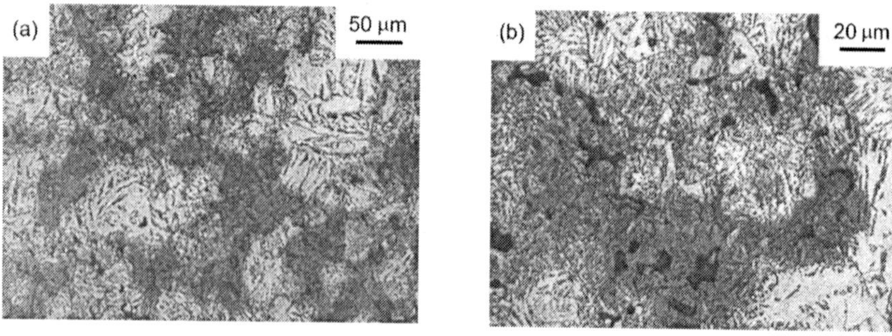

Fig. 11.41. Microstructures at two magnifications of Mn-containing steel: (Fe–0.3Mo)–1.5Mn–0.37C sintered at 1120°C for 60 min [278].

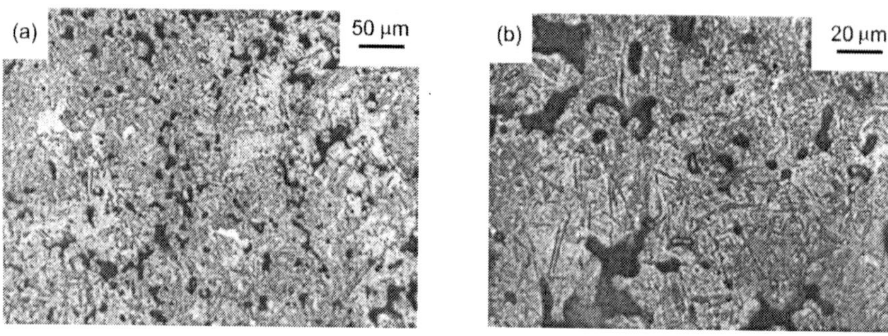

Fig. 11.42. Microstructures at two magnifications of Mn-containing steel: (Fe–1.5Cr–0.2Mo)–1.5Mn–0.4C sintered at 1200°C for 60 min [278].

(~0.45°C/min). It is noteworthy that the hardest transformation product is mostly located surrounding the residual porosity while the core of the Fe-based particles transforms to bainite of distinct morphology according to the chemical composition of the prealloyed powders. The microstructures are heterogeneous even after sintering at 1200°C.

Figure 11.43 is included here as an example to illustrate the characteristics of the microstructure of a wide range of sintered steels obtained from powder mixtures containing nickel in admixed form. In the present case the microstructure belongs to PM steel obtained from a powder mixture of prealloyed Fe–0.3Mo with additions of 4% Ni–0.6% C that was sintered at 1120°C during 15 min followed by cooling at 0.17°C/min. The microstructure in this case is constituted by a mixture of Fe–Mo grains that transform to coarse and fine bainite (labelled Bf) engulfed by Ni-rich areas (labelled B) that may transform to martensite in sufficiently Ni-diluted regions. An

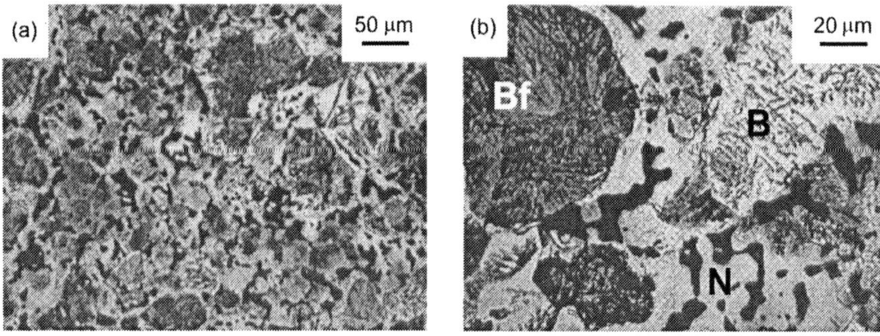

Fig. 11.43. Microstructure at two magnifications of PM steel containing Fe–0.3Mo–4Ni–0.6C at 1120°C for 15 min and followed by slow cooling at 0.17°C/min at two magnifications [278].

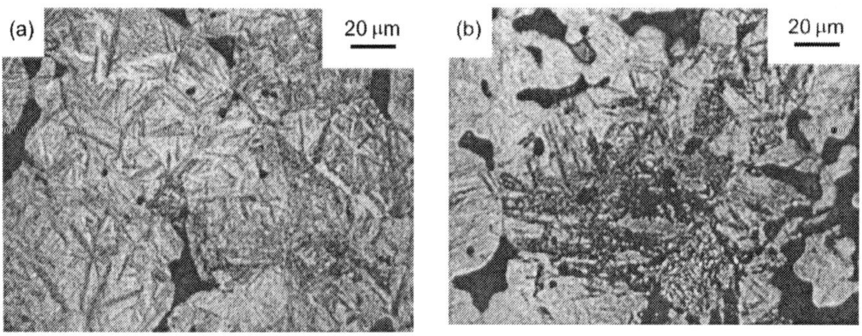

Fig. 11.44. Microstructures of PM steels containing 0.6% C, sintered in $90N_2$–$9H_2$–$1CH_4$ atmosphere at 1240°C for 30 min. Fully prealloyed grade (a), and with admixed Mn and Ni (b) [278].

important consequence of the behaviour of Ni is that it does not only produce grain refinement through the fragmentation of Fe-based particles but also causes a noticeable modification of pore morphology and acts activating material transport.

As shown by the micrographs shown in Fig. 11.44, the microstructure of the sintered prealloyed grade is chemically homogeneous and consists of a mixture of martensite and small amounts of lower bainite after cooling. In contrast, the specimen containing additions of Mn and Ni shows a larger amount of bainite.

Because of the diffusion routes followed and the relative volume diffusion coefficients between Mn, Ni and Fe steels obtained from the powder blends containing Mn and/or Ni in the admixed form inevitably lead to heterogeneous microstructures within the temperature–time combinations typically used in industrial practice. When Mn and Ni are added together the final microstructure is constituted by 'Mn/Ni'-rich regions engulfing Mn/Ni-free grains corresponding to the core of the Fe-based powder particles used.

11.5.1.2. Properties of Mn–Cr–Mo steels based on prealloyed powders containing different Cr and Mn contents

[279] analysed the evaporation during sintering of Fe–Mn–Cr compacts from prealloyed powders. The samples were compacted at 600 MPa from prealloyed powders coded A, D, G, I with low and high manganese content, mixed with 0.5% HWC lubricant. Powder A is virtually plain iron, with 0.15% Mn, while D is with about 0.6% Mn and I contains about 1.9% Mn. Material G is a complex prealloy steel with 1.4% Cr, 0.7% Mn and some Ni and Mo. Dewaxing of the samples was carried out at 600°C in N_2 for 30 min. The dewaxed bodies were then sintered at three temperatures (1120, 1250 and 1300°C) and for different times (1 h and 4 h isothermally) in a laboratory furnace.

Analysis of the Mn content through X-ray fluorescence (XRF) was carried out at different locations at the upper and lower surfaces, at the upper supper surface after polishing off 0.1 mm, also in the core. Table 11.31 shows the composition of the prealloyed powders with different manganese and chromium content, used as reference.

Compared to the surface, the polished face in material D shows lower manganese content after sintering for 1 h but higher one after sintering for 4 h except for sintering at 1120°C. In material A, with a low alloy content, there are not any differences in Mn content except at 1300°C, 1 h, Figs. 11.45 and 11.46.

For 4 h sintering at temperature of 1250°C and 1300°C, the Mn content in the surface is lower than in the polished surface and the core. In regard to the high vapour pressure of manganese, it was to be expected that the Mn amount in the surface is lower than in the core section, Fig. 11.47.

Table 11.31. Composition of prealloyed powder (determined on pre-sintered compacts by XRF)

Powder	Mn [%]	Cr [%]
A	0.15	0.07
D	0.58	0.19
G	0.68	1.42
I	1.87	0.12

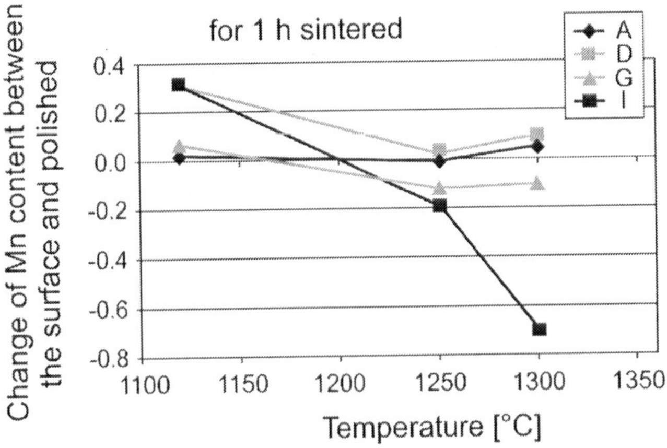

Fig. 11.45. Difference of Mn concentration between the upper surface and upper surface-polished for the different applied sintering temperatures for 1 h (positive value: Mn content is higher at the sintered than at the polished surface) [279].

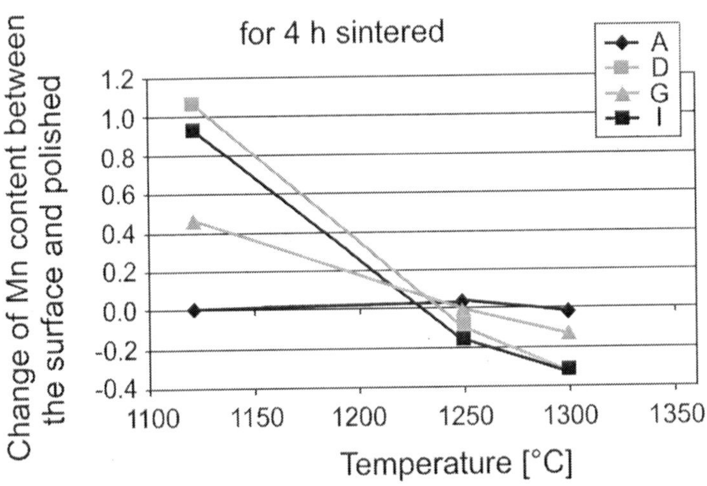

Fig. 11.46. Difference of Mn concentration between the upper surface and upper surface-polished for the different applied sintering temperatures for 4 h [279].

In Fig. 11.48 at 1 h sintering time relatively similar results were obtained – some differences in powders D and A – but all these materials show the same trend as those sintered at 1250°C and 1300°C for 4 h.

The chemical analysis of the sintered compacts through XRF has shown that for the material A, which is virtually plain iron, the Mn content is independent of their location. For the Mn prealloyed

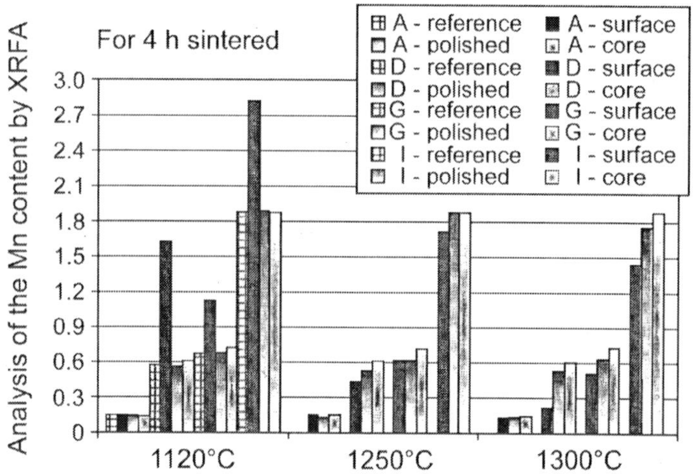

Fig. 11.47. Mn content (by XRF) at upper surface, upper surface polished and core for 4 h polished and core for 1 h [279].

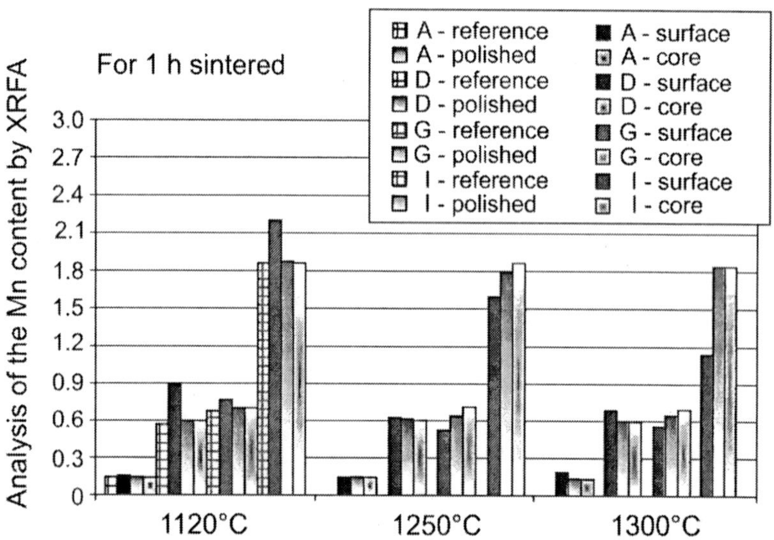

Fig. 11.48. Mn content (by XRF) at upper surface polished and core for 1 h [279].

powder D and especially I, progressive enrichment of Mn was detected at the upper surface after sintering at 1120°C, which is, however, surely an artefact, caused by preferential oxidation of Mn at the surface. When sintering at higher temperatures this enrichment was not observed; in particular after sintering at 1300°C there was a considerable Mn loss. The Cr–Mn prealloyed steel G was slightly sensitive to Mn evaporation compared to, for example, D which has about the same Mn content, indicating that Cr might slow down Mn evaporation. Generally, however, Mn evaporation has to be taken into consideration and isothermal sintering times should be kept short.

11.5.2. Influence of sintering temperature on the properties of prealloyed PM steel containing Cr, Mo and Mn

The following study demonstrates the mechanical properties obtained with prealloyed powder grades containing Cr, Mo and Mn after application of different press-sinter process routes [280].

Three different water-atomised powder grades were used as test materials, Table 11.32. The oxygen content of the powders were in the range 0.0–0.15% O. Powders A and B were mixed with 0.75% graphite for sintering at 1120°C and 0.8% graphite for sintering at 1250°C for both sintering temperatures. Powder C was mixed with 0.8% C for both sintering temperatures. The lubricant amount was 0.6% in all mixes.

Standard tensile test specimens and unnotched impact test specimens were produced with two different compaction pressures (600/800 MPa). A heated tool (60°C) was used for the highest pressure. Specimens of both density levels were sintered in a laboratory mesh belt furnace at 1120°C for 30 min in the $95N_2/5H_2$ atmosphere. High-temperature sintering at 1250°C for 30 min in the $95N_2/5H_2$ atmosphere was done in a laboratory batch furnace using the specimens compacted at 600 MPa. The cooling rate was in the range 0.5–1°C/s in both sintering processes. An overview of the applied process routes is given in Table 11.33.

The basic characteristics of the sintered specimens are given in Table 11.34.

The ultimate tensile strength (UTS) of the materials increases as much with higher density as with higher sintering temperature, Fig. 11.49. Material C has the highest UTS values, with 714 MPa after process 1 and close to 900 MPa after processes 2 and 3.

Table 11.32. Nominal chemical compositions of investigated powder grades

Grade	Fe [%]	Cr [%]	Mo [%]	Mn [%]
A	base	0.8	–	0.4
B	base	1.8	–	–
C	base	1.5	0.2	–

Table 11.33. Process routes applied in the test mixes

Process route	Compaction	Sintering
1	600 MPa	1120°C, 30 min, $95N_2/5H_2$
2	800 MPa, heated tool	1120°C, 30 min, $95N_2/5H_2$
3	600 MPa	1250°C, 30 min, $95N_2/5H_2$

Table 11.34. Carbon and oxygen content and sintered density (SD) for powders A/B/C mixed with 0.75–0.8% graphite after different process routes

Grade	A			B			C		
Process route	1	2	3	1	2	3	1	2	3
C [%]	0.66	0.68	0.67	0.68	0.69	0.67	0.74	0.75	0.69
O [%]	0.04	0.06	0.01	0.04	0.06	0.01	0.05	0.07	0.01
SD [g cm^{-3}]	7.04	7.26	7.06	7.03	7.23	7.08	6.98	7.19	7.06

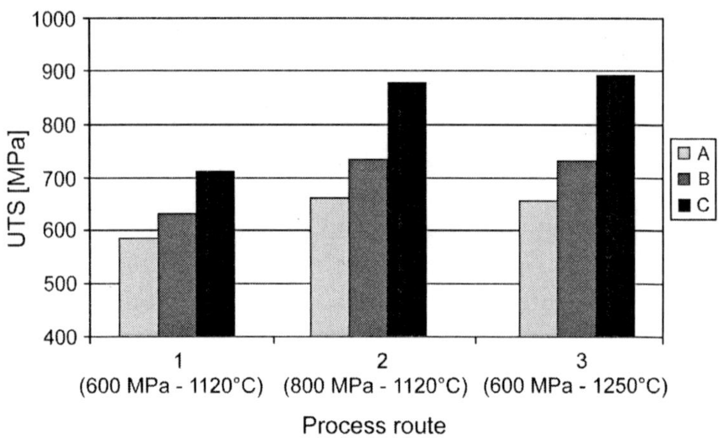

Fig. 11.49. Tensile strength values obtained for powders A/B/C mixed with 0.75–0.8% graphite after different process routes [280].

Correspondingly, the UTS values increase with change in process mode (1→2/3) from 620 to around 735 MPa for material B and from 586 MPa to around 660 MPa for material A. The yield strength (YS) of the materials is enhanced with the change in the process route, Fig. 11.50. However, the YS values are higher after process 3 than after process 2.

Material A has high elongation (A) in the range 3.2–4.0% after all three process routes, as shown in Fig. 11.51. Also material B has high elongation values. Material C has lower elongation than the other

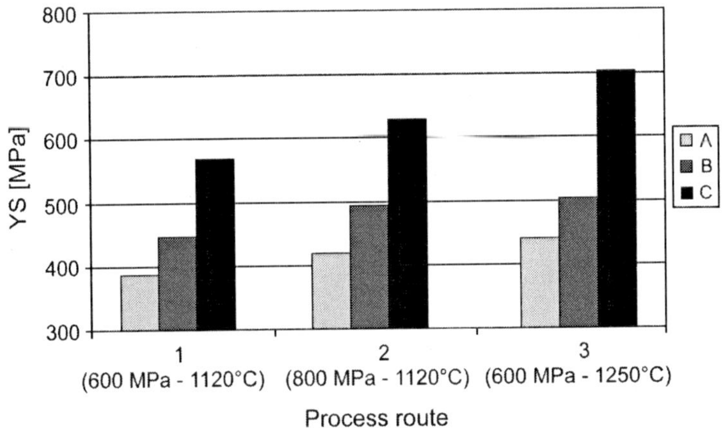

Fig. 11.50. Yield strength values obtained for powders A/B/C mixed with 0.75–0.8% graphite after different process routes [280].

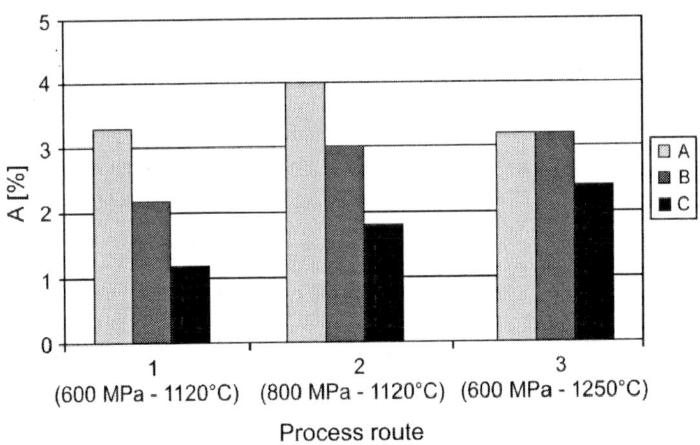

Fig. 11.51. Elongation values obtained for powders A/B/C mixed with 0.75–0.8% graphite after different process routes [280].

materials after all process routes, with the maximum value of 2.4% after process 3. After process 1, the impact strength (IE) value of 19 J was obtained, Fig. 11.52. After process 2, the impact strength was considerably higher for materials A (IE = 30 J) and B (IE = 28 J) but only slightly higher for material C (IE = 23 J). Process 3 resulted to high and equivalent IE values in the range 30–33 J for all materials.

High mechanical performance is achievable with cost effective prealloyed powder grades after conventional press-sinter process routes. The investigated materials provide different property profiles with regard to the combination of strength and ductility.

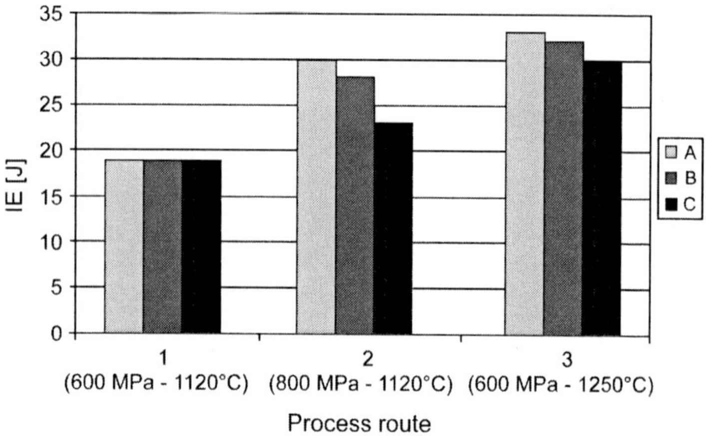

Fig. 11.52. Impact energy values obtained for powders A/B/C mixed with 0.75–0.8% graphite after different process routes [280].

11.5.3. Properties of manganese steels based on CrL and CrM prealloyed powders

The aim of the present study was to establish an economical route of producing low-alloy nitrogen-free structural steels, and second, to substitute Ni, classified as the class 3 carcinogenic and allergenic element, with Mn [281].

Prealloyed Astaloy CrL and Astaloy CrM were used as the base powders. 3% Mn in the form of medium-carbon ferromanganese, and 0.3% ultrafine graphite powder was added to the base powders in order to prepare two mixtures Fe–3% Mn–1.5% Cr–0.2% Mo–0.3% C or Fe–3% Mn–3% Cr–0.5% Mo–0.3% C based on Astaloy CrL and Astaloy CrM powders. The mixed powders were compacted in a die with lubricated walls. The compacts with approximately the same density were prepared for the TRS specimens and dogbone tensile strength bars. Sintering was carried out in a dry (10 ppm moisture) atmosphere with different H_2 to N_2 ratios in a laboratory tube furnace at 1120°C and 1250°C for 60 min, employing slow (1.4°C/min) and convective (65°C/min) cooling. The sintered density of tensile strength specimens of both steels was in the range of 6.84 to 6.91 g cm^{-3} and of transverse rupture strength specimens in the range of 7.00 to 7.09 g cm^{-3} sintered at both temperatures. Sintering was carried out in a semi-closed container. Table 11.35 lists the chemical compositions of the atmospheres used for sintering of the tested steels.

Table 11.35. Sintering atmospheres

Designation of atmosphere	Chemical composition
A0	100% H_2
A1	75% H_2–25% N_2
A2	25% H_2–75% N_2
A3	5% H_2–95% N_2
A4	100% N_2

For each chemical composition of steels, 5 batches were sintered in each (A0–A4) atmosphere and tempered at 200°C for 60 and 240 min.

The mechanical properties of Fe–Mn–Mo–Cr–C steels sintered at 1120°C, conventionally and rapidly cooled and tempered are presented in Tables 11.36 and 11.37.

As shown in Table 11.36, for higher Cr and Mn concentrations high-temperature sintering in nitrogen has to be employed to improve strength properties. The recorded values for UTS, elongation and TRS after sintering are higher than that after the low-temperature process. There are no evident percentage differences between sintering in the H_2-rich atmosphere and pure N_2. For lower Cr and Mo concentrations, slightly better properties were recorded after sintering in the H_2-rich atmosphere than in nitrogen, irrespective of sintering temperature. Table 11.37 summarises the mechanical properties of the investigated steels after tempering.

As shown in Figs. 11.53 and 11.54, the specimens with lower Cr and Mo content, sintered at 1120°C in H_2 and N_2 show a bainitic/martensitic structure, whereas those containing more Cr and Mo contained mainly a martensitic/austenic structure. Figures 11.55 and 11.56 shown for comparison the microstructures of the tested steels not tempered and tempered.

The strength results showed a strong dependence on sintering atmosphere (H_2, N_2), especially for tempered steels. A comparison of the tensile strength, elongation and impact values found quite good agreement. The effect of using a higher sintering temperature than 1120°C on the mechanical properties is evident from the results. The tensile and transverse rupture strength showed a measurable effect of the sintering temperature and tempering time. Increasing Cr content from 1.5% to 3% and/or tempering time offered improvement in the properties under the given conditions. The specimens sintered at 1250°C possessed higher mechanical properties, irrespective of the

Table 11.36. As sintered properties of Fe–Mn–Cr–Mo–C steels cooled at 1.4°C/min (mean values)

Chemical composition	Sintering temperature, [°C], and atmosphere		UTS [MPa]	$R_{0.2}$ [MPa]	A [%]	TRS [MPa]	KC [J]	HV 30
Fe–1.5% Cr–0.2% Mo–3% Mn–0.3% C	1120	A0	662	408	1.8	1460	5.6	231
		A1	512	455	1.3	872	4.1	190
		A2	438	361	0.9	1024	2.7	207
		A3	400	342	0.8	814	2.8	247
		A4	374	285	0.7	909	2.6	229
	1250	A0	629	453	1.5	1514	11.2	222
		A1	602	440	1.6	1226	5.7	260
		A2	510	462	1.1	1214	5.8	249
		A3	456	484	0.8	1208	3.9	235
		A4	466	440	1.2	970	3.9	225
Fe–3.0% Cr–0.5% Mo–3% Mn–0.3% C	1120	A0	598	541	0.9	1104	5.1	322
		A1	584	498	1.1	969	4.0	255
		A2	532	451	1.1	1048	3.9	302
		A3	515	386	1.0	972	3.5	341
		A4	495	479	1.0	975	3.9	353
	1250	A0	774	546	1.8	1425	7.7	254
		A1	816	587	2.5	1469	7.2	326
		A2	672	569	1.5	1145	4.7	345
		A3	572	526	1.1	1026	4.8	351
		A4	670	–	2.0	1056	3.7	362

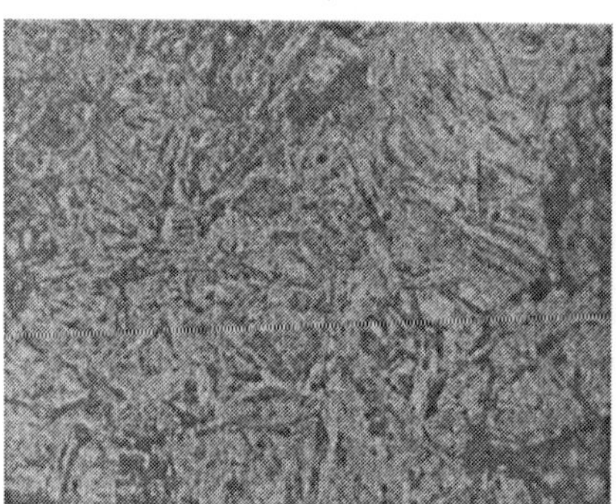

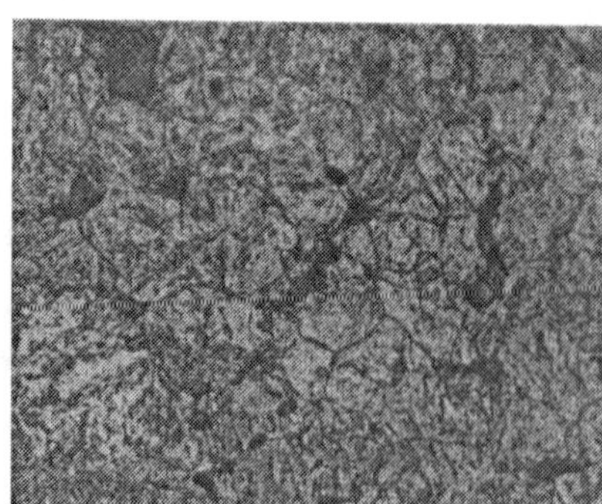

Fig. 11.53 (left). Micrograph of Astaloy CrM-based steel sintered at 1120°C in H_2; austenite and martensite, ×500 [281].

Fig. 11.54 (right). Micrograph of Astaloy CrM-based steel sintered at 1120°C in N_2; austenite/martensite/bainite, ×500 [281].

Table 11.37. Mechanical properties of Fe–Mn–Cr–Mo–C steels sintered at 1120°C, rapidly cooled at 65°C/min and tempered at 200°C

Chemical composition/sintering atmosphere/tempering time [min]			UTS [MPa]	$R_{p0.2}$ [MPa]	A [%]	TRS [MPa]	KC [J]	HV 30
Fe–1.5% Cr–0.2% Mo–3% Mn–0.3% C	A0	0	545	–	1.0	1171	6.0	198
		60	554	481	0.8	1216	3.4	245
		240	625	553	0.6	1163	5.2	263
	A1	0	514	507	0.8	825	4.1	222
		60	439	–	0.6	1017	3.0	262
		240	662	457	1.1	1045	3.69	210
	A2	0	634	517	1.2	1003	4.1	342
		60	662	608	1.0	1049	3.6	318
		240	608	554	0.9	1018	7.0	280
	A3	0	654	498	1.1	805	4.8	317
		60	542	481	0.8	1020	3.0	284
		240	506	405	0.7	912	2.5	278
	A4	0	495	450	0.58	765	4.3	315
		60	541	481	0.8	1191	3.4	274
		240	544	508	0.8	1082	3.4	298
Fe–3.0% Cr–0.5% Mo–3% Mn–0.3% C	A0	0	–	–	–	1041	4.6	389
		60	439	357	0.6	1195	5.8	305
		240	687	635	1.2	1096	5.5	291
	A1	0	447	–	0.7	702	3.0	206
		60	563	–	0.8	1030	3.3	258
		240	535	474	0.8	1260	3.5	310
	A2	0	642	537	1.1	689	5.5	400
		60	645	560	1.1	1105	5.0	317
		240	575	517	0.9	1035	4.0	309
	A3	0	416	–	0.5	828	4.1	372
		60	648	537	1.7	982	3.3	323
		240	621	577	2.0	941	3.2	307
	A4	0	695	492	1.9	399	2.5	395
		60	706	535	2.5	855	4.1	320
		240	700	587	2.5	1131	4.1	319

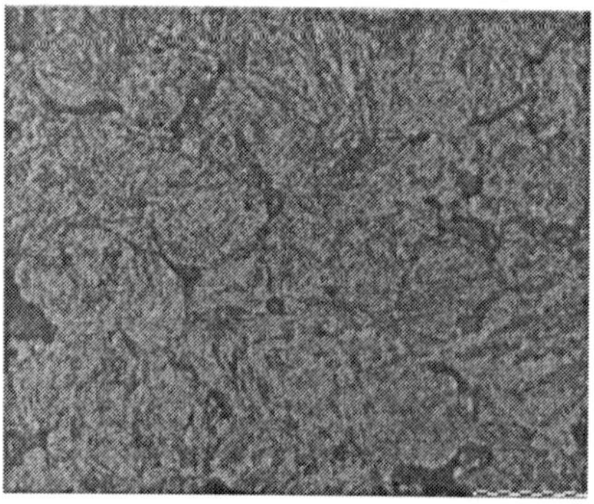
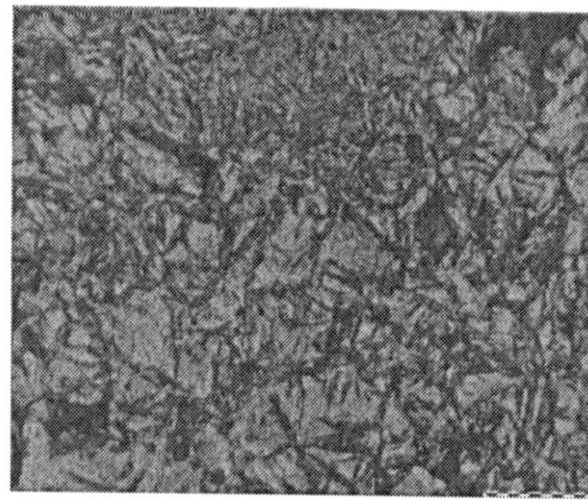

Fig. 11.55 (left). Micrograph of Astaloy CrM-based steel sintered at 1120°C in N_2, not tempered; austenite, martensite, acicular bainite, ×500 [281].
Fig. 11.56 (right). Micrograph of Astaloy CrM-based steel sintered at 1120°C in N_2, tempered at 200°C for 60 min; austenite, martensite, ×500 [281].

H_2–N_2 ratio in the furnace atmosphere. The tensile and transverse rupture strengths were approximately directly proportional to the H_2 content in the sintering atmosphere.

In [282] investigations were carried into the effect of the sintering atmosphere with different ratios of hydrogen and nitrogen with a dew point of –60°C on the properties of Fe–Cr–Mn–Mo–C steels based on AstCrL and AstCrM prealloyed powders sintered at 1120°C and 1250°C for 60 min. 3% of manganese was added in the form of a medium-carbon ferromanganese powder, and 0.3% carbon as an ultrafine graphite powder. To improve the local dew point of the microatmosphere and to minimise the loss of manganese due to volatisation, sintering was carried out in a semi-closed stainless steel container. Table 11.38 summarises the mechanical properties of the investigated steels after sintering.

Taking into account the tensile strength values in relation to the composition of the atmosphere, sintering in 100% H_2 did not result in the highest strength values. These were obtained at different H_2 and N_2 ratios. In can be seen that in a number of cases the highest strength values were recorded at sintering in the 5% H_2–95% N_2 and also 100% N_2 atmosphere. The mechanical properties after sintering at 1250°C were higher than after sintering at 1120°C.

Table 11.39 gives the carbon and oxygen contents after sintering for both material systems. According to these data, the highest oxygen content was recorded for the steels sintered in the A0 atmosphere and the lowest for the steels sintered mostly in the A3 and A4 atmospheres.

Table 11.38. Mechanical properties of sintered Fe–Mn–Cr–Mo–C steels sintered at 1120°C and 1250°C in atmospheres with different ratio of H_2 and N_2 with a dew point of –60°C. A0 = 100% H_2, A1 = 75% H_2–25% N_2, A2 = 25% H_2–75% N_2, A3 = 5% H_2–95% N_2, A4 = 100% N_2

Chemical composition / sintering atmosphere		UTS [MPa]	TRS [MPa]	Impact toughness [J]	HV30
1120°C					
CrL+3Mn +0.3C	A0	545	1171	6.04	198
	A1	514	825	4.09	222
	A2	634	1003	4.11	342
	A3	654	805	4.84	317
	A4	495	765	4.26	315
CrM+3Mn+0.3C	A0	–	1041	4.59	389
	A1	447	702	3.00	206
	A2	642	689	5.48	400
	A3	416	828	4.05	372
	A4	595	399	2.47	395
1250°C					
Crl+3Mn +0.3C	A0	573	1354	13.97	252
	A1	594	1214	11.35	253
	A2	622	1530	7.72	220
	A3	799	1519	5.23	296
	A4	822	1428	10.35	281
CrM+3Mn +0.3C	A0	858	1349	7.85	314
	A1	781	1501	8.23	306
	A2	886	1610	7.85	331
	A3	901	1572	13.00	334
	A4	802	1675	14.07	328

11.5.4. Properties of sintered steels based on Cr and Cr–Mn prealloyed powders with various Cr and Mn content

Two kinds of water-atomised powders, one prealloyed with 1.8% Cr (powder A) and the other prealloyed with 0.8% Cr and 0.4% Mn (powder B), were used as test materials. The water-atomised steel powder grade Astaloy Mo (powder C), which is prealloyed with 1.5% Mo, was included in the study as reference material. The chemical compositions of the investigated powders are given in Table 11.38 [283].

Table 11.39. Oxygen and carbon content of PM Fe–Mn–Cr–Mo–C steels

Sintering temperature and atmosphere		AstCrL+3Mn+0.3C		AstCrM+3Mn+0.3C	
		C [wt.%]	O_2 [wt.%]	C [wt.%]	O_2 [wt.%]
1120°C	A0	0.287	0.587	0.303	0.507
	A1	0.266	0.331	0.277	0.310
	A3	0.289	0.460	0.304	0.390
	A4	0.274	0.338	0.276	0.333
1250°C	A0	0.221	0.252	0.218	0.343
	A1	0.208	0.140	0.221	0.161
	A3	0.190	0.157	0.228	0.290
	A4	0.191	0.165	0.214	0.280

Each powder was mixed with 0.8% graphite and a lubricant (0.6% Kenolube or 0.5% Lube E). Two premixes (one for each lubricant) were made for all three kinds of powder. Standard tensile test specimens and unnotched impact test specimens with three different density levels were produced by cold compaction with 600 MPa compaction pressure, compaction at 800 MPa with heated tool (80°C), and double pressing with compaction pressures 700/900 MPa, respectively. The specimens were pre-sintered at 750°C for 30 minutes in the $75H_2/25N_2$ atmosphere between the two compaction steps in the double pressing operation. Chemistry of the tested specimens is listed in Table 11.40, and specimen green densities obtained for the different materials are presented in Table 11.41.

Test specimens of all material types and density levels were sintered for 30 min at three different temperatures (1120/1200/1250°C). Sintering at 1120°C was done in a laboratory mesh belt furnace, while sintering at 1200°C and 1250°C was carried out in a laboratory batch furnace. All specimens from high-temperature sintering were resintered in the mesh belt furnace (1120°C for 30 min) in order to ensure that the same cooling rate after sintering was applied for all materials. This cooling rate was between 0.5°C/s and 1°C/s in the temperature range 800–300°C. High-purity

Table 11.40. Chemistry of investigated prealloyed powder grades

Powder	Fe [%]	Cr [%]	Mn [%]	Mo [%]	O [%]	C [%]
A	Base	1.80	0.10	0.02	0.14	< 0.01
B	Base	0.82	0.39	0.03	0.13	< 0.01
C	Base	0.03	0.13	1.53	0.09	< 0.01

Table 11.41. Green densities of tensile test specimens after different pressing operations. (CC = cold compaction, HTC = heated tool compaction, DP = double pressing)

Material	Powder	Graphite [%]	Green density [g cm^{-3}]		
			CC – 600 MPa	HTC – 800 MPa	DP – 700/900 MPa
A-08C	A	0.8	6.99	7.23	7.54
B-08C	B	0.8	7.05	7.28	7.54
C-08C	C	0.8	7.06	7.31	7.54

90N_2/10H_2 gas mixture (dew point –40°C) was used in all sintering processes.

The bulk oxygen content in A and B material with density of ~7.0–7.3 g cm^{-3} was minimal (0.01–0.04 %) and the highest one (0.06–0.08%) was in the materials with the density over 7.5 g cm^{-3}. The lowest bulk oxygen content was determined in the steels sintered at 1250°C.

Mechanical properties. The influence of specimen density and sintering temperature on the ultimate tensile strength (UTS) and yield strength (YS) of the sintered materials is given by graphs a – d in Fig. 11.57. For material A-08C, the UTS increases from 650 MPa at a density of 7.0 g cm^{-3} to 960 MPa at a density of 7.5 g cm^{-3} after sintering at 1120°C. The UTS values of A-08C are 5–10% higher at each density level when the sintering temperature was raised from 1120°C to 1200°C. The same gain in UTS was achieved when the sintering temperature was raised from 1200°C to 1250°C, except at the highest density level where UTS is just above 1000 MPa in both cases. The YS of material A-08C is about 490 MPa at the lowest density level (7.0–7.1 g cm^{-3}) for all three sintering temperatures. At the next density level (7.2–7.3 g cm^{-3}), the YS is 535–540 MPa after sintering at 1120/1200°C and 580 MPa after sintering at 1250°C. For the highest specimen densities (7.5–7.6 g cm^{-3}), the YS of A-08C is 615 MPa after sintering at 1120°C and in the range 660–680 MPa after sintering at 1200/1250°C.

The UTS of material B-08C increases from 595 MPa at a density of 7.1 g cm^{-3} to 830 MPa at a density of 7.5 g cm^{-3} after sintering at 1120°C. The YS of material B-08C is 420–440 MPa at a density of 7.1 g cm^{-3}, 480–500 MPa at a density of 7.3 g cm^{-3}, and 505–540 MPa at a density of 7.5 g cm^{-3}.

The reference material C-08C shows UTS values between 550 MPa (7.1 g cm^{-3}) and 770 MPa (7.5 g cm^{-3}) after sintering at

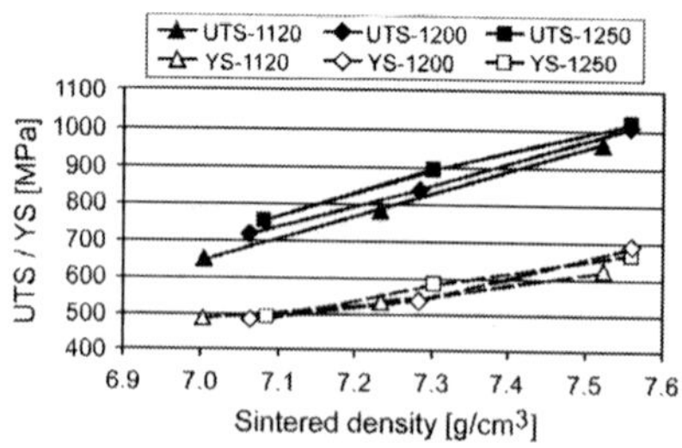

a) UTS and YS for material A-08C

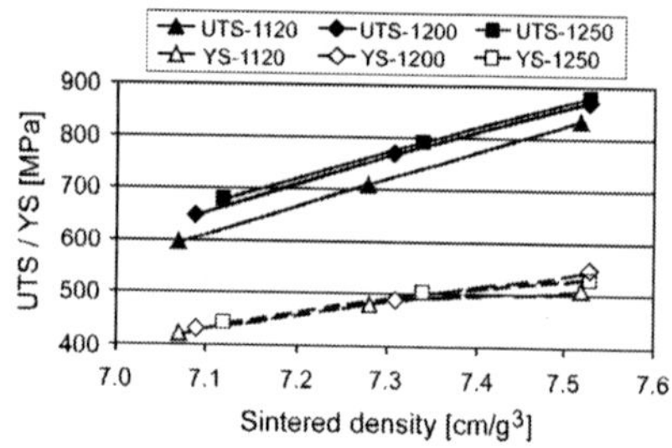

b) UTS and YS for material B-08C

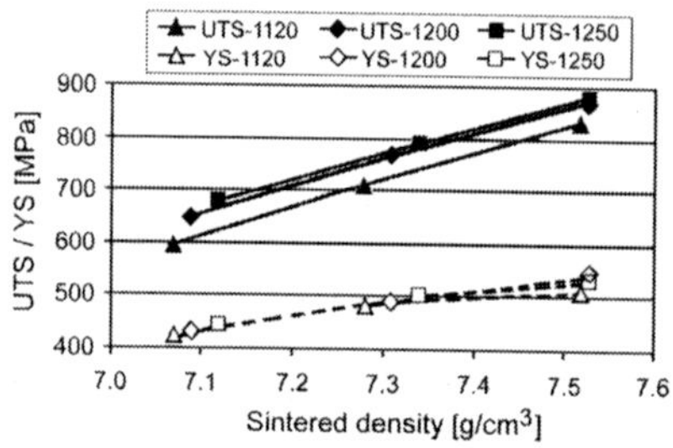

c) UTS and YS for material C-08C

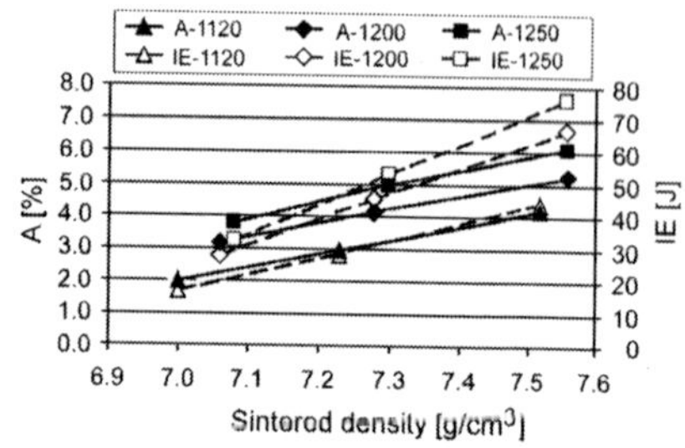

d) A and IE for material A-08C

Fig. 11.57. Mechanical properties obtained after sintering for 30 min at 1120/1200/1250°C in $90N_2/10H_2$ atmosphere (UTS = ultimate tensile strength, YS = yield strength, A = elongation, IE = impact energy) [283].

1120°C. The YS of material C-08C is similar (450–475 MPa) for all sintering temperatures at a density of 7.1 g cm^{-3}. At a density of 7.3 g cm^{-3}, there was no significant difference in YS (520–530 MPa) after sintering at 1120°C and 1200°C, while YS has increased to 550 MPa after sintering at 1250°C. The YS of C-08C is 560–595 MPa at the highest density level (7.5 g cm^{-3}) and the highest value is reached after sintering at 1200°C.

Specimen hardness values obtained after sintering at 1120°C are presented in Fig.11.58. Material A-08C has the highest hardness of the three materials, with values between 184 HV10 at a density of 7.0 g cm^{-3} and 263 HV10 at a density of 7.5 g cm^{-3}. Corresponding hardness ranges for the other materials are 167–220 HV10 (B-08C) and 176–216 HV10 (C-08C) in the density interval 7.1–7.5 g cm^{-3}.

According to [284], when sintering FeMn and Cr prealloyed powders in inert atmospheres, on the basis of the dilatometry tests the reduction of the surface oxides tends to be shifted to markedly higher temperatures than of Fe–C or conventionally alloyed PM steel grades. In Fe–C the reduction starts at about 700°C while in

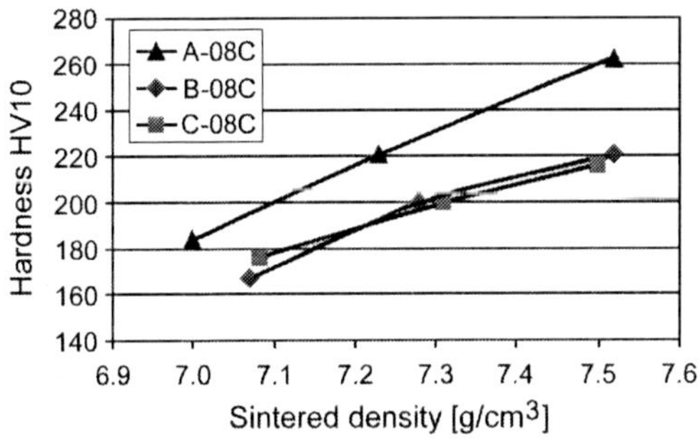

Fig. 11.58. Specimen hardness values HV 10 obtained after sintering for 30 min at 1120°C in $90N_2/10H_2$ atmosphere in dependence on sintered density [283].

the prealloyed grades there is a first peak at even slightly lower temperatures only if the alloy element content is moderate, with the peak completely disappearing at >1.5% Mn, and the reduction of the remaining oxides then starts at 1100°C, in both cases the removal of surface oxides being the first major process. The reduction is not complete at temperatures up to 1300°C. Sintering in high-purity H_2 results in partial reduction at 400°C, with H_2O being formed as the product, but removal of the remaining oxides requires more or less the same temperatures as in inert atmospheres, and the reaction product is mainly CO_2, thus confirming that at lower temperatures carbon is a reducing agent in any sintering atmosphere. This has been considered when defining the desired carbon level in the steels, since in Cr- and Mn-alloyed grades, the carbon loss during sintering also in H_2 will be significant.

For industrial processes, sintering of both Cr- and Mn-prealloyed steel powder compacts at significantly high temperature, typically >1200°C, is recommended to ensure effective deoxidation and thus sound interparticle bonding. Mn evaporation is less pronounced if Mn prealloying is employed compared to admixing, and there is also less risk of Mn heterogeneity in the sintered specimen.

In [285] is stated that the low-alloyed sintered steels, which use as alloying elements Cu, Ni and Mo, also make use of the elements Cr and Mn and partly also V, Si and B. Higher reactivity of these elements in relation to the sintering atmosphere, first of all the higher oxygen affinity, has expressive consequences for sintering and heat treatment. At sintering of the structural parts from Cr-prealloyed powders the removal of oxide layers on the particles and the formation of the stable interparticle necks in an inert atmosphere, in which C acts as reduction agent, as well at sintering in the H_2-containing atmosphere are more extensive than as with the unalloyed steels; this requires markedly higher temperatures, which in the region technically usual sintering temperatures can be lower. Actually, at these new sintered steel types for which high sintering temperature is notably advantageous, at which the mechanical properties in the case of high green densities are markedly improved. When manganese is used as an alloying element, high vapour pressure at sintering is important and supports the homogenisation of Mn in the steel matrix at sintering of admixed steel systems. The loss of Mn on the surface of the parts at industrial sintering appears less critical than at laboratory sintering. In some cases nitrogen in the atmosphere also plays an active role at sintering through the stabilisation of the nitrides, like V in the retardation of homogenisation or B which results in full blocking of the sintering activating effect.

In study [286], concerned with sintering in different reducing atmospheres, the role of manganese in relation to the investigated factors is also investigated.

Manganese forms more stable oxides than chromium and the reduction is therefore greater. But manganese due to its physical–metallurgical properties is extremely interesting because through the combination of its intense effect on the hardenability of the steel (higher Grossman factor) and the actual low price it has been favoured fo many years. However, up to present time the introduction of the powder metallurgy Mn-steels in the industry has not been successful although the scientific and technological works of manganese started many years ago and the results were published.

The doubt was that the additions of manganese in the form of elemental powder or high-alloyed Fe–Mn master alloys, due to their highly hardening effect, should have a worsening influence on the compressibility of the compacts and, therefore, the addition of manganese as an alloying element has not been introduced in the industrial practice. It is here necessary to mention the specific property of manganese – high manganese vapour pressure by which manganese differs substantially from other alloying elements. High manganese vapour pressure through its vapour forms the self cleaning/getter effect for the sintering atmosphere and by this makes it possible to sinter admixed manganese steels. However, it is also known that here there is a risk of the formation of low-melting phases with wall materials for industrial sintering furnaces. It is obviously that no PM parts producer can tolerate the risk from the defect of the wall material because there is insufficient information about it at present.

Part III

12

Sintering of manganese steels in low- and high-purity atmospheres: Results and evaluation

Initial research studies, concerned with manganese as the alloying element of sintered iron-based steels, were perfected in the 50s as a significant inlet of manganese in powder metallurgy alloys. In all subsequent studies concerned with manganese, the mechanical and other properties of sintered manganese steels obtained in the subsequent 20 years were similar to those of the initial materials and the sintering conditions without taking into account the type and purity of the protective atmospheres. At the beginning of the 70s, special attention started to be paid to the research of sintered manganese steels because of the increase of the price of nickel. Since then, as a result of the larger and larger number of research studies and more detailed analysis of the sintered Fe–Mn system, the majority of thermodynamic studies started to stress the high affinity of manganese for oxygen and, consequently, difficulties in the reduction of its oxides. For this reason, it became necessary to sinter manganese steels in high-purity atmospheres, with the purity which can not be obtained especially in the industrial production of sintered components, for example, with a dew point up to –70°C. This caused and still causes fear of the oxidation of particles of the manganese carrier during sintering in Fe-Mn compacts, especially already during preheating in the atmospheres without low dew points.

The aim of the studies carried out so far has been and is partial replacement of nickel and copper with manganese in the production of sintered alloyed steels, also because of the considerably higher ferrite hardening effect of manganese in comparison with other alloying elements. On the basis of the currently available results and properties of the sintered manganese steels, it has been possible to form an independent group of manganese-alloyed steels based on iron and prealloyed powders, supplementing the available sintered alloyed material systems. Another reason for the interest in the application of manganese as an alloying element for alloying sintered steels is its very low price, especially in the form of various types of ferromanganese also in comparison with other types of master alloys and, in particular, the price of nickel and copper.

The mechanical and other properties of the manganese steels, obtained in the studies concerned with manganese and published so far, are similar to or higher than those of the steels alloyed with conventional alloying elements.

This is so despite the fact that in the sintering of the manganese steels under greatly different sintering conditions, including sintering atmospheres and their purity, although according to the given thermodynamic data none of the atmospheres used in all works presented in this monograph fulfilled the requirements for the purity to protect manganese against oxidation or to reduce MnO formed during the sintering process regarding the thermodynamics mainly during preheating up to the $\alpha - \gamma$ phase transformation temperature. The atmosphere with the dew point of –30°C could not to fulfil this requirement during the entire sintering process. The dew point at the hold at 1150°C should be below –50°C if only isothermal sintering would be carried out. The atmosphere with the dew point of –70°C has oxidising character below 930°C, *i.e.* during the substantial part of the preheating stage. Manganese begins to oxidise in air at ~400°C.

12.1. Basic thermodynamic and physical characteristics of manganese in term of its use for alloying of sintered steels

All requirements concerning the formation and reduction of manganese oxides during the sintering of mixed manganese steel according to the Richardson–Ellingham diagram are related to MnO only. The basic reaction of manganese and oxygen expresses the equation:

Table 12.1. Equilibrium data of Mn/MnO in O_2 (partial oxygen pressure) gas mixture of H_2/H_2O and corresponding water contents; d. p. – dew point

	Temperature [°C]								
	600	700	800	900	1000	1100	1150	1200	1400
P_{O2} [Pa]	10^{-38}	10^{-34}	10^{-30}	10^{-26}	10^{-24}	10^{-22}	10^{-21}	10^{-20}	10^{-16}
H_2/H_2O	1×10^{8}	1×10^{7}	3×10^{6}	7×10^{5}	2×10^{5}	9×10^{4}	5.0×10^{4}	4.0×10^{4}	2.7×10^{3}
P_{H2O} [Pa]	0.01	0.05	0.13	1.06	2.35	2.06	3.89	10.10	37.67
d. p.[°C]	–102	–90	–80	–75	–60	–57	–50	–40	–30
H_2O vol. [%]	1×10^{-6}	9×10^{-6}	5.2×10^{-5}	1.4×10^{-4}	1.1×10^{-3}	1.3×10^{-3}	2.0×10^{-3}	3.8×10^{-3}	3.7×10^{-2}

$$2Mn + O_2 = 2MnO. \qquad (1)$$

MnO, green in colour, is reducible only with solid carbon at the temperature higher than 1280°C and also neither with endothermic gas and dissociated ammonia nor with the nitrogen-based atmospheres, at different dew points, the formed oxides can be reduced or further oxidation of manganese or chromium can be prevented.

All considerations concerning the reducibility of manganese oxide (MnO) during the sintering of mixed manganese steels can be realised by equilibrium conditions for the reaction in the presence of oxygen and in the H_2/H_2O mix shown in Table 12.1.

It is possible to deduce from these data why manganese as a PM alloying element has not found up to the present time application in the production of sintered parts. These thermodynamic data were and are in most cases considered as a decisive factor against manganese in spite of the results obtained at the actual time of sintering of manganese steels.

The thermodynamic requirements for the purity of a sintering atmosphere should be not constant in the entire sintering temperature range. The total preheating period and the cooling stage under laboratory and especially under industrial sintering conditions need the atmosphere with higher purity than that for isothermal sintering.

The preheating period in the sintering process is that in which the dewaxing of the compacts occurs at which the lubricant melts and evaporates followed by burn-off requiring a humid atmosphere at the temperature below ~600°C. On the other hand, the Mn-containing parts require for elimination the oxidation of the manganese carrier particles in the compacts an atmosphere with an extremely low dew point, for instance –102°C at 600°C, whis is not realisable in practice. Beside this, in the preheating period interparticle necks start to form in the compacts together with alloying of base powders with all alloying elements, including manganese, despite the fact that in the given thermodynamic conditions of sintering the manganese particles in the compacts should have been, according to equation (1), coated with the almost irreducible oxide MnO. Despite this, as mentioned in all the previously published studies, sintering of the compacts with the addition of manganese particles already during preheating resulted in the formation of interparticle necks and in the initial alloying of the particles of the main powders with manganese. This was reflected in the tensile strength values after sintering at 875°C, as shown in Fig. 5.32. Otherwise, if the oxides or chemisorbed oxygen on the surface of the powder particles were not reduced, no necks could form between the particles during sintering.

All the previously investigated processes of sintering of manganese steels have been successful, irrespective of the sintering conditions and the initial material used as well as the type of manganese carrier grade. This is the result of the fact that manganese belongs to the crystalline materials which exhibit significantly higher vapour pressure than, for example, Ni, Cu, Mo, Cr, Si and P and attain the barometric pressure at lower temperatures than those at which they melt. Consequently, at heating under the pressure of 1000 hPa they proceed directly into the gaseous state termed as sublimation (sublimation point). The formation of gaseous manganese by sublimation from the manganese carrier at sintering is therefore the consequence of the physical property of manganese demonstrated in high vapour pressure in dependence on temperature.

Table 12.2 gives the manganese vapour pressure values and sublimation rates of maganese. In Table 12.3 are listed the calculated isothermal sublimation times from the cubic Mn particles of three sizes and the manganese vapour volume formed from 1 g of manganese. Both data in dependence on temperature.

The calculation refers to the geometrical surface of the particles of elemental manganese and to the temperature range of 400°C to 1250°C covering the temperature ranges at which the oxidation of manganese in air starts (400°C), and most sintering processes are perfected up to 1250°C, regarding the melting point of manganese of 1244°C. The longer sublimation time compared to that for pure manganese can be expected by using ferromanganese or some special Mn-containing master alloys as manganese carriers due to their lower chemical activity (lower Mn content).

Table 12.2. Manganese vapour pressure and sublimation rate of manganese in dependence on temperature

Temperature [°C]										
400	500	600	700	800	900	1000	1100	1150	1200	1250
Manganese vapour pressure [Pa]										
$4.0\cdot10–^{10}$	$2.3\cdot10^{-7}$	$2.9\cdot10^{-5}$	$1.3\cdot10^{-3}$	$2.9\cdot10^{-2}$	0.38	3.2	20.0	45.0	99.0	188.0
Sublimation rate [$g\cdot cm^2\cdot s^{-1}$]										
$1.0\cdot10^{-14}$	$2.6\cdot10^{-11}$	$2.2\cdot10^{-9}$	$4.0\cdot10^{-9}$	$8.7\cdot10^{-8}$	$9.8\cdot10^{-7}$	$8.1\cdot10^{-6}$	$4.8\cdot10^{-5}$	$1.03\cdot10^{-5}$	$2.2\cdot10^{-4}$	$4.0\cdot10^{-4}$

Table 12.3. Calculated isothermal rates and time of manganese sublimation from the cubic elemental manganese particles 10, 20 and 50 mm in size and volume of manganese vapour generated from 1 g of manganese in dependence on temperature

Temperature [°C]		400	500	600	700	800	900	1000	1100	1150	1200	1250
Time of sublimation [s]	10 μm	$2.4\cdot10^{10}$	$4.6\cdot10^{7}$	$3.8\cdot10^{5}$	$8.7\cdot10^{3}$	$4.2\cdot10^{2}$	30.3	4.1	0.9	0.6	0.15	0.1
	20 μm	$4.8\cdot10^{10}$	$9.1\cdot10^{7}$	$7.6\cdot10^{5}$	$1.7\cdot10^{4}$	$8.2\cdot10^{2}$	66.7	8.2	1.8	1.5	0.3	0.2
	50 μm	$1.2\cdot10^{11}$	$21.3\cdot10^{8}$	$1.9\cdot10^{6}$	$4.4\cdot10^{4}$	$2.1\cdot10^{3}$	$1.7\cdot10^{2}$	20.4	4.5	3.5	0.8	0.5
Vapour volume [l]		1.00	1.15	1.30	1.45	1.60	1.75	1.90	2.1	2.1	2.2	2.5

It is evident from the given data that the manganese vapour volume, generated by sublimation from Mn particles in a compact, is considerably higher than the pore volume, *i.e.* the manganese vapour fills the total open volume in the compact already in early stages of sublimation. In a real system, manganese as a manganese carrier is added as a milled powder and should be used in the size in the range from <1 μm up to max. ~45 μm. In many works presented in previous chapters the manganese carrier particles with the size of <40 μm were succesfully used. The sublimation of all manganese particles starts at some temperature which is ~400°C. Since the manganese particle size distribution is markedly wide, the sublimation occurs very quickly from the lowest particles in size up to the end of sublimation from the largest particles. Sublimation is affected by the time and physical–chemical characteristics of the individual particles of the manganese carrier and the heating rate of the compacts.

The effect of manganese vapour formed during sintering of the compacts with the addition of manganese manifests itself in the following processes. When the oxygen potential of the atmosphere exceeds the equilibrium values for oxidation, in this case of manganese, the excess of oxygen in the sintering atmosphere causes differences in the manganese gradient and the oxygen in the atmosphere reacts with the very active gaseous form of manganese due to its high affinity for oxygen in a very active gaseous form for oxygen. It means that in such cases, only in the vapour form, this demonstrates the high affinity of manganese for oxygen, not in the solid state.

The reaction of manganese vapour with moisture lowers the oxygen content in the atmosphere and, consequently, its oxygen potential to the equilibrium value given by the Mn/MnO line in the Richardson–Ellingham diagram, Fig. 2.4, as a result of the reduction – 'self-cleaning/protection' effect of gaseous manganese for sintering in the H_2/H_2O atmosphere according to the equation:

$$Mn_{(g)} + H_2O = MnO + H_2. \qquad (2)$$

The product of this reaction is a very fine dispersed green MnO powder which is transported away by the flowing atmosphere from the work space (tube and other form of sintering furnaces). The solid surface of the sintered Fe–Mn specimens is by this protected from any oxidation. In addition to this, MnO is a characteristic phase of variable composition. At temperatures higher than 1000°C MnO is

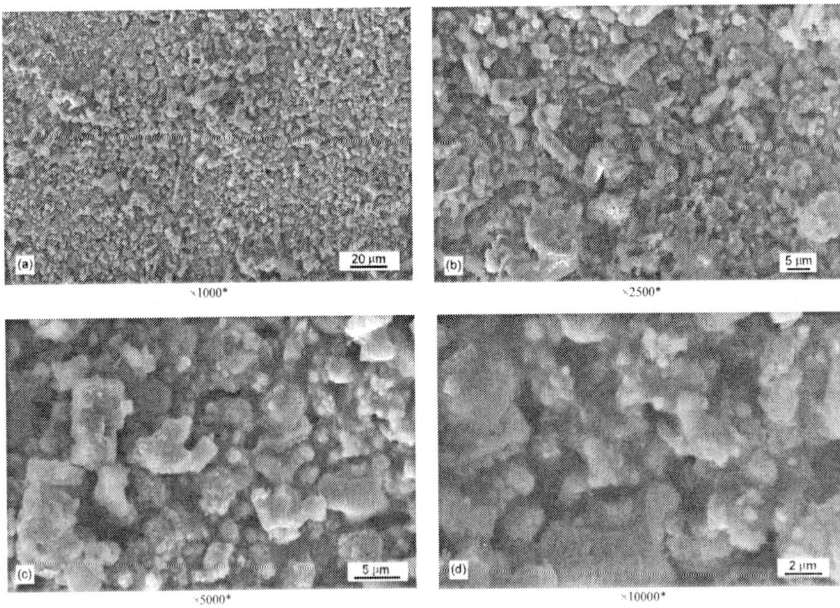

Fig. 12.1. MnO formed during laboratory sintering of Fe–3Cr–0.5Mo–(0,1,2,3) Mn–0.24C steels in dissociated ammonia with dew point −30°C. SEM. *original magnification.

capable of dissolving a notable overflow of oxygen and is chemically very active and involatile and this contributes to the decrease of the oxygen comntent in the flowing atmosphere over the reaction zone. By this reaction manganese in the gaseous form clearly demonstrates its high affinity for oxygen.

It is necessary to complete that the manganese vapour in the pores according to mentioned reaction firstly reduces the oxide films and/or chemisorbed oxygen covering the starting iron particles and the starting atmosphere filling the pores. This is the starting determining process which enables the formation of the interparticle necks during sintering among the metallic clean iron particles in a compact and by this to form by sintering a solid body from a dispersed body. The alloying of the base iron particles also starts during this sintering process. This process is decisive in solid phase (Fe) – gas phase (Mn vapour) alloying of sintered Mn-containing steels. During the preheating period the manganese addition in the compacts stabilises the γ-phase and decreases the diffusivity.

The formation of solid green MnO oxides is confirmed by Fig. 12.1. This oxide was detected on the bottom side of a steel plate which partly covered the steel box in the samples to be sintered were place, as shown in Fig. 8.28. MnO oxide due to its green colour was detected also on the walls of the industrial steel box containing the sintered manganese-containing samples.

As can be seen, the solid MnO particles do not have uniform size and form. At condensation they formed clusters on the steel plate.

The following section presents the results indicating clearly the effect of the purity of the sintering atmosphere on the chemical composition of sintered samples, *i.e.* the samples in which analysis showed the content of manganese and carbon after sintering. In cases in which these elements were not analysed, in addition to the sintering conditions with the atmospheres with the different H_2/N_2 ratio and different dew points, the resultant mechanical properties of the sintered steels are also presented to evaluate the effect of the appropriate characteristics of the protective atmosphere on the values of these properties of the sintered manganese steels.

This review of the properties of the sintered manganese-alloyed steels should contribute to evaluating the effect of low to high values of the dew point of the atmosphere on all the results presented in this monograph and also on the results of the studies not mentioned in this book. The tables include the numbers of tables from the original studies and the highest values of the strength and other mechanical properties obtained under the given treatment conditions.

12.2. Chemistry and mechanical properties of manganese containing steels sintered in H/N atmospheres with different dew points

The following tables present the content of manganese and carbon in the sintered state of the manganese steels prepared on the basis of Fe and of the prealloyed Cr and Mo powders with emphasis on the content of these elements after sintering in the atmospheres with a dew point of –30 and –50°C. The highest values of the tensile strength, included with the appropriate data in the tables, confirm efficient sintering and alloying of the manganese steels.

12.2.1. Sintering in H/N atmospheres with dew points of –30°C and –50°C

a) Table 12.4 gives the manganese and carbon contents in the initial conditions and after sintering of manganese steels prepared on the

Table 12.4 (*Table 5.12*). As sintered density (ρ), manganese and carbon contents added and as sintered (%) in the steels. 600 MPa, laboratory sintering, 1120°C, 3 h, cracked ammonia (c.a.), dew point –30°C. Mn and C added in form of high-carbon ferromanganese. The Mn and C contents in mass %

Iron powder	Hametag (H)				RZ (R)			
Designation	H1	H2	H3	H4	R1	R2	R3	R4
ρ [g cm^{-3}]	7.26	7.19	7.12	7.08	7.08	7.08	6.95	6.93
Mn added	2.5	3.5	4.5	5.5	2.5	3.5	4.5	5.5
C added	0.20	0.28	0.36	0.44	0.20	0.28	0.36	0.44
C_c	0.14	0.15	0.20	0.26	0.11	0.16	0.28	0.33
Mn	2.2	3.2	4.2	4.5	2.4	3.3	4.1	4.8

Table 12.5 (*Table 5.15*). As sintered density and carbon (C_c) and manganese content (%) for Fe–(1–5)Mn–xC steels. SC100.26 iron powder, 600 MPa, laboratory sintering, 1120°C, 60 min, c. a., d. p. –30°C; Mn and C added in form of high-carbon ferromanganese (76% Mn, 6% C)

As sintered	Manganese [mass %]				
	1	2	3	4	5
	Carbon [mass %]				
	0.08	0.16	0.24	0.32	0.40
Density [g cm^{-3}]	6.98	7.01	6.92	6.86	6.81
Mn	0.99	2.01	2.93	3.34	4.39
C_c	0.10	0.17	0.21	0.23	0.32

basis of Hametag iron powder (H, eddy milled) and water-atomised iron powder RZ (R).

The loss of manganese in sintered compacts was ~0.3% for the steels with the manganese addition of 2.5% to 4.5% and 1% and 0.7% for steels with the addition of 5.5% Mn. Relatively marked decarburisation was observed (hold time 3 h).

The corresponding mechanical properties are shown in Fig. 5.23. In this case the highest tensile strength of 886 MPa was attained for the H3 steel and 670 MPa for the R4 steel.

b) Table 12.6 lists the values of as sintered density and carbon and manganese content of manganese steels alloyed as in Table 12.3 but prepared on the basis of SC100.26 iron powder.

The manganese loss was observed in the steels with 4 and 5% Mn addition. Modest decarburisation of the specimens increased with increasing starting carbon content.

Table 12.6 (*Table 8.2*). As sintered density (ρ), carbon (C_c) and Mn content of hybrid H1, H2 and K2 steel samples sintered above mentioned conditions; *Mn added in the form of medium-carbon ferromanganese

Steel	H1, 1250°C/30 min				H2, 1180°C/40 min				K2, 1200°C/30 min				
*Mn [%]	0	1	2	3	0	1	2	3	0	1	1.5	2	3
ρ_s [g cm^{-3}]	6.98	7.06	7.01	6.97	7.05	7.04	7.01	6.98	7.01	7.03	7.04	7.02	6.99
C_c [%]	0.21	0.21	0.22	0.22	0.20	0.21	0.22	0.23	0.20	0.22	0.23	0.20	0.20
Mn [%]	0.10	0.89	1.93	2.71	0.25	0.95	1.86	2.80	0.17	0.87	1.24	1.76	2.75

The diagrams showing the tensile strength values of the tested steels in dependence on manganese addition are shown in Fig. 5.33. The highest tensile strength of 589 MPa was obtained for the Fe–4Mn–0.32C steel.

c) Table 12.6 gives the as sintered carbon and manganese contents of the hybrid steel samples based on prealloyed Fe–3Cr–0.5Mo powder (coded here H) and of those based on prealloyed Fe–3Cr–0.3Mo–0.3V powder (coded here K2). The compacts were sintered under laboratory conditions (H1 at 1250°C for 30 min and K2 at 1200°C for 60 min) in cracked ammonia with a dew point of –30°C. H2 samples were sintered under industrial conditions at 1180°C for 40 min in the $70N_2/30H_2$ atmosphere with a dew point of –50°C. All mixtures contained initially 0.24% C.

The manganese and carbon contents in specimens sintered under the laboratory as well under industrial conditions in both types of the atmosphere with the dew point of –30°C and of –50°C were on the starting levels with relatively small manganese and carbon losses compared to the starting values.

The tensile strength of the corresponding sintered steels in dependence on the manganese addition is shown in Fig. 8.18. The highest tensile strength of 940 MPa in the whole range of performed tests was obtained for the K2–2Mn steel. The H2–2Mn steel sintered under industrial conditions reached the highest tensile strength of 806 MPa.

d) Table 12.7 gives the manganese and carbon contents after sintering of manganese steels alloyed with three manganese carriers for the study of their effect on the properties of steels based on prealloyed Fe–Cr–Mo (CrM) powder. The addition of graphite to the mixtures with electrolytic manganese and medium-carbon

Table 12.7 (*Table 8.6*). As sintered density (ρ),carbon (C_c) and manganese (Mn) content of hybrid Fe–3Cr–0.5Mo–xMn–0.24C steel sintered under laboratory conditions for 30 min at 1250°C, c. a., d. p. –30°C. Mn added as electrolytic manganese (EMn), medium-carbon ferromanganese (FeMn) and high-carbon ferromanganese (FeMnC)

Mn carrier		EMn			FeMn			FeMnC		
Mn added [%]	0	1	2	3	1	2	3	1	2	3
ρ_s [g cm^{-3}]	6.98	7.07	7.03	6.98	7.06	7.01	6.97	7.04	7.02	6.97
Cc [%]	0.21	0.22	0.23	0.23	0.24	0.22	0.23	0.22	0.23	0.23
Mn [%]	0.10	0.88	1.87	2.87	0.94	1.88	2.86	0.0.87	1.92	2.92

Table 12.8 (*Table 8.11*). As sintered density (ρ_s) (tensile bars), combined carbon (C_c) and manganese content (mass %) of hybrid Fe–3Cr–0.5Mo–(0,1,2,3)Mn–0.24C steels. Industrial sintering, 1180°C, $70N_2/30H_2$, d. p. –50°C. *Mn added

Mn carrier	No	EMn			FMn			FMnC		
*Mn [%]	0	1	2	3	1	2	3	1	2	3
ρ_s [g cm^{-3}]	7.05	7.04	7.00	6.97	7.04	7.01	6.98	7.04	7.03	7.02
C_c [%]	0.20	0.20	0.23	0.24	0.21	0.23	0.24	0.21	0.24	0.24
Mn [%]	0.25	0.88	1.94	2.98	0.98	1.96	2.91	0.87	1.92	2.92

ferromanganese was about 0.24% corresponding to that brought in to the mixtures with high-carbon ferromanganese. The mass of to be together in one box sintered samples was again ~1200 g.

The manganisation of the samples without manganese addition to 0.10% was determined. As sintered manganese and carbon contents in all samples were on the level of the starting values independently on the manganese carrier grade.

The corresponding tensile strength values are shown in Fig. 8.29. The highest tensile strength of 856 MPa was obtained for the Fe–3Cr–0.5Mo–2Mn–0.24C steel. The steel with 1% Mn addition reached the tensile strength of 745 MPa (Mn added as FeMn).

e) Table 12.8 gives the characteristics of the same steel specimens as listed in Table 12.7. In this case, they were sintered under industrial conditions in a pusher furnace. The samples with the mass of ~1200 g were placed in a steel box (330 × 330 × 80 mm) between the boxes with structural parts of other compositions being sintered.

According to these results, also after sintering in an industrial furnace neither decarburisation nor demanganisation of the specimens occurred without some effect of the manganese carrier grade, as follows from the table. On the contrary, in the samples without

Table 12.9 *(Table 8.12, 8.14).* As sintered density (r_s) and Mn and C_c contents for Fe–0.85Mo–(1,2,3)Mn–0.3C steels prepared as hybrid steel based on prealloyed Fe–0.85Mo powder (prealloy) and as mixed (premix) steels sintered under laboratory and industrial conditions. *Mn added

(a) 1200°C	Fe–0.85Mo (prealloy)				Fe–0.85Mo (premix)			
*Mn [%]	0 Mn	1 Mn	2 Mn	3 Mn	0 Mn	1 Mn	2 Mn	3 Mn
ρ_s [g cm^{-3}]	7.02	7.00	6.94	6.86	7.08	7.06	6.98	6.91
C_c [%]	0.22	0.17	0.23	0.22	0.22	0.21	0.23	0.23
Mn [%]	0.12	0.90	1.95	2.86	0.09	0.97	1.95	2.92

(b) 1180°C	Ast85Mo (prealloy)				Fe–0.85Mo (premix)			
*Mn [%]	0 Mn	1 Mn	2 Mn	3 Mn	0 Mn	1 Mn	2 Mn	3 Mn
ρ_s [g cm^{-3}]	7.00	6.96	6.90	6.84	7.02	7.03	6.96	6.90
C_c [%]	0.29	0.28	0.26	0.27	0.27	0.26	0.27	0.28
Mn [%]	0.11	0.97	1.90	2.86	0.08	1.07	2.12	2.86

Mn addition, an increase of the manganese content to 0.25% was determined.

The dependence of the tensile strength on the manganese addition is shown in Fig. 8.34. The highest strength of 806 MPa was obtained for the Fe–3Cr–0.5Mo–0.24C steel with the addition of 2% Mn. The steel with the addition of 1% Mn had the tensile strength of 745 MPa (Mn added as FeMn).

f) Table 12.9 gives the as sintered density values and manganese and carbon contents of the hybrid steel based on prealloyed Ast85 Mo powder and the premixed steels based on ASC100.29 iron powder with the addition of manganese of 0, 1, 2 and 3% in the form of medium-carbon ferromanganese and graphite in the amount of 0.24%. Both steel groups were sintered together under the laboratory and industrial conditions. Sintering occurred under: (a) laboratory conditions at 1200°C for 60 min in cracked ammonia with a dew point of –30°C, (b) under industrial conditions at 1180°C for 40 min in the $70N_2/30H_2$ atmosphere with a dew point of –50°C.

Manganisation of the specimens of both steel groups without manganese addition (0Mn) after sintering at both temperatures to 0.08–0.12% was determined. Minimal differences in the as sintered manganese and carbon contents compared to the starting ones for prealloyed and mixed steels were determined. Neither decarburisation

Table 12.10 (*Table 10.1*). Properties of sintered Fe–3Mn–0.8C steel prepared on the basis of NC100.24 iron powder in dependence sintering temperature and dew point of hydrogen. Mn added as medium-carbon ferromanganese, graphite added to 0.8% C

Sint. temp. [°C]	Dew point [°C]	Density [g cm⁻³]	Rm [MPa]	TRS [MPa]
1120	–40	6.52	452	976
	–50		492	1128
	–60		433	1243
1150	–40	6.58	385	1236
	–50		490	1136
	–60		477	1254
1180	–40	6.70	451	832
	–50		423	1007
	–60		583	1253
1200	–40	6.60	469	1240
	–50		527	1091
	–60		445	1016

nor demanganisation of the samples occurred in spite of different atmospheres and dew points applied (–30 and –50°C).

The tensile strength values of the specimens sintered at 1200°C are shown in Fig. 8.44 and of those sintered at 1180°C in Fig. 8.62. The steel based on the Mo prealloyed powder sintered at 1200°C showed the highest tensile strength value of 670 MPa and the premix steel of 535 MPa. The hybrid steel after sintering at 1180°C had the highest UTS value of 655 MPa and the premix steel 548 MPa. All with the manganese addition of 3%.

12.2.3. Sintering of manganese steels in atmospheres with different ratios of H and N with dew points up to –60°C

This chapter discusses the mechanical properties of manganese steels at which the as sintered manganese and carbon contents were not determined. The following tables list only the tensile strength and transverse rupture strength values in relation to the sintering conditions as done in the previous cases.

g) Table 12.10 gives the strength values of Fe–3Mn–0.8C steel sintered in hydrogen with the dew point of –40, –50 and –60°C. The specimens were compacted by a pressure of 600 MPa. Sintering

Table 12.11 (*Tables 10.9, 10.10*). Fracture and stress, Rm and TRS strengths of Fe–3Mn–0.8C sintered steel. 660 MPa, 1120°C and 1250°C, 60 min, hydrogen and nitrogen, d. p. –60°C

Sintering	Strength [MPa]					
	NC100.24		ABC100.30		ASC100.29	
	fracture	Stress	fracture	stress	fracture	stress
1120°C, H_2	542±62	1120±138	524±74	1073±111	466±52	959±141
1120°C, N_2	677±50	1357±113	603±74	1128±89	565±38	1057 ±152
1250°C, H_2	713±68	1337±150	650±76	1250±196	578±64	996±93
1250°C, N_2	732±53	1354±121	651±67	1244±184	617±55	1109±134

was performed under laboratory conditions at 1120, 1150, 1180 and 1200°C for 60 min. To prevent oxidation, the specimens were placed in a stainless steel semi-closed container and were removed only after the boat had cooled down to room temperature.

The effect of the dew point on the tensile strength values differed from the effect on the TRS values, without any marked effect of sintering temerature. The highest tensile strength values were obtained after sintering at 1120, 1150 and 1200°C in the atmosphere with a dew point of –50°C. In the case of the transverse rupture strength the relation to the dew point is demonstrated by different values. The highest tensile strength of 586 MPa and transverse rupture strength value of 1253 MPa were obtained after sintering at 1180°C in the atmosphere with a dew point of –60°C.

h) Table 12.11 lists the yield and tensile strength values of Fe–3Mn–0.8C sintered steels prepared on the basis of NC100.24, ABC100.30, and ASC100.29 iron powders. 0.8% of carbon as graphite and 3% of manganese as medium-carbon ferromanganese with the mean particle size ~15 μm were added to the powder mixtures. The tensile strength specimens were compacted at 660 MPa, followed by sintering in hydrogen and nitrogen in a laboratory furnace in a semi-closed container including a water-jacketed rapid convective cooling section. All specimens were tempered at 200°C for 1 h in argon.

The values of 0.2% offset yield, tensile and transverse rupture strengths were determined. The averaged results are presented in Tables 10.10 and 10.11.

The highest tensile strength values of the specimens prepared on the basis of three iron powder grades were recorded for the steels sintered in nitrogen and with small deviations also the

Table 12.12 *(Tables 11.22, 11.23).* Chemical analysis and tensile and transverse rupture strength values of Fe–1.5% Mn–1.5% Cr–0.25% Mo–0.4% C steel. 1120°C, 60 min, hydrogen and nitrogen, dew point –60°C

Sintering atmosphere	Carbon [%]	Oxygen [%]	Nitrogen [%]	UTS [MPa]	TRS [MPa]
H_2	0.445	0.401	0.004	586	1178
N_2	0.412	0.380	0.044	633	1293

transverse rupture strength values after sintering at both temperatures. Differences in tensile and transverse rupture strength values listed in this table are approximately on the same level as those given in Table 12.9. The effect of the base iron powder grades on the investigated mechanical properties of the steels was also demonstrated.

i) Table 12.12 gives the carbon, oxygen and nitrogen contents and tensile and transverse rupture strength values of hybrid Fe–1.5% Mn–1.5% Cr–0.25% Mo–0.4% C steel. The specimens based on prealloyed CrL powder were compacted at 600 MPa and laboratory sintered in a semi-closed container with a labyrinth seal at 1120°C for 60 min in hydrogen and nitrogen with a dew point of –60°C. The container did not have any ferromanganese added to it. Carbon was introduced as graphite and 1.5% manganese was added as medium-carbon ferromanganese with a particle size smaller than 20 mm. Sintered specimens were tempered at 200°C for 60 min.

The as sintered carbon content was on the level of the starting one. Lower carbon and oxygen contents after sintering in nitrogen were analysed. The nitrogen content after sintering in N_2 slightly increased. The tensile strength of 586 MPa for simtering in hydrogen and of 633 MPa for sintering in nitrogen sintering was obtained. Similarly, the transverse rupture strength was higher for sintering in nitrogen.

j) Table 12.13 presents the effect of the sintering atmosphere with different ratio of hydrogen and nitrogen with dew point –60°C on properties of hybrid Fe–Cr–Mn–Mo–C steels based on prealloyed CrL and CrM powders. 3% of manganese was added in the form of medium-carbon ferromanganese powder, and 0.3% carbon as graphite. Sintering was carried out in a semi-closed steel container at two temperatures.

After sintering at the temperature of 1120°C the highest tensile strength of 654 MPa of the samples prepared on the basis of CrL powder was obtained after sintering in the A3 atmosphere. The highest transverse rupture strength values at the same sintering

Table 12.13 *(Tables 11.37, 11.38)*. Mechanical properties of Fe–Mn–Cr–Mo–C steels sintered at 1120°C and 1250°C in atmospheres with different ratio of H_2 and N_2 with dew point –60°C. A0 = 100% H_2, A1 = 75% H_2–25% N_2, A2 = 25% H_2–75% N_2, A3 = 5% H_2–95% N_2, A4 = 100% N_2. Sintered density of the Rm specimens was 6.84 to 6.91 g cm^{-3} and of the TRS specimens 7.00 to 7.09 g cm^{-3}

Sintering temperature and atmosphere		CrL+3Mn+0.3C		CrM+3Mn+0.3C	
		Rm [MPa]	TRS [MPa]	Rm [MPa]	TRS [MPa]
1120°C	A0	545	825	–	1041
	A1	514	1218	447	702
	A2	634	1003	642	689
	A3	654	805	416	828
	A4	495	765	695	399
1250°C	A0	573	1354	858	1349
	A1	594	1214	781	1501
	A2	622	1530	886	1610
	A3	799	1519	901	1572
	A4	822	1428	862	1675

temperature were obtained after sintering in the A2 and A3 atmospheres. In the samples based on the CrM powder the highest tensile strength of 695 MPa was obtained after sintering in the A4 atmosphere and the lowest one the transverse rupture strength value what is hardly to explain in relation to the values attained at sintering in the A0 to A3 atmospheres.

After sintering at 1250°C the steel based on CrL powder showed the highest tensile strength value of 822 MPa after sintering in the A4 atmosphere. The steel based on CrM powder after sintering in A3 atmosphere showed the tensile strength value of 901 MPa. The highest transverse rupture strength value of 1675 MPa was obtained after sintering in the A4 atmosphere.

As can be seen, the steels based on the CrL and CrM powders after sintering at both temperatures in the atmospheres with different ratios of H_2 and N_2 did not show comparable strength properties. The strength properties of the tested steels were also affected by the powder grades used in the experiments and differing in the Cr and Mo contents.

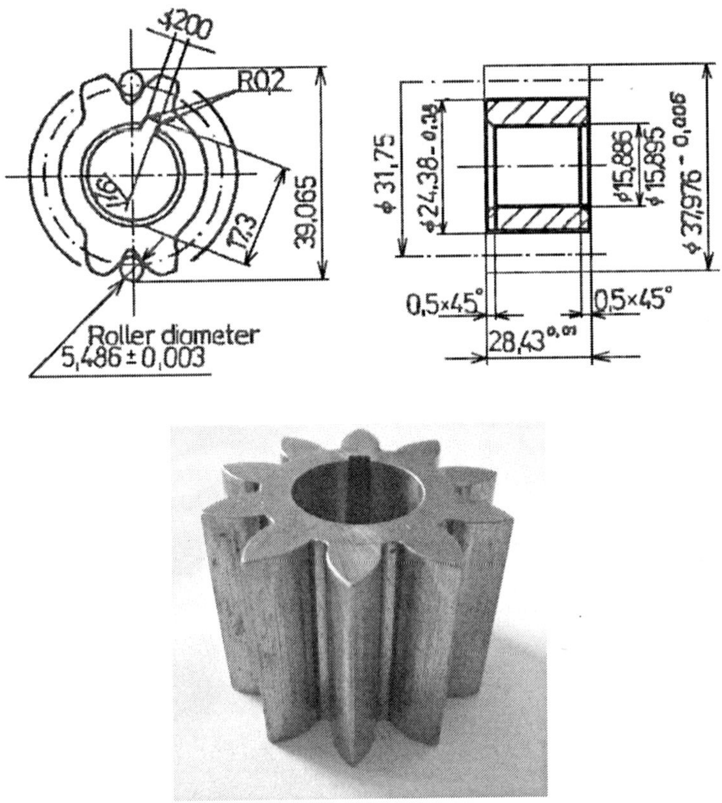

Fig. 12.2 (above) *(Fig. 5.54a).* Dimensions in mm of a spur gear for hydrogenerators; (bottom) Photo of the gear after testing during 5000 h *(Fig. 5.54b).*

12.2.2. Industrially sintered prototype structural parts prepared from manganese steel

This section describes components manufactured from manganese steels sintered in industrial furnaces in the H_2/N_2 atmospheres. As shown by the results, successful sintering was confirmed by the fact that sintering did not result in any demanganisation or decarburisation of the surface layers. This ensured the functional reliability of the components given by their technical and other properties required for application in engineering products. Taking into account the data on the minimum or no demanganisation of the sintered specimens, presented in the previous tables, shows that the other components, prepared from the manganese-alloyed steels, can be sintered in industrial furnaces without any risk of surface demanganisation or decarburisation.

Gears for hydrogenerators. The object of the interest was to prepare the gear shown in Fig. 12.2 used with the service pressure of 1.5 MPa with the aim to replace their production from wrought steel. The gears of the different Mn contents based on SC100.26 powder were compacted to the density of ~6.9 g cm^{-3} and industrially sintered for 40 min at 1180°C in the $75N_2/25H_2$ atmosphere with the dew point of –50°C. The sintered gears after final grinding to achieve dimensional stability were tested for the lifetime according to the requirements for the gears prepared from wrought steel. The as sintered density of the gears was 6.95 g cm^{-3} and hardness HV10/HRC 238/55.

Several Mn steels were investigated for this application. The required test lifetime of the gears in the hydrogenerators was 1000 h at the oil pressure of 3 MPa and a frequency 2800 rpm without damage. The tested gears were used as driven and the gear from hardened wrought steel as driving. The basic requirement was fulfilled for the gears from Fe–3Mn–0.5Mo–0.3C steel. The gears were further subjected to the tests under cyclic oil pressure changing every 10 s in the range of 0.7 to 3 MPa during 1000 h. This test condition was again fulfilled for the Fe–3Mn–0.5Mo–0.3C steel gear which was subjected further to long-time tests under cyclic pressure in the range of 0.7 to 5 MPa for 5000 h. The test was interrupted after 5000 h due to capacity reasons, and the as sintered Fe–3Mn–0.5Mo–0.3C steel gear was undamaged without the change of the dimensional tolerances.

The results prove that the gears for the hydrogenerators and similarly stressed components can be produced by the powder metallurgy method from the Fe–Mn–Mo–C steel as sintered only without some following heat treatment.

Components plate and cube. In the following are presented two structural components which were manufactured from the sintered steel based on Distaloy SE powder – plate and cube shown in Fig. 12.3. The aim of the study was to replace their production from the steel based on Distaloy SE powder by the manganese steel. The components and the bars for impact testing were prepared from 2, 3 and 4% Mn steels on the basis of SC100.26 iron powder and for comparison also those prepared from the DistSE powder were tested. Sintering of the components and specimens was carried out: (a) in a laboratory continuously furnace with manual shifting of the boxes (F) for 60 min at 1130°C in cracked ammonia with the dew point

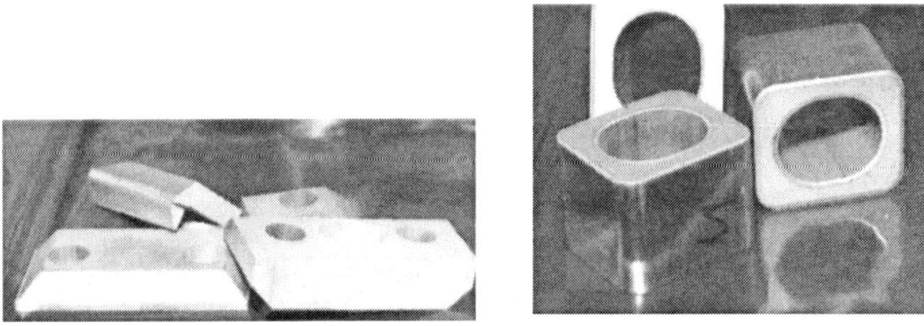

Fig. 12.3 (left) *(Fig.5.55a)*. Components *plate* (65 g), two undamaged and one broken at three point bending test, (right) *(Fig. 5.55b)* components *cube* (90 g).

Table 12.14 *(Table 5.29)*. Characteristic properties of the components *plate* and *cube*. DistSE – Fe–4% Ni–1.5% Cu–0.5% Mo, 2Mn – 2% Mn, 3Mn – 3% Mn, 4Mn – 4% Mn

(a) *plate*

Steel	HV 10	K [J]	Bending force [kN]	HV 10	K [J]	Bending force [kN]
	F furnace			C furnace		
DistSE	115	–	16.2	114	–	14.5
2Mn	125	12.0	18.4	135	16.5	15.0
3Mn	140	6.10	15.8	156	11.0	17.5
4Mn	210	4.40	12.2	200	10.0	21.0

Remark: The density of Mn-steel components was in the range 6.41–6.67 g cm^{-3}, the required hardness for FeNiCuMoC steel was 65 HB

(b) *cube*

Composition	HV 10	K [J]	Compression force [kN×10]	HV 10	K [J]	Compression force [kN×10]
	F furnace			C furnace		
DistSE	138	–	16.2	123	–	18.3
2Mn	140	12.0	18.4	132	17.0	10.2
3Mn	172	6.20	15.8	158	11.0	11.8
4Mn	210	4.10	12.2	202	10.2	12.0

Remark: The density of Mn steel components was in the range of 6.86 to 7.05 g cm^{-3}, the required hardness for FeNiCuMoC steel was 60 HB

of –23°C, and (b) in an industrial pusher furnace (C) for 60 min at 1100°C in the $70N_2/30N_2$ atmosphere with the dew point of –50°C.

The component *plate* was tested in the three-point bending test and the component *cube* by compression up to destruction. The

Fig. 12.4 *(Fig. 5.56).* Three types of the *rollers* in size of 22, 30 and 40 mm in dia. G = 14 g for 22 mm, 105 g for 30 mm, 170 g for 40 mm.

results, supplemented by the hadness and impact energy values, are presented in Table 12.14.

The height of the component plate was 11.5 mm. As can be seen from this table, the bending force for the component plate after sintering in the C furnace was higher than that of the plate prepared on the basis of DistSE powder. The compression force for the cube components prepared from the manganese steel was higher compared with FeNiCuMo steel components.

Roller component for conveyors. The photo of the roller component developed from Mn–steel is shown in Fig. 12.4 as a possible replacement of their production from the Fe–4%Ni–1.5% Cu–0.5% Mo–0.6% graphite mixture (DistSE) and from wrought steel. The rollers from the manganese steel were prepared from the Fe–3.5% Mn–0.7% graphite mixture on the basis of ASC100.29 iron powder. The components were compacted at 600 MPa and sintered

Table 12.15. (*Table 5.30*). Tribological characteristics of the rollers prepared from Fe–3.5Mn–0.7C and FeNiCuMo (DistSE–0.6C) steels. *W* – wear, *m* – coefficient of friction. *T* – temperature in the contact area. Mn-roller: ρ = 6.89 g cm^{-3}, 170 HV 100; Ni-roller: ρ = 6.99 g cm^{-3}, 206 HV 10. Test load: 140 and 100 N, sliding rate 0.5 m·s^{-1}, sliding distance 1000 m; *mild steel – counterpart, **hardened steel – counterpart

Steel	*Load 100 N			*Load 140 N			**Load 140 N		
	W [mg]	*m*	*T* [°C]	W [mg]	*m*	*T* [°C]	*W* [mg]	*m*	*T* [°C]
DistSE–0.6C	344	0.40	63	596	0.41	54	2405	0.36	58
3.5Mn–0.7C	96	0.39	56	104	0.60	64	50	0.19	57

under the industrial conditions in a pusher furnace (1180°C/40 min, $70N_2/30H_2$, entry dew point −50°C).

The tribological tests of the Mn rollers and of those prepared from DistSE powder were performed by the disc (roller)–prism method at two loads. The results are listed in Table 12.15.

The *rollers* from Fe–3.5Mn–0.7C steel exhibited the lowest wear (96 mg) compared to FeNiCuMoC steel (344 mg), and significantly lower (50→2405 mg) when tested against a hardened counterpart. The rollers from the manganese steel were successfully tested by the user.

12.3. Explanation of the sintering processes of manganese steels occurring in atmospheres with different partial pressure of oxygen

It follows from the chemical and mechanical characteristics of sintered manganese steels listed in this chapter and in previous chapters in dependence on materials and processing conditions used that the efficiency of the sintering processes was high.

All properties were attained independently of some differences in the composition and the purity of the atmospheres and this proves that the manganese particles in the compacts were not oxidised and that manganese sublimation in dependence on temperature occurred during the whole sintering process ensuring alloying of base powder particles in the starting compacts.

The explanation of these features includes the explanation of the effect of the sintering atmospheres and of their dew points on the sintering and properties of manganese steels under which the sintered specimens/parts did not show any loss of manganese and carbon and under which the high mechanical properties values of the manganese steels were attained depending on all factors acting at sintering by the formation of a solid body from a dispersed body.

As follows from the previous data, the sintering processes occurred in the atmospheres independently of their compositions and dew points resulting in the effective sintering of manganese steels.

The explanation of the effect of the atmosphere characteristics under which effective sintering of manganese steels was conducted will help to clarify the effect of the purity of the atmospheres and by this of the partial pressures of oxygen.

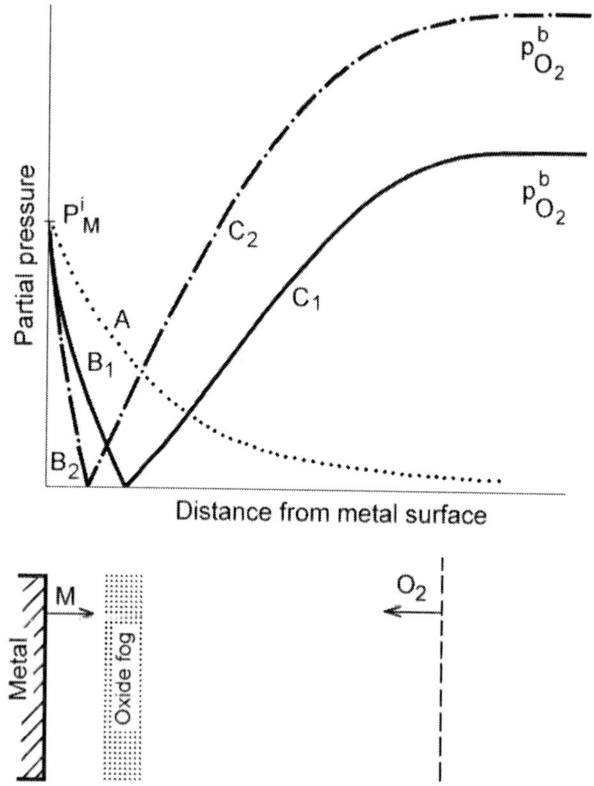

Fig. 12.5. Partial pressures of vaporising metal in flowing gas under isothermal conditions, shown as the function of the distance from the metal [287].

The following evaluation and explanation of the course of the sintering processes of manganese steels in low- and high-purity atmospheres is based on a study dealing with liquid copper at 1200°C and iron at 1600°C, *i.e.* on the mutual reaction of vaporisation of metal and flowing gas according to the following citation of Ref. 287 published also in [8, 288]:

'When metals are heated in cold flowing gases ($T_{gas} \ll T_{metal}$) the rates of vaporisation are greater than those predicted from the mass transfer coefficient. This is because the concentration gradient of vapour, which extends from the metal surface into the gas phase, is steepened by condensation of vapour in the temperature gradient which extends outward from the metal. This steepening is some three times greater than the calculated value, *i.e.* enhanced vapourisation occurs. A somewhat similar steepening can be obtained under isothermal conditions ($T_{gas} = T_{metal}$), if an involatile compound can

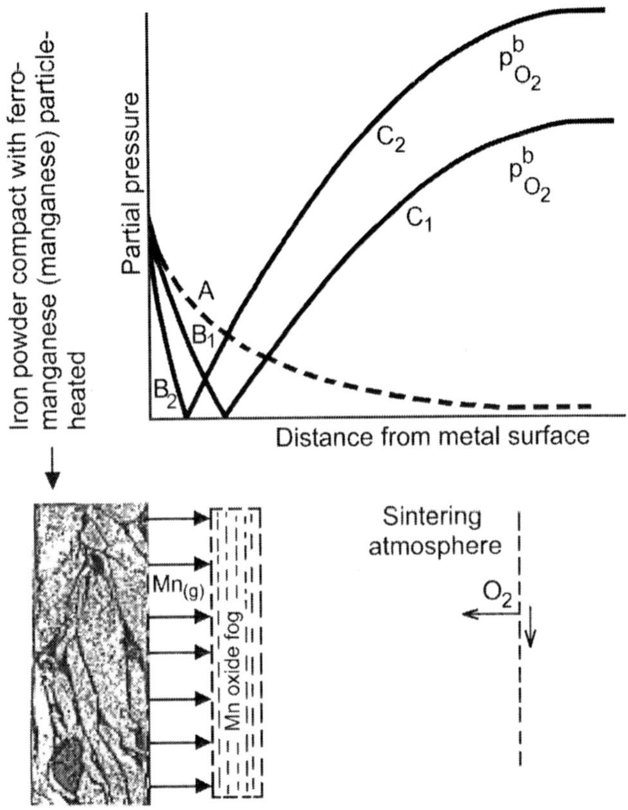

Fig. 12.6. Partial pressures of sublimating Mn in H_2/N_2 atmosphere containing oxygen under isothermal conditions as function of distance from Fe–Mn specimen/part surface [8,288].

be formed between the flowing gas and the vaporising metal. In the case of copper at say 1200°C, we might have the gradient A shown in Fig. 12.5 in the presence of a neutral gas. But when a low pressure of oxygen is admitted, the vapour is oxidised to solid particles of copper oxide, and the copper vapour pressure gradients are steepened as indicated by B_1 and the oxygen gradient might have the form C_1. The fluxes are represented schematically below the figure. If more oxygen is added, the gradients would be steepened to B_2 and C_2 and the outward flux of metal would become greater. Ultimately, as $p^!_{O2}$ is progressively increased, the rate of supply of oxygen to metal exceeds the maximum rate of supply of vapour – the free evaporation rate in vacuum. At that point the oxygen pressure at the drop surface $p^!_{O2}$ becomes high enough to produce free oxide on the metal surface. Vaporising then virtually ceases because the vapour pressure of Fe is so small'.

Figure 12.6 shows the same scheme as in Fig. 12.5 in which the metal–liquid copper specimen was replaced by a green Fe–Mn compact. The fluxes are represented schematically below the figures.

These schemes can be used to explain the effect of the reaction according to equation (2), *i.e.* $Mn_{(g)} - O$ occurring under isothermal sintering conditions, on the chemistry and mechanical properties of manganese steels. The condition for the realisation of this reaction must be the solid product – an involatile compound as listed in the previous citation. In the case of the reaction between the flowing atmosphere containing oxygen (H_2O vapour) and the manganese vapour a solid involatile compound MnO was formed as documented in Fig. 12.1. The condition for successful sintering of manganese steels under isothermal conditions according the previous citation was therefore fulfilled by the real formation of MnO making it possible to use the mentioned scheme for final explanation of the sintering processes for manganese steels.

In all sintering processes of manganese steels the controlling role is played by the manganese sublimation rate which depends on the manganese vapour pressure and is given by the equation:

$$\text{rate of sublimation} = p(\text{Mn})\cdot dc/dx,$$
$$(\log W = -3.358 + 0.5\log M + \log p - \log T)$$

where p(Mn) gradients are shown in Fig. 12.5 and 12.6 as the curves B_1 and B_2.

Taking into account the fact that this reaction was accompanied by the formation of MnO, Fig. 12.6 shows that for manganese the increase of the angle of inclination of the curve results in a steep manganese gradient given by the curves B_2 and B_1. Consequently, the reaction of the manganese vapour with oxygen, whose gradients determining the partial pressure of oxygen are given by the curves C_2 and C_1, takes place closer to the surface of the manganese sample, *i.e.* gradually from the lower gradient p_{O2} (C_1) at the lower manganese gradient B_1 in the direction to the higher gradient p_{O2} (C_2) and, therefore, to the steep gradient of manganese B_2. This means that the sublimation rate increases with increase of p_{O2} because at a constant value p_{Mn} is directly proportional to the corresponding manganese gradient. This corresponds in the case of copper to Fig. 12.5 and for the manganese vapour system to Fig. 12.6.

At sintering of Fe–Mn samples in the absolutely neutral atmosphere, sintering takes place in accordance with the gradient A which is the smallest and, consequently, the sublimation rate is the lowest. Therefore, the escape of manganese to the atmosphere decreases in intensity.

Also, at sintering of Fe–Mn samples in the atmospheres with increasing dew point, oxygen gradient C_1 gradually decreases and the appropriate manganese gradient B_1 becomes less steep until finally it changes to the form of the gradient A, as in the case of the neutral atmosphere. In this state, there is no longer reaction between the manganese vapour and oxygen, and only sublimation of manganese takes place in the samples or components during joint sintering in sintering boxes. This is caused by the fact that in the samples with the manganese additions and the samples with no addition, sintered together in the same box, the manganese content of the samples with no manganese addition increased. There was no decarburisation of the manganese steels containing carbon because the reaction of the manganese vapour with oxygen from the atmosphere stopped the reaction of carbon with oxygen already in the reaction zone.

This also applies without exception to the sintering of manganese steels in the atmosphere with a dew point of, for example, up to –70°C used for the dilatometric tests of the manganese steels described in chapter 11. The mechanical properties of the manganese steels, sintered in the atmospheres with different H_2/N_2 ratios and with the dew point up to –60°C, which were not affected by the composition of the atmosphere or by their purity, are presented in Tables 12.10–12.13.

All these data and considerations confirm that neither demanganisation nor decarburisation of the specimens occurred, as follows from Tables 12.6 and 12.9. The surface layers of the prototype structural components, sintered under the industrial conditions, shown in chapter 10.7, were not demanganised nor decarburised. Similarly, the prototype structural components, shown in Figs. 12.2 and 12.4, in which the surface layers were stressed by wet or dry friction at specific loading, *i.e.* the surfaces of which must be for functional reasons without demaganisation and decarburisation.

The above results show that the manganese steels can be sintered in the atmospheres with a considerably higher p_{O2} than the value permitted by the Richardson diagram, Fig. 2.4, as a result of the reaction of the manganese vapour with oxygen during sintering and the formation of MnO which, being the product of reaction according to equation (2), is the condition for the occurrence of this reaction and, consequently, successful sintering of the manganese steels. The resultant MnO is carried away by the flowing atmosphere away from the furnace space. As already mentioned, the absolutely neutral atmosphere would result in a minimum gradient and, consequently, the minimum loss of manganese – there would be only sublimation without oxidation and manganese would not be oxidised even if it did not sublimate. This shows that the manganese steels can also be sintered in pure nitrogen.

A suitable example of these efficient sintering processes of manganese steels is sintering in an industrial pusher furnace at a temperature of 1180°C in a $70N_2/30H_2$ atmosphere with a dew point of –50°C. In this case, the process was almost completely in equilibrium with the Ellingham–Richardson diagram (–55°C at 1100°C), so that this type of sintering did not cause any further reaction of the manganese vapour with oxygen. Consequently, there was practically no loss of manganese or carbon, Tables 12.6, 12.8, 12.9.

This result is also confirmed by the diagram of the industrial sintering furnace, shown in Fig. 8.33. In this case, the atmosphere flow was directed against the direction of displacement of the boxes with the specimens/components. The boxes with the specimens in the position of the gas inlet were already sintered, as confirmed by the results. These specimens showed no demanganisation or decarburisation, despite the fact that the atmosphere at entry of the boxes with the samples into the furnace could already have been contaminated because it travelled in the direction against the direction of movement of the boxes

Interaction between manganese and the sintering atmosphere. The previous results obtained in the evaluation of the effect of reactions between manganese and oxygen during sintering on the properties of the manganese steels are supplemented in the following citation by the results of the study of sintering of Fe–Mn samples in the atmosphere with a dew point of –20°C [289].

'According to the established thermodynamic data, Mn will oxidise at 1100°C if p_{H2O}/p_{H2} exceeds $4.77 \cdot 10^{-5}$. This implies that the dew point must lie below –55°C, which is a stringent requirement. Manufacturers of sintered steel parts prefer to use atmospheres with dew points low as of –20°C, in which $p_{H2O} = 125$ Pa, *i.e.* 26 times the equilibrium water vapour pressure. The oxidising atmosphere affects the sintering of iron–manganese powder mixtures by the following process.

The vapour pressure of manganese at 1100°C is 20 Pa and the rate of sublimation (in vaccum) is of the order of 1 mg/cm^2·s. A simple calculation shows that a supply of 320 cm^3 of sintering atmosphere (d. p. –20°C) to every cm^2 of manganese surface per second would be needed to oxidise all sublimating manganese. Further calculations, based on a model of the compact weighing 10 g and containing 6% Mn in form of spherical particles with 0.02 radius, lead to the conclusion that approx. 40 l/s of the atmosphere would have to react with the specimen's manganese, *i.e.* almost 150 m^3/h. This amount exceeds the total atmosphere flow, used in the experiments by three orders of magnitude and the volume, available for the reaction with manganese vapour inside the pores of the compact by an estimate of five to six orders of mangnitude. It can thus be concluded that the transport of manganese via the gas phase does not suffer perceptibly from an 'insufficiently' dry sintering atmosphere, and the volume fraction of the formed manganese oxide is totally negligible compared with the final porosity.

When high-carbon ferromanganese is used, the kinetics considerations are irrelevant, as a H_2/H_2O atmosphere will be thermodynamically reducing in the Mn–MnO system up to the zero dew point, due to its reaction with the carbon present.

Manganese is effectively transported through the gaseous phase from sites of high to lower activity during the sintering of iron–manganese powder mixtures.

The rate of sublimation of manganese far exceeds the rate of its oxidation by a hydrogen atmosphere (pure or diluted by inert gas) with a dew point of –20°C. This finding mitigates very considerably the stringent requirement for a –50°C dew point by purely thermodynamic considerations'.

12.4. Crucial results

From the detailed evaluation of the presented results concerning the properties of manganese containing steels sintered in low- and high-purity atmospheres following knowledge it follows that:

- sintering of manganese-containing steels can be efficiently carried out **in low-purity atmospheres**

- sintering of manganese-containing steels can be carried out **in nitrogen or, in nitrogen + 3–5% of hydrogen**
- at sintering of manganese-containing steels in low-purity atmospheres when the reaction of manganese vapour with oxygen occurs near the surfaces of to be sintered specimens/parts **only the products of this reaction, *i.e.* MnO and H_2 are transported by the flowing atmosphere away from the furnace space**
- at sintering of manganese-containing steels in high-purity atmospheres or in a neutral gas when the reaction of manganese with oxygen does not occur as documented also by no demanganisation and decarburisation of the industrially sintered specimens in the H_2/N_2 atmosphere with a dew point of –50°C, **the manganese vapour does not escape into the flowing atmosphere and by this the wall will not be affected by manganese vapour**.

12.5. Overview of materials and processes for preparation of manganese-alloyed steels with the highest strength properties

12.5.1. Materials and processing conditions

This section summarises the materials and processing conditions used for the manganese-alloyed sintered steels.

In research and development of sintered manganese-alloyed steels, they were prepared on the basis of the following materials and mostly under the following processing conditions. The powders used were: eddy milled, air-atomised, base sponge and water-atomised iron powder grades, low-manganese containing prealloyed powders, manganese–copper and manganese–nickel prealloyed powders, diffusion Fe–Mn–C prealloyed powders and prealloyed Fe–3Cr–0.5Mo, Fe–1,5Cr–0.25Mo, Fe–(1,3)Cr–0.3Mo–0.3V and Fe–0.85Mo and Fe–1.5Mo powders. Manganese carriers included pyrolusite, electrolytic manganese, ferromanganese grades, and many master alloys with different composition including Mn–Cr and Mn–Mo and carbide phase powder systems prepared under various manufacturing conditions. The specimens were single cold compacted to the densities from ~6.5 g cm^{-3} to about 7.0 g cm^{-3}, triple compacted to about 7.5 g cm^{-3} and powder forged and extruded to the full density of the specimens. The sintering of manganese steels was carried out at the temperatures of ~800°C up to 1300°C in dissociated ammonia and in H_2/N_2 atmospheres with the dew points of –23°C up to –70°C. In some studies, to prevent oxidation of manganese particles in the compacts, special measures were used as, for example, sintering in semi-closed containers. The effect of the sintering atmosphere with different ratios from 100% hydrogen to 100% nitrogen was the objective of the investigations. The manganese steels were sintered under the laboratory as well under industrial conditions, including the sintering of the structural components. The manganese steels were prepared as Fe–Mn–C with the addition of manganese up to 5% and of carbon in the amount up to 0.8% and as hybrid-based on Cr–Mo, Mo and Cr–Mo–V prealloyed powders, and as high manganese-alloyed austenitic steels. Liquid-phase sintering of manganese steels with the addition of liquid phase-forming elements Sn, Zn, B and P was performed. Silicon was also used for alloying of manganese steels. Sinter hardening, heat treatment and tempering of manganese steels were the objective of the investigations. The mechanical, toughness, fatigue endurance, rolling contact fatigue endurance and friction properties of the laboratory as well of industrially sintered specimens and of the structural components were tested.

12.5.2. Highest tensile (UTS) and transverse rupture strength (TRS) values of manganese-containing steels attained under corresponding conditions

Fe–Mn–C steels

a) The Fe–4.5Mn–0.4C steel based on ASC100.29 powder sintered at 875°C in cracked ammonia with a dew point of –30°C attained the UTS value of 210 MPa and the Fe–4Mn–0.23C steel prepared on the basis of SC100.26 powder sintered at 1120°C in cracked ammonia with a dew point of –30°C attained the UTS value of 590 MPa. The same steel sintered at 1280°C attained the UTS value of 960 MPa. The Fe–3Mn–0.3C steel compacted at 800 MPa after sintering 1160°C for 75 min attained the UTS value of 710 MPa. The Fe–3Mn–03C steel compacted at 800 MPa after sintering at 1160°C for 75 min attained the UTS of 710 MPa. The Fe–3Mn–0.8C steels prepared on the basis of NC100.24 iron powder were sintered at four sintering temperatures in hydrogen with dew points of –40°C, –50°C and –60°C. The UTS value of 492 MPa and TRS value of 1128 MPa were attained by the steel sintered at 1120°C in hydrogen with a dew point of –50°C. The steel after sintering at 1150°C in hydrogen with a dew point of –50°C attained the UTS value of 490 MPa and the TRS value of 1136 MPa. The UTS value of 583 MPa and the TRS value of 1007 MPa were obtained for the steel sintered at 1180°C in hydrogen with a dew point of –60°C. The steel after sintering at 1200°C in hydrogen with a dew point of –50°C showed the UTS value of 527 MPa and the TRS value of 1091 MPa.

b) The Fe–3Mn–0.8C steel specimens prepared on the basis of NC100.24, ABC100.30 and ASC100.29 iron powders were sintered at 1120°C and 1250°C in hydrogen and in nitrogen with a dew point of –60°C, followed by rapid cooling in the water-jacketed section. The steel sintered at 1120°C in hydrogen showed the UTS value of 542 MPa and the TRS value of 1120 MPa and the UTS value of 677 MPa and TRS value of 1357 MPa after sintering in nitrogen. The steel sintered at 1250°C in hydrogen showed the UTS value of 713 MPa and the TRS value of 1337 MPa and that sintered in nitrogen the UTS value of 732 MPa and the TRS value of 1354 MPa.

c) The Fe–3Mn–0.8C specimens were sintered under laboratory conditions at 1120, 1150, 1180 and 1200°C for 60 min in hydrogen with a dew point of –40, –50 and –60°C. To prevent oxidation, the specimens were set in a stainless steel semi-closed container. The steel sintered at 1120°C in hydrogen with a dew point of –50°C showed the UTS value of 492 MPa and TRS of 1128 MPa. The steel sintered at 1150°C in hydrogen with a dew point of –50°C reached the UTS value of 490 MPa and the TRS value of 1130 MPa. The steel sintered at 1180°C in hydrogen with a dew point of –60°C exhibited the UTS value of 583 MPa and the TRS value of 1253 MPa and that sintered at 1200°C in hydrogen with a dew point of –40°C attained the UTS value of 469 MPa and the TRS value of 1240 MPa.

d) The Fe–3Mn–0.8C steels prepared on the basis of NC100.24, ABC100.30, and ASC100.29 iron powders were sintered at 1120°C and 1250°C for 60 min in hydrogen and nitrogen with a dew point of –60°C in semi-closed containers including a water-jacketed rapid

Steel	State of the tested steels					
	As sintered		As sinter hardened		As heat-treated	
	UTS [MPa]	TRS [MPa]	UTS [MPa]	TRS [MPa]	UTS [MPa]	TRS [MPa]
Fe–2Mn–0.7C	530	1125	572	1148	640	1071
Fe–3Mn–0.3C	549	1011	602	1151	554	903
Fe–4Mn–0.7C	528	1058	540	1066	422	805

Steel	Sintering temperature [°C]	Cooling rate [°C/min]	UTS [MPa]	TRS [MPa]
Fe–3Mn–0.8C (C_c–0.6%)	1120	4.8	478	1270
		5.5	535	1000
Fe–3Mn–0.8C (C_c 0.5%)	1250	4.7	595	1330
		5.7	491	1090
Fe–4Mn–0.8C (C_c –0.6%)	1120	4.8	669	1021
		55	580	1097
Fe–4Mn–0.8C (C_c –0.5%)	1250	4.7	636	1467
		5.7	428	847

convective cooling section. All specimens were tempered at 200°C for 1 h in argon. The steel sintered at 1120°C in hydrogen showed the UTS value of 654 MPa and the TRS value of 1245 MPa and after sintering in nitrogen the UTS of 677 MPa and the TRS value of 1357 MPa were obtained. The steel sintered at 1250°C in hydrogen showed the UTS values of 713 MPa and the TRS value of 1337 MPa. The steel sintered at the same temperature in nitrogen reached the UTS value of 732 MPa and the TRS value of 1354 MPa. The steels based on ABC100.30 and ASC100.29 powder reached lower strength values.

e) The Fe–(2,3,4)Mn–(0.3,0.7)C steel specimens based on SC100.26 iron powder were industrially sintered at 1180°C in the $25H_2/75N_2$ atmosphere with a dew point of –55°C. The specimens were furnace cooled at ~10 K/min. Sinter hardening occurred after reheating of the specimens to 900°C and cooling in the range 900 – 400°C and then tempered at 200°C for 60 min. The specimens for heat treatment were austenitized at 850°C in the $25H_2/75N_2$ atmosphere, quenched in oil and tempered at 200°C for 60 min. The tensile and transverse rupture strength values of the investigated steels attained under the given treatment processes are listed in the table on the previous page.

f) The Fe–(3,4)Mn–0.8C steel compacts based on the NC100.24 iron powder were sintered at 1120 and at 1250°C for 60 min in hydrogen with a dew point of –60°C and cooled at two cooling rates. The results are listed in the table at the top of this page.

Hybrid steels

g) The hybrid Fe–3Cr–0.5Mo–2Mn–0.24C steel sintered at 1250°C for 30 min in cracked ammonia with a dew point of –30°C showed the UTS value of 850 MPa and TRS value of 1400 MPa. The same steel industrially sintered at 1180°C for 40 min in the $70N_2/30H_2$ atmosphere with a dew point of –50°C attained the UTS value of 806 MPa and the TRS value of 1300 MPa. The same steel laboratory-sintered with the manganese addition of 1% showed the UTS value of 745 MPa and the TRS value of 1275 MPa.

h) The hybrid Fe–3Cr–0.5Mo–2Mn–0.24C steel sintered at 1250°C for 30 min in cracked ammonia with a dew point of –30°C after sintering was subjected to sinter hardening by cooling from the temperature of 900 to 400°C at the rate of ~0.3°C s^{-1}.

The steel reached the UTS value of 989 MPa and the TRS value of 1548 MPa. The same base steel with the manganese addition of 1% reached the UTS value of 854 MPa and the TRS value of 1265 MPa.

i) In relation to the manganese carrier grade used, the hybrid Fe–3Cr–0.5Mo–0.24C steel with the 1% manganese addition in the form of electrolytic manganese attained the UTS value of 602 MPa and the TRS value of 1290 MPa. The same steel with the manganese addition of 1% in the form of medium-carbon ferromanganese attained the UTS value of 657 MPa and the TRS value of 1450 MPa. The steel with the manganese addition of 1% in the form of high-carbon ferromanganese attained the UTS value of 601 MPa and the TRS value of 1300 MPa. The values of both strengths with other additions of manganese were lower.

j) The hybrid Fe–3Cr–0.3Mo–0.3V–1.5Mn–0.24 C steel sintered at 1200°C in cracked ammonia with a dew point of –30°C attained the UTS value of 929 MPa and the TRS value of 1300 MPa.

k) The Fe–0.85Mo–3Mn–0.3C steels were prepared as hybrid based on Ast85Mo powder and as premixed based on SC100.26 iron powder. The steels were sintered at 1200°C in cracked ammonia with a dew point of –30°C under laboratory conditions and at 1180°C in an industrial furnace in the $70N_2/30H_2$ atmosphere with a dew point of –50°C. The hybrid steel sintered at 1200°C attained the UTS value of 670 MPa and the TRS value of 990 MPa and the premixed one the UTS value of 535 MPa and the TRS value of 985 MPa. The hybrid steel sintered at 1180°C attained the UTS value of 760 MPa and the TRS value of 1090 MPa and the premixed one of 535 MPa and the TRS value of 990 MPa.

l) The hybrid Fe–(1.5,3)Cr–(0.25,0.5)Mo–3Mn–0.3C steel specimens based on prealloyed CrL and CrM powders were sintered at 1120°C and 1250°C in the atmospheres with different ratios of H_2

and N_2 with a dew point of –60°C in semi-closed containers. The steel based on CrL powder sintered at 1120°C in $75H_2/25N_2$ reached the UTS value of 654 MPa and the TRS value of 805 MPa. The same steel sintered at 1250°C in nitrogen reached the UTS value of 822 MPa and the TRS value of 1428 MPa. The steel based on CrM powder sintered at 1120°C in nitrogen reached the UTS value of 695 MPa and the TRS value of 1399 MPa. The same steel after sintering at 1200°C in the $75H_2/25N_2$ atmosphere attained the UTS value of 901 MPa and the TRS value of 1572 MPa.

m) The hybrid as sinter hardened Fe–3Cr–0.5Mo–2Mn–0.24 C steel samples after sintering at 1250°C for 30 min in cracked ammonia cooled in a water-jacketed zone (cooling rate ~0.3°C s^{-1}) attained the UTS value of 989 MPa and the TRS value of 1548 MPa.

n) The following sintered structural components were ivestigated and tested for functional properties in laboratories as well by the potential users: gears tested for bend loading, spur gears for drilling machines, gears for hydrogenerators, oil pump rotors, conveyor rollers, cube and plate and powder forged roller bearing rings up to the outer diameter of 90 mm with the durability significantly higher than that required by the standards for roller bearings.

References

[1] ŠALAK, A.: *Ferrous Powder Metallurgy*. Cambridge Inter. Sci. Publ., Cambridge, UK, 1995.

[2] SCHATT, W.: *Sintering Processes (Sintervorgänge)*. VDI Verlag, Düsseldorf, 1992, in German.

[3] BOCCHINI, G.F.: In: Proc. Powder Metall. Summer School, Gothenburg, 1999, EPMA, Shrewsbury, p. 185.

[4] KUBASCHEWSKI, O., EVANS, E. and ALCOCK, C.B.: *Metallurgical Thermochemistry*. Pergamon Press, New York, 1974.

[5] SCHATT, W. and WIETERS, K.P.: *Powder Metallurgy Processing and Materials*. EPMA, Shrewsbury, 1997.

[6] ŠALAK, A., RUDNAYOVÁ, E. and SELECKÁ, M: Internal Nitriding of Sintered Iron in Nitrogen-Low Ammonia Atmospheres. *Powder Metalurgy*, Vol. 42, 1999, No. 1, p. 45.

[7] *Höganäs Handbook for Sintered Components*, Höganäs AB, Höganäs, Vol. 2, 1997.

[8] ŠALAK, A., SELECKÁ, M. and BUREŠ, R.: Manganese in Ferrous Powder Metallurgy, *Powder Metallurgy Progress*, Vol. 1, 2001, No. 1, p. 41.

[9] KIEFFER, F. and HOTOP, W.: *Sintered Iron and Sintered Steel (Sintereisen und Stahl)*. Springer Verlag, Wien, 1948, in German.

[10] BENESOVSKY, F. and KIEFFER, R.: Sintered Manganese Steels. *Berg- und Hüttenmännische Monatshefte*, Vol. 95, 1950, No. 8, p. 145, in German.

[11] KAZENAS, E.K. and CHIZHIKOFF, D.M.: *Pressure and Composition of the Vapour above the Oxides of Chemical Elements*. Nauka, Moscow, 1976, in Russian.

[12] GRETHE, K.: *Mangan - Grundlagen und technische Verwendung (Manganese – Base Characteristics and Use)*. Stahleisen, Düsseldorf, 1972, in German.

[13] TRETYAKOV, YU.D.: *Chemistry of Non-Stoichiometric Oxides*. Moscow University, Moscow, 1974, in Russian.

[14] GERMAN, R.M.: *Powder Metallurgy Science*. MPIF, Princeton, NJ, 1994.

[15] TENGZELIUS, J., GREK, S.E. and BLÄNDE, C.A.: Limitations and Possibilities in the Utilisation of Cr and Mn as Alloying Elements in High Strength Steels. In: Modern Dev. in Powder Metall., (Int. PM Conf., Washington), MPIF, Princeton, NJ, 1980, Vol. 13, p. 159; TENGZELIUS, J., BLÄNDE, C.A. and WASTENSON, G.: High Strength Sintered Steels Containing Chromium and Manganese. *Höganäs PM Iron Powder Information*, 1981.

[16] RIVLIN, V.G.: Critical Review of Constitution of Carbon–Chromium–Iron- and Carbon–Iron–Manganese Systems. *International Metals Reviews*, Vol. 29, 1984, No. 4, p. 299.

[17] RANKYN, W.J. and WYNNYCKYJ, J.R.: *Metallurgical and Materials Transactions B*, Vol. 28B, April, 1997, p. 307.

[18] BOCCHINI, G.F.: Influence of Controlled Atmospheres on the Proper Sintering of Carbon Steels. *Powder Metallurgy Progress*, Vol. 4, 2004, No. 1, p. 1.

[19] BRDIČKA, R. *Fundamentals of Physical Chemistry*. Přirodovědecké nakladatelství, Prague, 1952, in Czech.

[20] ŠALAK, A.: Manganese Sublimation and Carbon Ferromanganese Liquid Phase

Formation during Sintering of Premixed Manganese Steels. *Int. J. Powder Metall. Powder Technology,* Vol. 16, 1980, No. 4, p. 369.

[21] ŠALAK, A.: Manganese Vapor-Protection of Premixed Manganese Steels against Oxidation during Sintering. *PMI,* 1986, 4, p. 266.

[22] ŠALAK, A. and ĎURIŠIN, J.: Evaporation of Manganese in Solid State and its Effect on Sintering of Manganese Steels. *Pokroky Práškové Metalurgie (PPM-VUPM),* 1979, No. 2–3, p. 29, in Slovak.

[23] ŠALAK, A.: Sublimation and Evaporation of Manganese at Sintering of Manganese Steels Prepared from Mixed Powders. *Poroshkovaya Metallurgiya,* 1980, No. 10, p. 103, in Russian.

[24] ŠALAK, A.: Sublimation and Evaporation of Manganese. *Hutnické listy*, Vol. XXXV, 1980, No. 10, p. 724, in Slovak.

[25] ŠALAK, A.: Effect of the Temperature on the Sublimation and Condensation of Manganese Vapour. *Hutnické listy,* XXXIX, 1984, 2, p. 124, in Slovak.

[26] ŠALAK, A.: Sintered Manganese Steels, Part II: Manganese Evaporation during Sintering. *PMI,* Vol. 12, 1980, No. 2, p. 2.

[27] ŠALAK, A. and SELECKÁ, M.: Manganese as Alloying Element in Sintered Steels. In: (CD), Research and Technology Drivers for Future Low-alloyed PM Steels. Höganäs Seminar, TU, Vienna, Höganäs AB, Höganäs, 2005.

[28] ŠALAK, A.: High-Strength Manganese Steels. *Modern Development in Powder Metallurgy,* Vol. 13, MPIF Princeton, NY, (Int. PM Conf. Washington), 1980, p.183.

[29] ŠALAK, A.: Effect of Manganese Sublimation in the Atmosphere Containing Oxygen. In: VIth Fractography Conf. IMR SAV, Košice, 1981, p. 66, in Slovak.

[30] ŠALAK, A. and ĎURIŠIN, J.: Method for Removing the Oxygen from Gaseous Atmospheres. Patent CS (SK), AO 208295-82.

[31] ŠALAK, A.: Sublimation and Condensation of Manganese Vapour. *Practical Metallography,* Vol. 22, 1985, No. 1, p. 26.

[32] ŠALAK, A.: Effect of Fine Grain Structure of Iron Powder on Increase of the Properties of Sintered Manganese Steels. *Poroshkovaya Metallurgiya,* 1980, No. 11, p. 100, in Russian.

[33] ŠALAK, A.: Metallographic Study of Initial Stages of Sintering of Fe–Mn–System. In: IIIrd Metallography Symp., Štrbské Pleso, SVTS DT Košice, 1978, p. 63, in Slovak.

[34] ŠALAK, A.: Sintered Manganese Steel, Part I: Effect of Structure of Initial Iron Powders upon Mechanical Properties. *PMI,* Vol. 12, 1980, No. 1, p. 28.

[35] ŠALAK, A., SELECKÁ, M. and CHACHAĽÁK, M.: Effect of Recrystallisation on Structural and Technological Properties of the Hametag Iron Powder. *Kovové materialy (Metallic Materials),* Vol. 18, 1980, No. 1, p. 126, in Slovak.

[36] ŠALAK, A.: Structure of Manganese Alloyed Sintered Steels Part 2: Basis Powder Hametag Iron. *Practical Metallography,* Vol. 17, 1980, p. 390.

[37] ŠALAK, A.: Microstructure of Sintered Steel Alloyed with Manganese, Part 1: Base Powder Atomised Iron. *Practical Metallography,* Vol. 17, 1980, No. 6, p. 273.

[38] TOLBA-SALLAN, M., HENRY, G. and CIZERON, G.: Study of Diffusion Processes in Powder Iron–Manganese Systems. *METAUX,* 1981, No. 669, May, p. 61, in French.

[39] PARTHASARATHY, T.A. and SHEWMON, P.G.: Vapor Transport and DIGM in the Ni–Fe System. *Metallurgical Transaction A,* Vol. 14A, 1983, Dec., p. 2560.

[40] ADDA, Y.: *Diffusion in Solid Solutions.* Inst. Nat. des Sciences et Technique Nucleaires, Saclay, 1966, in French.

[41] ŠALAK, A.: Difusion Induced Grain Bouindary Migration at Alloying of Iron Powder by Manganese Vapour. *Kovové materialy (Metallic Materials)*, Vol. 27, 1989, No. 2, p. 159, in Slovak.
[42] ŠALAK, A.: Activated Alloying of Fe–Mn Powder Systems by Manganese Vapour during Sintering. *Science of Sintering*, Vol. 21, 1989, No. 3, p. 145.
[43] HILLERT, M.: On the Driving Force for Diffusion Grain Boundary Migration. *Acta Metallurgica*, Vol. 17, 1983, p. 237.
[44] CHONGMO, L.I. and HILLERT, M.: Diffusion Induced Grain Boundary Migration in Cu–Zn. *Acta Metallurgica*, Vol. 30, 1982, p. 1133.
[45] SEDLÁČEK, V., KRÁLIK, F. and ŠEJNOHA, R.: *Diffusion and Precipitation Processes in Metallic Systems*, Academia, Praha, 1968, in Czech.
[46] SHVINDLERMAN, L.S.: *Grain Boundary Migration in Metals*. 2nd ed., CRC Press, Boca Raton, FL, 2009.
[47] NAVARA, E.: Sintering of Iron Powder with an Addition of Ferromanganese. In: SINTERING-85, edited by G.C. Kuczynski and D.P. Uskokovic, Hayne Palmout III and M.M. Ristic, Plenum Publishing Corp., 1987, p. 343.
[48] ZAPF, G., HOFFMANN, G. and DALAL, K.: Effect of Additional Alloying Elements on the Properties of Sintered Manganese Steels. *PMI*, Vol. 8, 1975, No. 35, p. 214.
[49] ŠALAK, A.: *Manganese in Ferrous Powder Metallurgy*. DrSc. Thesis, TU, Košice, 1995, in Slovak.
[50] ŠALAK, A. and SELECKÁ, M.: Effect of Structural Activity of Iron Powder Hametag on the Process of its Diffusion Alloying with Manganese. In: IVth Metallography Symp., Štrbské Pleso, IMR SAS, Košice, Vol. 3, 1980, p. 129, in Slovak.
[51] ŠALAK, A., LEITNER, G. and HERMEL, W.: Properties of Short-time High-Frequency Sintered Powder Forged Manganese Steel Prepared of Atomised Fe-powder RZ. *Pokroky Práškové Metalurgie*, 1978, No. 4, p. 45, in Slovak.
[52] ŠALAK, A., LEITNER, G. and HERMEL, W: Properties of Induction Sintered Manganese Steel Prepared of Fe–powder Hametag. *Pokroky Práškové Metalurgie*, 1979, No. 2-3, p. 41, in Slovak.
[53] DANNINGER, H. PÖTTSCHACHER, R., BRADAC, S., ŠALAK, A. and SEYERKAMMER, J.: Comparison of Mn, Cr and Mo Alloyed Sintered Steels Prepared from Elemental Powders. *Powder Metallurgy*, Vol. 48, 2005, No.1, p. 23.
[54] WRONSKI, A,S., SELECKÁ, M. and STOYTCHEV, M.: Development of Powder Metallurgy Manganese Steels for Structural Components. NATO Project No. 972395 (1999-2003), Report April 2000, IMR SAS, Košice.
[55] GERMAN, R.M.: *Powder Metallurgy Science*, 2nd edition, MPIF, Princeton, NJ, 1994.
[56] DUDROVÁ, E. and ŠALAK, A.: Formation and Growth of Metal Bridges between Particles in Iron Compacts during Short Sintering Periods. *PMI*, Vol. 8, 1976, No. 3, p. 122.
[57] ŠALAK, A. and SELECKÁ, M.: Manganese Gas-Phase Alloying of Sintered Steels Analysed by Dilatometry: Effect of Carbon, base Powders and Manganese Carriers. *Powder Metallurgy Progress*, Vol. 8, 2008, No. 1, p. 7.
[58] SELECKÁ, M., ŠALAK, A., PIECZONKA, T. and CIAS, A.: Fracture, Microstructure and Microanalyses of Dilatometry Samples of Fe-Mn Steels. In: Deformation and Fracture in Structural PM Materials DF PM, Stará Lesná, IMR SAS, Košice, Vol. 1, 2002, p. 92.
[59] SELECKÁ, M., ŠALAK, A., PIECZONKA, T. and STOYTCHEV, M.: Solid-

Phase-Gas-Phase Sintering and Alloying of Powder Fe-Mn Steels Observed by Dilatometry. In: EURO PM2007 Int. Powder Metallurgy Congress & Exh., Toulouse, EPMA, Shrewsbury, Vol. 1, p. 47.

[60] ŠALAK, A. and SELECKÁ, M.: Sinter-Alloying and Properties of Manganese Steels – State of the Art. chapter 2, In: Bulk Materials Research, Technology and Applications. T. Frias and V. Maestas (eds.), NOVA Science Publishers, Inc., 2009, pp. 75-140.

[61] KLEIN, A.N., OBERACKER, R. and THÜMMLER, F.: Development of New High Strength Sintered Steels Containing Silicon and Manganese. *MPR*, 1984, June, p. 335.

[62] SELECKÁ, M., ŠALAK, A. and PIECZONKA, T.: Dilatometer Study on Solid-Phase Gas-Phase Sintering and Alloying of Powder Fe–Mn Steels. *STEEL GRIPS*, Vol. 6, 2008, No. 5, p. 355.

[63] CAMINHA, I.M.R. and ASSUNCAO, E.Z.R.: Sintering Behaviour and Micro-structural Evaluation of Iron Powder with Ferromanganese Addition. In:. PM 90 World Conf. on Powder Metallurgy, Wembley, The Institute of Metals, London, Vol. 3, 1990, p. 94.

[64] SELECKÁ, M., SIMKULET, V. and PARILÁK, Ľ.: Properties of Sintered Hybrid and Mixed Fe–0.85Mo–XMn-0.3C Steels. In: PM Auto 05 Int. Conf. on Powder Metallurgy for Automotive Parts, Isfahan, MPM Tehran, Iran, 2005, p. 309.

[65] KLEIN, A.N., OBERACKER, R. and THÜMMLER, F.: High Strength Si–Mn-Alloyed Sintered Steels. Microstructure-Properties. *PMI*, Vol 17, 1985, No. 1, p. 13.

[66] SULOWSKI, M.: Dilatometric Investigation of Fe–Mn–Cr–Mo PM Steels with Different Carbon Concentrations. *Powder Metallurgy Progress*, Vol. 8, 2008, No. 2, p. 151.

[67] PIECZONKA, T., MITCHELL, S.C., CIAS, A., SULOWSKI, M. and WRONSKI, A.S.: Dimensional Changes during Sintering of Manganese-Carbon Steel Compacts. *Inžynieria Materialowa*, Vol. XIX, 1998, No. 4, p. 1171.

[68] SELECKÁ, M. and ŠALAK, A.: Electrolytic Manganese and Ferromanganese Grades: Physical-Metallurgical and Technical Characteristics. *Powder Metallurgy Progress*, Vol. 9, 2009, No. 2, p. 65.

[69] UPADHYAYA, G.S.: *Manganese in Powder Metallurgy Alloys*. The Manganese Centre, Paris, 1985.

[70] KEPKA, M.: Content of Gases, of Non-Metallic Inclusions and of Trace Elements in Ferroalloys. *Hutník*, Vol. 23, 1978, No. 7, p. 256, in Czech.

[71] GASIK, D.N., IGNATYEV, V.S. and GASIK, M.I.: *Structure and Quality of Industrial Ferroalloys and of Alloying Elements*. Tekhnika, Kiev, 1975, in Russian.

[72] RASHEV, T.: *High Nitrogen Steels*. Publishing House of Bulgarian Academy of Sciences, Sofia, 1995.

[73] MOROZOV, AN.: *Hydrogen and Nitrogen in Steel*. SNTL Praha, 1953, in Czech.

[74] ŠALAK, A., KUCHÁR, J., DUDROVÁ, E. and RUDNAYOVÁ, E.: Saturation of Compacts with Nitrogen at Sintering in Nitrogen Containing Atmosphere. *Kovové Materiály (Metallic Materials)*, Vol. IX, 1971, No. 4, p. 345; Ibid. BISI Translation Service, no. 9678, April, 1976.

[75] ŠALAK, A., BENKE, V. and HRUBJAK, M.: *Material Research and Production Technology of Rolling Bearing Rings by Powder Metallurgy*. ZVL – Research Institute for Rationalisation of Bearing Production, Žilina, Final Report, 1998, in Slovak.

[76] ŠALAK, A., RUDNAYOVÁ, E. and SELECKÁ, M.: Properties of Short-Time in

Gas Nitrided Self-Lubricated Bearings. *Pokroky Práškové Metalurgie VUPM,* Vol. 34, 1996, No. 3, p. 43, in Slovak.

[77] ŠALAK, A. and SELECKÁ, M.: Mechanical and Friction Properties of Sintered (1-5)% Manganese Steel. In: 3rd Int. Conf. on Powder Metallurgy, TTMD, Ankara, 2002, p. 231.

[78] ŠALAK, A. and MIŠKOVIČ, V.: Properties of Sponge Iron Powder Manufactured by Reduction of Mill Scale. *Hutnické Listy*, 1972, No. 3, p. 159, in Slovak.

[79] ŠALAK, A.: Properties of Sintered Compacts based on Sponge Iron Powder Manufactured from Mill Scale. *Strojírenská Výroba*, 1972, No. 11, p. 817, in Slovak.

[80] ŠALAK, A., DUDROVÁ, E., MIŠKOVIČ, V. and RUDNAYOVÁ, E.: Low-Carbon Low-Alloy Sintered Steels with Bainitic Structure. In: IVth Int. Conf. on Powder Metallurgy in Czechoslovakia, Štrbské Pleso, IMR SAS, Košice, Vol. III, 1974, p. 101, in Slovak.

[81] ŠALAK, A.: Properties of Low-Carbon Low-alloyed Sintered Steel. *Strojírenství*, Vol. 26, 1976, No. 10, p. 619, in Slovak.

[82] ŠALAK, A. and SELECKÁ, M.: Sinter-Alloying and Properties of Manganese Steels – State of the Art. In: Advances in Condensed Materials and Materials Research, Vol. 7, chapter 3, pp. 100-171, Has Geelvinck and Sjaak Reynst, eds. NOVA Science Publishers, Inc., New York, 2010.

[83] PICKERING, F.B.: The Properties and Properties of Bainite in Steels. In: Transformation and Hardenability in Steels, symp. Climax Molybdenum Comp. of Michigan, 1971, p. 152.

[84] SELECKÁ, M. and ŠALAK, A.: Industrially Sintered Hybrid Low-Carbon 3Cr-0.5Mo–xMn Steels. *Int. J. Powder Metall.*, Vol. 46, 2010, No. 4, p. 29.

[85] LEITNER, G. and HERMEL, W.: Physical and Technological Preconditions for the Application of Inductive Rapid Heating Technique for Sintering of PM Parts. In: V[th] Int. Powder Metall. Conf. in ČSSR, Gottwadov (Zlin), IMR SAS Košice, Vol. I, 1978, p. 373, in German.

[86] LEITNER, G., HERMEL, W., ŠALAK, A. and DUDROVÁ, E.: The use of Induction Fast-Heating Technique for Sintering of PM Profiles and Parts. In: VIth Int. PM Tagung in DDR, Dresden,Vol. I, 1977, lecture 19, p. 9, in German.

[87] ŠALAK, A., LEITNER, G. and HERMEL, W.: Properties of Induction-Sintered Fe–Mn–C and Fe–Mn–Cu–C Steels as Sintered and as Sinter Forged. In: V[th] Int. PM Conf. in Poland, Poznaň, Vol. I, 1979, p. 393, in German.

[88] ŠALAK, A., LEITNER, G. and HERMEL, W.: Mechanical Properties, Structure and Fracture of Induction HF Sintered Manganese Steel Prepared on the Basis of Air Atomised Iron Powder RZ. *Pokroky Práškové Metalurgie*, 1978, No. 4, p. 31. in Slovak.

[89] ŠALAK, A., LEITNER, G. and HERMEL, W.: Properties of Induction-Sintered Fe-Mn–C and Fe–Mn–Cu–C Steel in the Sintered and Forged States. *PMI,* Vol. 13, 1981, No. 1, p. 21.

[90] NAVARA, E. and ŠLESÁR, M.: Theory of Eddy Communition of Metals. In: Problémy Práškovej Metalurgie, Slovak Academy of Sciences, Bratislava, 1963, p. 61, in Slovak.

[91 ŠALAK, A.: Properties and Sintering Conditions of Manganese Steels. In: UN Seminar, Minsk, 1985, lecture R6, in Russian.

[92] ŠALAK, A.: Properties of Sintered High Strength Steel. In: Int. Symp. Metallic Materials with High Mechanical-Physical Properties, SVTS DT Bratislava, Vol. 2, 1977, p. 100, in Slovak.

[93] ŠALAK, A.: Effect of Iron Powder Grade on Mechanical Properties of Manganese

and Manganese–Molybdenum Steels. In: 5th Int. Conf. on Powder Metallurgy in ČSSR, Gottwaldov (Zlin), SVTS DT Žilina, IMR SAS, Košice, Vol. I. 1978, p. 425, in German.

[94] ŠALAK, A.: Powder Forging – Basic Characteristics. In: Deformation Processes at Processing of Powder Materials and their Effect on Properties of the Products, Považská Bystrica, SVTS DT Žilina, 1980, p. 178, in Slovak.

[95] ŠALAK, A. and MIŠKOVIČ, V.: Properties of Sintered Hot Extruded Carbon and Manganese Steels. In: VIth Int. Powder Metall. Conf. in GDR, Dresden, Vol. 2, 1977, p. 33, in German.

[96] ŠALAK, A., MIŠKOVIČ, V. and MAÁR, K.: Mechanical Properties of Hot Extruded Sintered Carbon and Manganese Steels. *Zborník Vedeckých Prác VŠT Košice, Košice*, 1978, No. 2, p. 341, in Slovak.

[97] ŠALAK, A. and LAURINČÍK, J.: Effect of Particle Size, Heating Method and Lubricant Kind on Properties of Sintered Manganese Steel Prepared on the Basis of Hametag Iron Powder. *Pokroky Práškové Metalurgie*, 1983, No. 2, p. 21, in Slovak.

[98] ŠALAK, A. DUDROVÁ, E. and MIŠKOVIČ, V.: Properties of Sintered Carbon and with Manganese and Chromium Alloyed Steels. *Strojírenská Výroba*, Vol. 25, 1977, No. 7, p. 514, in Slovak.

[99] ŠALAK, A., DUDROVÁ, E. and MAÁR, K.: Mechanical Properties of Sintered Manganese and Chromium Steels. *Zborník Vedeckých Prác VŠT Košice*. Košice, 1978, No. 2, p. 325, in Slovak.

[100] ŠALAK, A. and MIŠKOVIČ, V.: Properties of Sintered Extruded Carbon and Manganese Steels. In: VIth Int. PM Tagung in DDR, Dresden, Vol. 2, 1977, lecture 33, p. 1, in German.

[101] ŠALAK, A., MIŠKOVIČ, V., MAÁR, K. and ŠPITÁL, J.: Effect of Hot Forming Methods and Conditions on Properties of Sintered Carbon and Manganese Steels. *Zborník Vedeckých Prác VŠT Košice*, Košice, 1979, No. 3, p. 259, in Slovak.

[102] ŠALAK, A.: Effect of Extreme Sintering Conditions upon Properties of Sintered Manganese Steels. *PMI*, Vol. 16, 1984, No. 6, p. 260.

[103] ŠALAK, A.: Evaluation Powder Activity used for Production of Manganese Steels according to Tensile Strength Value. *Poroshkovaya Metallurgiya*, 1984, No. 8, p. 58, in Russian.

[104] ŠALAK, A.: Evaluation of Iron Powder Activity by means of Tensile Strength Value of Sintered Manganese steel. *Pokroky Práškové Metalurgie*, 1983, No. 4, p. 3, in Slovak.

[105] ŠALAK, A.: Effect of Heating Stage at Sintering on Properties of Manganese steels. In: Seminary – Sintering of Powders and its Effect on Properties of Materials, Považská Bystrica, IMR SAS, Košice, 1983, p. 156, in Slovak.

[106] ŠALAK, A., SELECKÁ, M., PARILÁK, Ľ. and ČAJKA, A.: Effect of Manganese Donor on Properties of Fe–Cr–Mn–Mo–C Steels. In: PM2004 Powder Metallurgy World Congress & Exh., Vienna, H. Danninger and R. Ratzi eds., EPMA, Shrewsbury, Vol. 3, 2004, p. 237.

[107] DANNINGER, H., PÖTTSCHACHER, R., ŠALAK, A. and SEYERKAMMER, J.: Comparison of Manganese and Molybdenum as Alloying Metals in Sintered Steels. In: TECHNOLÓGIA'97 Int. Conf., STU Bratislava, Bratislava, Vol. 1, 1977, p. 16, in German.

[108] DANNINGER, H., PÖTTSCHACHER, R., JANGG, G., SEYRKAMMER, J. and ŠALAK, A.: Manganese Alloyed Sintered Steels Prepared from Elemental Powders. In: Powder Metall. World Congress, Paris, Vol. II, 1994, p. 879.

[109] ŠALAK, A., SELECKÁ, M. and PARILÁK, Ľ.: Mechanical Properties of

Sintered (1-4%) Manganese Steel. *Naučni Izvestija,* Vol. VII, 2000, June (DEFEKTOSKOPIJA´2000 NDT-2000), Sozopol, 2000, p. 94.

[110] CIAS, A., MITCHELL, S. and WRONSKI, A.S.: Processing, Microstructure and Properties of PM 0.6% C Manganese Steels. In: 1998 World PM Congress & Exh., Granada, EPMA, Shrewsbury, Vol. 3, 1998, p. 178.

[111] SINKA, V., ŠALAK, A. and SELECKÁ, M.: The Effect of Iron Powder Grade on Density and Bending Strength of Manganese Steels. In: DFPM'99 Deformation and Fracture in Structural PM Materials, Piešťany, IMR SAS, Košice, Vol. 2, 1999, p. 38.

[112] SINKA, V., SELECKÁ, M. and ŠALAK, A.: Influence of Iron Powder Grade and Manganese Carrier on the Homogeneity Modulus of Sintered Mn Steels. In: Third Int. Latin-American Conf. on Powder Metallurgy, Florianopolis, 2001, paper 11/06, p. 72.

[113] ŠALAK, A., SELECKÁ, M., ČAJKA, A. and FRANEK, Š.: Tribological Properties of Sintered Manganese Steel. In: Deformation and Fracture in PM Materials DF PM 2002, Stará Lesná, IMR SAS, Košice, Vol. 1, 2002, p. 285.

[114] ŠALAK, A., SELECKÁ, M. and DANNINGER H.: *Machinability of Sintered Steels.* CISP, Cambridge, UK, 2004; Viva Books, New Delhi, 2008.

[115] NAVARA, E.: The Investigation of Friction Properties of Various Sintered Steel Structures. In: *Problémy Práškové Metalurgie.* J. Kubelík, ed., Vydavateľstvo Slovenskej Akadémie Vied (SAV), Bratislava, 1964, p. 127, in Slovak.

[116] ŠALAK, A.: Gears for Hydrogenerators Manufactured from Mangaese Steel. *Strojírenská Výroba,* 1990, No. 12, p. 45, in Slovak.

[117] ŠALAK, A., DUDROVÁ, E. and MIŠKOVIČ, V.: Sintered Toothed Gears for Hydrogenerators Produced from Powder Mixed Manganese Steels. *Powder Metallurgy Science and Technology,* Vol. 3, 1992, No. 2, p. 26.

[118] WRONSKI, A. S., SELECKÁ, M. and STOYTCHEV, M.: Development of Powder Metallurgy Manganese Steels for Structural Components. NATO Project No. 972395 (1999-2003), Report April 2001, IMR SAS, Košice.

[119] SELECKÁ, M., ŠALAK, A. and JAKUBÉCZYOVÁ, D.: Mechanical and Tribological Characteristics of Sintered Manganese Steels. *Materials Science Forum,* Vol. 672 (2011), p. 59.

[120] ŠALAK, A.: Effect of Compacting Pressure and 0.5% Mo on Properties of Sintered Iron Powder and Manganese Steels. *Powder Metallurgy,* 1981, No. 2, p. 64.

[121] ŠALAK, A.: Effect of Copper and Phosphorus Additions on the Properties of Steels Alloyed with Manganese. In: 5th Int. Conf. on Powder Metallurgy, Central Section – Metalurgii Prószkow, Katowice, Poznaň, Vol. 1, 1979, p. 423, in German.

[122] ŠALAK, A.: Effect of Addition of Copper and Phosphorus on Character of Fracture of Sintered Manganese Steel. In: 5th State-wide Fractography Conf., Zlatá Idka, IMR SAS Košice, 1979, p. 69, in Slovak.

[123] ŠALAK, A.: Effect of Copper and Phosphorus Addition on the Properties of Manganese Alloyed Steels. In: VIth Int. Pulvermetall. Tagung in GDR, Dresden, Vol. I, 1997, p. 423, in German.

[124] KAREL, V.: Systematisation of Base Microfractographical Terms. In: Fractography'87, 9th Conf., Tatranské Matliare, TU Košice, IMR SAS, Košice, 1987, Vol. I, p. 119, in Slovak.

[125] ŠALAK, A.: Sintering of Manganese Steel Prepared from Mixed Powders at Presence of Liquid Phase of Additional Element. *Pokroky Práškové Metalurgie,* 1983, No. 4, p. 35, in Slovak.

[126] ŠALAK, A.: Effect of Copper and Phosphorus Addition on Properties of Manganese

Alloyed Steels. In: VIth Scientific Conf., VŠD Žilina, Vol. I, p. 423, in German.

[127] ŠALAK, A.: Properties of Forged Alloy Steels Prepared from Loose Bulk Sintered Mixed Powders. *Pokroky Práškové Metalurgie,* 1983, No. 4, p. 15, in Slovak.

[128] SELECKÁ, M., ŠALAK, A., DANNINGER, H. and PARILÁK, Ľ.: Properties of Liquid Phase Sintered Fe–Mn–B Alloys. In: 2000 Powder Metallurgy World Congress, Kyoto, JSPM, Part 1, 2000, p. 16.

[129] SELECKÁ, M., ŠALAK, A., PARILÁK, Ľ. and DANNINGER, H.: Tribological Properties of Boron Liquid Phase Sintered Fe–Mn–C Steels. In:. Euro PM2001, Nice, EPMA, Shrewsbury, Vol. 1, 2001, p. 368.

[130] SELECKÁ, M., ŠALAK, A. and DUDROVÁ, E.: Properties of Mixed Low-Alloyed Fe–Mn–Cr–Mo–C Steel Sintered at Presence of Boron Liquid Phase. *Pokroky Práškovej Metalurgie,* 1993, No. 3, p. 63, in Slovak.

[131] SELECKÁ, M., DANNINGER, H., ŠALAK, A. and BUREŠ, R.: Properties of Fe–Cr–Mo–V–(Mn)–(B) Sintered Steels. In: 6th Int. Conf. TECHNOLÓGIA`99, STU Bratislava, Bratislava, Vol. 1, p. 178, in Slovak.

[132] ŠALAK, A.: State of the Art of Powder Metallurgy in Czechoslovakia. In: 8th Int. Conf. on Powder Metallurgy in Czechoslovakia, Piešťany, IMR SAS, Košice, Vol. 1, 1992, p. 31–43, in German.

[133] SELECKÁ, M., ŠALAK, A., PARILÁK, Ľ. and DANNINGER, H.: Influence of Boron on Mechanical Properties of Fe–Cr–Mo-V Sintered Steels. *Nauchni Izvestiya,* (Defektoskopia NDT-2000), Vol. VII, 2000, No. 1, p. 100.

[134] ŠALAK, A., SELECKÁ, M., DANNINGER, H., ŠALAK, A., UNAMI, S. and PARILÁK, Ľ.: Properties of Boron Liquid Phase Sintered Fe–Cr–Mo–V Steels. In: Int. Conf. DF PM `99, Deformation nd Fracture in Structural PM Materials, Piešťany, IMR SAS, Košice, Vol. 1, 1999, p. 175.

[135] SELECKÁ, M., DUDROVÁ, E., BUREŠ, R., KABÁTOVÁ, M. and ŠALAK, A.: Effect of Boron on Microstructure of Selected Sintered Prealloyed Steels. *Pokroky Práškové Metalurgie,* Vol. 33, 1995, No. 1–2, p. 77, in Slovak.

[136] DUDROVÁ, E.: Effect of 0.2% B on Structure and Properties of Mixed Powder Fe–Cr–xMe–C Steel. In: CO MAT TECH `93, STU Trnava, 1993, p. 8, in Slovak.

[137] SELECKÁ, M., DANNINGER, H., ŠALAK, A., UNAMI, S. and PARILÁK, Ľ.: Properties of Boron Liquid Phase Sintered of Cr–Mo–V Alloyed Steels. In: Deformation and Fracture in Structural PM Materials, DF PM'99, Piešťany, IMR SAS Košice, Vol. 1, 1999, p. 75.

[138] SELECKÁ, M., ŠALAK, A. and DUDROVÁ, E.: The Influence of the Sintering Atmosphere on the Microstructure of Sintered Fe–B and Fe–Ni–Mo Materials in Continuous Process. In: Metallography'95, St. Lesná, TU Košice, 1995, p. 293.

[139] GRAF von MATUSCHKA, A.: Boriding (Borieren). Carl Hanser Verlag München, Wien, 1997, in German.

[140] STEWART, K.: Boronizing Protects Metals Against Wear. *Advanced Materials & Processes,* 1997, No. 3, p. 23.

[141] SINHA, A.K.: Boriding in Heat Treating. ASM Handbook, ASM International, Materials Park, OH, Vol. 4, 1991, p. 437.

[142] DEGER, M., RIEHLE, M. and SCHATT, W.: Increase of Wear Resistance of Sintered Steels through Chemical-Thermal Processing (Boriding). *Neue Hütte,* Vol. 17, 1972, No. 6, p. 341, in German.

[143] ŠALAK, A., SELECKÁ, M. and DANNINGER, H.: Manganese Assisted Pack Boriding of Sintered Manganese Steels. *Int. J. Powder Metall.,* Vol. 27, 2001, No. 4, p. 53.

[144] ŠALAK, A., SELECKÁ, M. and PARILÁK, Ľ.: Pack Boriding of PM Steels without

Halides as Activators. *Naučnyje Izvestija*, Vol. VII, 2000, No. 1, p. 120.

[145] HAWKINS, D. T. and HULTGREN, R.: *Constitution of Binary Alloys*. Metals Handbook, ASM, Metal Park, OH, 1976, Vol. 8, 8th edition, p. 284.

[146] ŠALAK, A., SELECKÁ, M, TRPČEVSKÁ, J. and DANNINGER, H.: Microanalysis of Alloying Elements in Manganese Assisted Pack Borided PM Steels. *Acta Metallurgica Slovaca*, Vol. 7, 2001, p. 365.

[147] ŠALAK, A., SELECKÁ, M., DANNINGER, H. and PARILÁK, Ľ.: Formation and Properties of Boride Layers on Sinetered Steels without Halides. In: 1st Int. Conf. on Physical Metallurgy and Fracture of Materials '99, Herľany, TU Košice, 1999, p. 172, in Slovak.

[148] ŠALAK, A., SELECKÁ, M. and DUDROVÁ, E.: Combined Sintering and Boriding of Powder Steels. *Metalurgia Proszków*, Vol. XXVIII, 1995, No. 1/2, p. 17.

[149] ŠALAK, A., DANNINGER H. and SELECKÁ, M.: Sinter Boriding of P/M Steels without Halides. *P/M Science & Technology Briefs*, Vol. 1, 1999, No. 1, p. 5.

[150] SELECKÁ, M., ŠALAK, A. and DANNINGER, H.: The Effect of Boron Liquid Phase Sintering on Properties of Ni, Mo, and Cr Alloyed Structural Steels. In: Int. Conf. New Developments on Technology (AMPT'01), Leganés, Madrid, Vol III, 2001, p. 1295.

[151] ŠALAK, A. and SELECKÁ, M.: New Aspects for Sinter Boriding of PM steels. *Powder Metallurgy Progress*, Vol. 2, 2002, No. 3, p. 161.

[152] Software: HSC Chemistry 4.1, Outokumpu Research Oy, Pori Finland, June 30, 1999.

[153] ŠALAK, A.: Properties of Sintered and Powder Forged Fe–Mn–Cr Steels Manufactured from Prealloyed Powders. *Pokroky Práškové Metalurgie*, 1985, No. 1, p. 25, in Slovak.

[154] ŠALAK, A., SELECKÁ, M. and PARILÁK, Ľ.: Some Propeties of Powder Forged Steel Based on Chromium Prealloyed Powders of High Oxygen Content. *J. Materials Processing Technology*, 143–144, 2003, p. 18.

[155] ŠALAK, A., SELECKÁ, M. and PARILÁK, Ľ.: Some Properties of Powder Forged Steel Based on Chromium Prealloyed Powders of High Oxygen Content. In: AMPT'01 Int. Conf. New Developments on Powder Technology, J.M. Torralba, ed., Leganés, Madrid, Vol. III, 2001, p. 1287.

[156] ŠALAK, A., SELECKÁ, M. and PARILÁK, Ľ.: Properties of Sintered Fe–Cr–Mo–V Steels. In: 6th Int. Conf. TECHNOLÓGIA'99, STU Bratislava, Bratislava, Vol. I, 1999, p. 184.

[157] ŠALAK, A., DANNINGER, H., SELECKÁ, M. and UNAMI, S.: Properties of Sintered Fe–Cr–Mo–V Steels. In: Deformation and Fracture in Structural PM Materials DF PM'99, Piešťany, IMR SAS Košice, Vol. 1, 1999, p. 178.

[158] ŠALAK, A., DANNINGER, H., SELECKÁ, M., OGURA, K. and UNAMI, S.: Properties of Sintered Fe–Cr–V Steels. In: PM Auto'99 The 1999 Int. Conf. on Powder Metall. for Automotive Parts. A. Arvand and H. Danninger, eds., Mashad Powder Metallurgy Co., MPM, Tehran, Isfahan, 1999, p. 177.

[159] BETZ, R.O.: Heavy-Duty PM Bearings Reach Full Production. *Iron Age*, 1976, No. 7, p. 35.

[160] ADAMS, R.J. and GLOVER, D.: Improved Bearings at Lower Cost via Powder Metallurgy. *Metal Progress*, 1977, No. 8, p. 29.

[161] ADAMS, R.J. and GLOVER, D.: Success Story for Powder Forged Bearings. *Metal Powder Report*, 1980, No. 1, p. 10.

[162] ŠALAK, A., BENKE, V. and HRUBJAK, M.: Manufacturing of Rolling Bearing Rings via Powder Metallurgy. In: Conf. Strojárska technológia a valivé ložiská'97,

Žilina, TU Žilina, 1997, p. 38, in Slovak.
[163] ŠALAK, A., KABÁT, M. and SINKA, V.: Effect of Sintering Temperatrure on Properties Powder Forged Fe–Mn–Cr Steels. In: 8th Conf. Fractography `85, Tatranské Matliare, TU Košice, 1985, p. 131, in Slovak.
[164] SINKA, V., VELES, P. and ŠALAK, A.: Effect of Sintering Atmosphere on Properties of Powder Forged Steel Sumiron 4100S. *Hutník,* Vol. 39, 1989, No. 8-9, p. 315, in Slovak.
[165] ŠALAK, A.: Properties of Powder Forged Oil Atomised 0.7Mn–1Cr–0.2Mo Steel. In: IX. Int. Powder Metallurgy Conf. in GDR, Dresden, Vol. 2, 1989, p. 243, in German.
[166] ŠALAK, A.,SELECKÁ, M. and DUDROVÁ, E.: Properties of Forged 0.7Mn-1.0Cr–0.2Mo Steel Produced from Oil-Dispersed Powder. *Transactions of the Technical University Košice*, Vol. 2, 1992, No. 2, p. 197.
[167] ŠALAK, A. and SELECKÁ, M.: Mechanical and Toughness Properties of Powder-Forged Low Alloyed Cr–Steels. In: 3rd Int. Powder Metallurgy Conf., Turkish PM Association, Ankara, 2002, p. 408.
[168] ŠALAK, A. and BENKE, V.: Powder Forged Rolling Bearing Races from Fe–Ni-Mo and Fe–Mn–Cr Powders. *PMI*, Vol. 22, 1990, No. 5, p. 9.
[169] ŠALAK, A. and SELECKÁ, M.: Manganese as an Alloy Element in Sintered Steels. In: Höganäs Chair Seminar 2005, Research and Technological Drivers for Future Low-Alloyed PM Steels, TU Vienna, 2005, CD.
[170] ŠALAK, A., ŠKOCH, V., JIRÁSEK, J. and ŠMEJKAL, M.: Manufacturing of Axisymetric Components by Powder Forging. In: Conf. Automation of Engineering Production IV. ČSVTS and VÚSTE Praha, 1986, p. 43, in Slovak.
[171] BENKE, V., ŠALAK, A. and KRES, D.: Effect of Density of Powder Preforgings on Deformational Force and Properties of Rotary Forgings. In: VIIth Int. Powder Metall. Conf. in ČSSR, Zlin, SVTS Žilina, 1987, p. 223, in Slovak.
[172] ŠALAK, A.: Manufacturing of Rolling Bearing Rings by Forging of Powder Steels. *Hutnické Listy*, Vol. XLV, 1990, No. 7, p. 483, in Slovak.
[173] ŠALAK, A., BENKE, V., KRES, D., BEŽILA, J. and VELES, P.: Powder Metallurgy in Manufacturing of Rolling Bearings. In: 1th Int. Symp. Interprogress – Metallurgy, SVTS, Bratislava, 1990, p. 9, in Slovak.
[174] ŠALAK, A.: Powder Forging of Rolling Bearing Races from Prealloyed Fe–Ni–Mo Powder. *Poroshkovaya Metallurghiya*, 1991, No. 11, p. 72, in Russian.
[175] SINKA, V., VELES, P. and ŠALAK, A.: Effect of Sintering Atmosphere on Properties of Powder Sumiron 4100S Steel. *Hutník*, Vol. 39, 1989, p. 325, in Slovak.
[176] ŠALAK, A. and SELECKÁ, M.: Contact Fatigue Endurance of Powder Forged Bearing Rings. In: Euro PM2001, Special Interest Seminar, Nice, EPMA, Shrewsbury, Vol. 1, 2001, p. 39.
[177] BENKE, V. and ŠALAK, A.: Optimalisation of Selected Factors on Forging Process of Powder Steels. In: VIIth Conf. on Powder Metallurgy in Poland, Polish Academy of Sciences, Krakow, Vol. II, 1988, p. 713, in Russian.
[178] ŠALAK, A. and SELECKÁ, M.: Manufacturing and Properties of Powder Forged Roller Bearing Rings. In: 1999 Int. Conf. on Powder Metallurgy for Automotive Parts PM AUTO '99, Mashad Powder Metallurgy Co., Tehran, Isfahan, 1999, p. 35.
[179] SELECKÁ, M. and ŠALAK, A.: Durability and Failure of Powder Forged Rolling Bearing Rings. *Wear*, 266, 1999, p. 47.
[180] SELECKÁ, M. and ŠALAK, A: Structural Features of Rolling Bearing Rings Damage at Contact Fatigue Loading Manufacured by Powder Forging. In: 7th

Metalography Conf., VUZ Bratislava, Vol. 1, 1989, p. 113, in Slovak.

[181] BILLY, J. and ŠALAK, A.: Microstructural Aspects of Pitting Origin on the Bearing Rings made by Powder Metallurgy. *Zborník Vedeckých Prác Vysokej Školy Technickej v Košiciach* (TU Košice), 1986, p. 279, in Slovak.

[182] BILLY, J. and ŠALAK, A.: Microstructural Changes in Rolling–Bearing Rings Manufactured by Powder Metallurgy. *Kovové Materiály (Metallic Materials)*, Vol. 23, 1965, No. 3, p. 277, in Slovak.

[183] BILLY, J. and ŠALAK, A.: Prediction of Durability of Rolling Bearing Specimens Produced via Powder Metallurgy trought Evaluation of Morphology of Fracture Surface. In: Conf. Fraktography'87, T. Matliare, SVTS DT Žilina, 1987, p. 217, in Slovak.

[184] ŠALAK, A.: Experimental Production of Metallic Powder by Dispersion with (diesel) Oil and Gases. *Hutnické Listy*, Vol. 45, 1990, No. 3, p. 185, in Slovak.

[185] ŠALAK, A, SELECKÁ, M. and PARILÁK, Ľ.: Mechanical Properties of Sintered Fe–Mn–Cr–Mo V Steels. In: 2000 Powder Metallurgy World Congress, Kyoto, Koji Kosuge and Hiroshi Nagai, eds., JSPM, Part 1, 2000, p. 9.

[186] ŠALAK, A, SELECKÁ, M. and PARILÁK, Ľ.: Properties od Hybrid Fe–Mn–Cr–C Sintered Steels. In: Eur. Congress & Exhibition on Powder Metallurgy, 2001, Nice, EPMA, Shrewsbury, Vol. 1, 2001, p. 251.

[187] ŠALAK, A. and SELECKÁ, M.: Fracture Behaviour of Sintered Hybrid Fe–Mn–Cr–Mo–V Steels. In: Int. Conf. Fractography 2000, Stará Lesná, IMR SAS Košice, 2000, p. 167.

[188] SELECKÁ, M. and ŠALAK, A.: Practure of Hybrid Sintered Fe-Cr-Mn-C Steel. In: Int. Conf. Fractography 2003, Stará Lesná, IMR SAS Košice, 2003, p. 116.

[189] ŠALAK, A. and SELECKÁ, M.: Effect of Manganese Addition and Sintering Conditions on Mechanical Properties of Low Carbon 3Cr Prealloyed Steels. *Materials Science Forum*, Vol. 672, 2011, p. 55.

[190] ŠALAK, A., SELECKÁ, M., PARILÁK, Ľ. and ČAJKA, A.: Effect of Manganese Donor on Properties of Fe–Cr–Mn–Mo–C Steels. In: PM2004 Powder Metallurgy World Congress & Exh., Vienna, H. Danninger and R. Ratzi, eds., EPMA, Shrewsbury, Vol.3, 2004, p. 237.

[191] ŠALAK, A. and SELECKÁ, M.: Effect of Manganese Content and Manganese Carrier on Properties of Sintered and Sinter Hardened Hybrid Fe–3Cr–0.5Mo–xMn–0.24C Steel. *Powder Metallurgy*, Vol. 51, 2008, No. 4, p. 327.

[191] ŠALAK, A. and SELECKÁ, M.: Effect of Manganese Content and Manganese Carrier on Properties of Sintered and Sinter Hardened Hybrid Fe–3Cr–0.5Mo–xMn-0.24C Steel. *Powder Metallurgy*, Vol. 51, 2008, No. 4, p. 327.

[192] ŠALAK, A. and SELECKÁ, M.: New Properties of Low-Carbon Sinter Hardened Fe–Cr–Mn–0.6C Steels. In:. PM Auto '05. 4th Int. Conf. on Powder Metallurgy for Automotive Parts. Isfahan, A. Arvand, H. Danninger, eds., Mashad Powder Metallurgy Co Tehran, Isfahan, 2005, p. 112.

[193] SELECKÁ, M. and ŠALAK, A.: Effect of Manganese Content and Mnganese Carrier on Properties of Hybrid Low Carbon Fe–3Cr–0.5Mo–xMn-C Steel Sintered under Industrial Conditions. In: PM Auto'09, 5th Int. Congress on Powder Metallurgy for Automotive Parts, A. Arvand, H. Danninger, eds., Mashad Powder Metallurgy Co. Tehran, Isfahan, 2008, p. 25.

[194] SELECKÁ, M. and ŠALAK, A.: Industrial Sintering of Hybrid Low-Carbon 3Cr-0.5Mo–xMn Steels. *I. J. Powder Metall.*, Vol. 46, 2010, No. 4, p. 29.

[195] SELECKÁ, M., SIMKULET, V. and PARILÁK, Ľ.: Properties of Sintered Hybrid Fe–0.85Mo–xMn–0.3C Steels. In: PM Auto'05, 4th Int. Conf. on Powder Metallurgy

for Automotive Parts. Isfahan, A. Arvand, H. Danninger, eds., Mashad Powder Metallurgy Co Tehran, Isfahan, 2005, p. 309.

[196] SELECKÁ, M. and SIMKULET, V.: Mechanical Properties of Fe–0.85Mo–Mn–C Steels. In: PM2004 Powder Metallurgy World Congress & Exhibition, Vienna, H. Danninger and R. Ratzi., eds., EPMA Shrewsbury, 2004, p. 237.

[197] SIMKULET, V. and SELECKÁ, M.: Microstructure and Fracture of Fe–0.85Mo-0.6C Steel. In: State-wide Fractography Conference, Stará Lesná, IMR SAV, Košice, 2003, p. 116, in Slovak.

[198] SIMKULET, V. and SELECKÁ, M.: Effect of Manganese on Fracture of Premix and Hybrid Fe–0.85Mo–xMn–0.3C Sintered Steel. *Powder Metallurgy Progress*, 2006, Vol. 6, 156-163.

[199] SELECKÁ, M., SIMKULET, V., KERESTI, R. and PARILÁK, Ľ.: Effect of Manganese on Properties of Sintered Fe-(1.5,0.85)Mo-0.5C Steel. In: EURO PM2005 Congress & Exhibition, Prague, EPMA, Shrewsbury, 2005, Vol. 1, p. 41.

[200] ŠALAK, A., SELECKÁ, M. and BUREŠ, R.: The Role of the Atmosphere for Sintering of Mn Containing Steels. Höganäs Chair Seminar, TU Vienna, 1999, CD.

[201] OFFERMANN, E.K.: *Mitt. Kohl.-Eisenforschung*, 1936, No. 1, p. 55, in German.

[202] KIEFFER, R. and HOTOP, W.: *Sintered Iron and Sintered Steel*. Springer Verlag, Wien, 1948, in German.

[203] BENESOVSKY, F. and KIEFFER, R.: Sintered Manganese Steels. *Berg- und Hüttenmänische Monatshefte*, Vol. 95, 1950, No. 8, p. 145, in German.

[204] WAKABAYASHI, S., MORIOKA, Y. and WATANABE, T.: *Sintered Manganese Steel*. Report of the Castings Research Laboratory, Waseda, Waseda University, Vol. 16, 1965, p. 11.

[205] VOLYNOVA, T.F., EMELYANOVA, I.Z. and SIDOROVA, I.B.: Pecularities of the Structure Formation in Powder Iron Manganese alloys at Deformation and Recrystalisation. *Poroskovaya Metallurghiya*, Part I, 1992, No. 5, p. 29, in Russian.

[206] VOLYNOVA, T.F., EMELYANOVA, I.Z. and SIDOROVA, I.B.: Sintering of Manganese Steels Alloyed with Manganese throught the Point Sources. *Poroskovaya Metallurgiya*, Part II, 1992, No. 6, p. 1, in Russian.

[207] WILSON, F.G. and KNOTT, B.R.: Effect of Particles Dispersion on the Behaviour of Two Austenitic Manganese Alloys. *I. J. Powder Metall. & Powder Technology*, Vol. 14, 1978, No. 2, p. 121.

[208] RADOMISELSKIY, I.D., MARTYUKHIN, I.D., GLAZKOV, V.M. and NOVOMEYSKIY, Yu.D.: *Sintered Structural Materials*. Ed. IPM UAN, Kiev, 1976, p. 88, in Russian.

[209] DOROFEYEV, Yu.G., ZHERDITSKIY, I.T. and KOLESNIKOV, V.A.: *Sintered Structural Materials*. Ed. IPM UAN, Kiev, 1976, p. 109, in Russian.

[210] MURAMATSU, Y.: Production of High Alloyed Steels by Powder Metallurgy Methods. *J. Japan. Inst. Metals*, Vol. 44, 1980, No. 12, p. 1341.

[211] ZAPF, G., HOFFMANN, G. and DALAL, K.: Effect of Additional Elements on the Properties of Sintered Manganese Steels. *PMI*, Vol.18, 1975. No. 35, p. 214.

[212] ZAPF, G. and DALAL, K.: *Modern Develop. in Powder Metall.*, MPIF, Princeton, NJ, Vol. 10, 1976, p. 129.

[213] HOFFMANN, G. and DALAL, K.: Development and Present Situation of Low Alloyed PM Steels Using MCM and MVM Master Alloys. *PMI*, Vol. 11, 1979, No. 4, p. 181.

[214] ZAPF, G., HOFFMANN, G. and DALAL, K.: Development of Sintered Steels with Optimum Yield Strength, Fatigue Strength and Notch Toughness Strength through

Powder Forging. *Zeitschrift f. Werkstoffkunde,* Part 1, 1975, No. 11, p. 384, in German.

[215] SCHLIEPPER, G. and THÜMMLER, F.: High Strength Heat-Treatable Sintered Steels Containing Manganese, Chromium, Vanadium and Molydenum. *PMI*, Vol. 11, 1979, No. 4, p. 172.

[216] BANERJEE, S., SCHLIEPPER, G, THÜMMLER, F. and ZAPF, G.: Liquid Phase Formation during Sintering of Low-Alloyed Steels with Carbide Master Alloy Additions. *PMI*, Vol. 13, 1980, No. 3, p. 143.

[217] RETELSDORF, H.J., FICHTE, R.M.: Manufacturing and Use of Base Powders of the Type MnCrMo. In: 4th eur. Symp. Powder Metall., Grenoble, EPMA, Shrewsburry, Vol. 1, 1975, 3-3, p. 1.

[218] ZHANG, Z., FRISK, K., SALWÉN, M. and SANDSTRÖM, R.: Mechanical Properties of Fe–Mo–Mn–Si–C Sintered Steels. *Powder Metallurgy*, Vol. 47, 2004, No. 3, p. 239.

[219] BEISS, P.: Mn–Cr-Alloyed Sintered Steels for High Loaded Structural Parts. Yearly Meeting Austria–Germany PM Association, Aachen, 2001, CD, in German.

[220] BEISS, P.: New Cr–Mn-Alloyed Sintered Steel for High-Performance Applications. In: Advances in Powder Metallurgy & Particulate Materials-2005, Int. Conf. on Powder Metallurgy & Particulate Materials, Montreal, MPIF, Princeton. NJ, Part 7, 2005, p. 22.

[221] BEISS, P.: Alloy Cost Optimisation of High Strength Mn-Cr-Mo Steels with Kerosene-Atomized Master Alloy. In: Powdermet 2008, Int. Conference on Powder Metallurgy & Particulate Materials, San Diego, MPIF, Princeton, 2006, CD.

[222] HAMIUDDIN, Md.: Increasing Mechanical Properties of Steels Containing Transition Metal; Carbide Master Alloy MCM and Phosphorus. *The I. J. of Powder Metall. & Powder Technology*, Vol. 20, 1984, No. 4, p. 343.

[223] HAMIUDDIN, M.D. and UPADHYAYA, G.S.: Effect of Transition Metal Carbides Master Alloys MCM on Sintering Iron and Iron–Phosphorus Powder Premix. *PMI*, Vol. 12, 1980, No. 2, p. 65.

[224] Anonymous: Master Alloys Help Produced High-Strength PM Steels. *MPR*, Vol. 16, 1981, No. 4, 187

[225] DOUGAN, M.J. and MORATŇ, J.: New Robust Mn-Based PM Steels. In: 2008 World Congress on Powder Metallurgy & Particulate Materials, Washington, MPIF, Princeton, 2008, p. 7-35.

[226] GARCIA, W., SAINZ, S. and CASTRO, F.: Tensile and Fatigue Properties of PM Steels with Additions of an Atomised Fe-Mn-C Master Alloy. In: 2008 World Congress on Powder Metallurgy & Particulate Materials, MPIF, Princeton, Washington, part 10, 2008, p. 231.

[227] NAPARA-VOLGINA, S.G.: Properties of Materials on the Basis of Powders Alloyed Prepared by Diffusion Saturation from Point Sources and Areas of their Application. *Poroshkovaya Metallurgiya*, 263, 1984, No. 11, p. 1, in Russian.

[228] Patent USA Nr. 4071354-1978 translated in Russian - New Processes and Materials in Powder Metallurgy. Ed. Metallurgiya, Moscow, 1983, p. 40, in Russian.

[229] MOCARSKI, S., HALL, R.A., CHERNEKOFF, D.A., YEAGER, A. and McHUGH, Ch.: Master Alloys to Obtain Premixed Hardenable Powder Metallurgy Steels. *Powder Metallurgy*, Vol. 39, 1996, No. 2, p. 130.

[230] FINDEISEN, F.: Sintered Steels Containing Manganese. In: Problems of Powder Metallurgy, Int. PM Conf., ed. J. Kubelík, SAV Bratislava, Smolenice, (IMR SAS Košice), 1964, p. 189, in Slovak.

[231] FEDORCHENKO, I.M. and SKOROKHOD, V.V.: Theory and Practice of

Sintering.*Poroshkovaya Metallurgiya*, 1967. No. 10. p. 29.

[232] Anonymous: *MPR*, Vol. 16, 1981, No. 4, p. 187.

[233] HUPPMANN, W.J., KAYSSEP, W. and ESPER, F.J.: Sintering of Iron–Manganese with Tin Additions. In: 4th European symposium for Powder Metallurgy, Grenoble, Vol. 1, 1975, paper 3–7, p. 1.

[234] HUPPMANN, WJ., KAYSSEP, W. and ESPER, FJ.: Sintering of Iron–Manganese with Addition of Tin. *Zeitschrift f. Werkstofftechnik*, 1976, No. 7, p. 190, in German.

[235] CHATTERJEE, S.K. and MACKAY, C.A.: Mechanical Properties of Sintered Iron-Tin–Manganese Alloys. *Powder Metallurgy*, 1980, No. 4, p. 183.

[236] FISCHMEISTER, H. and DRAR, H.: Zinc as Alloying Element in Cu–Mn–Steel Powder Forged Produced by Liquid-Solid Alloying. *PMI*, Vol. 19, 1977, No. 3, p. 114.

[237 BENDEREVA, E., TRAYKOVA, I., SIMEONOVA, R., KANEV, G., YAKIMOV, K., DOBREV, R. and PETCHENYAKOV, I.: Investigation on Effect of Type of Boron Additive on Structure Formation during Sintering of High Manganese PM Steel. In: PM2001 Congress & Exh., Nice, EPMA Shrewsbury, 2001, Part 4, p. 246.

[238] SULOWSKI, M. and CIAS, A.: Effect of Processing Variables on Mechanical Properties of Sintered Manganese Steels Fe–3%Mn–0.8%C. *Inżynieria Materialowa*, Vol. XIX, 1988, No. 4, p. 1179.

[239] DUDROVÁ, E., KABÁTOVÁ, M., PARILÁK, Ľ. and KOŠČ, E.: Microstructure and Properties of Fe–3Mn–Cr Sintered Steels. *Metalurgija*, Vol. 40, 2001, No. 4, p. 187.

[240] DUDROVÁ, E., KABÁTOVÁ, M. and BIDULSKÝ, R.: Effect of Sintering Conditions on Microstructure and Mechanical Properties of Fe–3Mn–0.7C Sintered Steel. In: Materiál v Inžinierskej Praxi 2002 (Materials in Engineering Practice), Herľany, Metallurgical Department of TU Košice, 2002, p. 79.

[241] CIAŚ, A. and WRONSKI, A.S.: Comparison of Mechanical Properties of Manganese Steels Based on Sponge and Atomised Iron Powders. *Powder Metallurgy Progress*, Vol. 7, 2007, No. 2, p. 80.

[242] CIAS, A. and WRONSKI, A.S.: Comparison of Mechanical Properties of Manganese Steels of the same Chemical Composition Based on Sponge and Atomised Iron Powders. *Powder Metallurgy Progress*, Vol. 8, 2008, No. 2, p. 76.

[243] SULOWSKI, M.: How Processing Variables Influence Mechanical Properties of PM Mn Steels? *Powder Metallurgy Progress*, Vol. 7, 2007, No. 2, p. 80.

[244] SULOWSKI, M. and CIAS, A.: The Effect of Cooling Rate on the Structure and Mechanical Properties of Sinter-Hardened Fe-(3–4)Mn–0.8C PM Steels. In: Eur. Congress & Exh. on Powder Metallurgy, Valencia, EPMA, Shrewsbury, Vol. 1, 2003, p. 453.

[245] CIAS, A., MITCHELL, SC., WATTS, A. and WRONSKI, AS.: Microstructure and Mechanical Properties of Sintered (2-4)Mn-(0.6-0.8)C Steels. *Powder Metallurgy*, Vol. 42, 1999, No. 3, p. 227.

[246] JALILIZIYAEIAN, M., GIERL, Ch. and DANNINGER, H.: Influenece of Manganese on the Sintering Behaviour of Atomized Iron Powder. In: PM Auto 5th Int. Congress on Powder Metallurgy for Automotive Parts, Mashad Powder Metallurgy Co., Tehran, Isfahan, 2008, p. 85.

[247] ZENDRON, M., MOLINARI, A. and GIRARDINI, L.: Hardenability Study of Low Alloy Sintered Steels Containing Manganese. *Powder Metallurgy Progress*, Vol. 6, 2006, No. 3, p. 111.

[248] SULOWSKI, M. and CIAS, A.: The Effect of Tempering Temperature on the Structure and Mechanical Properties of Sinter-Hardened Manganese Steels. In: 3rd

APP/051/19

Int. Powder Metalurgy Conf., Ankara, Gazi University, TTMD, 2002, p. 582.

[249] CIAS, A. and MITCHEL, S.C.: Processing and Ductile–Brittle Transitions in PM Manganese Steels. *Powder Metallurgy Progress*, Vol. 5, 2005, No. 2, p. 82.

[250] BIDULSKÝ, R., ACTIS GRANDE, M.A., HRYHA, E. and BIDULSKÁ, J.: The Effect of Manganese on Microstructure Development and Diffusion Characteristics in Admixed Fe–Mn–C Sintered Steels. In: PM2010 World Congress & Exhibition, EPMA, Shrewsbury, Florence, 2010, CD.

[251] PIECZONKA, T., MITCHELL, S.A., CIAS, A., SULOWSKI, M. and WRONSKI, A.S.: Dimensional Changes during Sintering of Manganese–Carbon Steel Compacts. *Inżynieria Materialowa*, Vol. XIX, 1998, No. 4, p. 1171.

[252] HRYHA, E. and DUDROVÁ, E.: The Sintering Behaviour of Fe–Mn–C Powder System, Correlation between Thermodynamics and Sintering Process, Mn Distribution, and Microstructure. *Materials Science Forum*, 534-536, 2007, p. 761.

[253] HRYHA, E., NYBORG, L., DUDROVÁ, E. and BENGSTSSON, S.: Sintered Steels Alloyed with Manganese: Effect of Alloying Mode. In: PM2010 World Powder Metallurgy Congress & Exhibition, Florence, EPMA, Shrewsbury, 2010, CD.

[254] HRYHA, E., NYBORG, L., DUDROVÁ, E. and BENGTSSON, S.: Brittleness of Structural PM Steels Admixed with Manganese Studied by Advanced Electron Microscopy and Spectroscopy. *Powder Metallurgy Progress*, Vol. 8, 2008, No. 2, p. 109.

[255] CIAS, A. and SULOWSKI, M.: Influence of Powders and Processing on Mechanical Properties of Sintered Manganese Steel for Structural Parts. In: Int. Conf. DF PM '99, Deformation and Fracture in Structural PM Materials, Piešťany, IMR SAS, Košice, 1999, p. 112.

[256] CIAS, A. and STOYTCHEV, M.: Powder Metallurgy Innovative Processing of Manganese Steel Gears. In: 4th Int. Powder Metallurgy Conf., Sakaraya University, Sakaraya, TTMD, 2005, p. 350.

[257] CIAS, A.: Processing and Mechanical Properties of Sintered Manganese Steel Gears. *Powder Metallurgy Progress*, Vol. 5, 2005, No. 4, p. 147.

[258] CARNAL, A., SICRE, J., CAMPOS, M. and TORRALBA, J.M.: Mn Additions Influence on Strain Micro-mechanism in As-Sintered 316L. *Powder Metallurgy Progress*, Vol. 5, 2005, No. 2, p. 73.

[259] NADJAFI, A., SADREMALI, J. and HOSSEINI, SR.: Mechanical Properties Study of Low Alloy Sintered Steels Containing Mn. In: PM Auto 2008 7th Int. Conf. on Powder Metallurgy for Automotive Parts. Mashad Powder Metallurgy Co, Tehran, Isfahan, 2008, p. 1.

[260] ROMANSKI, A. and CIAS, A.: Properties of Mo-Alloyed Sintered Manganese Steel. *Inżynieria Materialowa*, Vol. XIX, 1998, No. 4, p. 1175.

[261] CIAS, A., STOYTCHEV, M. and WRONSKI, A.S.: Processing, Strength, Ductility and Toughness of Sintered Mn-(Mo)-Steels. Advances in Powder Metallurgy & Particulate Materials – 2001 (Int. Conf., New Orleans), B. Eisen, S. Kassam, eds., Part 10, 2001, p. 141.

[262] YOUSEFFI, M., MITCHELL, S.C., WRONSKI, A.S. and CIAS, A.: Sintering, Microstructure, and Mechanical Properties of PM Manganese–Molybdenum Steels. *Powder Metallurgy*, Vol. 43, 2000, No. 4, p. 353.

[263] CIAS, A., SULOWSKI, M., MITCHELL, S.C. and WRONSKI, A.S.: Sinter Hardening of Fe–Mn–C Steels. In: PM2001 Congress & Exhibition, Nice, EPMA, Shrewsbury, 2001, Part 4, p. 246.

[264] CIAS, A., SULOWSKI, M., MITCHELL, S.C. and WRONSKI, A.S.: Sinter

Hardening of Fe–Mn–C Steels. In: PM2001, EPMA, Shrewsbury, Nice, Vol. 4, 2001, p. 246.

[265] PIECZONKA, T., STOYTCHEV, M. and MITCHELL, S.C.: Microstructure and Dimensional Changes during Sintering of ASC100.29 and NC100.24 Höganäs Grade Powders with Manganese and Molybdenum Additions. In: PM2001, EPMA, Shrewsbury, Nice, Vol. 1, 2001, p. 316.

[266] LINDSLEY, B., SCHADE, CH. and RUTZ, H.: High Performance Manganese Containing PM Steels. In: 2001 Int. Conf. on Powder Metallurgy & Particulate Materials, MPIF Princeton. NJ, Washington, 2008, Part 7, p.17. World Congress & Exhibition, Florence, EPMA Shrewsbury, 2010, CD.

[266] SUCIU, C., ARGHIR, G., BRANDUSAN, L. and PETEAN, I.: Microstructure of the Fe–FeMn Transition Zone. *Powder Metallurgy Progress*, Vol. 11, 2011, No. 1–2, p.153.

[267] LINDSLEY. B. and JAMES, B.: PM Steels that Contain Manganese. In: PM 2010 World Congress & Exhibition, Florence, EPMA, Shrewsbury, 2010, CD.

[268] DUDROVÁ, E., KABÁTOVÁ, M., BIDULSKÝ, R. and WRONSKI, A.S.: Industrial Processing, Microstructure and Mechanical Properties of Fe-(2–4)Mn-0.85Mo–(0.3–0.7)C Sintered Steel. *Powder Metallurgy*, Vol. 47, 2004, No. 2, p. 181.

[269] MITCHELL, S.C., WRONSKI, A.S., CIAS, A. and STOYTCHEV, M.: Microstructure and Mechanical Properties of Mn–Cr–Mo–C Steels Sintered at >1140°C. In: Advances in Powder Metallurgy and Particulate Materials PM² Tech'99, Vancouver, MPIF, Princeton, NJ, 2009, p. 7–129.

[270] MOTOOKA, S., KUROISHI, N., HARA, A. and FUKURAWA, N.: Strength and Ductility of Mn–Cr Sintered Steel. In: P/M in Europe Int. Powder Metallurgy Conference, Florence, Vol. III, 1982, p. 445.

[271] CIAS, A., MITCHELL, SC. and WRONSKI, A.S.: Mechanical Properties of Chromium PM Steel Sintered in Technical Nitrogen. In: PM2004 Powder Metallurgy World Congress & Exhibition, Vienna, EPMA, Shrewsbury, 2004, Vol. 2, p. 5.

[272] MITCHELL, S.C., WRONSKI, A.S. and CIAS, A.: Development of Sintered Mn-(Mo)–(Cr)–C Steels. In: PM Conf., AMTOI, Gdansk, 2001, p. 1.

[273] CAMPOS, M., SANCHEZ, D. and TORRALBA, J.M.: Sintering Behaviour Improvement of a Low Cr–Mo Prealloyed Powder Steel through Mn Additions and others Liquid Phase Promoters. In: AMPT '01 Int. Conf. on Advanced Materials and Processing Technologies, Leganés, Madrid, Vol. 3, 2001, p. 1265.

[274] MITCHELL, S.C., WRONSKI, A.S., GEORGIEV, J and STOYTCHEV, M.: Mechanical Properties of Mn–Cr–Mo Steels Nitrogen Sintered from Hydrocabon-Coated Powders. In: PM2004 Powder Metallurgy World Congress & Exhibition, Vienna, EPMA, Shrewsbury, Sintering, 2004, p. 235.

[275] PIECZONKA, T., MITCHELLS, S.C., STOYTHCHEV, M. and KAZIOR, J.: Dimensional Changes during Sintering of Astaloy CrM Höganäs Powder with Carbon and Manganese Additions. In: PM² Tech 2001 Advances in Powder Metallurgy & Particulate Materials – 2001, New Orleans, MPIF, Princeton, Part 10, 2001, p. 141.

[276] PIECZONKA, T., MITCHELLS, S.C., STOYTHCHEV, M. and KOWALCZYK, M.: Liquid Phase Sintering of Fe–3Cr–1Mn–0.5Mo High Carbon Steels. In: Advances in Powder Metallurgy & Particulate Materials – 2002, Orlando, MPIF, Princeton, 2002, p. 11.

[277] SHIVANATH, R., JONES, P.K. and LAWCOCK, R.: On the Synergies of High Temperature Sintering and Alloy Development for High Endurance P/M Powertrain

Components. In: PM²TEC`96 World Congress on Powder Metallurgy and Particulate Materials, MPIF, Princeton, Vol. 4, 1996, p. 427.

[278] GARCIA, W., SAINZ, S. and CASTRO, F.: Influence of Mn and Ni on the Mechanical Properties of Newly Developed PM Steels. In: PM2010 World Congress & Exhibition, Florence, EPMA, Shrewsbury, 2010, CD.

[279] JALILIZIYAEIAN, M., GIERL, Ch., ZBIRAL J. and DANNINGER, H.: Manganese Evaporation during Sintering of Fe-Mn-Cr Compacts from Prealloyed Powder. In: EURO PM2008 Int. Powder Metallurgy Congress & Exhibition, Mannheim, EPMA, Shrewsbury, Vol. 1, 2008, p. 115.

[280] BERGMAN, O. and BENGTSSON, S.: Influence of Sintering Temperature and Component Density on the Properties of Prealloyed PM Steel Grades Containing Cr, Mo and Mn. In: EURO PM2009 Congress & Exhibition, Copenhagen, EPMA, Shrewsbury, 2009, CD.

[281] SULOWSKI, M. and CIAS, A.: Microstructure and Properties of Cr–Mn Structural Steels Sintered in a Microatmosphere. In: PM2010 World Congress & Exh. Florence, EPMA, Shrewsbury, 2010, CD.

[282] SULOWSKI, M. and CIAS, A.: The Effect of Sintering Atmosphere on the Structure and Mechanical Properties of Sintered Fe–Mn–Cr–Mo–C Structural Steels. *Powder Metalllurgy Progress,* Vol. 11, 2011, No. 1–2, 123.

[283] BERGMAN, O. and NYBORG, L.: Evaluation of Sintered Properties of PM Steels based on Cr and Cr–Mn Prealloyed Steel Powders. *Powder Metallurgy Progress,* Vol. 10, 2010, No. 3, p. 1.

[284 DANNINGER, H., YALILIZIYAEIAN, M., GIERL, Ch. and BENGSTSSON, S.: Chemical Reaction during Sintering of Mn and Mn–Cr Prealloyed Steels in Inert Reducing Atmospheres. *Materials Science Forum,* Vol. 672, 2011, p. 293.

[285] DANNINGER, H.: Alloying Systems for PM structural Parts. Consequences for Sintering in PM and in Science and Practice. In: Symposium, Hagen, Fachverband Pulvermetallurgie, Hagen, Vol. 22, 2006, p. 21, in German.

[286] GIERL, Ch., DANNINGER, H. and de ORO CALDERÓN, R.: Characterisation of Sintering Processes in Compacts on Iron Basis through Process Analytics via Thermal Analytical Methods. In: Symposium Modern Developments in Powder Metallurgy, Hagen, Fachverband Pulvermetallurgie, Hagen, Vol. 27, 20011, p. 46, in German.

[287] RICHARDSON, F.D.: *Physical Chemistry of Steels in Metallurgy.* Vol. 2, 1974, London, New York, Academic Press.

[288] ŠALAK, A. and SELECKÁ, M.: Adverse Effect of High-purity Atmospheres on Sintering of Manganese Steels. *Powder Metallurgy,* Vol. 53, 2010, No. 4, p. 285.

[289] NAVARA, E.: Alloying of Sintered Steel with Manganese. In: 6th Int. Powder Metallurgy Conference in ČSSR, Brno, IMR SAS Košice, Košice, Vol. 1, 1982, p. 143.

Index